Basic Arithmetic

Basic Arithmetic

3rd Edition

ROBERT G. MOON
ARTHUR H. KONRAD
GUS KLENTOS
JOSEPH NEWMYER, JR.
all, Fullerton College

Merrill, an imprint of
Macmillan Publishing Company
New York

Collier Macmillan Canada, Inc.
Toronto

Maxwell Macmillan International Publishing Group
New York Oxford Singapore Sydney

This book was set in Times Roman and Trade Gothic
Cover Design: Tony Faiola
Text Design: Ben Shriver
Production Coordination: Ben Shriver

Printed in the United States of America

Library of Congress Catalog Card Number: 83–62899
International Standard Book Number: 0–675–20136–5
Printed in the United States of America
PRINTING 7 8 9 **YEAR** 3 4 5

Preface

The first and second editions of *Basic Arithmetic* were created and tested over a fourteen-year period at Fullerton College, Fullerton, California. The authors' primary purpose was to create a program combining a text with audio tapes, whereby students could understand and learn basic arithmetic. The response to this program from teachers and students alike has been most gratifying.

This third edition of *Basic Arithmetic* has the same purpose as the first two editions. This edition is based on the classroom experience of the authors and many other arithmetic teachers at numerous colleges across the country. The main features of the third edition are as follows:

- ☐ The units are divided into six modules. Each module includes a Practice Test with answers so that students may review for a major test.
- ☐ To obtain a clearer presentation of the topics, the audio tapes have been redone.
- ☐ The number of supplementary problems has been increased. The answers to the odd-numbered supplementary problems are included at the back of the text. Answers to the even-numbered problems are in the Instructor's Manual.
- ☐ More applied problems have been inserted throughout the text.
- ☐ A new unit covers equation solving and ratio and proportion. Problem solving by means of equations is used extensively.
- ☐ Addition and subtraction have been combined into a single unit (Unit 2).
- ☐ The units on percent are divided into two modules. The first module deals with basic operational procedures and interest. The second module contains optional applications and divides these applications into the two main categories of percent decrease and percent increase.
- ☐ The unit on interest is subdivided into two parts, each with appropriate review exercises and supplementary problems. Part I covers simple interest, credit accounts, and the Truth-In-Lending Act. Part II covers compound interest and the use of a table, extended to include higher interest rates.
- ☐ Three units are now devoted to the metric system. The first encourages students to "think metric." The second unit has students working entirely within the metric system, making use of a chart to simplify movement of the decimal point. Two additional prefixes have been introduced: micro and mega. Metric–English conversions are performed in the third unit.

- ☐ Metric units of length, volume, and mass are used in problems involving measurement.
- ☐ The Pythagorean Theorem is introduced and applied in the measurement module.

Robert G. Moon
Arthur H. Konrad
Gus Klentos
Joseph Newmyer, Jr.

To the Student

The Audio-Tutorial Approach

You are about to embark on a new adventure in learning arithmetic, namely the audio-tutorial method. This method utilizes a book designed to correlate with audio tapes. You will read through a unit and at the same time listen to an explanation of the problems and theory. Also, throughout each unit are study exercises for you to work to periodically check your progress.

This technique for the teaching of arithmetic was developed, tested, and used at Fullerton College, Fullerton, California. It has been received with widespread student approval since its inception. In general, students have indicated that this technique greatly improved study habits, resulting in a better understanding of mathematics with a commensurate higher grade.

In the study of mathematics, most students are accustomed to the traditional lecture-textbook method where they read a certain section in the text, go to class for a lecture on the material, and then are left to try to work the exercises. The frustrations of this situation are many. First, most students have difficulty reading a mathematics textbook; second, once the classroom lecture is over, the chalkboard is erased and the explanatory lecture is lost forever; and third, many students have great difficulty working the homework assignment five or six hours after the lecture.

The audio-tutorial method of studying mathematics is an attempt to remedy the defects of the traditional lecture-textbook method. In the audio-tutorial approach, the lecture and other explanations are put on audio tape. The chalkboard illustrations are put in the text. Therefore, each student has a permanent record of the material generally presented in the class. The advantages of this system are numerous. You may go through the lecture at your own rate and review any part of the lecture as often as desired by simply reversing the recorder and turning back in the text.

Format of the Text

There are 38 units contained in the text. Each unit has been divided into the following parts:

(a) Frames
(b) Study Exercises
(c) Review Exercises
(d) Solutions to Review Exercises
(e) Supplementary Problems
(f) Solutions to Study Exercises

Frames. Each unit consists of 15 to 40 frames. Each frame has a "frame number" located at the upper right-hand corner. The frames are to be studied as you listen to the commentary on the audio tape. Remember to watch the "frame numbers" to keep the tape synchronized with the text.

Study Exercises. Some of the frames are devoted to study exercises. When you come to a study exercise, turn off the recorder. The exercises should be worked and your answers checked before you turn the recorder on and proceed to the next frame. If you do poorly on a study exercise, repeat that part of the unit.

Review Exercises. When you come to the end of the frames, you should turn off the recorder and do the review exercises. Check your answers with the ones immediately following. The review exercises are extremely important, in that they will be a measure of your understanding of the entire unit. If you don't do well on the review exercises, go back and review the entire unit.

Solutions to Review Exercises. The solutions to the review exercises immediately follow the review exercises. They are, for the most part, detailed solutions. Be sure to compare your work carefully with the details of these solutions. Many of your mistakes can be understood and corrected in this section.

Supplementary Problems. Immediately following solutions to review exercises is a section of supplementary problems. There are answers for the odd-numbered problems at the end of your text, and your instructor may wish to use the others for homework problems or for quizzes. However, you are encouraged to try as many as you wish. Remember, in mathematics, *practice* is the key to success.

Solutions to Study Exercises. At the end of each unit you will find the detailed solutions to the study exercises which were located amongst the frames. Before you begin each unit with the recorder, you will want to locate and mark this section for ready reference. Remember, as you proceed through the unit, you will be asked to work study exercises and to check your results with this section.

Study Techniques

- ☐ The recorder may be stopped at any time to give more time for analysis of a frame.
- ☐ The recorder may be reversed to review a particular portion of a unit.
- ☐ Cumulative practice tests with answers are spaced periodically throughout the text. These practice tests each cover one module and should be worked before each examination in the course, since they contain questions similar to those you will encounter on the examination itself.
- ☐ Before an examination, review all of the units involved, examine the review exercises, and work the practice test.
- ☐ Before the final examination, read over each frame and rework the review exercises and practice tests.
- ☐ Remember that mathematics pyramids. That is, it builds on itself. Don't get behind. In fact, it is a good idea to try to stay a unit ahead.
- ☐ Many students have indicated that they had tremendous success by doing a "constant review"—that is, constantly going back and redoing earlier units.

Contents

CONTENTS

CONTENTS

CONTENTS

CONTENTS

MODULE

1

Operating with Whole Numbers

UNIT

1 Whole Numbers and the Place Value System

OBJECTIVES (1)

By the end of this unit you should be able to:

1. SHOW HOW PLACE VALUE AND MULTIPLES OF TEN ARE USED TO NAME WHOLE NUMBERS BY WRITING THEM IN EXPANDED FORM.
2. READ NUMERALS REPRESENTING WHOLE NUMBERS.
3. WRITE NUMERALS REPRESENTING WHOLE NUMBERS.
4. ROUND WHOLE NUMBERS TO A SPECIFIED DEGREE OF ACCURACY.

Whole Numbers (2)

Whole numbers may be used to count objects within a collection or to denote an order.

Example 1: Counting objects in a collection.

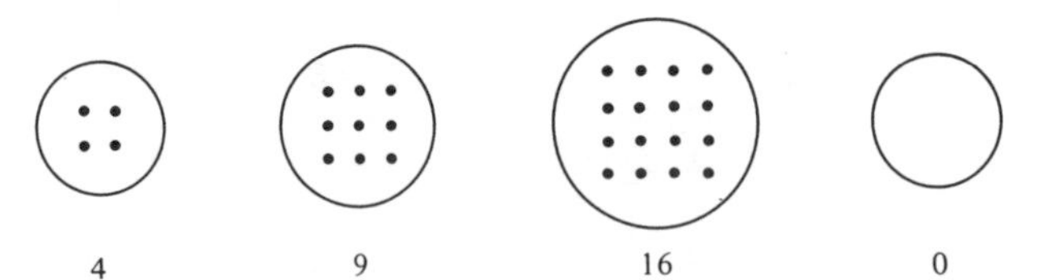

Example 2: Denoting an order.

a, b, c, d, e, f, g, h, i, j, k, l, m, n, o, p, q, r, s, t, u, v, w, x, y, z

a is the first letter of the alphabet.
e is the fifth letter of the alphabet.
z is the twenty-sixth letter of the alphabet.

The whole numbers are represented by the following symbols called *numerals*: 0, 1, 2, 3, 4, 5, 6, 7, 8, 9, 10, 11, 12, 13, 14, 15, . . .

Base Ten System (3)

Because our primitive ancestors counted on their fingers, we have a *base ten* or *decimal* system. The symbol for ten is *10*. It means *one group of ten plus no units.*

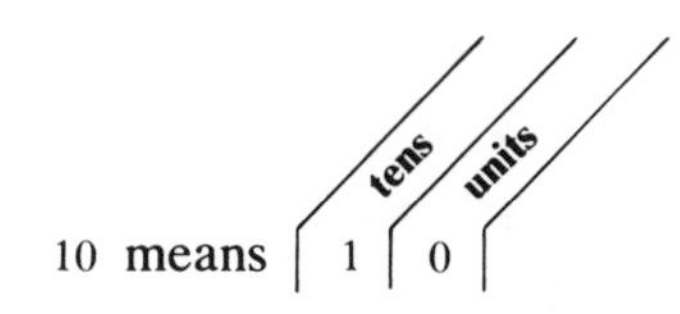

Basic Digits (4)

Each numeral representing a whole number may be composed of any of the following *basic digits*: 0, 1, 2, 3, 4, 5, 6, 7, 8, 9.

Examples:

1. 23 **2.** 456 **3.** 9,178 **4.** 631,205,479

Place Value (5)

When the basic digits are placed together to form symbols for whole numbers, each digit has a *place value*. Each place value is a power of ten: units, tens, hundreds, thousands, and so on.

Expanded Form (6)

Each whole number can be written in expanded form, where it is written as a sum in terms of the place values.

Example 1: 23

23 means *2 tens plus 3 units.*

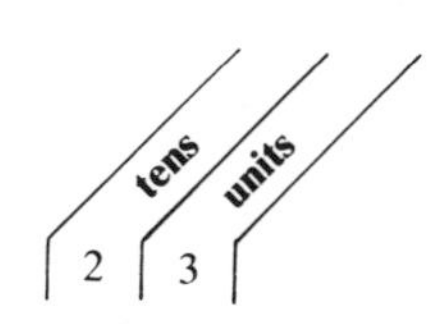

The place value of the digit 2 is *tens*.
The place value of the digit 3 is *units*.
Expanded form: 2 tens + 3 units

(7)

Example 2: 456

456 means *4 hundreds plus 5 tens plus 6 units.*

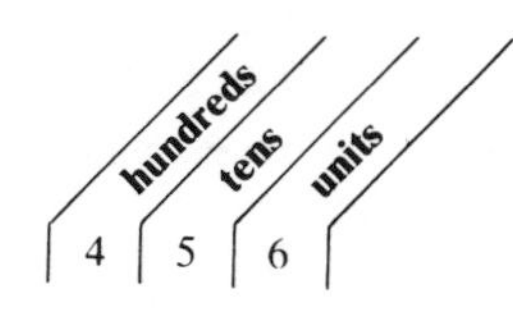

The place value of the digit 4 is *hundreds*.
The place value of the digit 5 is *tens*.
The place value of the digit 6 is *units*.
Expanded form: 4 hundreds + 5 tens + 6 units

Example 3: 9,178 **(8)**

9,178 means *9 thousands plus 1 hundred plus 7 tens plus 8 units.*

thousands	hundreds	tens	units
9	1	7	8

The place value of the digit 9 is *thousands.*
The place value of the digit 1 is *hundreds.*
The place value of the digit 7 is *tens.*
The place value of the digit 8 is *units.*

Expanded form: 9 thousands + 1 hundred + 7 tens + 8 units

Example 4: 631,205,479 **(9)**

This numeral means:

hundred millions	ten millions	millions	hundred thousands	ten thousands	thousands	hundreds	tens	units
6	3	1	2	0	5	4	7	9

Place Value Chart **(10)**

Memorize this chart:

trillions	hundred billions	ten billions	billions	hundred millions	ten millions	millions	hundred thousands	ten thousands	thousands	hundreds	tens	units

Each place is ten times the one to its immediate right.

Study Exercise One **(11)**

1. 8,975 means eight ________ plus ________ hundreds plus seven ________ plus ________ units.
2. Use the chart, if necessary, to determine the place value of each digit in the following numeral: 2,308,471,659,624
3. Write a numeral having 2 in the tens' place, 8 in the thousands' place, 3 in the units' place, 0 in the hundreds' place, 7 in the ten thousands' place, 2 in the millions' place, and 9 in the hundred thousands' place.
4. Write each of these whole numbers in expanded form.
   (a) 49 **(b)** 6,325 **(c)** 12,907

Reading Numerals Representing Whole Numbers (12)

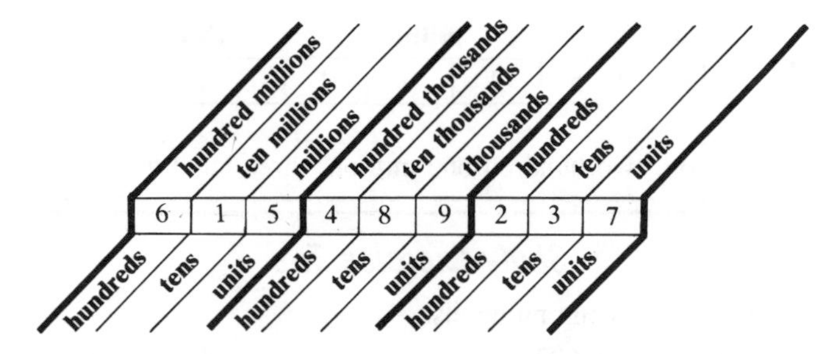

Six hundred fifteen million, four hundred eighty-nine thousand, two hundred thirty-seven.

In summary, to read a numeral representing a whole number, follow these steps:

Step (1): Separate the numeral into groups of three going from right to left.
Step (2): Read each of the resulting three-digit numerals, being sure to attach the correct labels.

Using Commas (13)

Commas are used to separate billions, millions, thousands, and units.

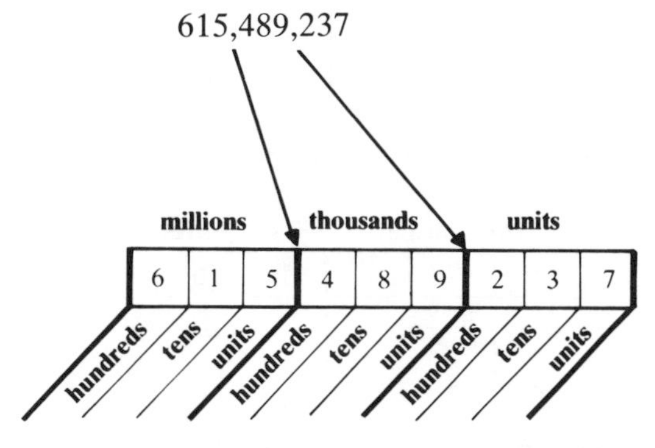

Six hundred fifteen million, four hundred eighty-nine thousand, two hundred thirty-seven.

Note: Two-digit numerals are hyphenated, such as eighty-nine and thirty-seven.

More Examples (14)

1. 121,648

thousands			units		
1	2	1	6	4	8

One hundred twenty-one thousand, six hundred forty-eight.

2. 37,502

thousands			units		
	3	7	5	0	2

Thirty-seven thousand, five hundred two.

More Examples (Continued)

3. 6,025,201,310,007

trillions			billions			millions			thousands			units		
		6	0	2	5	2	0	1	3	1	0	0	0	7

Six trillion, twenty-five billion, two hundred one million, three hundred ten thousand, seven.

Study Exercise Two (15)

Write in words the names for the following numerals:

1. 512 **2.** 1,602 **3.** 35,011

4. 615,901,002 **5.** 2,007,500,400 **6.** 9,245,063,106,002

Writing Numerals Representing Whole Numbers (16)

Example 1: Two thousand, three hundred forty-five.

Remember commas are used to separate thousands from units.

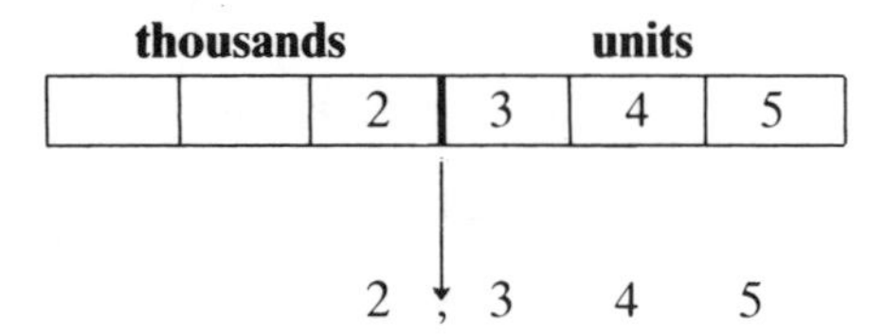

Example 2: Five hundred twenty-six thousand, one hundred sixty-three.

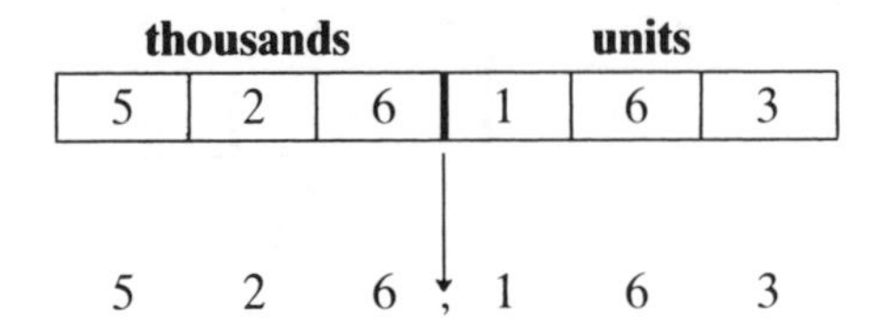

Example 3: Thirty-one million, four hundred eighty-two thousand, nine hundred twenty-eight.

millions			thousands			units		
	3	1	4	8	2	9	2	8

3 1 , 4 8 2 , 9 2 8

Example 4: Three hundred million.

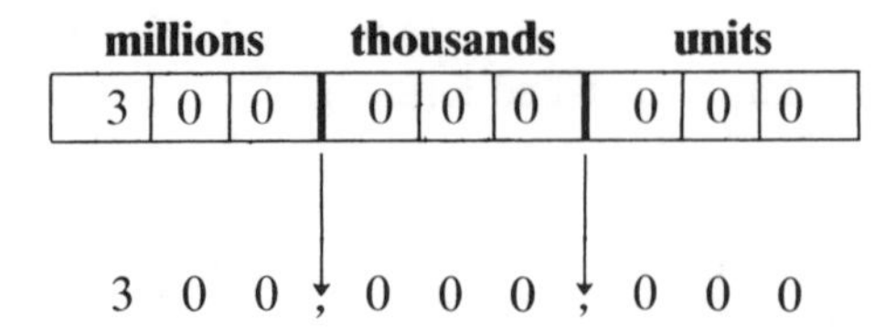

Writing Numerals Representing Whole Numbers (Continued)

Example 5: Nine billion, four million, six hundred five thousand, fourteen.

billions			millions			thousands			units		
		9	0	0	4	6	0	5	0	1	4

9 ; 0 0 4 ; 6 0 5 ; 0 1 4

Study Exercise Three (17)

A. Turn on the recorder. These exercises will be given orally.

B. Use the ten basic digits to write numerals for the following whole numbers:

8. Three hundred four.
9. Four thousand, nine hundred eleven.
10. Forty thousand, six hundred two.
11. Eight hundred fifty-five thousand, one.
12. Twenty-five million.
13. Thirty-two million, one thousand, nine hundred ten.
14. Three billion, four hundred twelve million, six hundred twenty-two thousand, three hundred ninety-five.
15. Twelve trillion, eighty-nine billion, four-hundred twenty-six million, one hundred five thousand, two hundred fifty-one.

Rounding Whole Numbers (18)

When great accuracy is not required, whole numbers may be rounded to a given place value by replacing certain digits with zeros.

Example 1: Rounding 4,681 to the nearest hundred, we have 4,700.

Example 2: Rounding 4,621 to the nearest hundred, we have 4,600.

Procedure for Rounding

Step (1): Locate the place value to which you are rounding and draw a box around this digit.
Step (2): Examine the digit which is to the immediate right of the box.

a. If this digit is 5, 6, 7, 8, or 9, then *increase the digit in the box by one, and replace all digits to the right with zeros.* (If the digit in the box is nine, it can be raised one by writing a zero and increasing the digit on its left by one.)

b. If this digit is 0, 1, 2, 3, or 4, then *leave the digit in the box as is and replace all digits to the right with zeros.*

Examples: (19)

1. Round 63,295 to the nearest hundred.
Step (1): 6 3,[2]9 5
Step (2): 6 3, 3 0 0

2. Round 398 to the nearest ten.
Step (1): 3[9]8
Step (2): 4 0 0

3. Round 67,973 to the nearest hundred.
Step (1): 6 7,[9]7 3
Step (2): 6 8, 0 0 0

4. Round 63,235 to the nearest hundred.
Step (1): 6 3,[2]3 5
Step (2): 6 3, 2 0 0

Examples (Continued)

5. Round 2,586,913 to the nearest hundred thousand.
   Step (1): 2,[5]8 6, 9 1 3
   Step (2): 2, 6 0 0, 0 0 0

6. Round 5,214,692 to the nearest million.
   Step (1): [5],2 1 4, 6 9 2
   Step (2): 5, 0 0 0, 0 0 0

Study Exercise Four (20)

A. Round each of the following whole numbers to the indicated place value:

1. 628 to the nearest ten.
2. 628 to the nearest hundred.
3. 628 to the nearest thousand.
4. 3,864,950 to the nearest million.
5. 3,864,950 to the nearest ten thousand.
6. 3,864,950 to the nearest hundred.

B. In the following paragraph, round all whole numbers to the nearest hundred thousand:

7. The assets of Valley Savings and Loan are $75,672,452 for the current fiscal year. Last year the assets were $61,971,412. The increase in assets has been $13,701,040.

REVIEW EXERCISES (21)

A. Give the place value for each digit in the following numerals:

1. 6,201
2. 5,906,871,234

B. Write in words the name for each of the following numerals:

3. 16
4. 496
5. 3,002
6. 46,950
7. 100,000
8. 3,000,000,000
9. 3,000,000
10. 9,070,021,004
11. 2,673,332

C. Write the numeral for each of the following whole numbers:

12. Twenty-two.
13. Five hundred two.
14. Sixteen thousand, nine hundred twelve.
15. Fifty thousand, four hundred.
16. Twenty million.
17. Three trillion, eighty-five million.
18. Two hundred sixty-five million, four hundred twenty-one thousand, two hundred eighty-one.

D. Round off 4,619,745 to the nearest:

19. Million.
20. Hundred thousand.
21. Thousand.
22. Hundred.
23. Ten.

E. Fill in the blanks:

24. 21,034 means two _________ plus _________ thousand plus _________ hundreds plus three _________ plus _________ units.

F. Write each of these numerals in expanded form:

25. 86
26. 309
27. 5,160
28. 29,468
29. 5,000
30. 98,061

Solutions to Review Exercises (22)

1.

Digit	Place Value
6	Thousands
2	Hundreds
0	Tens
1	Units

2.

Digit	Place Value
5	Billions
9	Hundred millions
0	Ten millions
6	Millions
8	Hundred thousands
7	Ten thousands
1	Thousands
2	Hundreds
3	Tens
4	Units

Solutions to Review Exercises (Continued)

3. Sixteen.
4. Four hundred ninety-six.
5. Three thousand, two.
6. Forty-six thousand, nine hundred fifty.
7. One hundred thousand.
8. Three billion.
9. Three million.
10. Nine billion, seventy million, twenty-one thousand, four.
11. Two million, six hundred seventy-three thousand, three hundred thirty-two.
12. 22
13. 502
14. 16,912
15. 50,400
16. 20,000,000
17. 3,000,085,000,000
18. 265,421,281
19. 5,000,000
20. 4,600,000
21. 4,620,000
22. 4,619,700
23. 4,619,750
24. 21,034 means two ten thousands plus one thousand plus zero hundreds plus three tens plus four units.
25. 8 tens + 6 units
26. 3 hundreds + 0 tens + 9 units
27. 5 thousands + 1 hundred + 6 tens + 0 units
28. 2 ten thousands + 9 thousands + 4 hundreds + 6 tens + 8 units
29. 5 thousands + 0 hundreds + 0 tens + 0 units
30. 9 ten thousands + 8 thousands + 0 hundreds + 6 tens + 1 unit

---

SUPPLEMENTARY PROBLEMS

A. Give the place value for each digit that is underlined:

1. $\underline{3}0{,}2\underline{0}8$
2. $\underline{5}6\underline{3}{,}4\underline{9}1$
3. $\underline{5}{,}0\underline{6}3{,}\underline{1}4\underline{9}$
4. $\underline{2}0{,}\underline{5}6\underline{0}$

B. Write each of these numerals in expanded form:

5. 93
6. 2,041
7. 12,832
8. 460,215

C. Write, in words, names for the following numerals:

9. 56
10. 506
11. 5,006
12. 20,316
13. 200,316
14. 4,953,629
15. 95,000,000
16. 9,500,000,000
17. 2,673,498,212
18. 302,000,104
19. 635,902,040
20. 7,654,503

D. Write numerals for the following whole numbers:

21. Three thousand, four hundred twenty-nine.
22. Thirty thousand, four hundred twenty-nine.
23. Eight hundred fifty-six million, two hundred fifteen thousand, one hundred forty-one.
24. Three billion, four million, two thousand, five.
25. Thirteen million.
26. One billion.
27. Five hundred thousand.
28. Two million, one hundred forty-one thousand, six hundred thirty-one.
29. One hundred thousand, six hundred two.

E. Round off the whole number 7,936,429 to the nearest:

30. Million.
31. Hundred thousand.
32. Ten thousand.
33. Thousand.
34. Hundred.
35. Ten.

F. Round each whole number in the following paragraphs to the nearest million:

36. The 1960 census showed that the United States population was 179,323,175 people. The 1970 census showed there were 204,821,693 people. This represents an increase of 25,498,518 people.
37. A corporation showed yearly receipts of $67,692,456. Its yearly expenditures were $35,449,991. This gives a gross profit of $32,242,465.
38. In 1980 the United States had a wheat reserve of 248,650,000 bushels. In 1975 the wheat reserve had declined to 212,470,000 bushels. This represents a decrease in the wheat reserve of 36,180,000 bushels.

SUPPLEMENTARY PROBLEMS (Continued)

G. When writing a check, the amount must be filled in both as a numeral and in words. For these imaginary checks write in the amount in words. (Real checks also have a decimal portion which will be studied later.)

39.

James Smart
Anaheim, CA

Aug. 15 19 86 — 286

Pay to the order of Computers Limited $ 6,072

_______________________________ *Dollars*

2nd National Bank of Anaheim — James Smart

40.

JANE GORDON
FARMVILLE, IA

May 6 19 87 — 803

Pay to the order of Farmville Hardware **$** 13,680

_______________________________ DOLLARS

Iowa National Bank — Jane Gordon

41.

Ramon Montoya
Silver City, TX

Sept. 2 19 85 — 511

Pay to the order of Western Saddle Co. **$** 896

_______________________________ *Dollars*

SILVER CITY *savings and loan* — Ramon Montoya

Solutions to Study Exercises (11A)

Study Exercise One (Frame 11)

1. 8,975 means eight thousands plus nine hundreds plus seven tens plus five units.

2.

Digit	Place Value
2	Trillions
3	Hundred billions
0	Ten billions
8	Billions
4	Hundred millions
7	Ten millions
1	Millions
6	Hundred thousands
5	Ten thousands
9	Thousands
6	Hundreds
2	Tens
4	Units

3. 2,978,023

4. **(a)** 4 tens + 9 units **(b)** 6 thousands + 3 hundreds + 2 tens + 5 units
(c) 1 ten thousand + 2 thousands + 9 hundreds + 0 tens + 7 units

Study Exercise Two (Frame 15) (15A)

A. **1.** Five hundred twelve.
2. One thousand, six hundred two.
3. Thirty-five thousand, eleven.
4. Six hundred fifteen million, nine hundred one thousand, two.
5. Two billion, seven million, five hundred thousand, four hundred.
6. Nine trillion, two hundred forty-five billion, sixty-three million, one hundred six thousand, two.

Study Exercise Three (Frame 17) (17A)

B. **8.** 304 **9.** 4,911 **10.** 40,602
11. 855,001 **12.** 25,000,000 **13.** 32,001,910
14. 3,412,622,395 **15.** 12,089,426,105,251

Study Exercise Four (Frame 20) (20A)

A. **1.** *Step (1):* 6[2]8
Step (2): 6 3 0
2. *Step (1):* [6]2 8
Step (2): 6 0 0
3. *Step (1):* [0]6 2 8
Step (2): 1 0 0 0
4. *Step (1):* [3], 8 6 4, 9 5 0
Step (2): 4, 0 0 0, 0 0 0
5. *Step (1):* 3, 8[6]4, 9 5 0
Step (2): 3, 8 6 0, 0 0 0
6. *Step (1):* 3, 8 6 4,[9]5 0
Step (2): 3, 8 6 5, 0 0 0

B. **7.** $75,700,000; $62,000,000; $13,700,000

UNIT

2 Addition and Subtraction of Whole Numbers

OBJECTIVES (1)

By the end of this unit you should be able to:

1. IDENTIFY THE ADDENDS AND THE SUM IN AN ADDITION PROBLEM.
2. ADD WHOLE NUMBERS USING THE CONCEPT OF CARRYING.
3. CHECK YOUR ANSWER TO AN ADDITION PROBLEM.
4. IDENTIFY THE COMMUTATIVE AND ASSOCIATIVE PROPERTIES OF ADDITION.
5. IDENTIFY THE TERMS *MINUEND, SUBTRAHEND,* AND *DIFFERENCE* IN A SUBTRACTION PROBLEM.
6. USE ADDITION AS A CHECK FOR SUBTRACTION.
7. PERFORM SUBTRACTION USING THE CONCEPT OF BORROWING.

Forms of Addition (2)

Horizontal Form

$4 + 2 = 6$

4 and 2: *addends*; 6: *sum*

Vertical Form

$$\begin{array}{r} 4 \\ +2 \\ \hline 6 \end{array}$$

4 and 2 ← *addends*; 6 ← *sum*

A plus sign is used to symbolize addition.
The numbers which are added are called *addends*.
The answer or total is called the *sum*.

Addition Facts (3)

This table contains the facts you must know in order to perform addition successfully. If you do not know these facts, you must memorize them.

Addition Facts (Continued)

Example: $5 + 7 = 12$

+	0	1	2	3	4	5	6	7	8	9
0	0	1	2	3	4	5	6	7	8	9
1	1	2	3	4	5	6	7	8	9	10
2	2	3	4	5	6	7	8	9	10	11
3	3	4	5	6	7	8	9	10	11	12
4	4	5	6	7	8	9	10	11	12	13
5	5	6	7	8	9	10	11	12	13	14
6	6	7	8	9	10	11	12	13	14	15
7	7	8	9	10	11	12	13	14	15	16
8	8	9	10	11	12	13	14	15	16	17
9	9	10	11	12	13	14	15	16	17	18

Interchanging Addends (The Commutative Property) (4)

In an addition problem the addends may be interchanged, but the sum remains the same.

Example 1: Horizontal form.

$$5 + 4 = 9$$
$$4 + 5 = 9$$

Example 2: Vertical form.

$$\begin{array}{r} 5 \\ +4 \\ \hline 9 \end{array} \qquad \begin{array}{r} 4 \\ +5 \\ \hline 9 \end{array}$$

Thus, because the addends may be interchanged, addition is said to be *commutative*.

Addition Involving Three or More Addends (The Associative Property) (5)

To evaluate a problem with three or more addends, we may add in any order and the sum will remain the same. For example, consider the following two cases where we evaluate $2 + 3 + 4$.

Horizontal Form — **Vertical Form**

Case (1):

$$\begin{aligned} 2 + 3 + 4 &= \underbrace{(2 + 3)}_{5} + 4 \\ &= 5 + 4 \\ &= 9 \end{aligned} \qquad \begin{array}{r} \left.\begin{array}{r} 2 \\ 3 \end{array}\right\} \boxed{5} \\ 4 \\ \hline 9 \end{array}$$

Case (2):

$$\begin{aligned} 2 + 3 + 4 &= 2 + \underbrace{(3 + 4)}_{7} \\ &= 2 + 7 \\ &= 9 \end{aligned} \qquad \begin{array}{r} 2 \\ \left.\begin{array}{r} 3 \\ 4 \end{array}\right\} \boxed{7} \\ \hline 9 \end{array}$$

Thus, because we may add in any order, addition is said to be *associative*.

Study Exercise One (6)

A. Fill in the blanks:

1. In the addition problem $5 + 3 = 8$

5 and 3 are called the _________ and 8 is called the _________.

2. In the addition problem

$$\begin{array}{r} 6 \\ +3 \\ \hline 9 \end{array}$$

the addends are _________ and _________. The sum is _________.

3. Addition is said to be _________ because the addends may be interchanged but the sum remains the same.

4. Addition is said to be _________ because we may add in any order and the sum will remain the same.

B. Evaluate the following. You may add in any order you wish.

5. $3 + 6$ **6.** $0 + 2$ **7.** $4 + 1 + 2$

8. $3 + 0 + 8$ **9.** $6 + 7 + 1$ **10.** $2 + 3 + 6 + 9$

Review of Place Value Notation and Expanded Form (7)

Example 1: Change 38 to place value notation by using expanded form:

Solution:

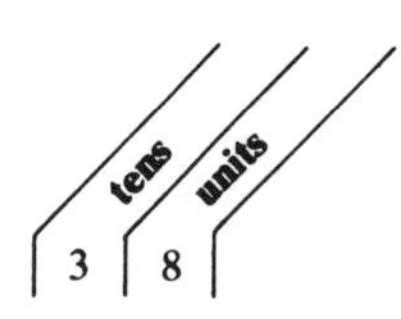

Line (a):

Line (b): 3 tens + 8 units.

Example 2: Change 58,423 to place value notation by using expanded form:

Solution:

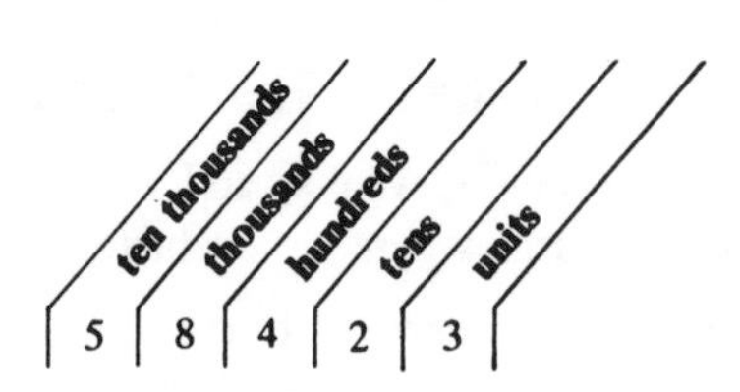

Line (a):

Line (b): 5 ten thousands + 8 thousands + 4 hundreds + 2 tens + 3 units.

Study Exercise Two (8)

Change each of the following to place value notation by using expanded form.

1. 15 **2.** 129 **3.** 4,923 **4.** 23,050

Addition Using Place Value Notation (9)

Example 1: 32 + 15

Solution: Write in vertical form, change to expanded form, and add the corresponding columns.

Step (1): Write in vertical form.

$$\begin{array}{r} 32 \\ +15 \\ \hline \end{array}$$

Step (2): Change to expanded form.

$$\begin{array}{r} 32 \\ +15 \\ \hline \end{array} \quad \text{means} \quad \begin{array}{l} 3 \text{ tens} + 2 \text{ units} \\ 1 \text{ ten} \ + 5 \text{ units} \\ \hline \end{array}$$

Step (3): Add the corresponding columns.

$$\begin{array}{l} 3 \text{ tens} + 2 \text{ units} \\ 1 \text{ ten} \ + 5 \text{ units} \\ \hline 4 \text{ tens} + 7 \text{ units} \end{array}$$

Step (4): 4 tens + 7 units is 47.

Example 2: 4,205 + 361 + 2,131 (10)

Solution:

Step (1): Write in vertical form.

$$\begin{array}{r} 4{,}205 \\ 361 \\ 2{,}131 \\ \hline \end{array}$$

Step (2): Change to expanded form.

$$\begin{array}{r} 4{,}205 \\ 361 \\ 2{,}131 \\ \hline \end{array} \quad \text{means} \quad \begin{array}{r} 4 \text{ thousands} + 2 \text{ hundreds} + 0 \text{ tens} + 5 \text{ units} \\ 3 \text{ hundreds} + 6 \text{ tens} + 1 \text{ unit} \\ 2 \text{ thousands} + 1 \text{ hundred} \ + 3 \text{ tens} + 1 \text{ unit} \\ \hline \end{array}$$

Step (3): Add the corresponding columns.

$$\begin{array}{r} 4 \text{ thousands} + 2 \text{ hundreds} + 0 \text{ tens} + 5 \text{ units} \\ 3 \text{ hundreds} + 6 \text{ tens} + 1 \text{ unit} \\ 2 \text{ thousands} + 1 \text{ hundred} \ + 3 \text{ tens} + 1 \text{ unit} \\ \hline 6 \text{ thousands} + 6 \text{ hundreds} + 9 \text{ tens} + 7 \text{ units} \end{array}$$

Step (4): 6 thousands + 6 hundreds + 9 tens + 7 units is 6,697

A Shorter Method (11)

Write in vertical form and add the corresponding place value columns:

Example 1: 32 + 15 (Compare with Frame 9):

Step (1):

$$\begin{array}{r} 32 \\ +15 \\ \hline \end{array}$$

$$\downarrow\downarrow$$

Step (2):

$$\begin{array}{r} 32 \\ +15 \\ \hline 47 \end{array}$$

Example 2: 4,205 + 361 + 2,131 (Compare with Frame 10): **(12)**

Step (1):

```
 4,205
   361
 2,131
 -----
 ↓ ↓↓↓
```

Step (2):

```
 4,205
   361
 2,131
 -----
 6,697
```

Checking Addition Problems **(13)**

Since addition is commutative and associative, the sum may be checked by adding the corresponding place value columns from bottom to top.

Example:

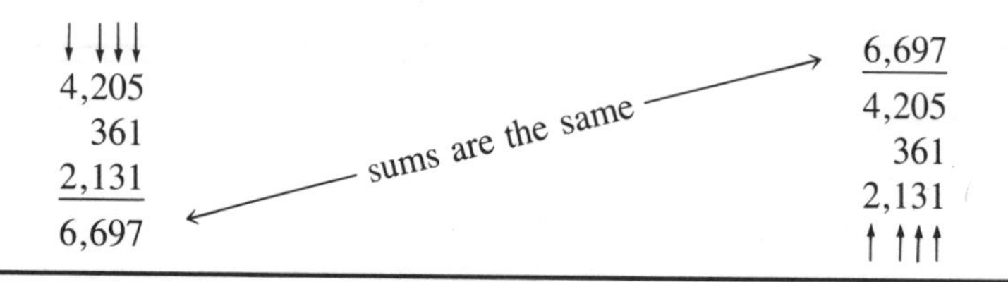

Study Exercise Three **(14)**

A. Find the sum for each of these problems by converting to expanded form:

Example: 62 + 35 **Solution:** 6 tens + 2 units
3 tens + 5 units
9 tens + 7 units or 97

1. 17 + 21
2. 54 + 23 + 20
3. 131 + 2,644 + 5,013
4. 53,074 + 125 + 2,700 + 31,100

B. Find the sum for each of these problems by using the shortcut. Also, check your answers by adding from bottom to top.

Example: 21 + 34 + 132 **Solution:**

```
  21
  34
 132
 ---
 187
```

Check:

```
 187
 ---
  21
  34
 132
 ↑↑↑
```

5. 12 + 33
6. 62 + 25 + 12
7. 203 + 1,132 + 6,344
8. 60,281 + 106 + 3,211 + 24,401

Addition Involving Carrying **(15)**

Example: 65 + 28

Line (a):

```
 65
 28
 --
  ↓
```

Addition Involving Carrying (Continued)

Line (b): Since 5 + 8 = 13 we carry the "1" to the tens' column. Leave 3 in the units' column.

Line (c):

```
 1
 65
 28
 --
  3
```

Line (d):

```
 1
 65
 28
 --
 93
```

More Examples (16)

Example 1: 34 + 18

Solution:

```
 1  ←—— space for carrying
 34
 18
 --
 52
```

Example 2: 342 + 279

Solution:

```
 11  ←—— space for carrying
 342
 279
 ---
 621
```

Example 3: 56,214 + 39,412 + 15,609

Solution:

```
 121  1  ←—— space for carrying
  56,214
  39,412
  15,609
 -------
 111,235
```

Example 4: 29 + 6 + 1,024 + 15 + 638

Solution:

```
    13   ←—— space for carrying
    29
     6
 1,024
    15
   638
 -----
 1,712
```

Study Exercise Four (17)

Find the sum for each of the following and check your answers by adding from bottom to top:

1. 78 + 16
2. 78 + 44
3. 296 + 459
4. 79,468 + 15,291 + 36,982 + 98,512
5. 12 + 6,915 + 298 + 3,001 + 79 + 907

Using Addition to Solve Applied Problems (18)

Addition is used when a problem asks for a sum or a total.

Example: A college drama club gave four performances of their play. Attendance opening night was 2,306. The second night 2,112 people attended. On the third night 2,215 people attended. And the attendance for the last performance was 2,462. What was the total attendance for the four performances?

Using Addition to Solve Applied Problems (Continued)

Solution: Add the attendance figures:

```
 2,306
 2,112
 2,215
 2,462
 -----
 9,095
```

The total attendance for the four performances was 9,095.

Study Exercise Five (19)

Solve the following applied problems:

1. Jim wishes to purchase a new stereo system. The AM/FM receiver is priced at $279. The turntable is priced at $129. The magnetic cartridge and diamond needle will cost $35. The air suspension speaker system is priced at $249 for the pair. Find the total price for the complete stereo system.

2. Find the total number of calories contained in this meal: coffee with cream and sugar, 95; rib roast, 260; green beans, 30; one slice of bread with butter, 170; one piece of iced chocolate cake, 445.

Subtraction (20)

Horizontal Form

$$\underset{\text{minuend}}{9} - \underset{\text{subtrahend}}{6} = \underset{\text{difference}}{3}$$

Vertical Form

$$\begin{array}{rl} 9 & \leftarrow \textit{minuend} \\ \underline{-6} & \leftarrow \textit{subtrahend} \\ 3 & \leftarrow \textit{difference} \end{array}$$

Check for Subtraction (21)

The difference plus the subtrahend equals the minuend.

	Problem	Check		Problem	Check
1.	$\begin{array}{r} 8 \\ \underline{-5} \\ 3 \end{array}$	$\begin{array}{r} 3 \\ \underline{+5} \\ 8 \end{array}$	2.	$8 - 5 = 3$	$3 + 5 = 8$
3.	$\begin{array}{r} 7 \\ \underline{-0} \\ 7 \end{array}$	$\begin{array}{r} 7 \\ \underline{+0} \\ 7 \end{array}$	4.	$7 - 0 = 7$	$7 + 0 = 7$
5.	$\begin{array}{r} 4 \\ \underline{-3} \\ 1 \end{array}$	$\begin{array}{r} 1 \\ \underline{+3} \\ 4 \end{array}$	6.	$4 - 3 = 1$	$1 + 3 = 4$

Study Exercise Six (22)

Subtract the following and check your answers:

1. $9 - 5$

2. $\begin{array}{r} 8 \\ \underline{-2} \end{array}$

3. $10 - 6$

4. $\begin{array}{r} 8 \\ \underline{-1} \end{array}$

5. $3 - 0$

6. $\begin{array}{r} 10 \\ \underline{-1} \end{array}$

7. $3 - 2$

8. $\begin{array}{r} 5 \\ \underline{-2} \end{array}$

Subtraction Using Place Value Notation (23)

Example 1: 96 − 32

Solution: Write in vertical form, change to expanded form, and subtract in the corresponding columns:

Step (1): Write in vertical form:

$$\begin{array}{r} 96 \\ -32 \\ \hline \end{array}$$

Step (2): Change to expanded form:

$$\begin{array}{r} 96 \\ -32 \\ \hline \end{array} \quad \text{means} \quad \begin{array}{r} 9 \text{ tens} + 6 \text{ units} \\ 3 \text{ tens} + 2 \text{ units} \\ \hline \end{array}$$

Step (3): Subtract in the corresponding columns:

$$\begin{array}{r} 9 \text{ tens} + 6 \text{ units} \\ 3 \text{ tens} + 2 \text{ units} \\ \hline 6 \text{ tens} + 4 \text{ units} \end{array}$$

Step (4): 6 tens + 4 units is 64

Step (5): Check the answer:

$$\begin{array}{r} 64 \\ +32 \\ \hline 96 \end{array}$$

Example 2: 6,765 − 523

Step (1): Write in vertical form:

$$\begin{array}{r} 6{,}765 \\ -523 \\ \hline \end{array}$$

Step (2): Change to expanded form:

$$\begin{array}{r} 6{,}765 \\ -523 \\ \hline \end{array} \quad \text{means} \quad \begin{array}{r} 6 \text{ thousands} + 7 \text{ hundreds} + 6 \text{ tens} + 5 \text{ units} \\ 5 \text{ hundreds} + 2 \text{ tens} + 3 \text{ units} \\ \hline \end{array}$$

Step (3): Subtract in the corresponding columns:

$$\begin{array}{r} 6 \text{ thousands} + 7 \text{ hundreds} + 6 \text{ tens} + 5 \text{ units} \\ 5 \text{ hundreds} + 2 \text{ tens} + 3 \text{ units} \\ \hline 6 \text{ thousands} + 2 \text{ hundreds} + 4 \text{ tens} + 2 \text{ units} \end{array}$$

Step (4): 6 thousands + 2 hundreds + 4 tens + 2 units is 6,242

Step (5): Check the answer:

$$\begin{array}{r} 6{,}242 \\ +523 \\ \hline 6{,}765 \end{array}$$

A Shorter Method (24)

Example 1: 96 − 32

Solution: Write in vertical form and subtract the digits in the corresponding place value columns:

Step (1): Write in vertical form:

$$\begin{array}{r} 96 \\ -32 \\ \hline \end{array}$$

A Shorter Method (Continued)

Step (2): Subtract the digits in the corresponding place value columns:

$$\begin{array}{r} 96 \\ -32 \\ \hline 64 \end{array}$$

Step (3): Check the answer:

$$\begin{array}{r} 64 \\ +32 \\ \hline 96 \end{array}$$

Example 2: 6,765 − 523

Step (1): Write in vertical form:

$$\begin{array}{r} 6{,}765 \\ -523 \\ \hline \end{array}$$

Step (2): Subtract the digits in the corresponding place value columns:

$$\begin{array}{r} 6{,}765 \\ -523 \\ \hline 6{,}242 \end{array}$$

Step (3): Check the answer:

$$\begin{array}{r} 6{,}242 \\ +523 \\ \hline 6{,}765 \end{array}$$

Study Exercise Seven (25)

A. Subtract the following by writing each numeral in expanded form. Check your answers by adding the difference to the subtrahend.

1. 58 − 25 **2.** 6,982 − 4,711 **3.** 28,976 − 17,005

B. Subtract the following using the shorter method. Check your answers by adding the difference to the subtrahend.

4. 74 − 21 **5.** 8,978 − 651 **6.** 58,976 − 31,760 **7.** 565,938 − 22,705

Subtraction Using Borrowing (26)

Example: 84 − 46

Long form: $\begin{array}{r} 84 \\ -46 \\ \hline \end{array}$

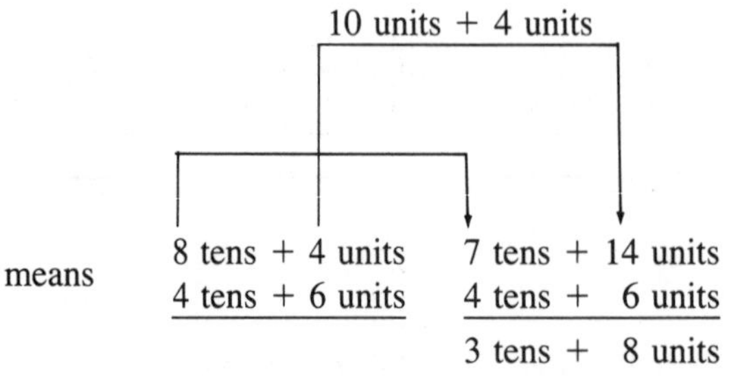

Short form: $\begin{array}{r} 84 \\ -46 \\ \hline \end{array}$ $\begin{array}{r} \overset{7}{\not 8}\,{}^{1}4 \\ 4\ 6 \\ \hline 3\ 8 \end{array}$

More About Borrowing (27)

In similar fashion we may borrow from any column since our place value system is such that each place value is 10 times the value on its right.

Example: 632 − 285

Solution:

```
   2         5 1 2
 63 ¹2       6  3 ¹2
−28 5      − 2  8  5
 ?? 7        3  4  7   (answer)
```

Check:

```
  11
  347
 +285
  632
```

More Examples (28)

Example 1: 4,347 − 1,478

Solution:

```
     3           2 1 3          3 1 2 1 3
 4,3 4¹7      4, 3  4¹7       4, 3  4¹7
−1,4 7 8     −1, 4  7 8      −1, 4  7 8
 ? ? ? 9      ?  ?  6 9       2, 8  6 9    (answer)
```

Check:

```
  1 11
  2,869
 +1,478
  4,347
```

Example 2: 900 − 265

Solution:

```
 8          8  9
 9¹00       9 ¹0¹0
−2 65      −2  6 5
 ? ??       6  3 5    (answer)
```

Check:

```
  11
  635
 +265
  900
```

Example 3: 26,050 − 9,768

Solution:

```
     4         5    4          5    9 1 4        1 5    9 1 4
 26,05¹0     26,¹05¹0      26, ¹0  5¹0      ¹2  6, ¹0  5¹0
−9,76 8     − 9, 76 8      − 9,  7  6 8      −    9,  7  6 8
 ? ? ? 2      ?  ?? 2       ??   2  8 2       1  6,   2  8 2   (answer)
```

Check:

```
  11 11
  16,282
 +9,768
  26,050
```

Study Exercise Eight (29)

Subtract the following and check your answers:

1. 517 − 279 **2.** 4,681 − 3,022 **3.** 1,684 − 255

4. 4,200 − 1,892 **5.** 3,000 − 1,784

6.
```
  48,000
 −19,736
```
7.
```
  40,300
 −21,856
```

Uses of Subtraction (30)

Subtraction is used when we wish to find:

1. the difference between two quantities,
2. how much greater the larger quantity is than the smaller, or
3. how much is left when a quantity is taken away.

Some Examples (31)

Example 1: The odometer on Harry's car registered 21,491 miles before he left on a trip. Upon his return the odometer registered 22,078 miles. How many miles did he travel on the trip?

Solution: Find the difference of 22,078 and 21,491:

```
   1             1  9
  2~~2~~,¹078     2~~2~~,¹~~0~~¹78
 −2 1 , 491     −2 1 , 4 91
 ----------     -----------
          7          5 87
```

The trip was 587 miles.

Example 2: Jim's annual salary is $9,750. Bob's annual salary is $12,225. How much greater is Bob's salary than Jim's?

Solution:
```
  1  1 1
 1~~2~~, ~~2~~¹25
 − 9 , 7 50
 ----------
   2 , 4 75
```

Bob's salary is $2,475 greater than Jim's.

Example 3: Mr. Jones had 1,500 feet of wire and used 378 feet. How many feet did he have left?

Solution:
```
    4   9
 1,~~5~~¹~~0~~¹0
 −  3  7 8
 ---------
 1, 1  2 2
```

Mr. Jones had 1,122 feet of wire left.

Study Exercise Nine (32)

1. Ted averaged 221 pins in his bowling league. George averaged 204 pins. Ted's average is how much greater than George's?
2. A contractor agrees to construct a building for $82,075. His materials and labor cost him $68,980. How much is his profit?

Study Exercise Nine (Continued)

3. Dick bought a new tire guaranteed for 40,000 miles. He drove 26,750 miles and the tire was worn out. According to the guarantee, how many more miles should the tire have lasted?
4. On January 1 a salesperson's odometer registered 13,642 miles. On February 1st the odometer registered 16,571 miles. How many miles did the salesperson drive for the month of January?

REVIEW EXERCISES (33)

1. In the addition problem

$$32 + 89 = 121$$

the addends are _________ and _________. The sum is _________.

2. In the addition problem

$$\begin{array}{r} 302 \\ +956 \\ \hline 1{,}258 \end{array}$$

302 and 956 are called _________ and 1,258 is called the _________.

3. Addition is said to be _________ because we may add in any order and the sum will remain the same. Example: $(4 + 2) + 3 = 9$ and $4 + (2 + 3) = 9$.
4. Addition is said to be _________ because the addends may be interchanged but the sum remains the same. Example: $8 + 2 = 10$ and $2 + 8 = 10$.
5. Change 29,058 into place value notation by using expanded form.
6. Add the following by changing to expanded form: 602 + 10,111 + 3,272.

Find the sum for each of the following problems; be sure to check your answers:

7. $9 + 8 + 2 + 1 + 0 + 5$
8. $19 + 23 + 56 + 7 + 14$
9. $2{,}619 + 586 + 4{,}012 + 89$
10. $51{,}612 + 92{,}915 + 21{,}713$
11. $\begin{array}{r} 10{,}026 \\ 15 \\ 5{,}948 \\ 36{,}712 \\ 4{,}893 \\ \hline \end{array}$
12. $\begin{array}{r} 2{,}698{,}512 \\ 5{,}207{,}060 \\ 318{,}945 \\ \hline \end{array}$
13. In the following subtraction problems, name the minuend, subtrahend, and difference:

(a) $9 - 7 = 2$

(b) $\begin{array}{r} 486 \\ -297 \\ \hline 189 \end{array}$

14. Explain how addition may be used to check a subtraction problem.
15. Perform the following subtraction by writing in expanded form: 5,691 − 2,380

Perform the following subtractions and check your answers:

16. $38 - 21$
17. $4{,}695 - 2{,}374$
18. $2{,}829 - 706$
19. $71 - 58$
20. $800 - 138$
21. $3{,}865 - 2{,}984$
22. $305{,}060 - 68{,}459$
23. $\begin{array}{r} 2{,}915{,}698 \\ -879{,}409 \\ \hline \end{array}$
24. $\begin{array}{r} 2{,}000{,}000 \\ -70{,}007 \\ \hline \end{array}$

Solve the following problems by using addition or subtraction.

25. A used car shows 38,924 miles on the odometer. The car is under warranty by the manufacturer for 50,000 miles. For how many more miles is the car under warranty?

REVIEW EXERCISES (Continued)

26. Bob's annual salary in his present job is $11,250. He is offered a new job paying an annual salary of $13,825. How much of a salary increase would Bob obtain if he accepted the new job?
27. An electrician had a 600-foot roll of electrical wire. In the process of wiring a room he used 237 feet. How much wire was left on the roll?
28. Janice decided to trade her car in on a later model costing $1,395. The dealer allowed her a $500 "trade-in" allowance. Exclusive of tax and license, how much extra would Janice have to pay for the later model?
29. What is the total seating capacity of a football stadium if it contains 5,200 box seats, 10,426 reserved seats, and 19,964 general admission seats?

Solutions to Review Exercises (34)

1. The addends are 32 and 89. The sum is 121.
2. 302 and 956 are called addends and 1,258 is called the sum.
3. Associative.
4. Commutative.
5. 2 ten thousands + 9 thousands + 0 hundreds + 5 tens + 8 units
6.
6 hundreds + 0 tens + 2 units
1 ten thousand + 0 thousands + 1 hundred + 1 ten + 1 unit
3 thousands + 2 hundreds + 7 tens + 2 units
1 ten thousand + 3 thousands + 9 hundreds + 8 tens + 5 units or 13,985

7.
9
8
2
1
0
5
25

8.
2
19
23
56
7
14
119

9.
1 22
2,619
586
4,012
89
7,306

10.
2 1
51,612
92,915
21,713
166,240

11.
12 12
10,026
15
5,948
36,712
4,893
57,594

12.
1 121 1
2,698,512
5,207,060
318,945
8,224,517

13. (a) Minuend 9; subtrahend 7; difference 2
(b) Minuend 486; subtrahend 297; difference 189
14. The difference plus the subtrahend must equal the minuend.
15. 5 thousands + 6 hundreds + 9 tens + 1 unit
2 thousands + 3 hundreds + 8 tens + 0 units
3 thousands + 3 hundreds + 1 ten + 1 unit or 3,311
16. 17
17. 2,321
18. 2,123
19. 13
20. 662
21. 881
22. 236,601
23. 2,036,289
24. 1,929,993
25. 11,076 miles left under warranty.
26. $2,575 increase in annual salary.
27 363 feet left on the roll.
28. $895 extra to trade cars.
29. 35,590 seats

SUPPLEMENTARY PROBLEMS

1. In this addition problem

$$121 + 236 = 357$$

121 and 236 are called _________ and 357 is called the _________.

2. In this addition problem

$$\begin{array}{r} 189 \\ +28 \\ \hline 217 \end{array}$$

the addends are _________ and _________. The sum is _________.

3. 12 + 8 = 20 and 8 + 12 = 20. This example illustrates that addition is _________.
4. (6 + 7) + 2 = 15 and 6 + (7 + 2) = 15. This example illustrates that addition is _________.
5. Change 302,465 into place value notation by using expanded form.
6. Add the following by changing to expanded form: 1,302 + 456 + 41.

Find the sum for each of the following problems; be sure to check your answers:

7. 2 + 4 + 0 + 5
8. 23 + 6 + 1 + 59 + 4
9. 25 + 65 + 183
10. 5,809 + 3,095 + 12,649
11. 53,941 + 68,011 + 92,349 + 129,814
12. 951 + 82 + 6,950 + 4 + 82,981
13. $\begin{array}{r} 129 \\ 20{,}692 \\ 3{,}413 \\ 12{,}971 \\ 15{,}604 \\ \hline \end{array}$
14. $\begin{array}{r} 369 \\ 81 \\ 2{,}040 \\ 54{,}985 \\ 408 \\ \hline \end{array}$
15. $\begin{array}{r} 21 \\ 326 \\ 5{,}029 \\ 85 \\ 647 \\ 9{,}087 \\ \hline \end{array}$
16. $\begin{array}{r} 1{,}295{,}642 \\ 5{,}468{,}711 \\ 2{,}905{,}829 \\ 6{,}002{,}193 \\ \hline \end{array}$
17. Study the following problems: $\begin{array}{r} 834 \\ +58 \\ \hline 892 \end{array}$ $\begin{array}{r} 562 \\ -86 \\ \hline 476 \end{array}$

Which of the above numbers is:
(a) A sum?
(b) An addend?
(c) A difference?
(d) A subtrahend?
(e) A minuend?

18. Perform the following subtraction by writing each number in expanded form: 5,963 − 402

Perform the following subtractions and check your answers:

19. 5,869 − 3,143
20. 7,639 − 3,528
21. 12,484 − 6,597
22. 60,502 − 28,767
23. 7,000 − 2,513
24. 73,008 − 49,679
25. 3,205,612 − 1,951,946

SUPPLEMENTARY PROBLEMS (Continued)

Solve the following problems using addition and/or subtraction:

26. Disneyland had the following attendance for each day of the week: Sunday, 31,362; Monday, 28,795; Tuesday, 23,647; Wednesday, 26,004; Thursday, 19,793; Friday, 20,782; Saturday, 36,271. What was the total attendance for the week?

27. Before Jose left on a trip, the odometer of his automobile read 48,652 miles. At the end of the trip the odometer reading was 50,028 miles. How far did Jose travel on his trip?

28. Subtract 695 from the sum of 219 and 1,142.

29. Fred had $110 and then he spent $25 for theater tickets, $7 for a corsage, and $6 for gasoline. How much money does he have left?

30. A motion picture theater has 500 seats. 386 tickets were sold for a performance. How many seats were vacant?

31. A contractor agreed to build a house for $87,225. Her costs were as follows: labor, $36,120; materials, $40,495; insurance, $1,150. How much was her profit?

32. From the sum of 3,492 and 9,065, subtract the sum of 1,096 and 5,607.

33. What is the total seating capacity of a baseball stadium if it contains 3,800 box seats, 11,048 reserved seats, and 17,928 general admission seats?

34. During a month a salesman bought the following amounts of gasoline for his car: 15 gallons, 12 gallons, 21 gallons, 9 gallons, 10 gallons, 8 gallons, 17 gallons, and 19 gallons. How many gallons of gasoline did he buy for the entire month?

35. The college library contains 1,065 magazines, 4,982 fiction books, 3,678 nonfiction books, and 1,074 reference books. What is the total number of magazines and books?

36. Jim decided to buy a new car having a base price of $8,143. He also wanted the following accessories: automatic transmission, $296; AM/FM radio, $206; interior decor trim, $63; exterior decor group, $38; deluxe wheel covers, $59; and convenience light group, $42. The delivery charge is $321, the sales tax is $250, and the license is $65. What is the total price of the new car?

Find the total "sticker" price for the following automobiles.

37. Datsun 300ZX

Base price	$10,990
GL Package	2,770
T-top	795
Dealer prep.	190
Total price =	☐

38. Dodge 400 Coupe

Base price	$8,043
Paint stripes	85
WSW Radial tires	157
AM/FM Radio	273
Air conditioning	645
Total price =	☐

39. Ford Mustang

Base price	$6943
Power steering	286
Power brakes	132
FM Stereo/tape	316
Undercoating	55
Delivery	175
Less Rebate	350
Total price =	☐

SUPPLEMENTARY PROBLEMS (Continued)

40. Find the total amount of the following bank deposit.

Please List Each Check Separately by Financial Institution Number

	DOLLARS	CENTS
CURRENCY	35	—
COIN	2	—
Checks by Financial Institution No.		
20-1299 1	185	—
30-1562 2	92	—
20-3059 3	206	—
30-4123 4	56	—
5		
6		
7		
8		
9		
10		
11		
12		
13		
Total Enter on front side.		

CURRENCY COUNT FOR FINANCIAL INSTITUTION'S USE ONLY

	x 1's		
	x 2's		
	x 5's		
	x 10's		
	x 20's		
	x 50's		
	x 100's		
	TOTAL $		

41. Find the total amount of this bank deposit. Note that some cash has been returned.

Please List Each Check Separately by Financial Institution Number

	DOLLARS	CENTS
CURRENCY		
COIN		
Checks by Financial Institution No.		
30-1892 1	21	—
30-1941 2	84	—
20-2031 3	106	—
30-1703 4	284	—
20-1911 5	6	—
50-1711 6	423	—
60-0021 7	895	—
8		
9		
10		
11		
Sub-total		
Less cash received	75	—
Total Enter on front side.		

CURRENCY COUNT FOR FINANCIAL INSTITUTION'S USE ONLY

	x 1's		
	x 2's		
	x 5's		
	x 10's		
	x 20's		
	x 50's		
	x 100's		
	TOTAL $		

42. The following lists show the itemized deductions for two families' income taxes. Determine the total deduction for each family.

Type of Deduction	Jones	Santos
Medical	$ 352	$ 283
Taxes	1898	1085
Interest	3044	3968
Contributions	618	736
Casualty Losses	236	382
Miscellaneous	517	961
Total		

SUPPLEMENTARY PROBLEMS (Continued)

According to the Internal Revenue Service, the following table lists the average deductions by adjusted gross income for the 1981 tax year. For exercises 43-52, compute the total deductions for the given income category.

	Adjusted Gross Income	Medical, Dental	Taxes	Contri- butions	Interest	Casualty, Theft	Total Deductions	
43.	$ 10,000–$ 12,000	$1,226	$1,005	$ 594	$2,201	$1,350	________	43.
44.	$ 12,000–$ 16,000	$1,078	$1,205	$ 564	$2,401	$1,524	________	44.
45.	$ 16,000–$ 20,000	$ 741	$1,393	$ 593	$2,608	$1,138	________	45.
46.	$ 20,000–$ 25,000	$ 664	$1,719	$ 615	$2,782	$ 980	________	46.
47.	$ 25,000–$ 30,000	$ 604	$1,997	$ 666	$3,004	$ 929	________	47.
48.	$ 30,000–$ 40,000	$ 543	$2,492	$ 808	$3,342	$ 898	________	48.
49.	$ 40,000–$ 50,000	$ 562	$3,258	$1,084	$3,901	$1,101	________	49.
50.	$ 50,000–$ 75,000	$ 727	$4,425	$1,539	$5,075	$1,261	________	50.
51.	$ 75,000–$100,000	$ 790	$6,516	$2,632	$7,286	$1,964	________	51.
52.	$100,000–$200,000	$1,167	$9,639	$4,605	$9,844	$2,390	________	52.

Solutions to Study Exercises (6A)

Study Exercise One (Frame 6)

A.
1. 5 and 3 are called the addends and 8 is called the sum.
2. The addends are 6 and 3. The sum is 9.
3. Commutative.
4. Associative.

B.
5. 9
6. 2
7. 7
8. 11
9. 14
10. 20

Study Exercise Two (Frame 8) (8A)

1. 1 ten + 5 units.
2. 1 hundred + 2 tens + 9 units.
3. 4 thousands + 9 hundreds + 2 tens + 3 units.
4. 2 ten thousands + 3 thousands + 0 hundreds + 5 tens + 0 units.

Study Exercise Three (Frame 14) (14A)

A.
1. 1 ten + 7 units
   2 tens + 1 unit
   3 tens + 8 units or 38
2. 5 tens + 4 units
   2 tens + 3 units
   2 tens + 0 units
   9 tens + 7 units or 97

Study Exercise Three (Frame 14) (Continued)

3.
1 hundred + 3 tens + 1 unit
2 thousands + 6 hundreds + 4 tens + 4 units
5 thousands + 0 hundreds + 1 ten + 3 units
7 thousands + 7 hundreds + 8 tens + 8 units or 7,788

4.
5 ten thousands + 3 thousands + 0 hundreds + 7 tens + 4 units
1 hundred + 2 tens + 5 units
2 thousands + 7 hundreds + 0 tens + 0 units
3 ten thousands + 1 thousand + 1 hundreds + 0 tens + 0 units
8 ten thousands + 6 thousands + 9 hundreds + 9 tens + 9 units or 86,999

B. 5. **Solution:** 12 + 33 = 45 **Check:** 45 = 12 + 33

6. **Solution:** 62 + 25 + 12 = 99 **Check:** 99 = 62 + 25 + 12

7. **Solution:** 203 + 1,132 + 6,344 = 7,679 **Check:** 7,679 = 203 + 1,132 + 6,344

8. **Solution:** 60,281 + 106 + 3,211 + 24,401 = 87,999 **Check:** 87,999 = 60,281 + 106 + 3,211 + 24,401

Study Exercise Four (Frame 17) (17A)

1. (carry 1) 78 + 16 = 94

2. (carry 1) 78 + 44 = 122

3. (carries 1 1) 296 + 459 = 755

4. (carries 3 2 2 1) 79,468 + 15,291 + 36,982 + 98,512 = 230,253

5. (carries 2 2 3) 12 + 6,915 + 298 + 3,001 + 79 + 907 = 11,212

Study Exercise Five (Frame 19) (19A)

1. (carries 1 3) 279 + 129 + 35 + 249 = 692 The complete stereo system costs $692.

2. (carries 3 1) 95 + 260 + 30 + 170 + 445 = 1,000 The meal contained 1,000 calories.

Study Exercise Six (Frame 22) (22A)

	Answer	Check
1.	4	4 + 5 = 9
2.	6	6 + 2 = 8
3.	4	4 + 6 = 10
4.	7	7 + 1 = 8
5.	3	3 + 0 = 3
6.	9	9 + 1 = 10
7.	1	1 + 2 = 3
8.	3	3 + 2 = 5

Study Exercise Seven (Frame 25) (25A)

A. 1.

5 tens + 8 units	33
−2 tens + 5 units	+25
3 tens + 3 units or 33	58

2.

6 thousands + 9 hundreds + 8 tens + 2 units	2,271
−4 thousands + 7 hundreds + 1 ten + 1 unit	+4,711
2 thousands + 2 hundreds + 7 tens + 1 unit or 2,271	6,982

3.

2 ten thousands + 8 thousands + 9 hundreds + 7 tens + 6 units	11,971
−1 ten thousand + 7 thousands + 0 hundreds + 0 tens + 5 units	+17,005
1 ten thousand + 1 thousand + 9 hundreds + 7 tens + 1 unit or 11,971	28,976

B.

	Problem	Check		Problem	Check
4.	74	53	**5.**	8,978	8,327
	−21	+21		−651	+651
	53	74		8,327	8,978
6.	58,976	27,216	**7.**	565,938	543,233
	−31,760	+31,760		−22,705	+22,705
	27,216	58,976		543,233	565,938

Study Exercise Eight (Frame 29) (29A)

	Problem	Check		Problem	Check
1.	517	238	**2.**	4,681	1,659
	−2 7 9	+279		−3,0 2 2	+3,022
	2 3 8	517		1,6 5 9	4,681
3.	1,684	1,429	**4.**	4,200	2,308
	−2 5 5	+255		−1,8 9 2	+1,892
	1,4 2 9	1,684		2,3 0 8	4,200
5.	3,000	1,216	**6.**	48,000	28,264
	−1, 7 8 4	+1,784		−1 9,7 3 6	+19,736
	1, 2 1 6	3,000		2 8,2 6 4	48,000
7.	40,300	18,444			
	−2 1, 8 5 6	+21,856			
	1 8, 4 4 4	40,300			

Study Exercise Nine (Frame 32) (32A)

1.
```
  221
 -204
   17
```
Ted's average is 17 pins greater than George's.

2.
```
  82,075
 -68,980
  13,095
```
The contractor made a profit of $13,095.

3.
```
  40,000
  26,750
  13,250
```
The tire should have lasted 13,250 miles longer.

4.
```
  16,571
 -13,642
   2,929
```
The salesperson drove 2,929 miles for the month of January.

UNIT

3 Multiplication of Whole Numbers

OBJECTIVES (1)

By the end of this unit you should be able to:

1. STATE HOW MULTIPLICATION IS RELATED TO ADDITION.
2. IDENTIFY THE FACTORS AND THE PRODUCT IN A MULTIPLICATION PROBLEM.
3. MULTIPLY BY POWERS OF TEN.
4. MULTIPLY WHOLE NUMBERS USING THE CONCEPT OF CARRYING.
5. IDENTIFY THE COMMUTATIVE AND ASSOCIATIVE PROPERTIES OF MULTIPLICATION.

Multiplication (2)

Several different-looking symbols are used to represent the operation of multiplication. Each of these examples means 3 times 5.

Example 1: 3×5

Example 2: $3 \cdot 5$

Example 3: $(3)(5)$

Multiplication means repeated addition.

Example 4: 2×4 means $4 + 4$ or 8

Example 5: 2×4 also means $2 + 2 + 2 + 2$ or 8

Example 6: $3 \cdot 2$ means $2 + 2 + 2$ or 6

Example 7: $3 \cdot 2$ also means $3 + 3$ or 6

Example 8: $(4)(1)$ means $1 + 1 + 1 + 1$ or 4

Example 9: $(4)(1)$ also means 4

Multiplication Facts (3)

This table contains the facts you must know in order to perform multiplication successfully. If you do not know these facts, you must memorize them.

Example: $4 \cdot 7 = 28$

×	0	1	2	3	4	5	6	7	8	9
0	0	0	0	0	0	0	0	0	0	0
1	0	1	2	3	4	5	6	7	8	9
2	0	2	4	6	8	10	12	14	16	18
3	0	3	6	9	12	15	18	21	24	27
4	0	4	8	12	16	20	24	28	32	36
5	0	5	10	15	20	25	30	35	40	45
6	0	6	12	18	24	30	36	42	48	54
7	0	7	14	21	28	35	42	49	56	63
8	0	8	16	24	32	40	48	56	64	72
9	0	9	18	27	36	45	54	63	72	81

Study Exercise One (4)

Find the value of each multiplication in two ways, using addition:

1. 2×3 **2.** $3 \cdot 5$ **3.** $(2)(6)$ **4.** 4×5 **5.** $6 \cdot 3$

Forms of Multiplication (5)

Horizontal Forms

$3 \times 2 = 6$

or

$3 \cdot 2 = 6$

or

$(3)(2) = 6$

factors (3)(2), *product* 6

Vertical Form

$$\begin{array}{r} 3 \\ \times 2 \\ \hline 6 \end{array}$$

3, 2 ← *factors*; 6 ← *product*

The numbers which are multiplied are called *factors*.
The answer is called the *product*.

Interchanging Factors (The Commutative Property) (6)

In a multiplication problem the factors may be interchanged, but the product remains the same.

Example 1: Horizontal form.

$3 \cdot 4 = 12$
$4 \cdot 3 = 12$

Interchanging Factors (The Commutative Property) (Continued)

Example 2: Vertical form.

$$\begin{array}{r} 5 \\ \times 2 \\ \hline 10 \end{array} \qquad \begin{array}{r} 2 \\ \times 5 \\ \hline 10 \end{array}$$

Thus, because the factors may be interchanged, multiplication is said to be *commutative.*

Multiplication Involving Three or More Factors (The Associative Property) (7)

To evaluate a problem having three or more factors, we may multiply in any order and the product will remain the same. For example, consider the following two cases where we evaluate $2 \times 3 \times 4$.

Case (1): $2 \times 3 \times 4 = \underbrace{(2 \times 3)} \times 4$
$= 6 \times 4$
$= 24$

Case (2): $2 \times 3 \times 4 = 2 \times \underbrace{(3 \times 4)}$
$= 2 \times 12$
$= 24$

Because we may multiply in any order, multiplication is said to be *associative.*

Study Exercise Two (8)

A. Fill in the blanks:

1. In the multiplication problem $6 \times 5 = 30$, 6 and 5 are called ________ and 30 is called the ________.

2. In the multiplication problem

$$\begin{array}{r} 7 \\ \times 2 \\ \hline 14 \end{array}$$

the factors are ________ and ________. The product is ________.

3. Multiplication is said to be ________ because the factors may be interchanged but the product remains the same.

4. Multiplication is said to be ________ because we may multiply in any order and the product will remain the same.

B. Evaluate the following (you may multiply in any order you wish):

5. 2×6 **6.** $3 \cdot 7$ **7.** $2 \cdot 9$ **8.** $(4)(5)$
9. $2 \cdot 5 \cdot 3$ **10.** $3 \times 2 \times 4$ **11.** $(5)(2)(2)$ **12.** $2 \cdot 3 \cdot 2 \cdot 3$

Multiplication by Zero (9)

1. 0×3 or 3×0 means $0 + 0 + 0$ or 0
2. 2×0 or 0×2 means $0 + 0$ or 0
3. $8 \times 0 = 0$
4. $0 \times 8 = 0$

If any number is multiplied by 0, the product is 0.

Multiplication by One (10)

1. 2×1 or 1×2 means $1 + 1$ or 2
2. 1×4 or 4×1 means $1 + 1 + 1 + 1$ or 4
3. $7 \times 1 = 7$
4. $1 \times 7 = 7$

If any number is multiplied by 1, the product is the original number.

Study Exercise Three (11)

Find the product for each of the following:

1. 6×0
2. 0×6
3. $8 \cdot 1$
4. $1 \cdot 8$
5. 32×1
6. $(32)(0)$
7. $\begin{array}{r} 5 \\ \times 0 \\ \hline \end{array}$
8. $\begin{array}{r} 1 \\ \times 2 \\ \hline \end{array}$
9. $\begin{array}{r} 12 \\ \times 0 \\ \hline \end{array}$
10. $\begin{array}{r} 169 \\ \times 1 \\ \hline \end{array}$
11. $\begin{array}{r} 169 \\ \times 0 \\ \hline \end{array}$
12. $(1)(1)$
13. 0×0

Multiplication by Ten (12)

1. 2×10 means $10 + 10$ or 20
2. 3×10 means $10 + 10 + 10$ or 30
3. $6 \times 10 = 60$
4. $9 \times 10 = 90$
5. $10 \times 10 = 100$
6. $100 \times 10 = 1{,}000$

Rule for Multiplying by Ten (13)

When multiplying by 10, the product may be obtained by attaching a zero to the right of the other factor.

Examples:

1. $6 \times 10 = 60$

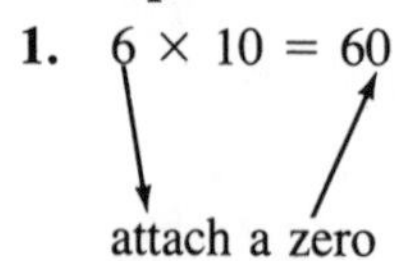

2. $10 \times 10 = 100$
attach a zero

3. 132 → attach a zero
×10
1,320

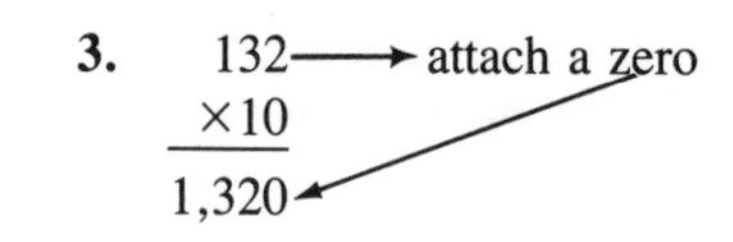

Multiplication by One Hundred (14)

1. 3×100 means $100 + 100 + 100$ or 300
2. 2×100 means $100 + 100$ or 200
3. $10 \times 100 = 1{,}000$
4. $100 \times 100 = 10{,}000$
5. $1{,}000 \times 100 = 100{,}000$

Rule for Multiplying by One Hundred (15)

When multiplying by 100, the product may be obtained by attaching *two zeros* to the right of the other factor.

Examples:

1. $7 \times 100 = 700$

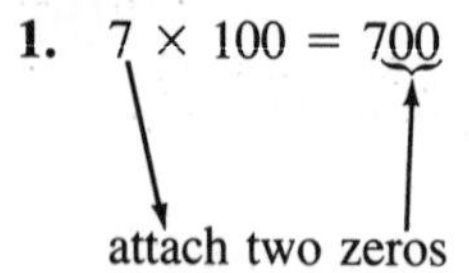

2. $1{,}000 \times 100 = 100{,}000$

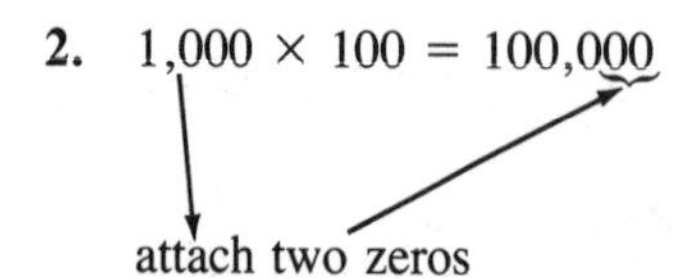

3. 1,026 → attach two zeros
×100
102,600

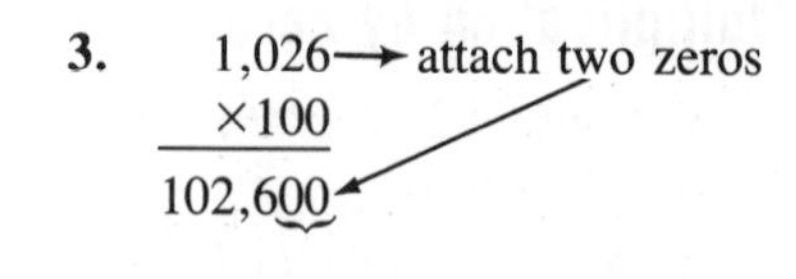

Study Exercise Four (16)

A. Multiply each of the following by 10:

1. 2
2. 5
3. 10
4. 10,000
5. 635
6. 630
7. 1,295
8. 100,000

***Study Exercise Four* (Continued)**

B. Multiply each of the following by 100:

9. 6	**10.** 9	**11.** 100	**12.** 1,000
13. 763	**14.** 760	**15.** 2,642	**16.** 100,000

Powers of Ten (17)

1	unit
10	ten
100	one hundred
1,000	one thousand
10,000	ten thousand
100,000	one hundred thousand
1,000,000	one million

Rule for Multiplying by Powers of Ten (18)

When multiplying by a power of ten, attach the number of zeros that follow the one.

Examples:

1. 63 × 1,000 = 63,000
attach three zeros

2. 4,397
×100
439,700
attach two zeros

3. 925 × 10,000 = 9,250,000
attach four zeros

4. 100 × 100,000 = 10,000,000
attach five zeros

5. 93 × 1 = 93
attach no zeros

Study Exercise Five (19)

A. Multiply each of the following by 1,000:

1. 8	**2.** 230	**3.** 10	**4.** 100

B. Multiply each of the following by 100,000:

5. 6	**6.** 50	**7.** 10	**8.** 1,000

C. How many zeros would be attached to the other factor if you multiplied by:

9. One hundred.	**10.** Ten.	**11.** One unit.
12. Ten thousand.	**13.** One thousand.	**14.** Ten million.

Multiplication Involving Place Values (20)

1. Units × units = units	1 × 1 = 1
2. Units × tens = tens	1 × 10 = 10
3. Units × hundreds = hundreds	1 × 100 = 100
4. Tens × units = tens	10 × 1 = 10
5. Tens × tens = hundreds	10 × 10 = 100
6. Tens × hundreds = thousands	10 × 100 = 1,000

Multiplication Involving Place Values (Continued)

7.	Hundreds × units = hundreds	$100 \times 1 = 100$
8.	Hundreds × tens = thousands	$100 \times 10 = 1{,}000$
9.	Hundreds × hundreds = ten thousands	$100 \times 100 = 10{,}000$

Multiplication of 21 × 4 (Long Form) (21)

Write each factor in place value notation by using expanded form. Multiply each term of one factor times each term of the other factor.

Step (1): Change to expanded form:

$$\begin{array}{r} 21 \\ \underline{\times 4} \end{array} \quad \text{means} \quad \begin{array}{r} 2 \text{ tens} + 1 \text{ unit} \\ \underline{4 \text{ units}} \end{array}$$

Step (2): Multiply each term of one factor times each term of the other factor:

$$\begin{array}{r} 2 \text{ tens} + 1 \text{ unit} \\ \underline{4 \text{ units}} \\ 8 \text{ tens} + 4 \text{ units or } 84 \end{array}$$

Multiplication of 21 × 4 (Short Form) (22)

$$\begin{array}{r} 21 \\ \underline{\times 4} \\ 84 \end{array}$$

Multiplication of 211 × 32 (Long Form) (23)

Step (1): Change to expanded form:

$$\begin{array}{r} 211 \\ \underline{\times 32} \end{array} \quad \text{means} \quad \begin{array}{r} 2 \text{ hundreds} + 1 \text{ ten} + 1 \text{ unit} \\ \underline{3 \text{ tens} + 2 \text{ units}} \end{array}$$

Step (2): Multiply each term of the upper factor by each term of the lower factor:

$$\begin{array}{r} 2 \text{ hundreds} + 1 \text{ ten} + 1 \text{ unit} \\ \underline{3 \text{ tens} + 2 \text{ units}} \\ \left.\begin{array}{r} 4 \text{ hundreds} + 2 \text{ tens} + 2 \text{ units} \\ \underline{6 \text{ thousands} + 3 \text{ hundreds} + 3 \text{ tens}} \end{array}\right\} \textit{partial products} \\ 6 \text{ thousands} + 7 \text{ hundreds} + 5 \text{ tens} + 2 \text{ units}\} \textit{ sum of partial products} \end{array}$$

Multiplication of 211 × 32 (Short Form) (24)

$$\begin{array}{rr} & 211 \\ & \underline{\times 32} \\ \textit{line (a)} \rightarrow & 422 \\ \textit{line (b)} \rightarrow & \underline{633} \\ \textit{line (c)} \rightarrow & 6752 \end{array} \quad \begin{array}{l} \text{(a), (b): } \textit{partial products} \\ \text{(c): } \textit{sum of partial products} \end{array}$$

The rightmost digit of each partial product must occupy the same place value position as the corresponding digit of the lower factor.

Study Exercise Six (25)

A. Multiply the following using the long form where each factor is written in expanded form:

1. 121 × 32　　**2.** (211)(42)

***Study Exercise Six* (Continued)**

B. Multiply the following using the short form:

3. 121 ×32 **4.** 112 ×22 **5.** 201 ×42

6. 311 × 21 **7.** 412 · 12 **8.** (1,012)(123)

Working With Zeros in Multiplication (26)

Example 1: 1,231 × 102

```
  1,231        1,231
   ×102         ×102     insert one zero
  -----        -----     and start next
  2 462        2 462     partial product
  00 00       123 10 <-
 123 1        -------
 -------      125,562
 125,562
```

Example 2: 12,131 × 2,001

```
   12,131         12,131
   ×2,001         ×2,001     insert two zeros
  -------        --------    and start next
   12 131         12 131     partial product
   000 00      24 262 00 <-
  0 000 0      ----------
 24 262        24,274,131
 ----------
 24,274,131
```

Example 3: 23,112 × 3,000

```
    23,112          23,112
    ×3,000          ×3,000    insert three
 ---------      ----------    zeros
    00 000      69,336,000 <-
    000 00
   0 000 0
 69 336
 ----------
 69,336,000
```

Study Exercise Seven (27)

Multiply each of the following by the short form in two ways: (a) keep all zeros in partial products; (b) delete as many zeros for partial products as possible.

1. 2,131 ×201 **2.** 2,131 ×2,001

3. 21,311 × 1,020 **4.** 12,112 × 4,000

Multiplication in Short Form with Carrying (28)

Example 1: 457 × 8

```
   45  <---- carrying space for units' multiplier
  457
   ×8
 -----
 3,656
```

Multiplication in Short Form with Carrying (Continued)

Example 2: 457 × 28

```
  1 1  ←—— carrying space for tens' digit
  4 5  ←—— carrying space for units' digit
  457
  ×28
 3 656
 9 14
12,796
```

More Examples (29)

1. 3,529 × 56

```
  2 14
  3 15
  3,529
    ×56
 21 174
176 45
197,624
```

2. 826 × 294

```
     1
    25
    12
   826
  ×294
 3 304
 74 34
165 2
242,844
```

3. 50,695 × 3,050

```
   2 21
   13 42
   50,695
  ×3,050
 2 534 750
152 085 0
154,619,750
```

Checking Answers to Multiplication Problems (30)

Because multiplication is commutative, you can check your work by interchanging the factors. The product should remain the same.

Example: 492 × 67

```
 Problem        Check
                    2
   5 1              6
   6 1              1
   492             67
   ×67           ×492
 3 444            134
 29 52           6 03
 32,964          26 8
                32,964
```

Study Exercise Eight (31)

Multiply the following and check your answers by interchanging the factors:

1. 23×7 **2.** $296 \cdot 78$ **3.** $(304)(203)$

4. $2,396 \times 259$ **5.** $50,406 \times 2,080$

Written Problems Using Multiplication (32)

Example 1: A company employs 237 people at an average salary of $938 a month. How much is the company's total monthly payroll?

Solution: Multiply 938 by 237:

$$\begin{array}{r} {\scriptstyle 1} \\ {\scriptstyle 12} \\ {\scriptstyle 25} \\ 938 \\ \underline{\times 237} \\ 6\ 566 \\ 28\ 14\ \ \\ \underline{187\ 6\ \ \ \ \ } \\ 222,306 \end{array}$$

The total monthly payroll is $222,306.

Example 2: On the average each person will use 60 gallons of water per day. How many gallons of water will a city use per day if its population is 600,000?

Solution: Multiply 600,000 by 60.

$$\begin{array}{r} 600,000 \\ \underline{\times 60} \\ 36,000,000 \end{array}$$

The city will use 36,000,000 gallons of water per day.

Study Exercise Nine (33)

Solve the following written problems:

1. Ed's salary is $1,246 per month. How much is his salary for one year?

2. A hydroelectric plant can generate 675,000 kilowatts per day. How many kilowatts of electrical energy can it generate in a year?

REVIEW EXERCISES (34)

A. Find the value of each multiplication two ways, using addition:

1. 3×4 **2.** $2 \cdot 5$ **3.** $(2)(3)$

B. Fill in the blanks:

4. In this multiplication problem, $6 \times 8 = 48$, the factors are ________ and ________. The product is ________.

5. In this multiplication problem

$$\begin{array}{r} 12 \\ \underline{\times 6} \\ 72 \end{array}$$

12 and 6 are called ________ and 72 is called the ________.

REVIEW EXERCISES (Continued)

6. Multiplication is said to be ________ because we may multiply in any order and the product will remain the same. Example: $(2 \cdot 4) \cdot 5 = 40$ and $2 \cdot (4 \cdot 5) = 40$.

7. Multiplication is said to be ________ because the factors may be interchanged and the product remains the same. Example: $3 \cdot 7 = 21$ and $7 \cdot 3 = 21$.

C. Find the product for each of the following:

8. $2 \times 4 \times 3$ **9.** $6 \cdot 2 \cdot 2$ **10.** $(2)(3)(4)(2)$ **11.** 56×0
12. 56×1 **13.** $75 \cdot 10$ **14.** $75 \cdot 100$ **15.** $75 \cdot 1{,}000$
16. $75 \cdot 10{,}000$ **17.** 0×1 **18.** $(3{,}958)(1{,}000)$ **19.** $(1{,}000)(1{,}000)$

D. Multiply the following using the long form where each factor is written in expanded form:

20. 24×12 **21.** 411×21

E. Multiply the following using the short form and check your answers:

22. 52×3 **23.** $(201)(24)$ **24.** $(318)(30)$ **25.** $(318)(300)$
26. $2{,}968 \times 43$ **27.** $50{,}208 \times 2{,}005$ **28.** $(318{,}659)(3{,}258)$

F. Solve the following written problems:

29. If a certain carpet sells for $23 a square yard installed, how much will it cost, less tax, to carpet a house containing 267 square yards?

30. If an automobile averages 26 miles per gallon of gas, how many miles were driven on a trip, if 155 gallons of gas were used?

Solutions to Review Exercises (35)

A. 1. $3 + 3 + 3 + 3 = 12$ or $4 + 4 + 4 = 12$
2. $5 + 5 = 10$ or $2 + 2 + 2 + 2 + 2 = 10$
3. $3 + 3 = 6$ or $2 + 2 + 2 = 6$

B. 4. The factors are 6 and 8. The product is 48.
5. 12 and 6 are called factors and 72 is called the product.
6. Associative.
7. Commutative.

C. **8.** 24 **9.** 24 **10.** 48 **11.** 0
12. 56 **13.** 750 **14.** 7,500 **15.** 75,000
16. 750,000 **17.** 0 **18.** 3,958,000 **19.** 1,000,000

D. 20.

```
                2 tens + 4 units
                1 ten  + 2 units
                ----------------
                4 tens + 8 units
2 hundreds + 4 tens
--------------------------------
2 hundreds + 8 tens + 8 units or 288
```

21.

```
            4 hundreds + 1 ten  + 1 unit
                         2 tens + 1 unit
            ----------------------------
            4 hundreds + 1 ten + 1 unit
8 thousands + 2 hundreds + 2 tens
----------------------------------------
8 thousands + 6 hundreds + 3 tens + 1 unit or 8,631
```

22. 156 **23.** 4,824 **24.** 9,540 **25.** 95,400
26. 127,624 **27.** 100,667,040 **28.** 1,038,191,022

F. 29. $267 \times 23 = 6{,}141$; $6,141 to carpet the house.
30. $155 \times 26 = 4{,}030$; 4,030 miles were driven on the trip.

SUPPLEMENTARY PROBLEMS

A. Find the value of each multiplication two ways, using addition:

1. $5 \cdot 6$ **2.** 4×2 **3.** $(3)(4)$

B. Fill in the blanks:

4. In this multiplication problem, $10 \times 31 = 310$, 10 and 31 are called _________ and 310 is called the _________.

5. In this multiplication problem,

$$\begin{array}{r} 35 \\ \underline{\times 8} \\ 280 \end{array}$$

the factors are _________ and _________. The product is _________.

6. $7 \times 9 = 63$ and $9 \times 7 = 63$. This example illustrates that multiplication is _________.

7. $(2 \times 5) \times 3 = 30$ and $2 \times (5 \times 3) = 30$. This example illustrates that multiplication is _________.

8. Repeated addition is called _________.

C. Find the product for each of the following:

9. $3 \times 2 \times 4$ **10.** $2 \times 2 \times 3 \times 3$ **11.** $4 \cdot 4 \cdot 2$
12. $5 \cdot 1$ **13.** 0×5 **14.** 1×5
15. 5×0 **16.** $2{,}695 \times 0$ **17.** $2{,}695 \times 1$
18. $2{,}695 \times 10$ **19.** $2{,}695 \times 100$ **20.** $2{,}695 \times 10{,}000$
21. $(100)(10)$ **22.** $(10)(10)$ **23.** $(10{,}000)(10{,}000)$

D. Multiply the following using the long form where each factor is written in expanded form:

24. 123×3 **25.** $2{,}113 \times 122$

E. Multiply the following using the short form and check your answers:

26. 32×9 **27.** $3{,}076 \times 38$ **28.** $2{,}529 \times 205$
29. $(40{,}105)(3{,}050)$ **30.** $(695)(400)$ **31.** $(23{,}960)(8{,}671)$

F. Solve the following applied problems:

32. Becky earns $1,150 per month. How much will she earn in one year?

33. ABC Department Stores has 56 employees. Each employee works 40 hours per week and earns $8 per hour. How much is the weekly payroll?

34. The Smiths own 385 shares of stock worth $28 per share. How much is their stock worth?

35. Find the cost of 275 feet of chain link fence if each foot costs $4.

36. Each month Art saves $175 from his paycheck. How much will he save in three years?

37. If building costs run $53 a square foot, how much would it cost to build a home containing 2,250 square feet?

38. Vincent buys a color TV, agreeing to pay $24 per month for 24 months. What is the total cost of the TV?

39. Kim agrees to purchase a used car for $1,000 down and $86 per month for three years. What is the total cost of the car? (Don't forget to add in the amount of the down payment.)

40. A certain shoe company manufactures and sells 3,785,000 pairs of shoes per year. What total revenue does the company earn if the average sales price of a pair of shoes is $18?

41. A secretary can type an average of 95 words per minute. How many words can he type in one hour?

SUPPLEMENTARY PROBLEMS (Continued)

42. A gallon of paint is advertised to cover 550 square feet. How many square feet will 15 gallons cover?
43. Find the cost of four automobile tires if each tire costs $65.
44. Find the distance a car can travel on 23 gallons of gasoline if it averages 28 miles per gallon.
45. Determine the distance a car travels in 5 hours if it averages 55 miles per hour.
46. A jet airliner flies at an average speed of 585 miles per hour. How far will it travel in 6 hours?
47. If a ream of typing paper contains 500 sheets, how many sheets are contained in 12 reams?
48. On a certain road map 1 inch represents 25 miles. What is the actual distance between two cities if they are 5 inches apart on the map?
49. A theater contains 23 rows of seats. If each row has 18 seats, what is the seating capacity of the theater?
50. If the speed of light is 186,284 miles per second, how far would light travel in 2 hours?

Solutions to Study Exercises (4A)

Study Exercise One (Frame 4)

1. $2 + 2 + 2 = 6$ or $3 + 3 = 6$
2. $3 + 3 + 3 + 3 + 3 = 15$ or $5 + 5 + 5 = 15$
3. $2 + 2 + 2 + 2 + 2 + 2 = 12$ or $6 + 6 = 12$
4. $4 + 4 + 4 + 4 + 4 = 20$ or $5 + 5 + 5 + 5 = 20$
5. $6 + 6 + 6 = 18$ or $3 + 3 + 3 + 3 + 3 + 3 = 18$

Study Exercise Two (Frame 8) (8A)

A.
1. 6 and 5 are called factors and 30 is called the product.
2. The factors are 7 and 2. The product is 14.
3. Commutative.
4. Associative.

B.
5. 12 | 6. 21 | 7. 18 | 8. 20
9. 30 | 10. 24 | 11. 20 | 12. 36

Study Exercise Three (Frame 11) (11A)

1. 0 | 2. 0 | 3. 8 | 4. 8 | 5. 32
6. 0 | 7. 0 | 8. 2 | 9. 0 | 10. 169
11. 0 | 12. 1 | 13. 0

Study Exercise Four (Frame 16) (16A)

A.
1. $2 \times 10 = 20$
2. $5 \times 10 = 50$
3. $10 \times 10 = 100$
4. $10{,}000 \times 10 = 100{,}000$
5. $635 \times 10 = 6{,}350$
6. $630 \times 10 = 6{,}300$
7. $1{,}295 \times 10 = 12{,}950$
8. $100{,}000 \times 10 = 1{,}000{,}000$

B.
9. $6 \times 100 = 600$
10. $9 \times 100 = 900$
11. $100 \times 100 = 10{,}000$
12. $1{,}000 \times 100 = 100{,}000$
13. $763 \times 100 = 76{,}300$
14. $760 \times 100 = 76{,}000$
15. $2{,}642 \times 100 = 264{,}200$
16. $100{,}000 \times 100 = 10{,}000{,}000$

Study Exercise Five (Frame 19) (19A)

A.
1. $8 \times 1{,}000 = 8{,}000$
2. $230 \times 1{,}000 = 230{,}000$
3. $10 \times 1{,}000 = 10{,}000$
4. $100 \times 1{,}000 = 100{,}000$

B.
5. $6 \times 100{,}000 = 600{,}000$
6. $50 \times 100{,}000 = 5{,}000{,}000$
7. $10 \times 100{,}000 = 1{,}000{,}000$
8. $1{,}000 \times 100{,}000 = 100{,}000{,}000$

Study Exercise Five (Frame 19) (Continued)

C. **9.** Two. **10.** One.
11. None. **12.** Four.
13. Three. **14.** Seven.

Study Exercise Six (Frame 25) (25A)

A. **1.**

```
                 1 hundred + 2 tens + 1 unit
                             3 tens + 2 units
                 ----------------------------
                2 hundreds + 4 tens + 2 units
3 thousands + 6 hundreds + 3 tens
---------------------------------------------
3 thousands + 8 hundreds + 7 tens + 2 units or 3,872
```

2.

```
                 2 hundred + 1 ten  + 1 unit
                             4 tens + 2 units
                 ----------------------------
                4 hundreds + 2 tens + 2 units
8 thousands + 4 hundreds + 4 tens
---------------------------------------------
8 thousands + 8 hundreds + 6 tens + 2 units or 8,862
```

B.

```
3.    121        4.    112        5.    201
      ×32              ×22              ×42
      ---              ---              ---
      242              224              402
     3 63             2 24             8 04
    -----            -----            -----
    3,872            2,464            8,442

6.    311        7.    412        8.    1,012
      ×21              ×12                ×123
      ---              ---                ----
      311              824               3 036
     6 22             4 12              20 24
    -----            -----             101 2
    6,531            4,944           -------
                                     124,476
```

Study Exercise Seven (Frame 27) (27A)

```
1.    2,131         2,131       2.    2,131          2,131
      ×201          ×201             ×2,001         ×2,001
      -----         -----            ------         ------
      2 131         2 131             2 131          2 131
      00 00        426 20             00 00       4 262 00
     426 2        -------             000 0     ----------
    -------       428,331           4 262        4,264,131
    428,331                       ---------
                                  4,264,131

3.    21,311        21,311      4.    12,112         12,112
     ×1,020        ×1,020            ×4,000         ×4,000
     ------        ------            ------         ------
     00 000        426 220           00 000     48,448,000
     426 22       21 311 0           000 00
    0 000 0     ----------          0 000 0
   21 311       21,737,220         48 448
 ----------                      ----------
 21,737,220                      48,448,000
```

Study Exercise Eight (Frame 31) (31A)

```
                                     64              1
                                     74              7
       2                                             4
1.    23           7           2.   296             78
      ×7         ×23                ×78           ×296
      ---        ---                ---           ----
      161         21              2 368            468
                  14              20 72           7 02
                 ---             ------           15 6
                 161             23,088         ------
                                                23,088
```

Study Exercise Eight (Frame 31) **(Continued)**

3.

$$\begin{array}{r} {\scriptstyle 1} \\ 304 \\ \times 203 \\ \hline 912 \\ 60\ 80 \\ \hline 61{,}712 \end{array} \qquad \begin{array}{r} {\scriptstyle 1} \\ 203 \\ \times 304 \\ \hline 812 \\ 60\ 90 \\ \hline 61{,}712 \end{array}$$

4.

$$\begin{array}{r} {\scriptstyle 11} \\ {\scriptstyle 1\ 43} \\ {\scriptstyle 3\ 85} \\ 2{,}396 \\ \times 259 \\ \hline 21\ 564 \\ 119\ 80 \\ 479\ 2 \\ \hline 620{,}564 \end{array} \qquad \begin{array}{r} {\scriptstyle 11} \\ {\scriptstyle 12} \\ {\scriptstyle 58} \\ {\scriptstyle 35} \\ 259 \\ \times 2{,}396 \\ \hline 1\ 554 \\ 23\ 31 \\ 77\ 7 \\ 518 \\ \hline 620{,}564 \end{array}$$

5.

$$\begin{array}{r} {\scriptstyle 1} \\ {\scriptstyle 3\ 4} \\ 50{,}406 \\ \times 2{,}080 \\ \hline 4\ 032\ 480 \\ 100\ 812\ 0 \\ \hline 104{,}844{,}480 \end{array} \qquad \begin{array}{r} {\scriptstyle 4} \\ {\scriptstyle 3} \\ {\scriptstyle 4} \\ 2{,}080 \\ \times 50{,}406 \\ \hline 12\ 480 \\ 832\ 00 \\ 104\ 000 \\ \hline 104{,}844{,}480 \end{array}$$

Study Exercise Nine (Frame 33) **(33A)**

1.

$$\begin{array}{r} {\scriptstyle 1} \\ 1{,}246 \\ \times 12 \\ \hline 2\ 492 \\ 12\ 46 \\ \hline 14{,}952 \end{array}$$

Ed's salary is $14,952 per year.

2.

$$\begin{array}{r} {\scriptstyle 21} \\ {\scriptstyle 43} \\ {\scriptstyle 32} \\ 675{,}000 \\ \times 365 \\ \hline 3\ 375\ 000 \\ 40\ 500\ 00 \\ 202\ 500\ 0 \\ \hline 246{,}375{,}000 \end{array}$$

246,375,000 kilowatts may be generated in a year.

UNIT

4 Division of Whole Numbers

OBJECTIVES (1)

By the end of this unit you should be able to:

1. GIVE THREE SYMBOLS THAT ARE USED TO REPRESENT DIVISION.
2. IDENTIFY THE TERMS *DIVIDEND, DIVISOR, QUOTIENT* AND *REMAINDER* IN A DIVISION PROBLEM.
3. CHECK DIVISION BY USING MULTIPLICATION AND ADDITION.
4. SHOW THAT DIVISION BY ZERO IS IMPOSSIBLE.
5. PERFORM LONG DIVISION USING THE CONCEPTS OF TRIAL DIVISOR, TRIAL DIVIDEND, AND TRIAL QUOTIENT.
6. USE DIVISION TO SOLVE APPLIED PROBLEMS.

Three Symbols for Division (2)

$$6 \div 3 \qquad 3\overline{)6} \qquad \frac{6}{3}$$

Each of these is read 6 *divided by* 3.

Parts of a Division Problem (3)

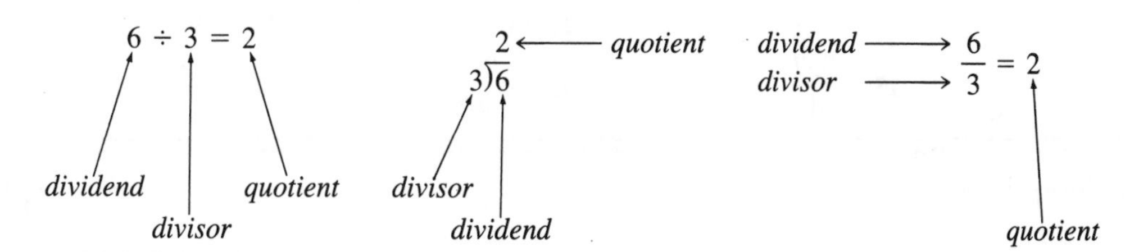

The Division Check (4)

1. $6 \div 3 = 2$ since $3 \times 2 = 6$
2. $3\overline{)6}$ with quotient 2 since $3 \times 2 = 6$

The Division Check (Continued)

3. $\frac{6}{3} = 2$ since $3 \times 2 = 6$

In a division problem, the *divisor times the quotient must equal the dividend.*

Properties of Division (5)

1. Any number divided by 1 produces the original number.

Example: $6 \div 1 = 6$ since $1 \times 6 = 6$

2. The divisor may *never* be 0.

Example: $5 \div 0$ does not represent a number because there is no quotient that can be multiplied times zero to produce 5.

3. If the dividend is zero and the divisor is not zero, then the quotient is zero.

Example: $0 \div 6 = 0$ since $6 \times 0 = 0$

More Examples (6)

1. $7 \div 1 = 7$
2. $243 \div 1 = 243$
3. $0 \div 1 = 0$
4. $0 \div 8 = 0$
5. $8 \div 0 = ?$ impossible
6. $0 \div 0 = ?$ impossible

Study Exercise One (7)

A. For each division problem, give the dividend, divisor, and quotient:

1. $6 \div 2 = 3$ **2.** $16 \div 2 = 8$ **3.** $3\overline{)24}$ with quotient 8

4. $5\overline{)15}$ with quotient 3 **5.** $\frac{21}{3} = 7$

B. Find the quotient for each of the following. If the division cannot be performed, write "impossible." Check each of your answers using multiplication.

6. $15 \div 5$	**7.** $8\overline{)16}$	**8.** $\frac{27}{9}$	**9.** $3\overline{)21}$
10. $42 \div 7$	**11.** $6 \div 1$	**12.** $6 \div 0$	**13.** $0 \div 6$
14. $7 \div 7$	**15.** $\frac{957}{1}$	**16.** $\frac{0}{2{,}963}$	**17.** $\frac{2{,}963}{0}$

The Long Division Form (8)

Example 1: $8 \div 2$

Line (a): $2\overline{)8}$ with quotient □ ⟵ 2 times "what number" is 8?

Line (b): $2\overline{)8}$ with quotient [4], 8 written below ⟵ 2 times [4] is 8.

The Long Division Form (Continued)

Line (c):

```
   4
2)8
  8 ←— subtract
  0 ←— remainder (amount left over)
```

There are 4 two's in 8 and zero left over. Thus, 8 is *evenly divisible* by 2 because the remainder is zero.

Example 2: 9 ÷ 2 **(9)**

Line (a):

```
  □ ←— 2 times "what number" is 9?
2)9
```

Line (b):

```
 [4] ←— 2 times [4] is 8.
2)9
  8
```

Line (c):

```
   4
2)9
  8 ←— subtract
  1 ←— remainder (must be less than 2)
```

There are 4 two's in 9 and a remainder of 1 left over. Therefore, 9 ÷ 2 = 4 R1.

Checking Division Involving Remainders (10)

Multiply the quotient times the divisor and add the remainder. The result must equal the dividend.

Problem: 9 ÷ 2 = 4 R1

Check: 4 × 2 = 8 and 8 + 1 = 9

4 = *quotient*, 2 = *divisor*, 1 = *remainder*, 9 = *dividend*

More Examples (11)

1. 7 ÷ 2

Solution:

```
   3
2)7
  6
  1
```

Therefore, 7 ÷ 2 = 3 R1

Check: 3 × 2 = 6 and 6 + 1 = 7

2. 23 ÷ 4

Solution:

```
    5
4)23
  20
   3
```

Therefore, 23 ÷ 4 = 5 R3

Check: 5 × 4 = 20 and 20 + 3 = 23

More Examples (Continued)

3. 98 ÷ 10

Solution:

$$\begin{array}{r} 9 \\ 10\overline{)98} \\ \underline{90} \\ 8 \end{array}$$

Therefore, 98 ÷ 10 = 9 R8

Check: 9 × 10 = 90 and 90 + 8 = 98

Study Exercise Two (12)

A. For each of the following, indicate the dividend, divisor, quotient, and remainder:

1. $\begin{array}{r} 2 \\ 5\overline{)14} \\ \underline{10} \\ 4 \end{array}$ **2.** 14 ÷ 5 = 2 R4 **3.** $\begin{array}{r} 8 \\ 10\overline{)86} \\ \underline{80} \\ 6 \end{array}$ **4.** 86 ÷ 10 = 8 R6

B. Fill in the blanks:

5. In a division problem the remainder must be less than the __________.

6. To check a division problem having a remainder, the __________ is multiplied times the divisor and the __________ is added to this product. This result must equal the __________.

7. 15 is evenly divisible by 3 because the remainder is __________.

C. Perform the following divisions and check your answers:

8. 9 ÷ 6 **9.** 27 ÷ 4 **10.** $\frac{13}{3}$ **11.** $10\overline{)56}$ **12.** $9\overline{)72}$

Long Division Where the Divisor Is a Single Digit (13)

Example 1: 95 ÷ 4

Line (a): $4\overline{)95}$ with □ ⟵ first digit of quotient goes above the "9"; 4 is smaller than 9

Line (b):

$$\begin{array}{r} 2 \\ 4\overline{)95} \\ \underline{8} \\ 1 \end{array}$$

8 ⟵ 4 × 2; 1 ⟵ subtract

Line (c):

$$\begin{array}{r} 2 \\ 4\overline{)95} \\ \underline{8}\downarrow \\ 15 \end{array}$$

15 ⟵ bring down the "5"

Line (d):

$$\begin{array}{r} 23 \\ 4\overline{)95} \\ \underline{8} \\ 15 \\ \underline{12} \\ 3 \end{array}$$

12 ⟵ 4 × 3; 3 ⟵ *remainder*

Line (e): 95 ÷ 4 = 23 R3

Example 2: 137 ÷ 6 **(14)**

Line (a):

$$\begin{array}{r} \square \\ 6\overline{)1\,3\,7} \end{array}$$
← first digit of quotient goes above the "3"

6 is smaller than 13

Line (b):

$$\begin{array}{r} 2 \\ 6\overline{)137} \\ \underline{12} \\ 1 \end{array}$$
12 ← 6 × 2
1 ← subtract

Line (c):

$$\begin{array}{r} 2 \\ 6\overline{)137} \\ \underline{12}\downarrow \\ 17 \end{array}$$
17 ← bring down the "7"

Line (d):

$$\begin{array}{r} 22 \\ 6\overline{)137} \\ \underline{12} \\ 17 \\ \underline{12} \\ 5 \end{array}$$
12 ← 6 × 2
5 ← remainder

Line (e): 137 ÷ 6 = 22 R5

Study Exercise Three **(15)**

Perform the following by long division and check your answers:

1. 87 ÷ 3 **2.** 239 ÷ 8 **3.** $6\overline{)755}$

(16)

We have just learned to perform long division where the divisor was a single digit. We will now learn to perform long division where the divisor contains two or more digits. To do this we must examine the concepts of trial divisor, trial dividend, and trial quotient.

Trial Divisor—Trial Dividend—Trial Quotient **(17)**

Example 1: 78,469 ÷ 52

	Trial Divisor	Trial Dividend	Trial Quotient
Solution: $\begin{array}{r} \boxed{1} \\ 52\overline{)78,469} \end{array}$	5~~2~~ → 5	7~~8~~ → 7	7 ÷ 5 → 1

52 is smaller than 78

Example 2: 178,469 ÷ 52

	Trial Divisor	Trial Dividend	Trial Quotient
Solution: $\begin{array}{r} \boxed{3} \\ 52\overline{)178,469} \end{array}$	5~~2~~ → 5	17~~8~~ → 17	17 ÷ 5 → 3

52 is smaller than 178

Trial Division—Trial Dividend—Trial Quotient (Continued)

Example 3: 68,405 ÷ 223

```
                 [3] <----------------------------------------------+
Solution:   223)68,405      2̸2̸3 → 2      6̸8̸4 → 6      6 ÷ 2 → 3
            \_/ \_/
   223 is smaller than 684
```

Example 4: 16,840 ÷ 223

```
                  [8] <---------------------------------------------+
Solution:   223)16,840      2̸2̸3 → 2      1,6̸8̸4 → 16      16 ÷ 2 → 8
            \_/ \__/
   223 is smaller than 1,684
```

Checking the Trial Quotient (18)

Example 1: 78,469 ÷ 52

Solution:

```
     1
52)78,469
   52 <— 52 × 1
   26 <— difference is less than 52
```

Example 2: 178,469 ÷ 52

Solution:

```
      3
52)178,469
   156 <— 52 × 3
    22 <— difference is less than 52
```

Example 3: 68,405 ÷ 223

Solution:

```
        3
223)68,405
    66 9 <— 223 × 3
     1 5 <— difference is less than 223
```

Example 4: 16,840 ÷ 223

Solution:

```
         8  <— trial quotient is too large
223)16,840
    17 84 <— 223 × 8
          <— no difference can be obtained
```

Adjusting the Trial Quotient (19)

If the trial quotient is too large, adjust it by making it smaller.

```
        8 --+
223)16,840  |
    17 84   |
            |
        7 <-+
223)16,840
    15 61 <— 223 × 7
     1 23 <— difference is less than 223
```

Another Example (20)

Example 1: 4,268 ÷ 25

```
                   [2]<----------------------------------------------+
Line (a):  25)4,268          2̸5 → 2        4̸2 → 4        4 ÷ 2 → 2
   25 is smaller than 42

                    2     <--- trial quotient is too large
Line (b):     25)4,268
                  5 0     <--- 25 × 2
                          <--- no difference can be obtained
                    1     <--- adjust by making smaller
Line (c):     25)4,268
                  2 5     <--- 25 × 1
                  1 7     <--- difference is less than 25
```

Study Exercise Four (21)

A. In each of the following division problems, indicate the place value of the first digit of the quotient by placing a box above the corresponding digit of the dividend:

Example:

```
        □
65)53, 9 46
```

1. 6)825 **2.** 6)372 **3.** 56)983 **4.** 56)257

5. 73)69,582 **6.** 73)84,106 **7.** 438)972,154 **8.** 438)300,256

B. Find the trial divisor, trial dividend, and trial quotient for each of the following:

9. 7)936 **10.** 83)956 **11.** 83)1,026

12. 28)36,581 **13.** 28)87,643 **14.** 562)243,965

C. For each of the following, the trial quotient needs to be adjusted. Give the correct adjusted trial quotient:

15. 56)257 **16.** 89)50,641 **17.** 438)300,256

Long Division Where the Divisor Contains Two or More Digits (22)

Example 1: 842 ÷ 56

Solution:

```
Line (a):       15
Line (b):   56)842
Line (c):      56↓
Line (d):      282
Line (e):      280
Line (f):        2 <--- remainder
```

Therefore, 842 ÷ 56 = 15 R2

To check the answer, multiply 56 by 15 and add the remainder, 2

```
  56     840
 ×15      +2
 280     842
 56
 840
```

Example 2: 5,692,073 ÷ 734 **(23)**

Solution:

```
Line (a):                 7 754
Line (b):          734)5,692,073
Line (c):             5 138
Line (d):               554 0
Line (e):               513 8
Line (f):                40 27
Line (g):                36 70
Line (h):                 3 573
Line (i):                 2 936
Line (j):                   637 <-- remainder
```

Therefore, 5,692,073 ÷ 734 = 7,754 R637

To check the answer, multiply 734 by 7,754 and add the remainder, 637

```
    7,754      5,691,436
  ×   734    +       637
   31 016      5,692,073
  232 62
5 427 8
5,691,436
```

Study Exercise Five (24)

A. Fill in the boxes with the correct numbers to illustrate the following long division:

```
       □ 3 7 1
1. 39)9 2 , 5 0 3
      7 8
      1 4 □
      1 1 7
        2 8 □
      □ □ □
            7 □
            3 9
            □ □
```

B. Perform the following long divisions and check your answers:

2. 1,892 ÷ 7 **3.** 1,892 ÷ 73 **4.** 142)2,556

5. 67)4,796 **6.** 188,732 ÷ 887 **7.** 5,719,437 ÷ 37,149

Working With Zeros in Division (25)

Example 1: 1,425 ÷ 7

Long Form

	Long Form	
Line (a):	203	
Line (b):	7)1,425	
Line (c):	1 4 ↓ ↓	
Line (d):	02	
Line (e):	0 ↓	← 7 × 0
Line (f):	25	
Line (g):	21	
Line (h):	4	

Therefore, 1,425 ÷ 7 = 203 R4

Short Form

	Short Form
Line (a):	203
Line (b):	7)1,425
Line (c):	1 4 ↓ ↓
Line (d):	25
Line (e):	21
Line (f):	4

Therefore, 1,425 ÷ 7 = 203 R4

Example 2: 84,056 ÷ 42 **(26)**

Long Form

	Long Form	
Line (a):	2,001	
Line (b):	42)84,056	
Line (c):	84	
Line (d):	0 0	
Line (e):	0 ↓	42 × 0
Line (f):	05	
Line (g):	0 ↓	42 × 0
Line (h):	56	
Line (i):	42	
Line (j):	14	

Therefore, 84,056 ÷ 42 = 2,001 R14

Short Form

	Short Form
Line (a):	2,001
Line (b):	42)84,056
Line (c):	84
Line (d):	0 056
Line (e):	42
Line (f):	14

Therefore, 84,056 ÷ 42 = 2,001 R 14

Study Exercise Six (27)

A. Fill in the blanks for the following long form division:

1.

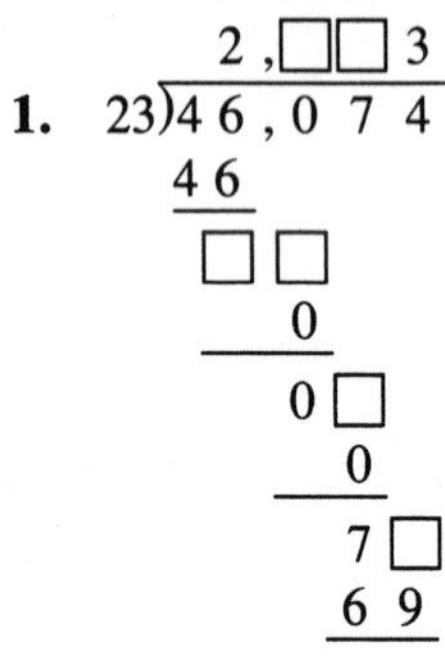

B. Fill in the blanks for the following short form division:

2.

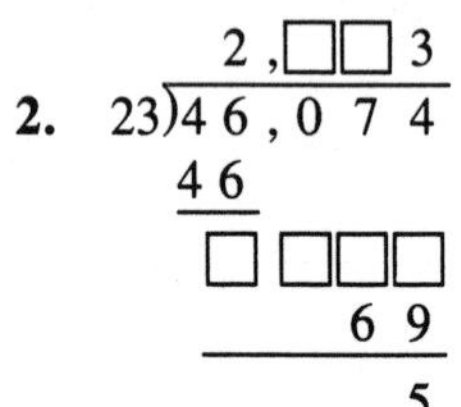

C. Divide the following using short form:

3. 1,535 ÷ 5 **4.** 54,146 ÷ 27 **5.** $\frac{42{,}182}{42}$ **6.** $\frac{13{,}225}{66}$ **7.** $59\overline{)11{,}807}$

Two Uses of Division (28)

1. Given a certain number of objects and a specified number of groups, division will tell us how many objects need to be placed in each group so that the groups will be the same size.

Example: There are 52 cards in a deck. If all the cards are to be dealt to 4 people in a bridge game, how many cards will each person receive?

Solution: There are 52 objects to be placed in 4 equal groups. Divide 52 by 4:

$$\begin{array}{r} 13 \\ 4\overline{)52} \\ \underline{4} \\ 12 \\ \underline{12} \\ 0 \end{array}$$

Each person will receive 13 cards.

2. Given a certain number of objects and a specified number of objects to be placed in a group, division will tell us how many groups are needed.

Example: At a certain college there are 450 students who wish to take an arithmetic course. If each classroom seats 30 students, how many classes should be offered?

Two Uses of Division (Continued)

Solution: There are 450 objects. Each group can hold 30. Therefore, divide 450 by 30:

$$\begin{array}{r} 15 \\ 30\overline{)450} \\ \underline{30} \\ 150 \\ \underline{150} \\ 0 \end{array}$$

15 groups or classes of arithmetic should be offered.

Study Exercise Seven (29)

Using division, solve the following word problems:

1. Juanita's annual salary is $11,100. How much is her monthly salary?
2. A building contractor estimates he needs 384 cubic yards of dirt fill. If a truck can carry 12 cubic yards of dirt, how many truckloads of fill will he need?
3. On his vacation, Harry traveled 3,290 miles and used 235 gallons of gasoline. On the average, how many miles did he travel for each gallon of gasoline?

REVIEW EXERCISES (30)

A. For the following problems name the dividend, divisor, quotient, and remainder:

1.
$$\begin{array}{r} 36 \\ 6\overline{)216} \\ \underline{18} \\ 36 \\ \underline{36} \\ 0 \end{array}$$

2.
$$\begin{array}{r} 248 \\ 43\overline{)10,674} \\ \underline{8\ 6} \\ 2\ 07 \\ \underline{1\ 72} \\ 354 \\ \underline{344} \\ 10 \end{array}$$

B. Fill in the blanks:

3. To check problem 1, 6 × __36__ = 216.

4. To check problem 2, 43 × __248__ = 10,664 and 10,664 + __10__ = 10,674.

C. Find the quotient and check your answer. If the division cannot be performed, write "impossible."

5. 16 ÷ 4 **6.** $\frac{25}{5}$ **7.** 21 ÷ 7 **8.** 0 ÷ 7

9. 7 ÷ 0 **10.** 12 ÷ 1 **11.** 72 ÷ 8 **12.** $6\overline{)42}$

13. $3\overline{)12}$ **14.** $\frac{15}{1}$ **15.** $\frac{0}{9}$ **16.** $\frac{9}{0}$

D. Perform the following long divisions and check your answers:

17. $6\overline{)95}$ **18.** $7\overline{)832}$ **19.** $42\overline{)8,875}$ **20.** 5,692 ÷ 21

21. 88,241 ÷ 44 **22.** $\frac{62,005}{17}$ **23.** $\frac{9,659}{123}$ **24.** 852,561 ÷ 272

25. 4,203 ÷ 21 **26.** $153\overline{)459,062}$ **27.** $38\overline{)17,328}$ **28.** $19\overline{)7,657}$

REVIEW EXERCISES (Continued)

E. Solve the following word problems:

29. A college scholarship fund contains $16,675. If each scholarship is to be worth $725, how many scholarships can be awarded?

30. A 120-foot block wall fence costs $720. What is the cost per running foot?

Solutions to Review Exercises (31)

A. **1.** Dividend is 216; divisor is 6; quotient is 36; remainder is 0
2. Dividend is 10,674; divisor is 43; quotient is 248; remainder is 10

B. **3.** $6 \times \underline{36} = 216$
4. $43 \times \underline{248} = 10{,}664$ and $10{,}664 + \underline{10} = 10{,}674$

C.

	Answer	Check		Answer	Check
5.	4	$4 \times 4 = 16$	**6.**	5	$5 \times 5 = 25$
7.	3	$7 \times 3 = 21$	**8.**	0	$7 \times 0 = 0$
9.	Impossible.		**10.**	12	$1 \times 12 = 12$
11.	9	$8 \times 9 = 72$	**12.**	7	$6 \times 7 = 42$
13.	4	$3 \times 4 = 12$	**14.**	15	$1 \times 15 = 15$
15.	0	$9 \times 0 = 0$	**16.**	Impossible.	

D.

	Answer	Check
17.	15 R5	$15 \times 6 = 90$ and $90 + 5 = 95$
18.	118 R6	$118 \times 7 = 826$ and $826 + 6 = 832$
19.	211 R13	$211 \times 42 = 8{,}862$ and $8{,}862 + 13 = 8{,}875$
20.	271 R1	$271 \times 21 = 5{,}691$ and $5{,}691 + 1 = 5{,}692$
21.	2,005 R21	$2{,}005 \times 44 = 88{,}220$ and $88{,}220 + 21 = 88{,}241$
22.	3,647 R6	$3{,}647 \times 17 = 61{,}999$ and $61{,}999 + 6 = 62{,}005$
23.	78 R65	$78 \times 123 = 9{,}594$ and $9{,}594 + 65 = 9{,}659$
24.	3,134 R113	$3{,}134 \times 272 = 852{,}448$ and $852{,}448 + 113 = 852{,}561$
25.	200 R3	$200 \times 21 = 4{,}200$ and $4{,}200 + 3 = 4{,}203$
26.	3,000 R62	$3{,}000 \times 153 = 459{,}000$ and $459{,}000 + 62 = 459{,}062$
27.	456	$456 \times 38 = 17{,}328$
28.	403	$403 \times 19 = 7{,}657$

E. **29.** 23 scholarships.
30. $6 per running foot.

SUPPLEMENTARY PROBLEMS

A. For the following problems name the dividend, divisor, quotient, and remainder:

1.
$$\begin{array}{r} 539 \\ 3\overline{)1{,}617} \\ \underline{1\ 5} \\ 11 \\ \underline{9} \\ 27 \\ \underline{27} \\ 0 \end{array}$$

2.
$$\begin{array}{r} 124 \\ 58\overline{)7{,}205} \\ \underline{5\ 8} \\ 1\ 40 \\ \underline{1\ 16} \\ 245 \\ \underline{232} \\ 13 \end{array}$$

B. Fill in the blanks:

3. To check problem 2, 58 × ________ = 7,192 and 7,192 + ________ = ________.

SUPPLEMENTARY PROBLEMS (Continued)

4. To check a division problem, the quotient is multiplied times the ________ and the ________ is added to this product.
5. A dividend is said to be evenly divisible by the divisor if the remainder is ________.

C. Find the following quotients and check your answers. If the division cannot be performed, write "impossible."

6. $\frac{28}{7}$ **7.** $\frac{30}{6}$ **8.** $\frac{9}{1}$ **9.** $0 \div 4$ **10.** $0 \div 0$

11. $4 \div 0$ **12.** $8\overline{)56}$ **13.** $6\overline{)18}$ **14.** $7\overline{)42}$

D. Perform the following long divisions and check your answers:

15. $3\overline{)423}$ **16.** $5\overline{)950}$ **17.** $\frac{720}{8}$

18. $6{,}642 \div 9$ **19.** $5{,}408 \div 27$ **20.** $329 \div 17$

21. $2{,}005 \div 13$ **22.** $\frac{5{,}212}{21}$ **23.** $98\overline{)1{,}205}$

24. $126\overline{)4{,}658}$ **25.** $231\overline{)46{,}251}$ **26.** $630{,}000 \div 21$

27. $2{,}400 \div 80$ **28.** $3{,}600 \div 60$ **29.** $20{,}504 \div 120$

30. $604\overline{)569{,}478}$ **31.** $236\overline{)5{,}652{,}058}$ **32.** $\frac{12{,}856}{301}$

33. $\frac{100{,}261}{92}$ **34.** $120\overline{)360{,}051}$ **35.** $765\overline{)865{,}042}$

E. Solve the following word problems:

36. A stock market investor has $1,725 which he wishes to invest in a stock costing $23 a share. Neglecting commissions and taxes, how many shares could he purchase?
37. A contractor builds a 1,450-square foot house for $76,850. What was the cost per square foot?
38. George bought a color television set for $425. If he paid $65 in cash and arranged to pay the balance in 12 equal monthly payments, what is the amount of each payment?
39. Jean earns a yearly salary of $11,700. What is her weekly salary?
40. Two cities are to be plotted on a map where the scale is 1 inch = 15 miles. If the cities are 345 miles apart, how many inches apart should they be plotted on the map?
41. A 3,500-calorie reduction is necessary to lose 1 pound of body weight. If a person reduces his calorie consumption by 700 calories per day, how long will he take to lose 1 pound?
42. Bob drives 980 miles on 35 gallons of gasoline. How many miles per gallon does his car obtain?
43. A medical patient needs to receive 300,000 units of penicillin in six doses. What is the amount of each dose?
44. A manufacturing company leases 4,200 square feet of floor space at an annual cost of $33,600. How much is the cost per square foot?
45. An 8-ounce box of cereal costs 96 cents. What is the cost per ounce?
46. A worker made $352 for 44 hours of work. How much did she make per hour?
47. A loan of $5,376 is to be repaid in 48 equal monthly installments. How much is each payment?
48. Thirteen sales people made total sales of $373,880. What was the average amount sold per person?

Solutions to Study Exercises

(7A)

Study Exercise One (Frame 7)

A.

	Dividend	Divisor	Quotient
1.	6	2	3
2.	16	2	8
3.	24	3	8
4.	15	5	3
5.	21	3	7

B.

	Quotient	Check
6.	3	$5 \times 3 = 15$
7.	2	$8 \times 2 = 16$
8.	3	$9 \times 3 = 27$
9.	7	$3 \times 7 = 21$
10.	6	$7 \times 6 = 42$
11.	6	$1 \times 6 = 6$
12.	Impossible.	
13.	0	$6 \times 0 = 0$
14.	1	$7 \times 1 = 7$
15.	957	$1 \times 957 = 957$
16.	0	$2{,}963 \times 0 = 0$
17.	Impossible.	

Study Exercise Two (Frame 12)

(12A)

A.

	Dividend	Divisor	Quotient	Remainder
1.	14	5	2	4
2.	14	5	2	4
3.	86	10	8	6
4.	86	10	8	6

B.
5. Divisor.
6. Quotient, remainder, dividend.
7. Zero.

C.

	Answer	Check
8.	1 R3	$1 \times 6 = 6$, $6 + 3 = 9$
9.	6 R3	$6 \times 4 = 24$, $24 + 3 = 27$
10.	4 R1	$4 \times 3 = 12$, $12 + 1 = 13$
11.	5 R6	$5 \times 10 = 50$, $50 + 6 = 56$
12.	8 or 8 R0	$8 \times 9 = 72$, $72 + 0 = 72$

Study Exercise Three (Frame 15)

(15A)

	Problem	Answer	Check
1.	$\begin{array}{r} 29 \\ 3\overline{)87} \\ \underline{6} \\ 27 \\ \underline{27} \\ 0 \end{array}$	29 R0	$29 \times 3 = 87$ and $87 + 0 = 87$
2.	$\begin{array}{r} 29 \\ 8\overline{)239} \\ \underline{16} \\ 79 \\ \underline{72} \\ 7 \end{array}$	29 R7	$29 \times 8 = 232$ and $232 + 7 = 239$
3.	$\begin{array}{r} 125 \\ 6\overline{)755} \\ \underline{6} \\ 15 \\ \underline{12} \\ 35 \\ \underline{30} \\ 5 \end{array}$	125 R5	$125 \times 6 = 750$ and $750 + 5 = 755$

Study Exercise Four (Frame 21) (21A)

A.

1. $6\overline{)825}$ (quotient box over 8)
2. $6\overline{)372}$ (quotient box over 7)
3. $56\overline{)983}$ (quotient box over 8)
4. $56\overline{)257}$ (quotient box over 7)
5. $73\overline{)69,582}$ (quotient box over 5)
6. $73\overline{)84,106}$ (quotient box over 1)
7. $438\overline{)972,154}$ (quotient box over 1)
8. $438\overline{)300,256}$ (quotient box over 2)

B.

	Trial Divisor		Trial Dividend		Trial Quotient	
9.	7		9		9 ÷ 7	1
10.	8~~3~~	8	9~~5~~	9	9 ÷ 8	1
11.	8~~3~~	8	10~~2~~	10	10 ÷ 8	1
12.	2~~8~~	2	3~~6~~	3	3 ÷ 2	1
13.	2~~8~~	2	8~~7~~	8	8 ÷ 2	4
14.	5~~62~~	5	2,4~~39~~	24	24 ÷ 5	4

C.

	Problem	Adjusted Trial Quotient
15.	$56\overline{)257}$ → 4; 224; 33	4
16.	$89\overline{)50,641}$ → 5; 44 5; 6 1	5
17.	$438\overline{)300,256}$ → 6; 262 8; 37 4	6

Study Exercise Five (Frame 24) (24A)

A. 1.

```
      [2] 3 7 1
39)9 2 , 5 0 3
   7 8
   1 4 [5]
   1 1 7
     2 8 [0]
    [2][7][3]
         7 [3]
         3 9
        [3][4]
```

B.

	Problem	Answer		Problem	Answer
2.	$7\overline{)1,892}$ → 270; 1 4; 49; 49; 02; 0; 2	270 R2	3.	$73\overline{)1,892}$ → 25; 1 46; 432; 365; 67	25 R67

Study Exercise Five (Frame 24) (Continued)

4.
```
      18
142)2,556
    1 42
    1 136
    1 136
        0
```
18 or 18 R0

5.
```
    71
67)4,796
   4 69
    106
     67
     39
```
71 R39

6.
```
     212
887)188,732
    177 4
     11 33
      8 87
      2 462
      1 774
        688
```
212 R688

7.
```
             153
37,149)5,719,437
       3 714 9
       2 004 53
       1 857 45
         147 087
         111 447
          35 640
```
153 R35,640

Study Exercise Six (Frame 27)

(27A)

A. 1.
```
    2 [0][0] 3
23)4 6 , 0 7 4
   4 6
   [0] [0]
        0
      0 [7]
          0
        7 [4]
        6 9
          5
```

B. 2.
```
    2 [0][0] 3
23)4 6 , 0 7 4
   4 6
   [0] [0][7][4]
            6 9
              5
```

C.

Problem	Answer	Problem	Answer
3. 5)1,535	307 or 307 R0	4. 27)54,146	2,005 R11
5. 42)42,182	1,004 R14	6. 66)13,225	200 R25
7. 59)11,807	200 R7		

3.
```
   307
5)1,535
  1 5
   035
    35
     0
```

4.
```
   2 005
27)54,146
   54
    0 146
      135
       11
```

5.
```
   1 004
42)42,182
   42
    0 182
      168
       14
```

6.
```
   200
66)13,225
   13 2
      025
        0
       25
```

7.
```
   200
59)11,807
   11 8
      007
        0
        7
```

Study Exercise Seven (Frame 29) **(29A)**

1.
$$\begin{array}{r} 925 \\ 12\overline{)11,100} \\ \underline{10\ 8} \\ 30 \\ \underline{24} \\ 60 \\ \underline{60} \\ 0 \end{array}$$

Juanita's monthly salary is $925.

2.
$$\begin{array}{r} 32 \\ 12\overline{)384} \\ \underline{36} \\ 24 \\ \underline{24} \\ 0 \end{array}$$

32 truckloads of fill.

3.
$$\begin{array}{r} 14 \\ 235\overline{)3,290} \\ \underline{2\ 35} \\ 940 \\ \underline{940} \\ 0 \end{array}$$

14 miles for each gallon of gasoline.

UNIT

5 Solving Verbal Problems by Reduction and Expansion

OBJECTIVES (1)

By the end of this unit you should be able to:

1. SOLVE REDUCTION PROBLEMS BY DIVISION.
2. SOLVE EXPANSION PROBLEMS BY MULTIPLICATION.
3. SOLVE REDUCTION-EXPANSION PROBLEMS BY USING BOTH DIVISION AND MULTIPLICATION.

Reduction (2)

Reduction means to change from *many to one* by using the operation of division.

Example 1: If 12 identical items cost a total of 60 cents, how much does 1 item cost? **(3)**

Solution: The total cost must be divided into 12 equal parts in order to find the cost of 1 item.

$$\begin{array}{r} 5 \\ 12\overline{)60} \\ \underline{60} \\ 0 \end{array}$$

Each item costs 5 cents.

Example 2: If 5 apples cost 70 cents, what is the price of 1 apple? **(4)**

Solution: The total cost must be divided into 5 equal parts.

$$\begin{array}{r} 14 \\ 5\overline{)70} \\ \underline{5\ } \\ 20 \\ \underline{20} \\ 0 \end{array}$$

Each apple costs 14 cents.

Example 3: An automobile can travel 325 miles on 25 gallons of gasoline. How far can it travel on 1 gallon of gasoline? **(5)**

Solution: The total mileage must be divided into 25 equal parts.

$$\begin{array}{r} 13 \\ 25\overline{)325} \\ \underline{25} \\ 75 \\ \underline{75} \\ 0 \end{array}$$

The automobile can travel 13 miles on 1 gallon of gasoline.

Example 4: How many identical articles will 1 dollar buy, if 15 dollars will buy 45 articles? **(6)**

Solution: Divide 45 into 15 equal parts.

$$\begin{array}{r} 3 \\ 15\overline{)45} \\ \underline{45} \\ 0 \end{array}$$

One dollar will buy 3 articles.

Study Exercise One **(7)**

Solve the following problems by reduction:

1. If 5 candy bars cost 80 cents, how much does 1 candy bar cost?
2. If 23 identical articles weigh a total of 1,863 pounds, what is the weight of 1 article?
3. A guided missile travels 5,280 feet in 6 seconds. How far does it travel in 1 second?
4. How many identical articles will 1 dollar buy if 21 dollars will buy 63 articles?

Expansion **(8)**

Expansion means to change from *one to many* by using the operation of multiplication.

Example 1: If 1 candy bar costs 7 cents, what is the cost of 12 candy bars? **(9)**

Solution: The cost of each is 7 cents, so the cost of 12 may be obtained by multiplying 12 times 7 cents.

$$\begin{array}{r} 12 \\ \underline{\times 7} \\ 84 \end{array}$$

The 12 candy bars cost a total of 84 cents.

Example 2: How far will a train travel in 60 seconds if it is traveling 88 feet each second? **(10)**

Solution: Multiply 88 feet by 60:

$$\begin{array}{r} 88 \\ \underline{\times 60} \\ 5{,}280 \end{array}$$

The train will travel 5,280 feet in 60 seconds.

Example 3: If a box weighs 17 pounds, what is the total weight of 9 such boxes? **(11)**

Solution: Multiply 17 pounds by 9:

$$\begin{array}{r} 17 \\ \times 9 \\ \hline 153 \end{array}$$

The total weight of 9 boxes is 153 pounds.

Study Exercise Two (12)

Solve the following problems by expansion:

1. If 1 pound of apples costs 29 cents, what is the cost for 3 pounds?
2. If 1 article weighs 87 pounds, find the total weight of 8 identical articles.
3. If a car can travel 16 miles on 1 gallon of gasoline, how far can it travel on 20 gallons of gasoline?
4. If the cost of 1 square yard of carpet is 13 dollars, what would be the cost to carpet a room having an area of 15 square yards?

Reduction and Expansion (13)

1. In doing reduction problems we "reduce to one" by using division.
2. In doing expansion problems we "expand to many" by using multiplication.

We will now combine these steps to solve problems involving both reduction and expansion.

Example 1: If 6 similar pencils cost 24 cents, what is the cost of 9 pencils? **(14)**

Solution:

Step (1): Organize the data:
6 pencils cost 24 cents.

Step (2): Reduce to the cost of 1:
$24 \div 6 = 4$
Each pencil costs 4 cents.

Step (3): Expand to cost for 9:
$4 \cdot 9 = 36$
The 9 pencils cost 36 cents.

Note: We begin with 6 pencils and end with 9 pencils. Therefore we divide by 6 and multiply by 9. Always divide by what you must change to one.

Example 2: If a car can travel 98 miles on 7 gallons of gasoline, find how far it can travel on 20 gallons of gasoline. **(15)**

Solution:

Step (1): Organize the data:
98 miles on 7 gallons of gasoline.

Step (2): Reduce to 1 gallon:
$98 \div 7 = 14$
The car will travel 14 miles on 1 gallon of gasoline.

Step (3): Expand to 20 gallons:
$(14)(20) = 280$
The car will travel 280 miles on 20 gallons of gasoline.

Example 3: Find the distance a car will travel in 7 seconds, if it travels 150 feet in 3 seconds. **(16)**

Solution:

Step (1): Organize the data:
150 feet in 3 seconds.

Step (2): Reduce to 1 second:
$150 \div 3 = 50$
The car will travel 50 feet in 1 second.

Step (3): Expand to 7 seconds:
$50 \times 7 = 350$
The car will travel 350 feet in 7 seconds.

Example 4: If 56 identical articles weigh 28 pounds, how many articles will weigh 7 pounds? **(17)**

Solution:

Step (1): Organize the data:
56 articles weigh 28 pounds.

Step (2): Reduce to 1 pound:
$56 \div 28 = 2$
2 articles weigh 1 pound.

Step (3): Expand to 7 pounds:
$(2)(7) = 14$
14 articles weigh a total of 7 pounds.

Study Exercise Three **(18)**

1. If 7 articles cost 28 cents, find the cost of 21 articles.
2. Find the distance a train will travel in 12 seconds if it travels 180 feet in 2 seconds.
3. If a car travels 105 miles on 7 gallons of gasoline, how far will it travel on 21 gallons?
4. If 60 identical articles weigh 15 pounds, how many articles will weigh 9 pounds?

REVIEW EXERCISES **(19)**

Solve the following problems using reduction and expansion:

1. If 1 article weighs 47 pounds, find the weight of 6 such articles.
2. If 8 identical articles weigh 120 pounds, find the weight of 1 article.
3. If 9 articles weigh 72 pounds, find the weight of 6 articles.
4. If 7 articles weigh 56 pounds, find the weight of 11 articles.
5. If a car travels 42 miles on 2 gallons of gasoline, how far will it travel on 5 gallons?
6. If a car travels 360 miles on 20 gallons of gasoline, how far will it travel on 13 gallons?
7. If 80 feet of blockwall fencing costs 480 dollars, how much will 120 feet cost?
8. If a rocket travels 8,800 feet in 5 seconds, how far will it travel in 60 seconds?
9. If an industrial plant produces 11,200 articles in 28 days, how many would it produce in 5 days?
10. Find the cost of 4 candy bars if 6 bars sell for 72 cents.
11. 84 identical articles weigh 21 pounds; how many articles will weigh 8 pounds?
12. 96 identical articles cost a total of 12 cents; how many articles may be purchased for 5 cents?

Solutions to Review Exercises (20)

1. $47 \times 6 = 282$
The 6 articles weigh 282 pounds.

2. $120 \div 8 = 15$
One article weighs 15 pounds.

3. $72 \div 9 = 8$
$8 \times 6 = 48$
The 6 articles weigh 48 pounds.

4. $56 \div 7 = 8$
$8 \times 11 = 88$
The 11 articles weigh 88 pounds.

5. $42 \div 2 = 21$
$21 \cdot 5 = 105$
The car will travel 105 miles on 5 gallons of gasoline.

6. $360 \div 20 = 18$
$(18)(13) = 234$
The car will travel 234 miles on 13 gallons of gasoline.

7. $480 \div 80 = 6$
$6 \times 120 = 720$
120 feet of fence will cost 720 dollars.

8. $8{,}800 \div 5 = 1{,}760$
$1{,}760 \times 60 = 105{,}600$
The rocket will travel 105,600 feet in 60 seconds.

9. $11{,}200 \div 28 = 400$
$400 \cdot 5 = 2{,}000$
2,000 articles would be produced in 5 days.

10. $72 \div 6 = 12$
$12 \times 4 = 48$
The 4 candy bars would cost 48 cents.

11. $84 \div 21 = 4$
$(4)(8) = 32$
32 articles would weigh a total of 8 pounds.

12. $96 \div 12 = 8$
$8 \times 5 = 40$
40 articles may be purchased for 5 cents.

SUPPLEMENTARY PROBLEMS

1. If 1 article costs 12 cents, find the cost of 8 identical articles.
2. If 7 articles cost 63 cents, find the cost of 1 article.
3. If 7 articles cost 35 cents, find the cost of 13 articles.
4. If 5 articles cost 75 cents, find the cost of 2 articles.
5. Find the time for a jet airliner to travel 5 miles if it travels 2 miles in 12 seconds.
6. Find the distance a car will travel in 12 seconds, if it travels 300 feet in 6 seconds.
7. If a car travels 60 miles on 4 gallons of gasoline, how far will it travel on 9 gallons?
8. If 96 identical objects weigh 48 pounds, how many objects weigh 16 pounds?
9. If it costs 600 dollars to drill a 200-foot well, how much will it cost to drill a 320-foot well?
10. If 15 identical boxes are required to package a total of 5 cubic feet, how many boxes will be needed to hold 7 cubic feet?
11. If 42 identical articles cost a total of 21 dollars, how many articles will cost 11 dollars?
12. If a car travels 121 miles on 11 gallons of gasoline, how far will it travel on 14 gallons?
13. If 12 articles weigh 36 pounds, find the weight of 20 articles.
14. If 50 feet of fencing costs 600 dollars, find the cost of 87 feet of fencing.
15. If a rocket travels 9,385 feet in 5 seconds, how far will it travel in 30 seconds?
16. Find the cost of 6 candy bars if 5 bars sell for 80 cents.
17. A manufacturing plant produces 582 articles in 3 days. How many articles can it produce in a week?
18. Find the time for a train to travel 18 miles if it travels 4 miles in 8 minutes.
19. If the water level in a tank rises 15 inches in 5 minutes, how far will it rise in 2 minutes?
20. A tree grew 18 inches in height in 270 days. At the same rate, how long did it take to grow 7 inches?
21. If Rick earns $660 in two weeks, how much will he earn in 9 weeks?
22. If Liz earns $1035 in 3 weeks, how much will she earn in 8 weeks?
23. If a manufacturing plant produces 3,520 items in 4 months, how many items should it be able to produce in one year?

SUPPLEMENTARY PROBLEMS (Continued)

24. If it costs $2,415,000 to construct 7 miles of highway, what is the cost per mile?

25. If 75 feet of slumpstone fencing costs $975, how much does 54 feet cost?

26. If a car travels 232 miles on 8 gallons of gasoline, how far will it travel on 11 gallons?

27. If an airplane travels 18 miles in 2 minutes, how far will it travel in 1 hour?

28. If 6 ounces of liquid soap costs 96 cents, what is the cost per ounce?

29. If 200 square feet of patio decking costs $800, find the cost of 1,000 square feet of decking.

30. Find the time for a car to travel 135 miles if it travels 315 miles in 7 hours.

Solutions to Study Exercises (7A)

Study Exercise One (Frame 7)

1. $80 \div 5 = 16$; each candy bar costs 16 cents.

2. $1{,}863 \div 23 = 81$; 1 article weighs 81 pounds.

3. $5{,}280 \div 6 = 880$; it travels 880 feet in 1 second.

4. $63 \div 21 = 3$; 1 dollar will buy 3 articles.

Study Exercise Two (Frame 12) (12A)

1. $29 \times 3 = 87$; 3 pounds costs 87 cents.

2. $87 \times 8 = 696$; 8 articles weigh 696 pounds.

3. $16 \cdot 20 = 320$; the car can travel 320 miles.

4. $(13)(15) = 195$; the cost of carpet would be 195 dollars.

Study Exercise Three (Frame 18) (18A)

2. $28 \div 7 = 4$
$4 \times 21 = 84$
The cost is 84 cents for 21 articles.

2. $180 \div 2 = 90$
$(90)(12) = 1{,}080$
The train will travel 1,080 feet in 12 seconds.

3. $105 \div 7 = 15$
$15 \times 21 = 315$
The car will travel 315 miles on 21 gallons.

4. $60 \div 15 = 4$
$4 \cdot 9 = 36$
36 articles will weigh a total of 9 pounds.

UNIT

6 Exponents, Perfect Squares, and Square Roots

OBJECTIVES (1)

By the end of this unit you should be able to:

1. IDENTIFY THE TERMS *BASE, EXPONENT, PERFECT SQUARE,* AND *SQUARE ROOT.*
2. EVALUATE EXPONENTIAL EXPRESSIONS.
3. FIND SQUARE ROOTS OF PERFECT SQUARES.

Factors (2)

Numbers related by multiplication are called *factors*.

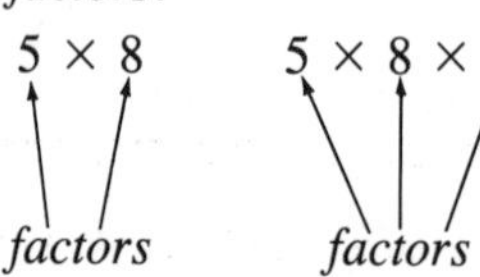

One number may be used as a factor several times. (3)

Example 1: $2 \times 2 \times 2$ **Example 2:** $3 \cdot 3 \cdot 3 \cdot 3 \cdot 3$

A Simpler Notation (4)

Example 1: $2 \times 2 \times 2 = 2^3$ **Example 2:** $3 \cdot 3 \cdot 3 \cdot 3 \cdot 3 = 3^5$

Exponential Notation (5)

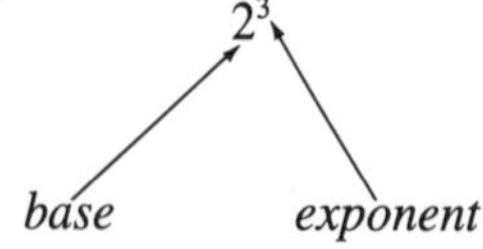

The base, 2, is used as a factor three times. The exponent indicates the number of times the base is used as a factor.

(6)

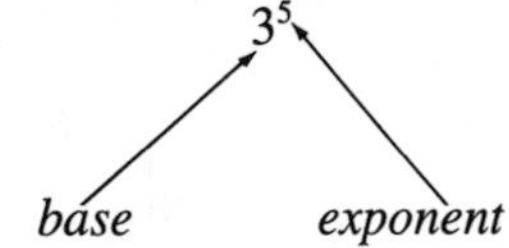

The exponent, 5, indicates that the base, 3, is used as a factor five times.

Reading Exponents

(7)

Example 1: 5^2 is read, *"5 to the second"* or "5 squared."

Example 2: 2^3 is read, *"2 to the third"* or *"2 cubed."*

Example 3: 7^5 is read, *"7 to the fifth."*

Example 4: $5^2 \times 2^3$ is read, *"5 squared times 2 cubed."*

Example 5: $3^7 \cdot 5^{12}$ is read, *"3 to the seventh times 5 to the twelfth."*

Study Exercise One

(8)

A. Indicate the base and exponent for each of the following:

1. 3^2 **2.** 2^3 **3.** 10^2 **4.** 5^7
5. 10^3 **6.** 3^{10}

B. Write a word statement describing how each of the following is read:

7. 7^2 **8.** 9^3 **9.** $7^2 \times 9^3$ **10.** $2^3 \cdot 3^5$

Changing to Exponential Notation

(9)

Example 1: $5 \times 5 = 5^2$

Example 2: $7 \cdot 7 \cdot 7 \cdot 7 = 7^4$

Example 3: $(10)(10)(10) = 10^3$

Example 4: $6 = 6^1$

Example 5: $3\times3\times5\times5\times5\times5=3^2\times5^4$

Example 6: $2\times2\times2\times5\times5\times7=2^3\times5^2\times7^1$

Sometimes exponential notation is called *exponential form.*

Study Exercise Two

(10)

Change each of the following to exponential notation:

1. 3×3 **2.** $5 \cdot 5 \cdot 5$ **3.** $2 \times 2 \times 2 \times 7 \times 7$
4. $5 \times 5 \times 5 \times 5 \times 11 \times 11 \times 13 \times 13 \times 13 \times 17$ **5.** $(2)(3)(3)(7)(7)(7)$

Evaluating Exponential Expressions

(11)

Evaluate means to find the value.

Example 1: 3^2

Solution:

Line (a): $3^2 = \underbrace{3 \times 3}$

Line (b): $3^2 = 9$

Example 2: 2^3

Solution:

Line (a): $2^3 = \underbrace{2 \times 2} \times 2$

Line (b): $2^3 = \underbrace{4 \times 2}$

Line (c): $2^3 = 8$

Evaluating Exponential Expressions (Continued)

Example 3: 3^4

Solution:

Line (a): $3^4 = \underbrace{3 \times 3}\times\underbrace{3 \times 3}$

Line (b): $3^4 = \underbrace{9 \quad \times \quad 9}$

Line (c): $3^4 = \quad 81$

Example 4: $2^3 \times 3^2$

Solution:

Line (a): $2^3 \times 3^2 = \underbrace{2 \times 2 \times 2}\times\underbrace{3 \times 3}$

Line (b): $2^3 \times 3^2 = \underbrace{8 \quad \times \quad 9}$

Line (c): $2^3 \times 3^2 = \quad 72$

Study Exercise Three (12)

Evaluate the following:

1. 5^2 **2.** 3^3 **3.** 2^2 **4.** 7^2
5. 1^3 **6.** 0^2 **7.** 10^2 **8.** 10^3
9. $2^3 \times 5^2$ **10.** $3^2 \times 5^1 \times 7^2$ **11.** $2^4 \times 3^2 \times 5^1$ **12.** $2^1 \times 3^2 \times 5^2$

Perfect Squares (13)

Any whole number is said to be a *perfect square* if it can be written using a whole number base with an exponent of 2.

Example:

$$25 = 5 \times 5 \quad \text{or} \quad 5^2$$

Therefore, 25 is a perfect square.

More Examples of Perfect Squares (14)

1. 9 is a perfect square because $9 = 3^2$
2. 16 is a perfect square because $16 = 4^2$
3. 36 is a perfect square because $36 = 6^2$
4. 100 is a perfect square because $100 = 10^2$

Study Exercise Four (15)

Decide which of the following are perfect squares. Then write using a base with an exponent of 2:

1. 25 **2.** 49 **3.** 12 **4.** 100
5. 10 **6.** 0 **7.** 1 **8.** 50
9. 64 **10.** 144

Square Roots of Perfect Squares (16)

A square root of a number is one of its two equal factors.

Example: A square root of 25 is 5 because $25 = 5 \times 5$.

equal factors

More Examples (17)

1. A square root of 9 is 3 because $9 = 3 \times 3$
2. A square root of 16 is 4 because $16 = 4 \times 4$
3. A square root of 1 is 1 because $1 = 1 \times 1$
4. A square root of 144 is 12 because $144 = 12 \times 12$

When a perfect square is written using a base with an exponent of 2, the square root is simply the base. **(18)**

Example: $25 = 5^2$

The square root is the base 5

Relationship of Squaring to Square Rooting **(19)**

Example 1: $5^2 = 25$ (*squaring*: $5^2 \rightarrow 25$; *square rooting*: $25 \rightarrow 5$)

Example 2: $7^2 = 49$ (*squaring*: $7^2 \rightarrow 49$; *square rooting*: $49 \rightarrow 7$)

Symbol for Square Root **(20)**

The symbol $\sqrt{25}$ says to find the square root of 25.

$$\sqrt{25} = 5$$

Study Exercise Five **(21)**

Find the following square roots:

1. $\sqrt{36}$ **2.** $\sqrt{49}$ **3.** $\sqrt{1}$ **4.** $\sqrt{0}$

5. $\sqrt{64}$ **6.** $\sqrt{100}$ **7.** $\sqrt{121}$

Table of Perfect Squares and Square Roots **(22)**

Perfect Square	Square Root	Perfect Square	Square Root
0	0	49	7
1	1	64	8
4	2	81	9
9	3	100	10
16	4	121	11
25	5	144	12
36	6	169	13

This table should be memorized!

REVIEW EXERCISES

(23)

A. Fill in the blanks:

1. Numbers related by multiplication are called ________.

2. For the expression 5^7, 5 is the ________ and 7 is the ________.

3. The expression 8^2 is read "8 to the ________" or "8 ________."

4. The expression 7^3 is read "7 to the ________" or "7 ________."

5. The expression 10^5 is read "10 to the ________."

B. Indicate the base and exponent for each of the following:

6. 1^2 **7.** 10^5 **8.** 5^{10} **9.** 7^8

C. Change each to exponential notation:

10. 3×3 **11.** $2 \cdot 2 \cdot 2 \cdot 2$

12. $(10)(10)(10)$ **13.** 5

14. $2 \times 2 \times 3 \times 3 \times 3$ **15.** $5 \times 7 \times 7 \times 11 \times 11 \times 11 \times 11$

D. Evaluate the following:

16. 4^2 **17.** 5^3 **18.** 6^1 **19.** 1^3

20. 7^3 **21.** 10^3 **22.** $2^2 \times 5^2$ **23.** $3^2 \times 5^1 \times 7^2$

E. Find the following square roots:

24. $\sqrt{4}$ **25.** $\sqrt{49}$ **26.** $\sqrt{100}$ **27.** $\sqrt{64}$

28. $\sqrt{81}$ **29.** $\sqrt{1}$ **30.** $\sqrt{0}$ **31.** $\sqrt{169}$

32. $\sqrt{144}$

Solutions to Review Exercises

(24)

A. **1.** Factors. **2.** Base, exponent. **3.** Second, squared.

4. Third, cubed. **5.** Fifth.

B.

	Base	Exponent
6.	1	2
7.	10	5
8.	5	10
9.	7	8

C. **10.** 3^2 **11.** 2^4 **12.** 10^3 **13.** 5^1

14. $2^2 \times 3^3$ **15.** $5^1 \times 7^2 \times 11^4$

D. **16.** 16 **17.** 125 **18.** 6 **19.** 1

20. 343 **21.** 1,000 **22.** 100 **23.** 2,205

E. **24.** 2 **25.** 7 **26.** 10 **27.** 8

28. 9 **29.** 1 **30.** 0 **31.** 13

32. 12

SUPPLEMENTARY PROBLEMS

A. Indicate the base and exponent for each of the following:

1. 2^8 **2.** 8^2 **3.** 1^5 **4.** 5^1

5. 10^2 **6.** 0^3 **7.** 3^4 **8.** 2^{12}

SUPPLEMENTARY PROBLEMS (Continued)

B. Change each to exponential notation:

9. 4×4
10. $3 \cdot 3 \cdot 3 \cdot 3 \cdot 3 \cdot 3$
11. 8
12. $2 \cdot 2 \cdot 2 \cdot 3 \cdot 3$
13. $(4)(4)(5)(5)(5)$
14. $2 \times 2 \times 3 \times 5 \times 5 \times 7 \times 7 \times 7$

C. Evaluate the following:

15. 2^4
16. 3^3
17. 5^1
18. 7^2
19. $2^2 \cdot 3^2$
20. $(2^3)(5^2)$
21. $2^1 \times 3^3 \times 7^2$
22. $2^2 \cdot 3^1 \cdot 5^3$
23. $10^2 \cdot 10^3$
24. $(2^1)(3^2)(5^2)$
25. $3^2 \times 5^2 \times 7^1$
26. $3^2 \cdot 5^1 \cdot 7^2 \cdot 11^1$

D. Find the following square roots:

27. $\sqrt{144}$
28. $\sqrt{100}$
29. $\sqrt{4}$
30. $\sqrt{36}$
31. $\sqrt{0}$
32. $\sqrt{121}$
33. $\sqrt{1}$
34. $\sqrt{81}$
35. $\sqrt{169}$
36. $\sqrt{64}$
37. $\sqrt{49}$
38. $\sqrt{9}$

Solutions to Study Exercises

(8A)

Study Exercise One (Frame 8)

A.

	Base	Exponent
1.	3	2
2.	2	3
3.	10	2
4.	5	7
5.	10	3
6.	3	10

B.
7. "7 to the second" or "7 squared."
8. "9 to the third" or "9 cubed."
9. "7 squared times 9 cubed."
10. "2 cubed times 3 to the fifth."

Study Exercise Two (Frame 10)

(10A)

1. 3^2
2. 5^3
3. $2^3 \times 7^2$
4. $5^4 \times 11^2 \times 13^3 \times 17^1$
5. $(2^1)(3^2)(7^3)$

Study Exercise Three (Frame 12)

(12A)

1. 25
2. 27
2. 4
4. 49
5. 1
6. 0
7. 100
8. 1,000
9. 200
10. 2,205
11. 720
12. 450

Study Exercise Four (Frame 15)

(15A)

1. Yes; 5^2
2. Yes; 7^2
3. No
4. Yes; 10^2
5. No
6. Yes; 0^2
7. Yes; 1^2
8. No
9. Yes; 8^2
10. Yes; 12^2

Study Exercise Five (Frame 21)

(21A)

1. 6
2. 7
3. 1
4. 0
5. 8
6. 10
7. 11

UNIT

7 Primes, Composites, and Prime Factoring

OBJECTIVES (1)

By the end of this unit you should be able to:

1. UNDERSTAND THE MEANING OF THE FOLLOWING TERMS: *EVENLY DIVISIBLE, EVEN NUMBERS, ODD NUMBERS, NATURAL NUMBERS, PRIME NUMBERS, AND COMPOSITE NUMBERS.*
2. USE THE DIVISIBILITY TESTS FOR 2, 3, 5, AND 10.
3. WRITE, IN EXPONENTIAL NOTATION, THE PRIME FACTORS OF A COMPOSITE NUMBER.

Evenly Divisible (2)

A number is said to be *evenly divisible* by another number if the remainder is zero.

Example 1: 34 is evenly divisible by 2

$$\begin{array}{r} 17 \\ 2\overline{)34} \\ \underline{2} \\ 14 \\ \underline{14} \\ 0 \end{array} \longleftarrow \text{remainder is zero}$$

Example 2: 21 is evenly divisible by 3

$$\begin{array}{r} 7 \\ 3\overline{)21} \\ \underline{21} \\ 0 \end{array} \longleftarrow \text{remainder is zero}$$

Example 3: 35 is evenly divisible by 5

$$\begin{array}{r} 7 \\ 5\overline{)35} \\ \underline{35} \\ 0 \end{array} \longleftarrow \text{remainder is zero}$$

Evenly Divisible (Continued)

Example 4: 90 is evenly divisible by 10

$$\begin{array}{r} 9 \\ 10\overline{)90} \\ \underline{90} \\ 0 \end{array} \longleftarrow \text{remainder is zero}$$

Example 5: 16 is *not* evenly divisible by 5

$$\begin{array}{r} 3 \\ 5\overline{)16} \\ \underline{15} \\ 1 \end{array} \longleftarrow \text{remainder is } \textit{not} \text{ zero}$$

Even Numbers (3)

Whole numbers which are evenly divisible by 2 are said to be *even numbers*.

0, 2, 4, 6, 8, 10, 12, 14, 16, 18, 20, 22, . . .

Note: 0 is an even number because

$$\begin{array}{r} 0 \\ 2\overline{)0} \\ \underline{0} \\ 0 \end{array} \longleftarrow \text{remainder is zero}$$

Odd Numbers (4)

Whole numbers which are not evenly divisible by 2 are said to be *odd numbers*.

1, 3, 5, 7, 9, 11, 13, 15, 17, 19, 21, . . .

Study Exercise One (5)

A. Answer yes or no:

1. Is 48 evenly divisible by 2?
2. Is 23 evenly divisible by 2?
3. Is 321 evenly divisible by 3?
4. Is 925 evenly divisible by 5?
5. Is 836 evenly divisible by 5?
6. Is 670 evenly divisible by 10?

B. Classify each of the following whole numbers as being even or odd:

7. 24 **8.** 38 **9.** 49 **10.** 0
11. 283 **12.** 2,005 **13.** 33,641 **14.** 531,738

Divisibility Tests (6)

A *divisibility test* is a quick method of looking at a number and deciding whether it is evenly divisible by a given number.

Divisibility Test for 2 (7)

Any whole number is evenly divisible by 2 if the last digit represents an even number.

Example 1: 5,794 is evenly divisible by 2
↑ last digit is even

Example 2: 8,645 is *not* evenly divisible by 2
↑ last digit is *not* even

Divisibility Test for 3 (8)

Any whole number is evenly divisible by 3 if the sum of its digits is evenly divisible by 3.

Example 1: 42 is evenly divisible by 3

$4 + 2 = 6$ sum of the digits is 6

Example 2: 7,062 is evenly divisible by 3

$7 + 0 + 6 + 2 = 15$ sum of the digits is 15

Example 3: 514 is not evenly divisible by 3

$5 + 1 + 4 = 10$ sum of the digits is 10

Study Exercise Two (9)

A. Is each of the following whole numbers evenly divisible by 2? Answer yes or no:

1. 16 **2.** 35 **3.** 146
4. 2,097 **5.** 1,570 **6.** 2,613,536

B. Is each of the following whole numbers evenly divisible by 3? Answer yes or no:

7. 51 **8.** 92 **9.** 132
10. 3,041 **11.** 6,936 **12.** 2,012,541

Divisibility Test for 5 (10)

Any whole number is evenly divisible by 5 if the last digit is either a 5 or a 0.

Example 1: 125 is evenly divisible by 5
↑ last digit is a 5

Example 2: 4,230 is evenly divisible by 5
↑ last digit is a 0

Example 3: 2,057 is *not* evenly divisible by 5
↑ last digit is neither 5 nor 0

Divisibility Test for 10 (11)

Any whole number is evenly divisible by 10 if the last digit is a 0.

Example 1: 230 is evenly divisible by 10
↑ last digit is a 0

Example 2: 504 is *not* evenly divisible by 10
↑ last digit is not a 0

Study Exercise Three (12)

A. Is each of the following whole numbers evenly divisible by 5? Answer yes or no:

1. 235 **2.** 452 **3.** 552 **4.** 1,960

B. Is each of the following whole numbers evenly divisible by 10? Answer yes or no:

5. 675 **6.** 200 **7.** 1,004 **8.** 123,590

***Study Exercise Three* (Continued)**

C. Examine each number and decide if it is evenly divisible by 2, 3, 5, or 10.

Example: 132 is divisible by 2 and 3. It is divisible by 2 because the last digit represents an even number. It is also divisible by 3 because the sum of the digits is 6 ($1 + 3 + 2 = 6$) and 6 is divisible by 3. The number 132 is not divisible by 5 or 10 because the last digit is neither 5 nor 0.

9. 135 **10.** 720 **11.** 4,233 **12.** 1,680
13. 2,802

Natural Numbers (13)

The *natural numbers* are all of the whole numbers except zero.
1, 2, 3, 4, 5, 6, 7, 8, 9, 10, 11, . . .

Prime Numbers (14)

A *prime number* is any natural number which is evenly divisible *only* by itself and one.
Remember, *evenly divisible* means that the remainder is zero. One is not considered to be a prime number.

Examples (15)

1. 2 is a prime number.
2. 6 is not a prime number, because 6 is evenly divisible by 2 and 3.
3. 11 is a prime number.
4. 15 is not a prime number, because 15 is evenly divisible by 3 and 5.

Composite Numbers (16)

A *composite number* is any natural number which is evenly divisible by a number *other* than one or itself.
No number is both prime and composite.

Examples of Composite Numbers (17)

1. 6 is composite because it is evenly divisible by 2 and 3.
2. 15 is composite because it is evenly divisible by 3 and 5.
3. 22 is composite because it is evenly divisible by 2 and 11.

Natural Numbers Separated Into Three Categories (18)

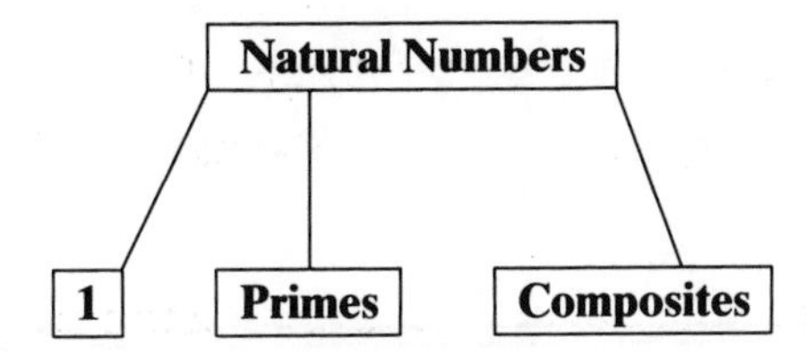

Given any natural number, it is either 1, prime, or composite.

Using the Divisibility Tests (19)

The divisibility tests may be used to help determine if a natural number is composite.

Example 1: 25 is composite because it is evenly divisible by 5.

Example 2: 42 is composite because it is evenly divisible by 2 and 3.

Example 3: 5,421 is composite because it is evenly divisible by 3.

Example 4: 77 is composite because it is evenly divisible by 7 and 11.

Study Exercise Four (20)

Classify each of the following as either prime or composite:

1. 2 **2.** 6 **3.** 7 **4.** 20 **5.** 16
6. 19 **7.** 141 **8.** 2,675 **9.** 17 **10.** 121

The Prime Numbers Less Than 30 (21)

$$2, 3, 5, 7, 11, 13, 17, 19, 23, 29$$

These should be memorized.

Prime Factoring (22)

Every composite number can be written as a product of prime factors.

Example: $6 = 2 \cdot 3$

composite (6) *prime factors* (2, 3)

The Prime Factorization of 30 (23)

Line (a): $2 \underline{| \; 30}$
Line (b): $3 \underline{| \; 15}$
Line (c): 5

$$30 = 2 \times 3 \times 5$$

The Prime Factorization of 90 (24)

Line (a): $2 \underline{| \; 90}$
Line (b): $3 \underline{| \; 45}$
Line (c): $3 \underline{| \; 15}$
Line (d): 5

$$90 = 2 \times \underbrace{3 \times 3} \times 5$$

Exponential Form: $90 = 2^1 \times 3^2 \times 5^1$

The Prime Factorization of 200 (25)

Line (a): $2\,\underline{|\;200}$
Line (b): $2\,\underline{|\;100}$
Line (c): $2\,\underline{|\;50}$
Line (d): $5\,\underline{|\;25}$
Line (e): 5

$$200 = \underbrace{2 \times 2 \times 2}_{} \times \underbrace{5 \times 5}_{}$$

Exponential Form: $200 = 2^3 \times 5^2$

The Prime Factorization of 1,911 (26)

Line (a): $3\,\underline{|\;1{,}911}$
Line (b): $7\,\underline{|\;637}$
Line (c): $7\,\underline{|\;91}$
Line (d): 13

$$1{,}911 = 3 \times \underbrace{7 \times 7}_{} \times 13$$

Exponential Form: $1{,}911 = 3^1 \times 7^2 \times 13^1$

Summary of Prime Factorization (27)

When performing this method of prime factorization, be sure to follow these steps precisely:

Step (1): Divide by the smallest prime number that goes in evenly. Continue dividing the resulting quotients until the division is no longer even.

Step (2): Divide by the next larger prime number that divides evenly. DO NOT SKIP AROUND! Test the primes in sequential order (smallest to largest). Keep dividing by the prime number until it no longer divides evenly, then move to the next larger prime number.

Step (3): Continue this process, moving through the primes sequentially, until the remainder is a prime number. Then, stop and write the primes in exponential notation.

Example: Give the prime factorization for 17,199.

Solution: Test the primes in sequential order. Do not skip around. The number 2 does not divide evenly, since 17,199 ends in an odd digit. 3 will divide evenly.

$3\,\underline{|\;17{,}199}$
$3\,\underline{|\;5733}$ ← Exhaust the division process
$3\,\underline{|\;1911}$
$7\,\underline{|\;637}$ ← 5 does not divide evenly, since the number does not end with a 5 or 0. So, try 7.
$7\,\underline{|\;91}$
13 ← Stop when the remainder is prime.

Thus, $17{,}199 = 3^3 \times 7^2 \times 13^1$

Study Exercise Five (28)

Use the preceding method to give the prime factorization for each of the following. Put your answers in exponential form:

1. 12 **2.** 24 **3.** 108 **4.** 180
5. 92 **6.** 2,205 **7.** 1,683

REVIEW EXERCISES (29)

A. Using the divisibility tests for 2, 3, 5, and 10, determine the numbers by which each of the following is evenly divisible:

1. 141 **2.** 570 **3.** 23,010
4. 5,162 **5.** 485 **6.** 1,620

B. Fill in the blanks:

7. The natural numbers are all of the whole numbers except _________.

8. *Evenly divisible* means that the remainder is _________.

9. No number is both prime and _________.

10. Given any natural number, it is either _________, _________, or composite.

C. Classify each of the following as either prime or composite:

11. 12 **12.** 23 **13.** 21
14. 49 **15.** 105 **16.** 7
17. 19 **18.** 3,111 **19.** 1,075

D. **20.** List all the prime numbers which are less than 30.
21. List all the composite numbers which are less than 30.
22. Is 1 a prime number?
23. Is 1 a composite number?

E. Give the prime factorizations for each of the following. Put your answers in exponential form:

24. 4 **25.** 9 **26.** 36
27. 35 **28.** 75 **29.** 63
30. 144 **31.** 174 **32.** 4,410

Solutions to Review Exercises (30)

A. **1.** 3 **2.** 2, 3, 5, 10 **3.** 2, 3, 5, 10
4. 2 **5.** 5 **6.** 2, 3, 5, 10

B. **7.** Zero. **8.** Zero. **9.** Composite.
10. One, prime.

C. **11.** Composite; divisible by 2 and 3. **12.** Prime.
13. Composite; divisible by 3. **14.** Composite; divisible by 7.
15. Composite; divisible by 3 and 5. **16.** Prime.
17. Prime. **18.** Composite; divisible by 3.
19. Composite; divisible by 5.

D. **20.** 2, 3, 5, 7, 11, 13, 17, 19, 23, 29
21. 4, 6, 8, 9, 10, 12, 14, 15, 16, 18, 20, 21, 22, 24, 25, 26, 27, 28
22. No.
23. No.

E. **24.** $4 = 2^2$ **25.** $9 = 3^2$ **26.** $36 = 2^2 \times 3^2$
27. $35 = 5^1 \times 7^1$ **28.** $75 = 3^1 \times 5^2$ **29.** $63 = 3^2 \times 7^1$
30. $144 = 2^4 \cdot 3^2$ **31.** $174 = 2^1 \cdot 3^1 \cdot 29^1$ **32.** $4{,}410 = 2^1 \times 3^2 \times 5^1 \times 7^2$

SUPPLEMENTARY PROBLEMS

A. Using the divisibility tests for 2, 3, 5, and 10, determine the numbers for which each of the following is evenly divisible:

1. 213 **2.** 720 **3.** 2,342
4. 93 **5.** 385 **6.** 2,820

B. Classify each of the following as being even or odd:

7. 0 **8.** 67 **9.** 132
10. 5,132 **11.** 13,040 **12.** 2,001

C. Classify each of the following as being prime or composite:

13. 2 **14.** 7 **15.** 9
16. 141 **17.** 29 **18.** 2,675

D. True or false:

19. 0 is a natural number.
20. 1 is a prime number.
21. 15 is a composite number.
22. 103 is evenly divisible by 3.
23. 4,263 is evenly divisible by 3.
24. 9,710 is evenly divisible by 2.
25. $2^2 \cdot 21^1$ is the prime factorization of 84 in exponential form.
26. The prime factors of 882 are $2 \times 3 \times 3 \times 7 \times 11$.
27. 72 is both prime and composite.
28. The following group contains all composite numbers: 6, 38, 15, 27, 100.

E. Give the prime factorization for each of the following. Put your answers in exponential form:

29. 14 **30.** 18 **31.** 21 **32.** 56
33. 252 **34.** 378 **35.** 1,155 **36.** 4,235
37. 2,275 **38.** 3,400 **39.** 1,000 **40.** 2,940

Solutions to Study Exercises

(5A)

Study Exercise One (Frame 5)

A. 1. Yes. **2.** No. **3.** Yes.
4. Yes. **5.** No. **6.** Yes.

B. 7. Even. **8.** Even. **9.** Odd.
10. Even. **11.** Odd. **12.** Odd.
13. Odd. **14.** Even.

(9A)

Study Exercise Two (Frame 9)

A. 1. Yes. **2.** No. **3.** Yes.
4. No. **5.** Yes. **6.** Yes.

B. 7. Yes. **8.** No. **9.** Yes
10. No. **11.** Yes. **12.** Yes.

(12A)

Study Exercise Three (Frame 12)

A. 1. Yes. **2.** No. **3.** No. **4.** Yes.
B. 5. No. **6.** Yes. **7.** No. **8.** Yes.
C. 9. 3, 5 **10.** 2, 3, 5, 10 **11.** 3 **12.** 2, 3, 5, 10
13. 2, 3

Study Exercise Four (Frame 20) **(20A)**

1. 2 is prime.
2. 6 is composite because it is evenly divisible by 2 and 3.
3. 7 is prime.
4. 20 is composite because it is evenly divisible by 2, 5, and 10.
5. 16 is composite because it is evenly divisible by 2.
6. 19 is prime.
7. 141 is composite because it is evenly divisible by 3.
8. 2,675 is composite because it is evenly divisible by 5.
9. 17 is prime.
10. 121 is composite because it is evenly divisible by 11.

Study Exercise Five (Frame 28) **(28A)**

1. $2 \underline{| 12}$; $2 \underline{| 6}$; 3

$12 = 2 \times 2 \times 3$

Answer: $12 = 2^2 \times 3^1$

2. $2 \underline{| 24}$; $2 \underline{| 12}$; $2 \underline{| 6}$; 3

$24 = 2 \times 2 \times 2 \times 3$

Answer: $24 = 2^3 \times 3^1$

3. $2 \underline{| 108}$; $2 \underline{| 54}$; $3 \underline{| 27}$; $3 \underline{| 9}$; 3

$108 = 2 \times 2 \times 3 \times 3 \times 3$

Answer: $108 = 2^2 \times 3^3$

4. $2 \underline{| 180}$; $2 \underline{| 90}$; $3 \underline{| 45}$; $3 \underline{| 15}$; 5

$180 = 2 \times 2 \times 3 \times 3 \times 5$

Answer: $180 = 2^2 \times 3^2 \times 5^1$

5. $2 \underline{| 92}$; $2 \underline{| 46}$; 23

$92 = 2 \times 2 \times 23$

Answer: $92 = 2^2 \times 23^1$

6. $3 \underline{| 2{,}205}$; $3 \underline{| 735}$; $5 \underline{| 245}$; $7 \underline{| 49}$; 7

$2{,}205 = 3 \times 3 \times 5 \times 7 \times 7$

Answer: $2{,}205 = 3^2 \times 5^1 \times 7^2$

7. $3 \underline{| 1{,}683}$; $3 \underline{| 561}$; $11 \underline{| 187}$; 17

$1{,}683 = 3 \times 3 \times 11 \times 17$

Answer: $1{,}683 = 3^2 \times 11^1 \times 17^1$

UNIT

8 Least Common Multiple

OBJECTIVES (1)

By the end of this unit you should be able to:

1. EXPLAIN THE MEANING OF THE FOLLOWING TERMS: *MULTIPLE*, *COMMON MULTIPLE*, AND *LEAST COMMON MULTIPLE*.
2. FIND THE LEAST COMMON MULTIPLE OF TWO OR MORE NUMBERS BY USING PRIME FACTORS AND EXPONENTS.
3. FIND THE LEAST COMMON MULTIPLE OF SOME NUMBERS BY USING ONE OF TWO SHORTCUTS.

Multiple (2)

The product of any natural number and 5 is said to be a *multiple* of 5.

Examples:

1. 15 is a multiple of 5 because $15 = 5 \times 3$
2. 30 is a multiple of 5 because $30 = 5 \times 6$
3. 5 is a multiple of 5 because $5 = 5 \times 1$

6, 9, and 27 are multiples of 3 because: **(3)**

1. $6 = 3 \times 2$
2. $9 = 3 \times 3$
3. $27 = 3 \times 9$

7, 14, 21, and 28 are multiples of 7 because: **(4)**

1. $7 = 7 \times 1$
2. $14 = 7 \times 2$
3. $21 = 7 \times 3$
4. $28 = 7 \times 4$

---

1. The *first* multiple of 7 is 7 × $\underline{1}$ or 7
2. The *second* multiple of 7 is 7 × $\underline{2}$ or 14
3. The *third* multiple of 7 is 7 × $\underline{3}$ or 21
4. The *fourth* multiple of 7 is 7 × $\underline{4}$ or 28
5. The *ninth* multiple of 7 is 7 × $\underline{9}$ or 63

(5)

1. The first five multiples of 2 are 2, 4, 6, 8, and 10.
2. The first five multiples of 3 are 3, 6, 9, 12, and 15.
3. The first five multiples of 9 are 9, 18, 27, 36, and 45.

(6)

Study Exercise One

(7)

A. Write the first six multiples for each of the following:

1. 4 **2.** 5 **3.** 8

4. 12 **5.** 25 **6.** 100

B. **7.** Find the third multiple of 6 **8.** Find the sixth multiple of 3

9. Find the tenth multiple of 7 **10.** Find the seventh multiple of 10

Common Multiples of 2 and 3

(8)

1. The first ten multiples of 2: 2, 4, 6, 8, 10, 12, 14, 16, 18, 20

common multiples

2. The first seven multiples of 3: 3, 6, 9, 12, 15, 18, 21

The first three *common multiples* of 2 and 3 are 6, 12, and 18

The common multiples of 2 and 3 are evenly divisible by both 2 and 3

(9)

Example: 6, 12, and 18 are common multiples of 2 and 3

Line (a): $2\overline{)6}$ = 3 $3\overline{)6}$ = 2

Line (b): $2\overline{)12}$ = 6 $3\overline{)12}$ = 4

Line (c): $2\overline{)18}$ = 9 $3\overline{)18}$ = 6

6, 12, and 18 are each evenly divisible by 2 and 3

Common Multiples of 8 and 12

(10)

1. The first six multiples of 8: 8, 16, 24, 32, 40, 48

common multiples

2. The first six multiples of 12: 12, 24, 36, 48, 60, 72

The first two common multiples of 8 and 12 are 24 and 48

The common multiples of 8 and 12 are evenly divisible by both 8 and 12 **(11)**

Example: 24 and 48 are common multiples of 8 and 12

Line (a): $8\overline{)24}^{\,3}$ $\quad 12\overline{)24}^{\,2}$

Line (b): $8\overline{)48}^{\,6}$ $\quad 12\overline{)48}^{\,4}$

24 and 48 are evenly divisible by 8 and 12

Study Exercise Two **(12)**

1. List the first six multiples of 9 and the first nine multiples of 6; then find the first three *common* multiples.
2. Find the first three common multiples of 10 and 15.
3. Find the first two common multiples of 2, 4, and 6.

The Least Common Multiple of 6 and 9 **(13)**

Multiples of 6: 6, 12, (18), 24, 30, (36), 42, 48, (54), . . .
Multiples of 9: 9, (18), 27, (36), 45, (54), . . .

Some common multiples of 6 and 9 are 18, 36, and 54
The *smallest* common multiple of 6 and 9 is 18
The smallest common multiple is called the *least common multiple* and is abbreviated *LCM*.
The LCM of 6 and 9 is 18

The least common multiple (LCM) of 6 and 9 is the smallest number which is evenly divisible by both 6 and 9 **(14)**

Example: The LCM of 6 and 9 is 18
18 is the smallest number which is evenly divisible by 6 and 9

$6\overline{)18}^{\,3}$ and $9\overline{)18}^{\,2}$

The Least Common Multiple (LCM) of 2, 4, and 6 **(15)**

Multiples of 2: 2, 4, 6, 8, 10, (12), 14, 16, 18, 20, 22, (24), . . .
Multiples of 4: 4, 8, (12), 16, 20, (24), 28, . . .
Multiples of 6: 6, (12), 18, (24), 30, 36, . . .

The smallest common multiple is 12
The LCM = 12

The LCM of 2, 4, and 6 is 12 **(16)**
Therefore, 12 is the smallest number which is evenly divisible by 2, 4, and 6

$2\overline{)12}^{\,6}$ $\quad 4\overline{)12}^{\,3}$ $\quad 6\overline{)12}^{\,2}$

Study Exercise Three (17)

1. By listing multiples of 10 and 15, find the least common multiple (LCM)
2. Find the LCM of 3, 6, and 4
3. Find the smallest number which is evenly divisible by 2, 3, and 4

Using Exponents to Find the LCM (18)

Example 1: Find the LCM of 60 and 72:

Solution:

Step (1): Write each number in prime factored form:

$$60 = 2 \times 2 \times 3 \times 5 \qquad 72 = 2 \times 2 \times 2 \times 3 \times 3$$

Step (2): Write in exponential form:

$$60 = 2^2 \times 3^1 \times 5^1 \qquad 72 = 2^3 \times 3^2$$

Step (3): Write each number that was used as a base, but write it *only once* (leave out repetitions):

$$2 \times 3 \times 5$$

Step (4): Attach the largest exponent used on each base from *Step (2)*:

$$\text{LCM} = 2^3 \times 3^2 \times 5^1$$

Step (5): Evaluate the exponential expression:

$$\text{LCM} = 8 \times 9 \times 5 \quad \text{or} \quad 360$$

Example 2: Find the LCM of 12 and 45: (19)

Solution:

Step (1): Write each number in prime factored form:

$$12 = 2 \times 2 \times 3 \qquad 45 = 3 \times 3 \times 5$$

Step (2): Write in exponential form:

$$12 = 2^2 \times 3^1 \quad 45 = 3^2 \times 5^1$$

Step (3): Write each number that was used as a base, but write it *only once* (leave out repetitions):

$$2 \times 3 \times 5$$

Step (4): Attach the largest exponent used on each base from *Step (2)*:

$$\text{LCM} = 2^2 \times 3^2 \times 5^1$$

Step (5): Evaluate the exponential expression:

$$\text{LCM} = 4 \times 9 \times 5 \quad \text{or} \quad 180$$

Example 3: Find the LCM of 15 and 77: (20)

Solution:

Step (1): Write each number in prime factored form:

$$15 = 3 \times 5 \qquad 77 = 7 \times 11$$

Example 3 (Continued)

Step (2): Write in exponential form:

$$15 = 3^1 \times 5^1 \qquad 77 = 7^1 \times 11^1$$

Step (3): Write each number that was used as a base, but write it only once (leave out repetitions):

$$3 \times 5 \times 7 \times 11$$

Step (4): Attach the largest exponent used on each base from *Step (2)*:

$$\text{LCM} = 3^1 \times 5^1 \times 7^1 \times 11^1$$

Step (5): Evaluate the exponential expression:

$$\text{LCM} = 1{,}155$$

Example 4: Find the LCM of 120, 126, and 900: **(21)**

Solution:

Step (1): Write each number in prime factored form:

$$120 = 2 \times 2 \times 2 \times 3 \times 5 \qquad 126 = 2 \times 3 \times 3 \times 7$$
$$900 = 2 \times 2 \times 3 \times 3 \times 5 \times 5$$

Step (2): Write in exponential form:

$$120 = 2^3 \times 3^1 \times 5^1 \qquad 126 = 2^1 \times 3^2 \times 7^1$$
$$900 = 2^2 \times 3^2 \times 5^2$$

Step (3): Write each number that was used as a base, but write it once (leave out repetitions):

$$2 \times 3 \times 5 \times 7$$

Step (4): Attach the largest exponent used on each base from *Step (2)*:

$$\text{LCM} = 2^3 \times 3^2 \times 5^2 \times 7^1$$

Step (5): Evaluate the exponential expression:

$$\text{LCM} = \underbrace{8 \times 9}_{} \times \underbrace{25 \times 7}_{}$$
$$\text{LCM} = 72 \times 175$$
$$\text{LCM} = 12{,}600$$

Study Exercise Four (22)

A. Use exponents to find the LCM:

1. Find the LCM of 12 and 10

2. Find the LCM of 40, 20, and 28

3. Find the LCM of 21 and 10

B. **4.** Find the smallest number which is evenly divisible by 15, 45, and 9

5. The following three numbers are given in prime factored form using exponents:

$$3{,}500 = 2^2 \times 5^3 \times 7^1$$
$$2{,}450 = 2^1 \times 5^2 \times 7^2$$
$$990 = 2^1 \times 3^2 \times 5^1 \times 11^1$$

Find the LCM. You may leave your answer in prime factored form with exponents.

Shortcuts For Finding the LCM (23)

Sometimes the LCM is simple enough to obtain by a shortcut. This happens in the following two cases.

Case (1): Numbers Having No Common Factors (24)

If numbers have no common factors, the LCM is simply their product.

Example 1: Find the LCM of 5 and 6.

Solution: 5 and 6 have no common factors. No whole number, other than one, will divide evenly into both 5 and 6. Thus, the LCM is $5 \cdot 6$, or 30.

Example 2: Find the LCM of 3, 7, and 10.

Solution: 3, 7, and 10 have no common factors. Thus, the LCM is $3 \cdot 7 \cdot 10$, or 210.

Case (2): Numbers Having Smaller LCM's (25)

Sometimes numbers have an LCM small enough to find by looking at the first three or four multiples of the largest number.

Example 1: Find the LCM of 6 and 8.

Solution: List the first three or four multiples of the larger number, 8. Check each multiple by dividing by 6. The LCM is the first multiple of 8 which is also evenly divisible by 6.

Multiples of 8: $1 \cdot 8 = 8$
$2 \cdot 8 = 16$
$3 \cdot 8 = 24 \longleftarrow$ divisible by 6.

Thus, the LCM of 6 and 8 is 24.

Example 2: Find the LCM of 10, 12, and 15.

Solution: The largest number is 15. List its first three or four multiples. Check to see if any are divisible by both 10 and 12.

Multiples of 15: $1 \cdot 15 = 15$
$2 \cdot 15 = 30$
$3 \cdot 15 = 45$
$4 \cdot 15 = 60 \longleftarrow$ divisible by 10 and 12.

Thus, the LCM of 10, 12, and 15 is 60.

A Word of Caution (26)

When finding the LCM, you may try these two shortcuts. If they do not apply or if they take too long, go back and use the original method of prime factoring and exponents.

Study Exercise Five (27)

Try to use a shortcut to do the following:

1. Find the LCM of 7 and 9.
2. Find the LCM of 20 and 10.
3. Find the LCM of 10 and 21.
4. Find the LCM of 6, 15, and 20.
5. Find the LCM of 5, 6, and 11.
6. Find the LCM of 4, 6, and 12.

REVIEW EXERCISES (28)

A. Fill in the blanks:

1. The product of any natural number and 7 is said to be a _________ of 7.
2. The second multiple of 8 is _________.
3. The fifth multiple of 9 is _________.
4. The first multiple of 3 is _________.
5. The first four multiples of 5 are: 5, _________, _________, 20.
6. The first six multiples of 7 are: 7, _________, 21, 28, _________, _________.

B. Find the first three *common* multiples for:

7. 2 and 5
8. 2, 4, 6

C.

9. Which of the three common multiples in question 7 is the least common multiple?
10. Which of the three common multiples in question 8 is the least common multiple?

D. Use exponents to find the LCM:

11. Find the LCM of 12 and 18
12. Find the LCM of 40 and 60
13. Find the LCM of 700, 490, and 196
14. The following three numbers are given in prime factored form using exponents:

$$4{,}875 = 3^1 \times 5^3 \times 13^1$$
$$450 = 2^1 \times 3^2 \times 5^2$$
$$10{,}584 = 2^3 \times 3^3 \times 7^2$$

Find the LCM. Leave your answer in prime factored form using exponents.

E. Use a shortcut to find the LCM:

15. Find the LCM of 5 and 12.
16. Find the LCM of 6 and 9.
17. Find the LCM of 2, 9, and 5.
18. Find the LCM of 6, 12, and 18.

Solutions to Review Exercises (29)

A. 1. Multiple. 2. 16 3. 45
4. 3 5. 10, 15 6. 14, 35, 42

B. 7. 2, 4, 6, 8, (10), 12, 14, 16, 18, (20), 22, 24, 26, 28, (30), . . .
5, (10), 15, (20), 25, (30), 35, 40, . . . Thus, the first three common multiples of 2 and 5 are 10, 20, and 30.

8. 2, 4, 6, 8, 10, (12), 14, 16, 18, 20, 22, (24), 26, 28, 30, 32, 34, (36), . . .
4, 8, (12), 16, 20, (24), 28, 32, (36), . . .
6, (12), 18, (24), 30, (36), . . . Thus, the first three common multiples of 2, 4, and 6 are 12, 24, and 36.

C. 9. LCM = 10 10. LCM = 12

D. 11. *Step (1):* $12=2\times2\times3$ $18=2\times3\times3$
Step (2): $12=2^2\times3^1$ $18=2^1\times3^2$
Step (3): 2×3
Step (4): $\text{LCM}=2^2\times3^2$
Step (5): LCM=36

12. *Step (1):* $40=2\times2\times2\times5$ $60=2\times2\times3\times5$
Step (2): $40=2^3\times5^1$ $60=2^2\times3^1\times5^1$
Step (3): $2\times3\times5$
Step (4): $\text{LCM}=2^3\times3^1\times5^1$
Step (5): LCM=120

Solutions to Review Exercises (Continued)

13. *Step (1):* $700=2\times2\times5\times5\times7$ $490=2\times5\times7\times7$
$196=2\times2\times7\times7$
Step (2): $700=2^2\times5^2\times7^1$ $490=2^1\times5^1\times7^2$
$196=2^2\times7^2$
Step (3): $2\times5\times7$
Step (4): $LCM=2^2\times5^2\times7^2$
Step (5): LCM=4,900

14. $LCM=2^3\times3^3\times5^3\times7^2\times13^1$

E. 15. 5 and 12 have no common factors. Thus, the LCM = $5 \cdot 12$ or 60.

16. Multiples of 9: $1 \cdot 9 = 9$
$2 \cdot 9 = 18 \longleftarrow$ divisible by 6
Thus, the LCM of 6 and 9 is 18.

17. 2, 9, and 5 have no common factors. Thus, the LCM = $2 \cdot 9 \cdot 5$ or 90.

18. Multiples of 18: $1 \cdot 18 = 18$
$2 \cdot 18 = 36 \longleftarrow$ divisible by 6 and 12
Thus, the LCM of 6, 12, and 18 is 36.

SUPPLEMENTARY PROBLEMS

A. Fill in the blanks:

1. The smallest number which is evenly divisible by two numbers is said to be the _______ _______ _______ of those numbers.
2. 35 is a multiple of 7 because _______ $\times 5 = 35$
3. The third multiple of 4 is _______.
4. The seventh multiple of 10 is _______.
5. The first multiple of 7 is _______.
6. The first four multiples of 6 are: 6, _______, 18, _______.
7. The first six multiples of 8 are: _______, 16, _______, _______, 40, _______.

B. Find the first three common multiples for:

8. 2 and 4
9. 2, 4, and 5

C. Use exponents to find the LCM:

10. 6 and 12
11. 60 and 18
12. 4 and 9
13. 4, 6, and 8
14. 2, 3, and 9
15. 12, 20, and 15
16. 4, 8, and 12
17. 50 and 20
18. 24 and 18
19. 126, 108, and 98
20. The following three numbers are given in prime factored form using exponents:

$$3{,}500 = 2^2 \times 5^3 \times 7^1$$
$$4{,}875 = 3^1 \times 5^3 \times 13^1$$
$$10{,}584 = 2^3 \times 3^3 \times 7^2$$

Find the LCM. Leave your answer in prime factored form using exponents.

D. If possible, use a shortcut to find the LCM:

21. 7 and 11
22. 2, 3, and 7
23. 4 and 6
24. 2, 3, and 8
25. 6, 8, and 12
26. 20 and 30
27. 9 and 15
28. 2, 4, and 8
29. 2, 3, and 15
30. 12, 15, and 5
31. 7, 9, and 11
32. 4, 8, and 16

Solutions to Study Exercises

(7A)

Study Exercise One (Frame 7)

A.
1. 4, 8, 12, 16, 20, 24
2. 5, 10, 15, 20, 25, 30
3. 8, 16, 24, 32, 40, 48
4. 12, 24, 36, 48, 60, 72
5. 25, 50, 75, 100, 125, 150
6. 100, 200, 300, 400, 500, 600

B.
7. 6 × 3 or 18
8. 3 × 6 or 18
9. 7 × 10 or 70
10. 10 × 7 or 70

(12A)

Study Exercise Two (Frame 12)

1. 9, (18), 27, (36), 45, (54)
   6, 12, (18), 24, 30, (36), 42, 48, (54)
   The first three common multiples of 9 and 6 are 18, 36, and 54
2. 10, 20, (30), 40, 50, (60), 70, 80, (90)
   15, (30), 45, (60), 75, (90)
   The first three common multiples of 10 and 15 are 30, 60, and 90
3. 2, 4, 6, 8, 10, (12), 14, 16, 18, 20, 22, (24)
   4, 8, (12), 16, 20, (24)
   6, (12), 18, (24), 30
   The first two common multiples of 2, 4, and 6 are 12 and 24

(17A)

Study Exercise Three (Frame 17)

1. 10, 20, (30), 40, 50, 60, 70, 80, 90, . . .
   15, (30), 45, 60, 75, 90, . . .
   The LCM of 10 and 15 is 30
2. 3, 6, 9, (12), 15, 18, 21, 24, 27, 30, . . .
   6, (12), 18, 24, 30, . . .
   4, 8, (12), 16, 20, 24, 28, 32, . . .
   The LCM of 3, 6, and 4 is 12
3. 2, 4, 6, 8, 10, (12), 14, 16, 18, 20, 22, 24, . . .
   3, 6, 9, (12), 15, 18, 21, 24, 27, 30, . . .
   4, 8, (12), 16, 20, 24, 28, . . .
   The LCM is 12. Therefore, 12 is the smallest number which is evenly divisible by 2, 3, and 4

(22A)

Study Exercise Four (Frame 22)

A.
1. *Step (1):* $12=2\times2\times3$ $10=2\times5$
   Step (2): $12=2^2\times3^1$ $10=2^1\times5^1$
   Step (3): $2\times3\times5$
   Step (4): $LCM=2^2\times3^1\times5^1$
   Step (5): $LCM=60$
2. *Step (1):* $40=2\times2\times2\times5$ $20=2\times2\times5$ $28=2\times2\times7$
   Step (2): $40=2^3\times5^1$ $20=2^2\times5^1$ $28=2^2\times7^1$
   Step (3): $2\times5\times7$
   Step (4): $LCM=2^3\times5^1\times7^1$
   Step (5): $LCM=280$
2. *Step (1):* $21=3\times7$ $10=2\times5$
   Step (2): $21=3^1\times7^1$ $10=2^1\times5^1$
   Step (3): $2\times3\times5\times7$
   Step (4): $LCM=2^1\times3^1\times5^1\times7^1$
   Step (5): $LCM=210$

B.
4. Find the LCM of 15, 45, and 9.
   Step (1): $15=3\times5$ $45=3\times3\times5$ $9=3\times3$
   Step (2): $15=3^1\times5^1$ $45=3^2\times5^1$ $9=3^2$
   Step (3): 3×5
   Step (4): $LCM=3^2\times5^1$
   Step (5): $LCM=45$
   45 is the smallest number which is evenly divisible by 15, 45, and 9.
5. $LCM=2^2\times3^2\times5^3\times7^2\times11^1$

Study Exercise Five (Frame 27) (27A)

1. LCM $= 7 \cdot 9$ or 63
2. Multiples of 20: $1 \cdot 20 = 20 \longleftarrow$ divisible by 10
   LCM $= 20$
3. LCM $= 10 \cdot 21$ or 210
4. Multiples of 20: $1 \cdot 20 = 20$
   $2 \cdot 20 = 40$
   $3 \cdot 20 = 60 \longleftarrow$ divisible by 6 and 15
   LCM $= 60$
5. LCM $= 5 \cdot 6 \cdot 11$ or 330
6. Multiples of 12: $1 \cdot 12 = 12 \longleftarrow$ divisible by 4 and 6.
   LCM $= 12$

Module 1 Practice Test

Units 1–8

1. Give the place value for the underlined digit: 26,<u>4</u>57,053
2. Write in words the name for 26,041,030
3. Write the numerals for eight hundred forty-two thousand, six hundred nine.
4. Round off the whole number 1,365,432 to the nearest ten thousand.
5. Add the following by writing each number in expanded form: 2,341 + 413 + 35
6. Perform the following addition and check your answer:

$$\begin{array}{r} 62{,}948 \\ 408{,}932 \\ \underline{9{,}756} \end{array}$$

7. Perform the following subtraction by writing each number in expanded form: 765 − 341
8. Perform the following subtraction and check your answer:

$$\begin{array}{r} 206{,}401 \\ \underline{-58{,}382} \end{array}$$

9. Multiply the following: $3{,}652 \times 1{,}000$
10. Multiply the following using the short form and check your answer: $6{,}048 \times 708$
11. List three symbols for showing that 15 is to be divided by 3
12. Which of the following divisions *cannot* be performed?
    (a) $20 \div 5$ (b) $0 \div 5$ (c) $5 \div 0$
13. Perform the following long division and check your answer: $32{,}607 \div 153$
14. State the divisibility test for 3
15. In the exponential expression 5^4, indicate the base and indicate the exponent.
16. Evaluate the following exponential expression: $2^3 \cdot 3^2 \cdot 5^2$
17. Evaluate: $\sqrt{81}$
18. List the first five prime numbers.
19. Give the prime factorization of 450
20. Give the fifth multiple of 12
21. Find the LCM of 120, 126, and 450
22. If 3 candy bars cost 96 cents, how much does one candy bar cost?
23. If a box weighs 21 pounds, what is the total weight of 8 identical boxes?
24. If a car can travel 155 miles on 5 gallons of gasoline, how far can it travel on 12 gallons?
25. If a manufacturing concern produces 6,000 articles in 5 days, how many articles will it produce in 21 days?

Answers to Module 1 Practice Test

1. Hundred thousands.
2. Twenty-six million, forty-one thousand, thirty.
3. 842,609
4. 1,370,000
5. 2 thousands + 7 hundreds + 8 tens + 9 units or 2,789
6. 481,636
7. 4 hundreds + 2 tens + 4 units or 424
8. 148,019

12. 3,652,000
10. 4,281,984
11. $15 \div 3$, $3\overline{)15}$, $\frac{15}{3}$

12. (c) $5 \div 0$
13. 213 R18
14. The sum of the digits is evenly divisible by 3
15. The base is 5 and the exponent is 4
16. 1,800
17. 9
18. 2, 3, 5, 7, 11
19. $2^1 \times 3^2 \times 5^2$
20. 60
21. 12,600
22. 32 cents
23. 168 pounds
24. 372 miles
25. 25,200 articles

MODULE

2

Operating with Fractions

UNIT

Introduction to Fractions

OBJECTIVES (1)

By the end of this unit you should be able to:

1. STATE THE TWO MEANINGS OF A FRACTION.
2. KNOW WHAT IS MEANT BY *NUMERATOR, DENOMINATOR* AND *TERMS* OF A FRACTION.
3. STATE THE FUNDAMENTAL PRINCIPLE OF FRACTIONS.

Part of a Whole (2)

If a pie is cut into 4 equal portions, then what part of the pie is each portion?

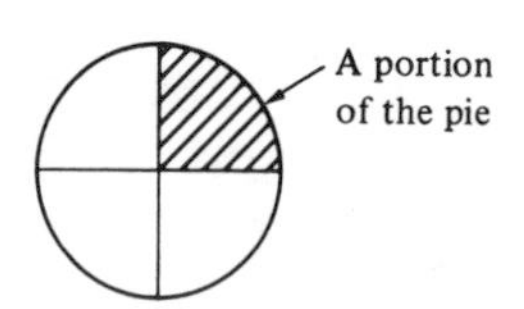

Comparing a Part to a Whole (3)

Four portions are equal to the whole pie. The portion is a part of the whole pie. Can you compare the part to the whole?

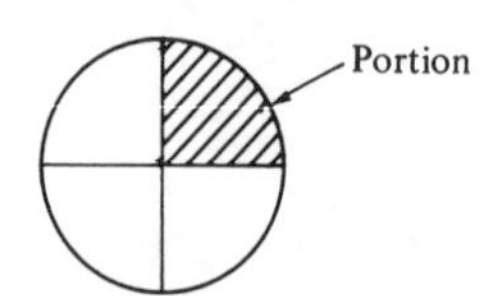

Fraction (4)

We will say that a portion of the pie is to the whole pie as 1 is to 4. The number idea that describes this comparison will be written $\frac{1}{4}$ and will be called a *fraction*.

The comparison of a portion to the whole is the idea of a fraction.

Comparing Parts to the Whole (5)

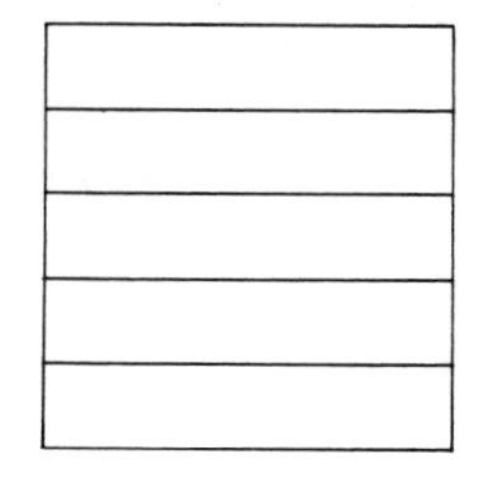

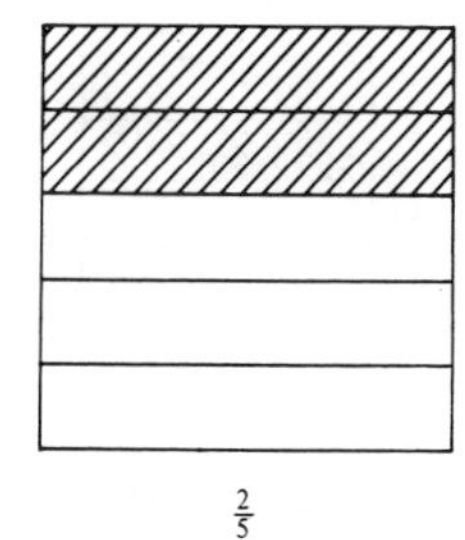

$\frac{2}{5}$

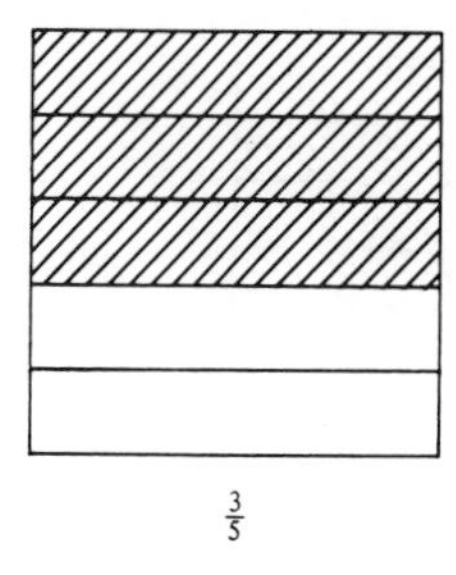

$\frac{3}{5}$

The square has been divided into 5 equal parts. Each part of the square is $\frac{1}{5}$. If two parts are taken, we have two-fifths and write it $\frac{2}{5}$. If three parts are taken, we have $\frac{3}{5}$.

A Second Meaning of a Fraction (6)

A fraction can also be thought of as an indicated division. Thus, $\frac{4}{2}$ means $4 \div 2$ and $\frac{2}{3}$ means $2 \div 3$.

Division of Zero (7)

Since division by zero is not allowed, no fraction can be written that would indicate division by zero. $\frac{5}{0}$ would mean $5 \div 0$ and is undefined.

Examples (8)

Which of the following are allowed? Answer yes or no:

1. $\frac{8}{0}$ **2.** $\frac{6}{1}$ **3.** $\frac{0}{5}$ **4.** $\frac{0}{0}$

Solution: **1.** No, since $\frac{8}{0}$ would have to mean $8 \div 0$ and division by zero is impossible.

2. Yes, since $\frac{6}{1}$ would mean $6 \div 1$ or 6

3. Yes, since $\frac{0}{5}$ would mean $0 \div 5$ or 0

4. No, since $\frac{0}{0}$ would have to mean $0 \div 0$ and division by zero is impossible.

Study Exercise One (9)

A. In the drawings below, write the fraction that the shaded part represents to the whole:

1.

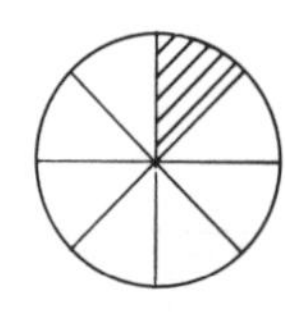

2.

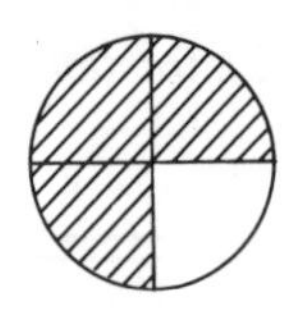

3.

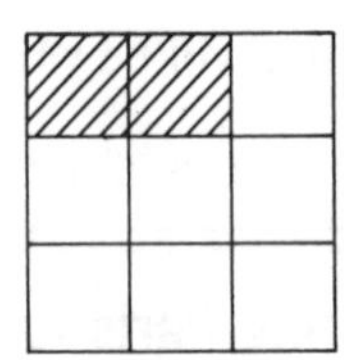

4.

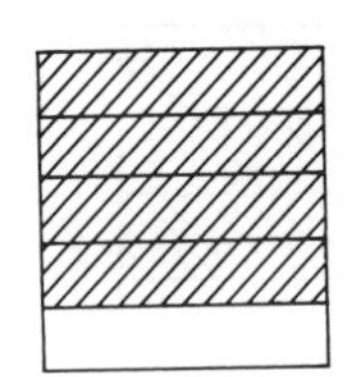

5.

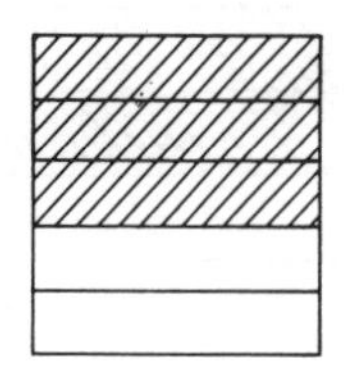

6.

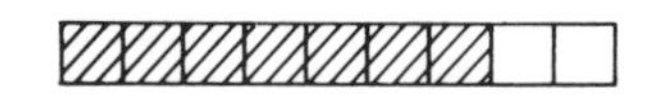

B. Write each of the following fractions as a division. Example: $\frac{3}{4} = 3 \div 4$

7. $\frac{6}{7}$ **8.** $\frac{2}{5}$ **9.** $\frac{1}{2}$

C. Indicate which of the following represent fractions. Answer yes or no.

10. $\frac{5}{5}$ **11.** $\frac{6}{0}$ **12.** $\frac{0}{6}$

13. $\frac{0}{0}$ **14.** $\frac{1}{0}$ **15.** $\frac{0}{1}$

Terms of a Fraction (10)

We have written fractions by writing one whole number over another. In the fraction $\frac{3}{5}$, the 3 and 5 are called *terms* of the fraction; the 3 is called the *numerator* and the 5 is called the *denominator.*

$$\frac{3}{5} \begin{array}{l} \longleftarrow \textit{numerator} \\ \longleftarrow \textit{denominator} \end{array}$$

Equivalent Fractions (11)

The same fraction may be written in many different forms.

In the diagrams below, $\frac{2}{4} = \frac{8}{16}, \frac{8}{16} = \frac{1}{2}, \frac{1}{2} = \frac{3}{6}, \frac{2}{4} = \frac{3}{6}$, etc.

1.

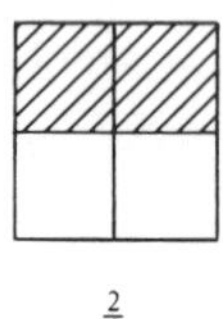

$\frac{2}{4}$

2.

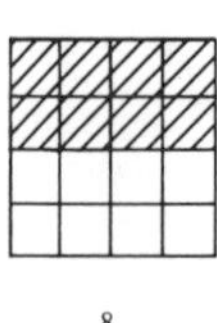

$\frac{8}{16}$

3.

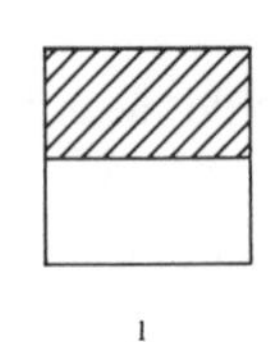

$\frac{1}{2}$

4.

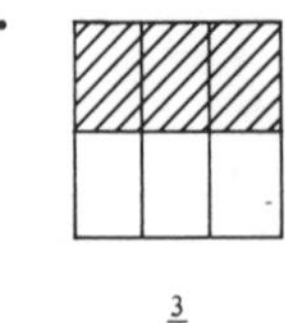

$\frac{3}{6}$

Thus, we say that $\frac{2}{4}, \frac{8}{16}, \frac{1}{2}, \frac{3}{6}$ are *equivalent* fractions.

Comparing Equivalent Fractions (12)

In the preceding frame we saw that:

$$\frac{1}{2} = \frac{2}{4} \quad \text{or} \quad \frac{1}{2} = \frac{3}{6} \quad \text{or} \quad \frac{1}{2} = \frac{8}{16}$$

In the statement $\frac{1}{2} = \frac{3}{6}$, observe that the numerator of the second fraction is three times the numerator of the first fraction and that the denominator of the second fraction is also three times the denominator of the first fraction. In other words, $\frac{1}{2}$ can be changed to the equivalent fraction $\frac{3}{6}$ by multiplying numerator and denominator each by 3.

The Fundamental Principle of Fractions—Part 1 (13)

We have discovered in Frame 12 a part of the basic rule of fractions which is sometimes called the Fundamental Principle of Fractions:

Multiplying both the numerator and denominator of a fraction by the same nonzero number does not alter the value of the fraction.

Examples (14)

1. We take the fraction $\frac{3}{5}$ and multiply numerator and denominator by the number 4.

$$\frac{3}{5} = \frac{3 \times 4}{5 \times 4} = \frac{12}{20}$$

2. We take $\frac{2}{7}$ and multiply numerator and denominator by 6.

$$\frac{2}{7} = \frac{2 \times 6}{7 \times 6} = \frac{12}{42}$$

3. We take $\frac{1}{5}$ and multiply numerator and denominator by 5.

$$\frac{1}{5} = \frac{1 \times 5}{5 \times 5} = \frac{5}{25}$$

Fundamental Principle of Fractions—Part 2 (15)

Since $\frac{1}{2} = \frac{3}{6}$, it is equally true that $\frac{3}{6} = \frac{1}{2}$. Observe that the numerator and denominator of the first fraction have each been divided by 3 to get the second fraction. We thus arrive at the second part of the basic rule of fractions:

Dividing both numerator and denominator of a fraction by the same nonzero number does not alter the value of the fraction.

Fundamental Principle of Fractions (16)

The two parts combined form the Fundamental Principle of Fractions:

Multiplying or dividing both numerator and denominator of a fraction by the same nonzero number does not alter the value of the fraction.

Examples (17)

1. We take the fraction $\frac{6}{10}$ and divide numerator and denominator by 2.

$$\frac{6}{10} = \frac{6 \div 2}{10 \div 2} = \frac{3}{5}$$

2. We take $\frac{20}{36}$ and divide numerator and denominator by 4.

$$\frac{20}{36} = \frac{20 \div 4}{36 \div 4} = \frac{5}{9}$$

Study Exercise Two (18)

Take the given fraction and find an equivalent fraction by multiplying or dividing numerator and denominator by the indicated number:

1. $\frac{1}{7}$; multiply by 2
2. $\frac{4}{6}$; divide by 2
3. $\frac{10}{12}$; multiply by 3
4. $\frac{10}{12}$; divide by 2
5. $\frac{8}{9}$; multiply by 4
6. $\frac{12}{21}$; divide by 3

Equality of Fractions (19)

Two fractions are said to be *equal* or *equivalent* if their *cross products* are equal.

$$\frac{12}{16} \times \frac{3}{4}$$

Since $12 \times 4 = 3 \times 16$, the fractions $\frac{12}{16}$ and $\frac{3}{4}$ are equal.

Changing the Form of a Fraction (20)

The Fundamental Principle may be used to change the terms of a fraction.

Problem: Change $\frac{1}{4}$ to an equivalent fraction with denominator 12.

$$\frac{1}{4} = \frac{?}{12}$$

Changing the Form of a Fraction (Continued)

Solution: Since $3 \times 4 = 12$, $\dfrac{1}{4} = \dfrac{?}{12}$ (denominator $\times 3$)

The denominator was multiplied by 3, so we must also multiply the numerator by 3.

$$\frac{1}{4} \overset{\times 3}{=} \frac{?}{12} \text{ which gives } \frac{1}{4} = \frac{3}{12}$$

Check: The cross products must be equal: $\underbrace{1 \times 12}_{12} = \underbrace{3 \times 4}_{12}$

More Examples (21)

Example 1: Supply the missing number: $\dfrac{4}{7} = \dfrac{?}{21}$

Solution: Since $3 \times 7 = 21$, $\dfrac{4}{7} \overset{\times 3}{=} \dfrac{?}{21}$ which gives $\dfrac{4}{7} = \dfrac{12}{21}$

Check: The cross products must be equal: $\underbrace{4 \times 21}_{84} = \underbrace{12 \times 7}_{84}$

Example 2: Supply the missing number: $\dfrac{12}{18} = \dfrac{?}{9}$

Solution: Since $18 \div 2 = 9$, $\dfrac{12}{18} \overset{\div 2}{=} \dfrac{?}{9}$ which gives $\dfrac{12}{18} = \dfrac{6}{9}$

Check: The cross products must be equal: $\underbrace{12 \times 9}_{108} = \underbrace{6 \times 18}_{108}$

Study Exercise Three (22)

A. Supply the missing term by using the Fundamental Principle and check your answer by examining the cross products:

1. $\dfrac{3}{10} = \dfrac{?}{30}$ **2.** $\dfrac{3}{4} = \dfrac{?}{12}$ **3.** $\dfrac{2}{3} = \dfrac{?}{27}$ **4.** $\dfrac{2}{5} = \dfrac{?}{25}$

5. $\dfrac{9}{15} = \dfrac{?}{5}$ **6.** $\dfrac{8}{32} = \dfrac{?}{8}$ **7.** $\dfrac{9}{18} = \dfrac{?}{6}$ **8.** $\dfrac{35}{60} = \dfrac{?}{12}$

9. $\dfrac{5}{8} = \dfrac{25}{?}$ **10.** $\dfrac{3}{10} = \dfrac{15}{?}$

Study Exercise Three **(Continued)**

B. True or false (check the cross products):

11. $\frac{3}{6} = \frac{1}{2}$ **12.** $\frac{2}{3} = \frac{36}{48}$ **13.** $\frac{12}{21} = \frac{4}{7}$ **14.** $\frac{21}{49} = \frac{3}{7}$

15. $\frac{6}{7} = \frac{48}{42}$ **16.** $\frac{7}{15} = \frac{21}{60}$ **17.** $\frac{12}{15} = \frac{4}{5}$ **18.** $\frac{7}{8} = \frac{28}{36}$

Writing Whole Numbers as Fractions (23)

A whole number can be written as a fraction having a denominator of 1.

Examples: $3 = \frac{3}{1}$, $5 = \frac{5}{1}$, and $8 = \frac{8}{1}$

A whole number can also be written as a fraction with any nonzero number as a denominator.

Examples: $3 = \frac{12}{4}$, $3 = \frac{15}{5}$, $5 = \frac{10}{2}$, and $8 = \frac{24}{3}$

Procedure (24)

To write a whole number as a fraction with a given number for a denominator, follow these steps:

Step (1): Write the whole number as a fraction having a denominator of 1.
Step (2): Use the Fundamental Principle to multiply numerator and denominator by the given number.

Example 1: Write 5 as a fraction with a denominator of 4.

Step (1): $5 = \frac{5}{1}$

Step (2): Multiply numerator and denominator by 4.

$$\frac{5 \times 4}{1 \times 4} = \frac{20}{4}$$

Thus, $5 = \frac{20}{4}$

Example 2: Write 8 as a fraction with a denominator of 8.

Step (1): $8 = \frac{8}{1}$

Step (2): Multiply numerator and denominator by 8.

$$\frac{8 \times 8}{1 \times 8} = \frac{64}{8}$$

Thus, $8 = \frac{64}{8}$

Zero Written as a Fraction (25)

Zero, when written as a fraction, can have any nonzero number for a denominator.

Example 1: Write zero as a fraction having a denominator of 2.

Step (1): $0 = \frac{0}{1}$

Step (2): Multiply numerator and denominator by 2.

$$\frac{0 \times 2}{1 \times 2} = \frac{0}{2}$$

Thus, $0 = \frac{0}{2}$

Example 2: Write zero as a fraction having a denominator of 5.

Step (1): $0 = \frac{0}{1}$

Step (2): Multiply numerator and denominator by 5.

$$\frac{0 \times 5}{1 \times 5} = \frac{0}{5}$$

Thus, $0 = \frac{0}{5}$

Study Exercise Four (26)

1. Write 7 as a fraction with a denominator of 1.
2. Write the whole number 6 as a fraction with a denominator of 3.
3. Write the whole number 8 as a fraction with a denominator of 7.
4. Write the number 0 as a fraction with a denominator of 4.
5. What whole number is equal to $\frac{72}{8}$?

REVIEW EXERCISES (27)

A. Write the fraction that the shaded part represents to the whole:

1.

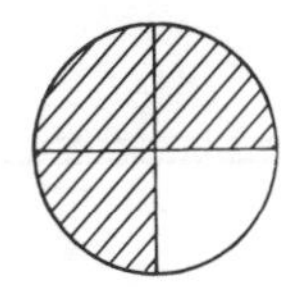

2.

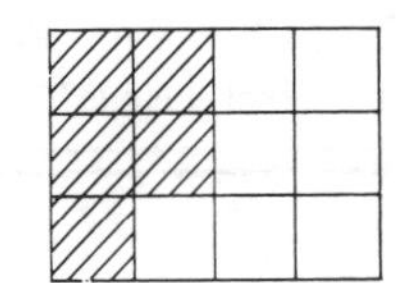

B. **3.** For the fraction $\frac{2}{3}$, we call 2 the ________ and 3 the ________.

4. The fractions $\frac{2}{7}$ and $\frac{2}{5}$ have the same ________.

5. Which of these is not a fraction: $\frac{5}{9}, \frac{1}{3}, \frac{4}{0}$?

REVIEW EXERCISES (Continued)

C. Find an equivalent fraction by using the Fundamental Principle as indicated:

6. $\frac{7}{9}$; multiply by 3 **7.** $\frac{16}{18}$; divide by 2 **8.** $\frac{12}{20}$; multiply by 4

D. Supply the missing term; then check your answer by examining the cross products:

9. $\frac{2}{3} = \frac{6}{?}$ **10.** $\frac{8}{20} = \frac{?}{5}$ **11.** $7 = \frac{14}{?}$ **12.** $0 = \frac{?}{7}$

E. True or false:

13. $\frac{3}{10} = \frac{6}{20}$ **14.** $\frac{24}{36} = \frac{8}{9}$

Solutions to Review Exercises

(28)

A. **1.** $\frac{3}{4}$ **2.** $\frac{5}{12}$

B. **3.** Numerator; Denominator. **4.** Numerator.

5. $\frac{4}{0}$ is not a fraction since $4 \div 0$ is undefined.

C. **6.** $\frac{7}{9} = \frac{7 \times 3}{9 \times 3} = \frac{21}{27}$ **7.** $\frac{16}{18} = \frac{16 \div 2}{18 \div 2} = \frac{8}{9}$

8. $\frac{12}{20} = \frac{12 \times 4}{20 \times 4} = \frac{48}{80}$

D. **9.** $\frac{2}{3} \overset{\times 3}{=} \frac{6}{?} = \frac{6}{9}$ (×3)

Check: $2 \times 9 = 3 \times 6$
$18 = 18$

10. $\frac{8}{20} \overset{\div 4}{=} \frac{?}{5} = \frac{2}{5}$ (÷4)

Check: $8 \times 5 = 20 \times 2$
$40 = 40$

11. $\frac{7}{1} \overset{\times 2}{=} \frac{14}{?} = \frac{14}{2}$ (×2)

Check: $7 \times 2 = 1 \times 14$
$14 = 14$

12. $\frac{0}{1} \overset{\times 7}{=} \frac{?}{7} = \frac{0}{7}$ (×7)

Check: $0 \times 7 = 1 \times 0$
$0 = 0$

E. **13.** True, since $3 \times 20 = 6 \times 10$

14. False, since $24 \times 9 \neq 8 \times 36$*

*The symbol $\neq$ means "is not equal to".

SUPPLEMENTARY PROBLEMS

A. Find the fractions suggested by the following drawings:

1.

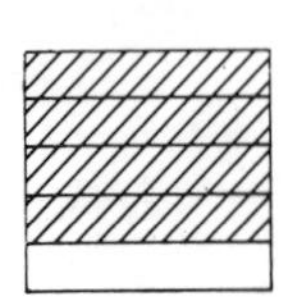

2.

3.

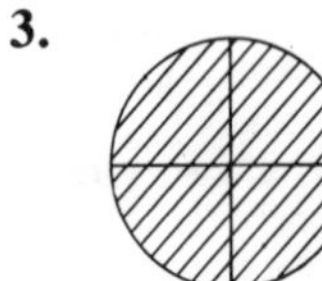

SUPPLEMENTARY PROBLEMS (Continued)

B. **4.** In writing $\frac{3}{5}$, we call 3 the _________ of the fraction and 5 the _________ of the fraction; we call 3 and 5 _________ of the fraction.

5. State the two meanings of a fraction.

6. State the Fundamental Principle of Fractions.

C. Find an equivalent fraction by using the Fundamental Principle.

7. $\frac{32}{40}$; divide by 8 **8.** $\frac{25}{120}$; divide by 5 **9.** $\frac{7}{35}$; divide by 7

10. $\frac{3}{8}$; multiply by 8 **11.** $\frac{7}{8}$; multiply by 5 **12.** $\frac{0}{4}$; multiply by 5

D. Supply the missing term; then check your answer by examining the cross products:

13. $\frac{5}{8} = \frac{?}{56}$ **14.** $\frac{8}{?} = \frac{48}{54}$ **15.** $\frac{1}{6} = \frac{12}{?}$

16. $\frac{?}{7} = \frac{33}{77}$ **17.** $\frac{36}{60} = \frac{?}{15}$ **18.** $\frac{1}{?} = \frac{11}{66}$

19. $\frac{?}{64} = \frac{7}{16}$ **20.** $\frac{1}{8} = \frac{?}{56}$ **21.** $\frac{?}{4} = \frac{0}{12}$

E. True or false:

22. $\frac{4}{5} = \frac{12}{15}$ **23.** $\frac{18}{21} = \frac{6}{7}$ **24.** $4 = \frac{80}{20}$

25. $\frac{7}{84} = \frac{42}{12}$ **26.** $0 = \frac{5}{0}$ **27.** $\frac{55}{80} = \frac{11}{16}$

F. **28.** A baseball team won 11 out of 17 games. Express as a fraction the part of the games won to the total number of games played.

29. If John is 16 years old and Bill is 17, express as a fraction John's age to Bill's age.

30. Write the whole number 12 as a fraction with denominator of 8.

Solutions to Study Exercises (9A)

Study Exercise One (Frame 9)

A. **1.** $\frac{1}{8}$ **2.** $\frac{3}{4}$ **3.** $\frac{2}{9}$

4. $\frac{4}{5}$ **5.** $\frac{3}{5}$ **6.** $\frac{7}{9}$

B. **7.** $6 \div 7$ **8.** $2 \div 5$ **9.** $1 \div 2$

C. **10.** Yes. **11.** No. **12.** Yes.

13. No. **14.** No. **15.** Yes.

Study Exercise Two (Frame 18) (18A)

1. $\frac{1 \times 2}{7 \times 2} = \frac{2}{14}$ **2.** $\frac{4 \div 2}{6 \div 2} = \frac{2}{3}$ **3.** $\frac{10 \times 3}{12 \times 3} = \frac{30}{36}$

4. $\frac{10 \div 2}{12 \div 2} = \frac{5}{6}$ **5.** $\frac{8 \times 4}{9 \times 4} = \frac{32}{36}$ **6.** $\frac{12 \div 3}{21 \div 3} = \frac{4}{7}$

Study Exercise Three (Frame 22) (22A)

A. 1. $\frac{3}{10} \overset{\times 3}{=} \frac{9}{30}$ (×3)

Check: $3 \times 30 = 9 \times 10$
$90 = 90$

2. $\frac{3}{4} \overset{\times 3}{=} \frac{9}{12}$ (×3)

Check: $3 \times 12 = 9 \times 4$
$36 = 36$

3. $\frac{2}{3} \overset{\times 9}{=} \frac{18}{27}$ (×9)

Check: $2 \times 27 = 18 \times 3$
$54 = 54$

4. $\frac{2}{5} \overset{\times 5}{=} \frac{10}{25}$ (×5)

Check: $2 \times 25 = 10 \times 5$
$50 = 50$

5. $\frac{9}{15} \overset{\div 3}{=} \frac{3}{5}$ (÷3)

Check: $9 \times 5 = 3 \times 15$
$45 = 45$

6. $\frac{8}{32} \overset{\div 4}{=} \frac{2}{8}$ (÷4)

Check: $8 \times 8 = 2 \times 32$
$64 = 64$

7. $\frac{9}{18} \overset{\div 3}{=} \frac{3}{6}$ (÷3)

Check: $9 \times 6 = 3 \times 18$
$54 = 54$

8. $\frac{35}{60} \overset{\div 5}{=} \frac{7}{12}$ (÷5)

Check: $35 \times 12 = 7 \times 60$
$420 = 420$

9. $\frac{5}{8} \overset{\times 5}{=} \frac{25}{40}$ (×5)

Check: $5 \times 40 = 25 \times 8$
$200 = 200$

10. $\frac{3}{10} \overset{\times 5}{=} \frac{15}{50}$ (×5)

Check: $3 \times 50 = 15 \times 10$
$150 = 150$

B. 11. True, since $3 \times 2 = 1 \times 6$

*12. False, since $2 \times 48 \neq 36 \times 3$

13. True, since $12 \times 7 = 4 \times 21$

14. True, since $21 \times 7 = 3 \times 49$

15. False, since $6 \times 42 \neq 48 \times 7$

16. False, since $7 \times 60 \neq 21 \times 15$

17. True, since $12 \times 5 = 4 \times 15$

18. False, since $7 \times 36 \neq 28 \times 8$

*The symbol $\neq$ means "is not equal to".

Study Exercise Four (Frame 26) (26A)

1. $7 = \frac{7}{1}$

2. $6 = \frac{6}{1} = \frac{6 \times 3}{1 \times 3} = \frac{18}{3}$

3. $8 = \frac{8}{1} = \frac{8 \times 7}{1 \times 7} = \frac{56}{7}$

4. $0 = \frac{0}{1} = \frac{0 \times 4}{1 \times 4} = \frac{0}{4}$

5. $\frac{72}{8}$ means $72 \div 8 = 9$; therefore $\frac{72}{8} = 9$

UNIT

10 Reducing Fractions

OBJECTIVES (1)

By the end of this unit you should be able to:

1. REDUCE FRACTIONS TO LOWEST TERMS BY THE FUNDAMENTAL PRINCIPLE.
2. REDUCE FRACTIONS TO LOWEST TERMS BY CANCELLING.
3. IDENTIFY MIXED NUMERALS AND IMPROPER FRACTIONS.
4. CHANGE IMPROPER FRACTIONS TO MIXED NUMERALS.

Review of Fundamental Principle of Fractions (2)

Recall that the Fundamental Principle of Fractions states that you may multiply or divide both numerator and denominator of a fraction by the same nonzero number without altering the value of the fraction.

Example 1: Consider $\frac{3}{7}$.

Multiply numerator and denominator by 3.

$$\frac{3}{7} \overset{\times 3}{=} \frac{9}{21} \quad (\times 3)$$

Example 2: Consider $\frac{20}{36}$.

Divide numerator and denominator by 4.

$$\frac{20}{36} \overset{\div 4}{=} \frac{5}{9} \quad (\div 4)$$

Expanding to Higher Terms (3)

By the Fundamental Principle, a fraction such as $\frac{4}{6}$ may be changed to an equivalent fraction by multiplying numerator and denominator by 3. Thus,

$$\frac{4}{6} = \frac{4 \times 3}{6 \times 3} = \frac{12}{18}$$

When the application of the Fundamental Principle results in a larger numerator and denominator, it is called *expanding to higher terms*.

Here are examples of fractions expanded to higher terms: **(4)**

Example 1: $\frac{3}{5}$; multiply by 5: $\frac{3 \times 5}{5 \times 5} = \frac{15}{25}$ thus $\frac{3}{5} = \frac{15}{25}$

Example 2: $\frac{1}{7}$; multiply by 2: $\frac{1 \times 2}{7 \times 2} = \frac{2}{14}$ thus $\frac{1}{7} = \frac{2}{14}$

Example 3: $\frac{5}{1}$; multiply by 3: $\frac{5 \times 3}{1 \times 3} = \frac{5}{3}$ thus $\frac{5}{1} = \frac{15}{3}$

Reduction to Lower Terms (5)

When the numerator and denominator of a fraction have a common factor, we can divide by the common factor and reduce the fraction to lower terms.

For example, $\frac{4}{6}$ can be reduced to lower terms since 4 and 6 contain a factor of 2.

$$\frac{4}{6} = \frac{4 \div 2}{6 \div 2} = \frac{2}{3}$$

When the application of the Fundamental Principle results in a smaller numerator and denominator, it is called *reducing to lower terms*.

Here are examples of fractions reduced to lower terms: **(6)**

Example 1: $\frac{12}{27}$; divide by 3: $\frac{12 \div 3}{27 \div 3} = \frac{4}{9}$ thus $\frac{12}{27} = \frac{4}{9}$

Example 2: $\frac{27}{36}$; divide by 9: $\frac{27 \div 9}{36 \div 9} = \frac{3}{4}$ thus $\frac{27}{36} = \frac{3}{4}$

Example 3: $\frac{30}{42}$; divide by 6: $\frac{30 \div 6}{42 \div 6} = \frac{5}{7}$ thus $\frac{30}{42} = \frac{5}{7}$

Study Exercise One (7)

A. Expand to higher terms by multiplying numerator and denominator by the indicated number:

1. $\frac{3}{7}$; multiply by 4 **2.** $\frac{4}{6}$; multiply by 3 **3.** $\frac{5}{8}$; multiply by 7

***Study Exercise One* (Continued)**

B. Reduce to lower terms by dividing numerator and denominator by the indicated number:

4. $\frac{9}{15}$; divide by 3 **5.** $\frac{12}{26}$; divide by 2 **6.** $\frac{4}{18}$; divide by 2

7. $\frac{14}{21}$; divide by 7 **8.** $\frac{30}{42}$; divide by 6 **9.** $\frac{4}{100}$; divide by 4

Reducing (8)

A fraction may be reduced only when both numerator and denominator have a common factor.

Example 1: The fraction $\frac{12}{27}$ can be reduced since 12 and 27 have a common factor of 3.

Example 2: The fraction $\frac{5}{8}$ can not be reduced since 5 and 8 have no common factor.

Fractions With More Than One Common Factor (9)

Many times the terms of a fraction have several common factors.

$\frac{16}{24}$ can be reduced to $\frac{8}{12}$: $\frac{16}{24} = \frac{16 \div 2}{24 \div 2} = \frac{8}{12}$

But, $\frac{8}{12}$ can be reduced to $\frac{4}{6}$: $\frac{8}{12} = \frac{8 \div 2}{12 \div 2} = \frac{4}{6}$

But, $\frac{4}{6}$ can be reduced to $\frac{2}{3}$: $\frac{4}{6} = \frac{4 \div 2}{6 \div 2} = \frac{2}{3}$

Thus, the fraction $\frac{16}{24}$ reduces to $\frac{2}{3}$.

Reducing to Lowest Terms (10)

If we reduce until we obtain an equivalent fraction where there are no common factors in the numerator and denominator, we say that fraction is *reduced to lowest terms*.

Answers in fractional form should always be reduced to lowest terms.

Fractions Reduced To Lowest Terms (11)

Example 1: Reduce $\frac{24}{28}$ to lowest terms:

Solution:

$$\frac{24}{28} = \frac{24 \div 2}{28 \div 2} = \frac{12}{14}$$

$$\frac{12}{14} = \frac{12 \div 2}{14 \div 2} = \frac{6}{7}$$

Thus, $\frac{24}{28}$ reduces to $\frac{6}{7}$

Reducing to Lowest Terms (Continued)

Example 2: Reduce $\frac{26}{39}$ to lowest terms:

Solution:

$$\frac{26}{39} = \frac{26 \div 13}{39 \div 13} = \frac{2}{3}$$

Thus, $\frac{26}{39}$ reduces to $\frac{2}{3}$

Study Exercise Two (12)

Reduce to lowest terms:

1. $\frac{15}{20}$ **2.** $\frac{18}{24}$ **3.** $\frac{36}{45}$

4. $\frac{28}{35}$ **5.** $\frac{8}{24}$ **6.** $\frac{18}{45}$

7. $\frac{30}{96}$ **8.** $\frac{54}{60}$ **9.** $\frac{17}{51}$

A Shortcut for Reducing Fractions (Cancellation) (13)

We can use prime factors of numbers to help us reduce a fraction.

To reduce a fraction:

Step (1): Write the numerator as a product of primes.
Step (2): Write the denominator as a product of primes.
Step (3): Divide numerator and denominator by their common factors. This is called *cancelling the common factors.*

Example: Reduce $\frac{6}{14}$ to lowest terms:

Solution:

Step (1): Write 6 as a product of primes: 2×3
Step (2): Write 14 as a product of primes: 2×7

$$\frac{6}{14} = \frac{2 \times 3}{2 \times 7}$$

Step (3): Cancel the common factors:

$$\frac{{}^{1}\not{2} \times 3}{{}^{1}\not{2} \times 7} = \frac{3}{7}$$

A Word About Cancellation (14)

Remember, cancellation is simply a shortcut for dividing numerator and denominator by the same nonzero number.

Example: $\frac{{}^{1}\not{2} \times 3}{{}^{1}\not{2} \times 7}$ means $\frac{6 \div 2}{14 \div 2}$ or $\frac{3}{7}$

Examples of Reducing Fractions By Cancelling (15)

Example 1: Reduce $\frac{12}{16}$ to lowest terms:

Solution:

Steps (1) and (2): Write the prime factors of the numerator and denominator:

$$12 = 2 \times 2 \times 3$$
$$16 = 2 \times 2 \times 2 \times 2$$

Step (3): Reduce by cancelling:

$$\frac{12}{16} = \frac{\cancel{2}^1 \times \cancel{2}^1 \times 3}{\cancel{2}_1 \times \cancel{2}_1 \times 2 \times 2}$$
$$= \frac{1 \times 1 \times 3}{1 \times 1 \times 2 \times 2}$$
$$= \frac{3}{4}$$

Example 2: Reduce $\frac{60}{84}$ to lowest terms:

Solution:

Steps (1) and (2): Write the prime factors of the numerator and denominator:

$$60 = 2 \times 2 \times 3 \times 5$$
$$84 = 2 \times 2 \times 3 \times 7$$

Step (3): Reduce by cancelling:

$$\frac{60}{84} = \frac{\cancel{2}^1 \times \cancel{2}^1 \times \cancel{3}^1 \times 5}{\cancel{2}_1 \times \cancel{2}_1 \times \cancel{3}_1 \times 7}$$
$$= \frac{1 \times 1 \times 1 \times 5}{1 \times 1 \times 1 \times 7}$$
$$= \frac{5}{7}$$

Example 3: Reduce $\frac{26}{39}$ to lowest terms:

Solution: Sometimes the cancellation process is done mentally as follows:

Line (a): $\frac{26}{39} = \frac{\cancel{26}^2}{\cancel{39}^3}$

Line (b): $= \frac{2}{3}$

Study Exercise Three (16)

Reduce to lowest terms:

1. $\frac{4}{22}$ **2.** $\frac{21}{36}$ **3.** $\frac{27}{36}$ **4.** $\frac{28}{40}$

5. $\frac{12}{48}$ **6.** $\frac{30}{36}$ **7.** $\frac{33}{165}$ **8.** $\frac{16}{100}$

Mixed Numerals (17)

A *mixed numeral* is a symbol for a number that contains both a whole number and a fraction. Some mixed numerals are:

$$2\tfrac{1}{3}, \quad 5\tfrac{2}{7}, \quad 1\tfrac{1}{2}$$

Reading Mixed Numerals (18)

$2\frac{1}{3}$ is read *two and one-third.*

$5\frac{2}{7}$ is read *five and two-sevenths.*

$1\frac{1}{2}$ is read *one and one-half.*

Meaning of a Mixed Numeral (19)

A mixed numeral such as $3\frac{2}{5}$ actually means $3 + \dfrac{2}{5}$. It is customary to omit the plus sign and merely write $3\frac{2}{5}$.

Question: What is the meaning of $6\frac{5}{7}$?

Answer: $6\frac{5}{7}$ means $6 + \dfrac{5}{7}$.

Improper Fractions (20)

A fraction whose numerator is equal to or greater than its denominator is called an *improper fraction*. Some improper fractions are:

$$\frac{7}{2} \quad \frac{8}{5} \quad \frac{3}{3} \quad \frac{12}{5}$$

Study Exercise Four (21)

1. Which of the following are mixed numerals?

$$7 \quad \frac{5}{5} \quad \frac{2}{3} \quad 4\tfrac{1}{3} \quad 8 \quad \frac{13}{3} \quad 6\tfrac{2}{5}$$

2. Which of the numbers in problem 1 are improper fractions?
3. How is $5\frac{7}{10}$ read?
4. What is the meaning of $5\frac{7}{10}$?

Changing Improper Fractions to Mixed Numerals (22)

Step (1): Divide to get a whole number quotient and remainder.
Step (2): Write the remainder over the divisor.
Step (3): Add the whole number quotient to the fraction found in *Step (2)*.
Step (4): Reduce the fraction if possible.

Example: Change $\dfrac{34}{5}$ to a mixed numeral.

Changing Improper Fractions to Mixed Numerals (Continued)

Solution:

Step (1): Divide:

$$34 \div 5 = 6 \text{ with remainder } 4: \quad \begin{array}{r} 6 \\ 5\overline{)34} \\ \underline{30} \\ 4 \end{array}$$

Step (2): Write the remainder over the divisor: $\frac{4}{5}$

Step (3): Add the quotient to the fraction: $6 + \frac{4}{5}$

Step (4): $\frac{34}{5} = 6\frac{4}{5}$

Additional Examples (23)

Example 1: Write $\frac{38}{7}$ as a mixed numeral:

Solution: Divide 38 by 7 and place the remainder over 7.

$$\begin{array}{r} 5 \\ 7\overline{)38} \\ \underline{35} \\ 3 \end{array}$$

Answer: $\frac{38}{7} = 5\frac{3}{7}$

Example 2: Write $\frac{20}{8}$ as a mixed numeral:

Solution: Divide 20 by 8 and place the remainder over 8

$$\begin{array}{r} 2 \\ 8\overline{)20} \\ \underline{16} \\ 4 \end{array}$$

$$\frac{20}{8} = 2\frac{4}{8}$$

Reduce the fraction $\frac{4}{8}$ to $\frac{1}{2}$

Answer: $\frac{20}{8} = 2\frac{1}{2}$

When you obtain an improper fraction as an answer, reduce and change to a mixed numeral.

Study Exercise Five (24)

Change to mixed numerals; be sure fractions are in lowest terms.

1. $\frac{19}{5}$ 2. $\frac{52}{8}$ 3. $\frac{16}{6}$ 4. $\frac{46}{8}$
5. $\frac{37}{3}$ 6. $\frac{84}{8}$ 7. $\frac{24}{9}$ 8. $\frac{102}{5}$

REVIEW EXERCISES (25)

A. Expand to higher or lower terms by multiplying or dividing numerator and denominator by the indicated number:

1. $\frac{11}{16}$; multiply by 3 2. $\frac{20}{32}$; divide by 4

B. Reduce to lowest terms by cancellation:

3. $\frac{12}{28}$ 4. $\frac{32}{48}$ 5. $\frac{15}{27}$ 6. $\frac{4}{40}$

C. Fill in the blanks:

7. A mixed numeral is a symbol for a number that contains both a ________ number and a ________.
8. An improper fraction is a fraction whose numerator is ________ or ________ its denominator.
9. $2\frac{1}{3}$ is read as ________.

D. Change to mixed numerals:

10. $\frac{26}{3}$ 11. $\frac{52}{5}$ 12. $\frac{11}{3}$ 13. $\frac{86}{6}$

Solutions to Review Exercises (26)

A. 1. $\frac{11}{16} = \frac{11 \times 3}{16 \times 3} = \frac{33}{48}$ 2. $\frac{20}{32} = \frac{20 \div 4}{32 \div 4} = \frac{5}{8}$

B. 3. $\frac{12}{28} = \frac{\cancel{12}^3}{\cancel{28}^7} = \frac{3}{7}$ 4. $\frac{32}{48} = \frac{\cancel{32}^2}{\cancel{48}^3} = \frac{2}{3}$ 5. $\frac{15}{27} = \frac{\cancel{15}^5}{\cancel{27}^9} = \frac{5}{9}$

6. $\frac{4}{40} = \frac{\cancel{4}^1}{\cancel{40}^{10}} = \frac{1}{10}$

C. 7. Whole, fraction 8. Greater than, equal to. 9. Two and one-third.

D. 10. $8\frac{2}{3}$ 11. $10\frac{2}{5}$ 12. $3\frac{2}{3}$ 13. $14\frac{1}{3}$

SUPPLEMENTARY PROBLEMS

A. Expand to higher terms by multiplying numerator and denominator by the indicated number:

1. $\frac{3}{4}$; multiply by 4 2. $\frac{5}{12}$; multiply by 5 3. $\frac{1}{6}$; multiply by 7
4. $\frac{5}{6}$; multiply by 2 5. $\frac{15}{16}$; multiply by 4 6. $\frac{5}{9}$; multiply by 8

SUPPLEMENTARY PROBLEMS (Continued)

B. Reduce to lower terms by dividing numerator and denominator by the indicated number:

7. $\frac{20}{24}$; divide by 2 **8.** $\frac{16}{48}$; divide by 16 **9.** $\frac{14}{21}$; divide by 7

10. $\frac{64}{80}$; divide by 4 **11.** $\frac{36}{64}$; divide by 4 **12.** $\frac{75}{108}$; divide by 3

C. Reduce to lowest terms:

13. $\frac{18}{24}$ **14.** $\frac{45}{75}$ **15.** $\frac{16}{36}$ **16.** $\frac{84}{108}$

17. $\frac{8}{20}$ **18.** $\frac{42}{60}$ **19.** $\frac{21}{28}$ **20.** $\frac{57}{76}$

21. $\frac{62}{93}$ **22.** $\frac{40}{200}$ **23.** $\frac{98}{105}$ **24.** $\frac{26}{91}$

25. $\frac{60}{84}$ **26.** $\frac{160}{640}$ **27.** $\frac{17}{68}$ **28.** $\frac{45}{195}$

D. 29. Select the mixed numerals from the following:

$\frac{3}{5}$ 8 $3\frac{1}{4}$ $\frac{5}{6}$ $2\frac{11}{13}$ $\frac{7}{8}$

30. Select the improper fractions from the following:

$\frac{1}{6}$ $\frac{7}{6}$ $\frac{3}{3}$ $\frac{2}{7}$ $\frac{5}{2}$ $\frac{8}{8}$

E. Change to a whole number or a mixed numeral:

31. $\frac{3}{2}$ **32.** $\frac{12}{5}$ **33.** $\frac{12}{3}$ **34.** $\frac{22}{7}$ **35.** $\frac{46}{16}$

36. $\frac{33}{9}$ **37.** $\frac{8}{8}$ **38.** $\frac{23}{16}$ **39.** $\frac{54}{12}$ **40.** $\frac{108}{75}$

Solutions to Study Exercises

(7A)

Study Exercise One (Frame 7)

A. 1. $\frac{3}{7} = \frac{3 \times 4}{7 \times 4} = \frac{12}{28}$ **2.** $\frac{4}{6} = \frac{4 \times 3}{6 \times 3} = \frac{12}{18}$ **3.** $\frac{5}{8} = \frac{5 \times 7}{8 \times 7} = \frac{35}{56}$

4. $\frac{9}{15} = \frac{9 \div 3}{15 \div 3} = \frac{3}{5}$ **5.** $\frac{12}{26} = \frac{12 \div 2}{26 \div 2} = \frac{6}{13}$ **6.** $\frac{4}{18} = \frac{4 \div 2}{18 \div 2} = \frac{2}{9}$

7. $\frac{14}{21} = \frac{14 \div 7}{21 \div 7} = \frac{2}{3}$ **8.** $\frac{30}{42} = \frac{30 \div 6}{42 \div 6} = \frac{5}{7}$ **9.** $\frac{4}{100} = \frac{4 \div 4}{100 \div 4} = \frac{1}{25}$

Study Exercise Two (Frame 12)

(12A)

1. $\frac{15}{20} = \frac{15 \div 5}{20 \div 5} = \frac{3}{4}$ **2.** $\frac{18}{24} = \frac{18 \div 6}{24 \div 6} = \frac{3}{4}$ **3.** $\frac{36}{45} = \frac{36 \div 9}{45 \div 9} = \frac{4}{5}$

4. $\frac{28}{35} = \frac{28 \div 7}{35 \div 7} = \frac{4}{5}$ **5.** $\frac{8}{24} = \frac{8 \div 8}{24 \div 8} = \frac{1}{3}$ **6.** $\frac{18}{45} = \frac{18 \div 9}{45 \div 9} = \frac{2}{5}$

7. $\frac{30}{96} = \frac{30 \div 6}{96 \div 6} = \frac{5}{16}$ **8.** $\frac{54}{60} = \frac{54 \div 6}{60 \div 6} = \frac{9}{10}$ **9.** $\frac{17}{51} = \frac{17 \div 17}{51 \div 17} = \frac{1}{3}$

Study Exercise Three (Frame 16) **(16A)**

1. $\frac{4}{22} = \frac{\not{2}^1 \times 2}{\not{2}^1 \times 11} = \frac{2}{11}$ or $\frac{4}{22} = \frac{\not{4}^2}{\not{22}^{11}} = \frac{2}{11}$
2. $\frac{21}{36} = \frac{\not{3}^1 \times 7}{2 \times 2 \times 3 \times \not{3}_1} = \frac{7}{12}$ or $\frac{21}{36} = \frac{\not{21}^7}{\not{36}_{12}} = \frac{7}{12}$
3. $\frac{27}{36} = \frac{\not{3}^1 \times \not{3}^1 \times 3}{2 \times 2 \times \not{3}_1 \times \not{3}_1} = \frac{3}{4}$ or $\frac{27}{36} = \frac{\not{27}^3}{\not{36}_4} = \frac{3}{4}$
4. $\frac{28}{40} = \frac{\not{2}^1 \times \not{2}^1 \times 7}{\not{2}_1 \times \not{2}_1 \times 2 \times 5} = \frac{7}{10}$ or $\frac{28}{40} = \frac{\not{28}^7}{\not{40}_{10}} = \frac{7}{10}$
5. $\frac{12}{48} = \frac{\not{2}^1 \times \not{2}^1 \times \not{3}^1}{\not{2}_1 \times \not{2}_1 \times 2 \times 2 \times \not{3}_1} = \frac{1}{4}$ or $\frac{12}{48} = \frac{\not{12}^1}{\not{48}_4} = \frac{1}{4}$
6. $\frac{30}{36} = \frac{\not{2}^1 \times \not{3}^1 \times 5}{\not{2}_1 \times 2 \times \not{3}_1 \times 3} = \frac{5}{6}$ or $\frac{30}{36} = \frac{\not{30}^5}{\not{36}_6} = \frac{5}{6}$
7. $\frac{33}{165} = \frac{\not{3}^1 \times \not{11}^1}{\not{3}_1 \times 5 \times \not{11}_1} = \frac{1}{5}$ or $\frac{33}{165} = \frac{\not{33}^1}{\not{165}_5} = \frac{1}{5}$
8. $\frac{16}{100} = \frac{\not{2}^1 \times \not{2}^1 \times 2 \times 2}{\not{2}_1 \times \not{2}_1 \times 5 \times 5} = \frac{4}{25}$ or $\frac{16}{100} = \frac{\not{16}^4}{\not{100}_{25}} = \frac{4}{25}$

Study Exercise Four (Frame 21) **(21A)**

1. $4\frac{1}{3}, 6\frac{2}{3}$
2. $\frac{5}{5}, \frac{13}{3}$
3. Five and seven-tenths.
4. $5 + \frac{7}{10}$

Study Exercise Five (Frame 24) **(24A)**

1. $\frac{19}{5} = 19 \div 5 = 3\frac{4}{5}$
2. $\frac{52}{8} = 52 \div 8 = 6\frac{4}{8} = 6\frac{1}{2}$
3. $\frac{16}{6} = 16 \div 6 = 2\frac{4}{6} = 2\frac{2}{3}$
4. $\frac{46}{8} = 46 \div 8 = 5\frac{6}{8} = 5\frac{3}{4}$
5. $\frac{37}{3} = 37 \div 3 = 12\frac{1}{3}$
6. $\frac{84}{8} = 84 \div 8 = 10\frac{4}{8} = 10\frac{1}{2}$
7. $\frac{24}{9} = 24 \div 9 = 2\frac{6}{9} = 2\frac{2}{3}$
8. $\frac{102}{5} = 102 \div 5 = 20\frac{2}{5}$

UNIT

11 Lowest Common Denominator and Comparison of Fractions

OBJECTIVES (1)

By the end of this unit you should be able to:

1. FIND THE LOWEST COMMON DENOMINATOR (LCD) OF A GROUP OF FRACTIONS.
2. COMPARE TWO OR MORE FRACTIONS TO SEE WHICH IS LARGEST.

Like and Unlike Fractions (2)

If two fractions have the same denominator, they are called *like fractions*. If two fractions have different denominators, they are called *unlike fractions*.

The fractions $\frac{1}{7}$ and $\frac{3}{7}$ are *like fractions* since they have the same denominator, seven. The fractions $\frac{2}{3}$ and $\frac{3}{4}$ are *unlike fractions* since they have different denominators.

Common Denominators (3)

It will be necessary to change unlike fractions to the same or *common denominator*. Let us change $\frac{2}{3}$ and $\frac{4}{7}$ to fractions with the same denominator.

We need a denominator which is a multiple of the two denominators 3 and 7. We take $3 \times 7 = 21$ to be the common denominator.

$$\frac{2}{3} = \frac{2 \times 7}{3 \times 7} = \frac{14}{21}$$

$$\frac{4}{7} = \frac{4 \times 3}{7 \times 3} = \frac{12}{21}$$

Thus, $\frac{2}{3}$ is renamed $\frac{14}{21}$ and $\frac{4}{7}$ is renamed $\frac{12}{21}$.

Finding Common Denominators (4)

Suppose we want to change $\frac{1}{6}$ and $\frac{3}{10}$ to fractions with the same denominator.

We could choose $6 \times 10 = 60$ as the common denominator.

$$\frac{1}{6} = \frac{1 \times 10}{6 \times 10} = \frac{10}{60}$$

$$\frac{3}{10} = \frac{3 \times 6}{10 \times 6} = \frac{18}{60}$$

However, there is a smaller number that would also work. That number is 30.

$$\frac{1}{6} = \frac{1 \times 5}{6 \times 5} = \frac{5}{30}$$

$$\frac{3}{10} = \frac{3 \times 3}{10 \times 3} = \frac{9}{30}$$

But how do you determine the *lowest* common denominator?

Lowest Common Denominator (LCD) (5)

In order to keep the numbers we work with as small as possible, we wish to use the lowest common denominator. We will refer to the lowest common denominator as the *LCD*.

Rule: The LCD of two fractions that do not have a common factor in their denominators is the product of the denominators.

Example: The LCD of $\frac{1}{2}$ and $\frac{2}{5}$ is 10, since 2 and 5 have no common factor.

Finding Like Fractions Using the Rule (6)

Example: Change $\frac{3}{5}$ and $\frac{1}{6}$ to fractions with the same denominator:

Solution: Since 5 and 6 do not have a common factor, the LCD is $5 \times 6 = 30$

$$\frac{3}{5} = \frac{3 \times 6}{5 \times 6} = \frac{18}{30}$$

$$\frac{1}{6} = \frac{1 \times 5}{6 \times 5} = \frac{5}{30}$$

Study Exercise One (7)

A. Find the LCD of each of the following groups of fractions:

1. $\frac{1}{3}$ and $\frac{5}{8}$ **2.** $\frac{3}{11}$ and $\frac{4}{7}$ **3.** $\frac{2}{5}$ and $\frac{4}{9}$ **4.** $\frac{3}{4}$ and $\frac{2}{5}$

B. Change the following to like fractions:

5. $\frac{2}{3}$ and $\frac{1}{5}$ **6.** $\frac{1}{4}$ and $\frac{1}{3}$ **7.** $\frac{3}{8}$ and $\frac{4}{9}$

LCD for Three or More Fractions (8)

Fractions do not always come in pairs. It may be necessary to find the LCD for three or more fractions. If none of the fractions have a common factor, the LCD is the product of all the denominators.

Example: Find the LCD of $\frac{2}{3}$, $\frac{3}{4}$, and $\frac{1}{5}$:

Solution: Since 3, 4, and 5 have no common factor, the LCD is $3 \times 4 \times 5 = 60$

$$\frac{2}{3} = \frac{?}{60} = \frac{40}{60} \quad (\times 20) \qquad \frac{3}{4} = \frac{?}{60} = \frac{45}{60} \quad (\times 15) \qquad \frac{1}{5} = \frac{?}{60} = \frac{12}{60} \quad (\times 12)$$

Method for Finding the LCD (9)

On many occasions denominators have common factors which make it difficult to simply see the LCD. Therefore, we need a method of finding that LCD.

Rule: The LCD of two or more fractions is the least common multiple (LCM) of their denominators.

Example: Find the LCD of $\frac{5}{8}$ and $\frac{1}{6}$:

Solution: Find the LCM of the denominators 8 and 6.

Step (1): Write each denominator in prime factored exponential form:

$$8 = 2^3$$
$$6 = 2^1 \times 3^1$$

Step (2): List each factor that was used as a base:

$$2, 3$$

Step (3): Use the largest exponent that occurs on each:

$$2^3, 3^1$$

Step (4): Multiply together:

$$2^3 \times 3^1 = 8 \times 3 = 24$$

Step (5): The LCD is 24

Note: This procedure is similar to the one discussed in Unit 8, Frames 18-21.

(10)

Another Example: Find the LCD of $\frac{1}{6}$ and $\frac{3}{16}$:

Solution:

Step (1): Write each denominator in prime factored exponential form:

$$6 = 2^1 \times 3^1$$
$$16 = 2^4$$

Another Example (Continued)

Step (2): List each factor that was used as a base:

$$2, 3$$

Step (3): Use the largest exponent that occurs on each:

$$2^4, 3^1$$

Step (4): Multiply together:

$$2^4 \times 3^1 = 16 \times 3 = 48$$

Step (5): The LCD is 48

Study Exercise Two (11)

Find the LCD of the fractions:

1. $\frac{1}{2}, \frac{2}{3}$, and $\frac{3}{5}$ (follow the example in frame 8)

2. $\frac{2}{3}$ and $\frac{7}{12}$ **3.** $\frac{13}{20}$ and $\frac{7}{12}$ **3.** $\frac{1}{2}$ and $\frac{11}{16}$

5. $\frac{11}{24}$ and $\frac{5}{8}$ **6.** $\frac{5}{6}$ and $\frac{7}{18}$ **7.** $\frac{2}{3}$ and $\frac{1}{6}$

The LCD of Three or More Fractions (12)

The LCD of three or more fractions is found in a similar fashion to the LCD of two fractions.

Example: Find the LCD of $\frac{1}{6}, \frac{1}{8}$, and $\frac{1}{9}$.

Solution: Find the LCM of the denominators 6, 8, and 9.

$$6 = 2^1 \times 3^1$$
$$8 = 2^3$$
$$9 = 3^2$$

The different factors are 2, 3

$$2^3, 3^2$$
$$2^3 \times 3^2 = 8 \times 9 = 72$$

Thus, the LCD is 72

Study Exercise Three (13)

Find the LCD of the fractions:

1. $\frac{1}{4}, \frac{1}{12}$, and $\frac{5}{6}$ **2.** $\frac{7}{15}, \frac{2}{25}$, and $\frac{1}{10}$

3. $\frac{3}{8}, \frac{3}{10}, \frac{3}{4}$, and $\frac{1}{2}$

Renaming Fractions (14)

Now that we can find the LCD, we should be able to expand fractions to equivalent fractions with the same denominator. That denominator should be the LCD.

Expanding Fractions Using the LCD (15)

Example: Change to equivalent fractions having the same denominator: $\frac{1}{6}, \frac{2}{9}$

Solution:

Step (1): Find the LCD:

$$6 = 2^1 \times 3^1; \qquad 9 = 3^2$$
$$\text{LCD} = 2^1 \times 3^2 = 18$$

Step (2): Expand each fraction:

$$\frac{1}{6} = \frac{1 \times 3}{6 \times 3} = \frac{3}{18}$$
$$\frac{2}{9} = \frac{2 \times 2}{9 \times 2} = \frac{4}{18}$$

Thus, $\frac{1}{6} = \frac{3}{18}$ and $\frac{2}{9} = \frac{4}{18}$

Study Exercise Four (16)

Change to equivalent fractions having the same denominator. Use the LCD:

1. $\frac{1}{16}$ $\frac{7}{12}$ **2.** $\frac{3}{4}$ $\frac{5}{6}$ **3.** $\frac{1}{3}$ $\frac{2}{15}$ $\frac{5}{18}$

Comparing Fractions with the Same Denominator (17)

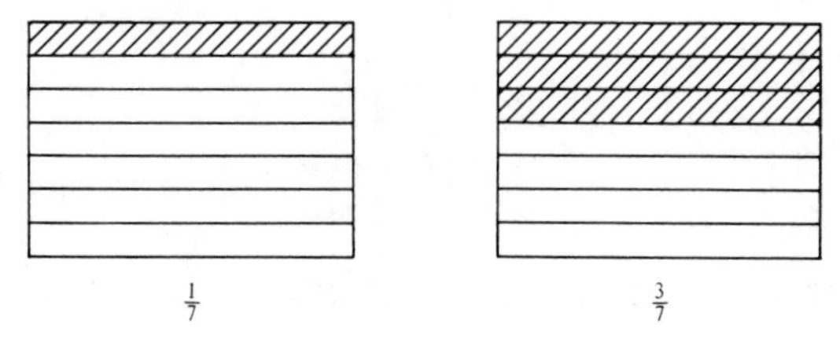

It is easy to compare two fractions with the same denominator. Which do you think is larger, $\frac{1}{7}$ or $\frac{3}{7}$?

Three parts is larger than 1 part out of 7, so $\frac{3}{7}$ is larger than $\frac{1}{7}$.

Example: Which is the largest: $\frac{4}{13}$, $\frac{7}{13}$, or $\frac{2}{13}$? **(18)**

Solution: Seven parts is larger than either 4 or 2 parts. Hence, $\frac{7}{13}$ is larger than either $\frac{4}{13}$ or $\frac{2}{13}$.

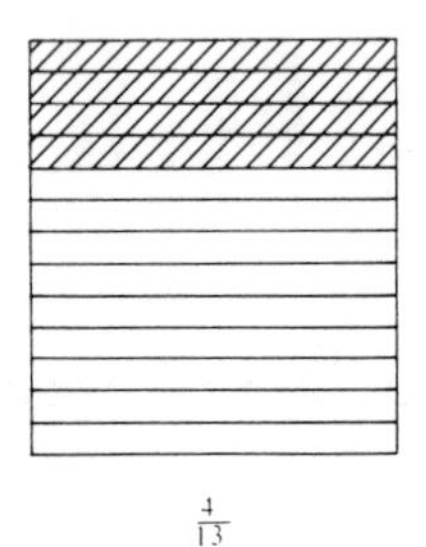

$\frac{4}{13}$

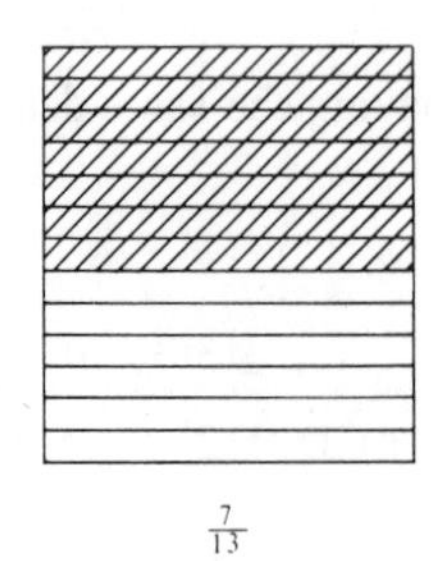

$\frac{7}{13}$

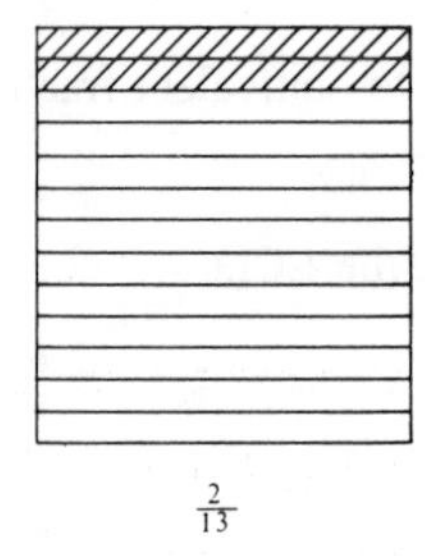

$\frac{2}{13}$

When comparing fractions having the *same denominator,* the largest fraction has the largest numerator.

Study Exercise Five (19)

1. Which is larger, $\frac{3}{4}$ or $\frac{2}{4}$?

For each group, find the largest fraction:

2. $\frac{2}{8}$ $\frac{5}{8}$ $\frac{6}{8}$

3. $\frac{7}{13}$ $\frac{6}{13}$ $\frac{5}{13}$

Comparing Fractions with Different Denominators (20)

Question: Which is larger $\frac{2}{3}$ or $\frac{3}{5}$?

Answer: It is difficult to compare 2 parts out of 3 with 3 parts out of 5.

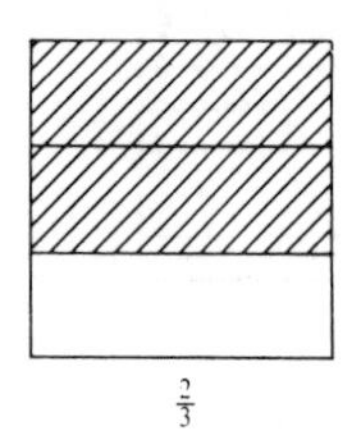

$\frac{2}{3}$

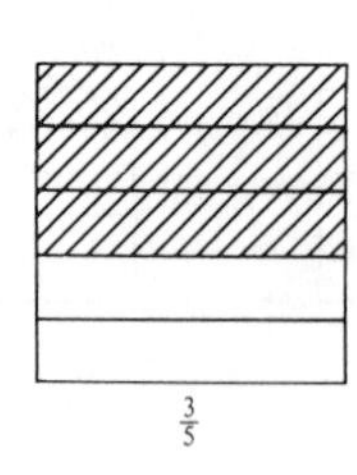

$\frac{3}{5}$

⟵ more parts are taken here but the parts are smaller

But if the denominators were the same, comparison would be easy. We will use the LCD and expand the denominators to 15.

$$\frac{2}{3} = \frac{2 \times 5}{3 \times 5} = \frac{10}{15}$$

$$\frac{3}{5} = \frac{3 \times 3}{5 \times 3} = \frac{9}{15}$$

Comparing Fractions with Different Denominators (Continued)

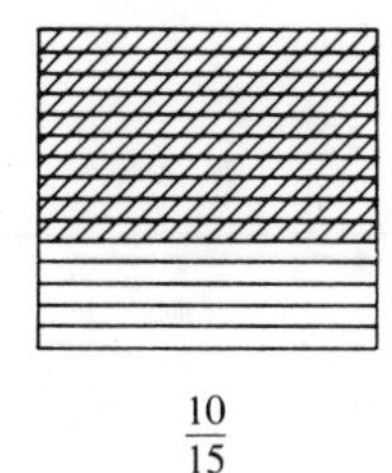
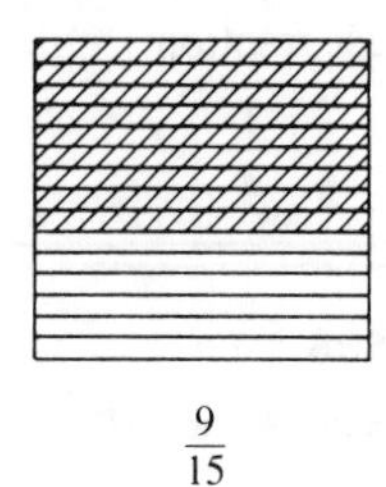

$\frac{10}{15}$ $\qquad$ $\frac{9}{15}$

$\frac{10}{15}$ is larger than $\frac{9}{15}$; therefore, $\frac{2}{3}$ is larger.

Comparing Unlike Fractions (21)

To compare unlike fractions, expand the fractions so they all have the same denominator, using the LCD. Then compare them as outlined in Frames 17 and 18.

Example: Which is larger, $\frac{1}{2}$ or $\frac{3}{5}$?

Solution: We will expand so each denominator is 10.

$$\frac{1}{2} = \frac{1 \times 5}{2 \times 5} = \frac{5}{10}$$

$$\frac{3}{5} = \frac{3 \times 2}{5 \times 2} = \frac{6}{10}$$

Since $\frac{6}{10}$ has a larger numerator than $\frac{5}{10}$, $\frac{6}{10}$ is larger than $\frac{5}{10}$. Therefore, $\frac{3}{5}$ is larger than $\frac{1}{2}$.

Study Exercise Six (22)

In each group, find the largest fraction:

1. $\frac{1}{4}$ $\frac{2}{3}$ **2.** $\frac{1}{6}$ $\frac{2}{9}$ **3.** $\frac{3}{5}$ $\frac{7}{10}$ **4.** $\frac{3}{4}$ $\frac{5}{6}$

5. $\frac{3}{5}$ $\frac{1}{10}$ $\frac{1}{6}$

REVIEW EXERCISES (23)

A. Find the LCD for each of the following groups of fractions:

1. $\frac{8}{9}$ $\frac{2}{5}$ **2.** $\frac{1}{8}$ $\frac{5}{12}$ **3.** $\frac{1}{4}$ $\frac{5}{6}$ $\frac{3}{10}$

4. $\frac{1}{8}$ $\frac{3}{20}$ **5.** $\frac{3}{8}$ $\frac{7}{18}$ **6.** $\frac{1}{9}$ $\frac{1}{3}$ $\frac{5}{36}$

B. Change to equivalent fractions having the same denominator. Use the LCD:

7. $\frac{1}{4}$ $\frac{1}{6}$ **8.** $\frac{5}{12}$ $\frac{7}{18}$ **9.** $\frac{1}{4}$ $\frac{3}{8}$ $\frac{5}{16}$

REVIEW EXERCISES (Continued)

C. In each group, find the largest fraction:

10. $\frac{3}{5}\ \frac{2}{5}$ **11.** $\frac{2}{5}\ \frac{1}{4}$ **12.** $\frac{3}{7}\ \frac{3}{8}$ **13.** $\frac{1}{9}\ \frac{1}{3}\ \frac{5}{36}$

Solutions to Review Exercises (24)

A. **1.** Since 9 and 5 have no common factor, the LCD = $9 \times 5 = 45$

2. $8 = 2^3, \quad 12 = 2^2 \times 3^1$
LCD = $2^3 \times 3^1 = 8 \times 3 = 24$

3. $4 = 2^2, \quad 6 = 2^1 \times 3^1, \quad 10 = 2^1 \times 5^1$
LCD = $2^2 \times 3^1 \times 5^1 = 4 \times 3 \times 5 = 60$

4. $8 = 2^3, \quad 20 = 2^2 \times 5^1$
LCD = $2^3 \times 5^1 = 8 \times 5 = 40$

5. $8 = 2^3, \quad 18 = 2^1 \times 3^2$
LCD = $2^3 \times 3^2 = 8 \times 9 = 72$

6. $9 = 3^2, \quad 3 = 3^1, \quad 36 = 2^2 \times 3^2$
LCD = $2^2 \times 3^2 = 4 \times 9 = 36$

B. **7.** $4 = 2^2, \quad 6 = 2^1 \times 3^1$
LCD = $2^2 \times 3^1 = 4 \times 3 = 12$

$\frac{1}{4} = \frac{1 \times 3}{4 \times 3} = \frac{3}{12}$

$\frac{1}{6} = \frac{1 \times 2}{6 \times 2} = \frac{2}{12}$

Thus, $\frac{1}{4} = \frac{3}{12}$ and $\frac{1}{6} = \frac{2}{12}$

8. $12 = 2^2 \times 3^1, \quad 18 = 2^1 \times 3^2$
LCD = $2^2 \times 3^2 = 4 \times 9 = 36$

$\frac{5}{12} = \frac{5 \times 3}{12 \times 3} = \frac{15}{36}$

$\frac{7}{18} = \frac{7 \times 2}{18 \times 2} = \frac{14}{36}$

Thus, $\frac{5}{12} = \frac{15}{36}$ and $\frac{7}{18} = \frac{14}{36}$

9. $4 = 2^2, \quad 8 = 2^3, \quad 16 = 2^4$
LCD = $2^4 = 2 \times 2 \times 2 \times 2 = 16$

$\frac{1}{4} = \frac{1 \times 4}{4 \times 4} = \frac{4}{16}$

$\frac{3}{8} = \frac{3 \times 2}{8 \times 2} = \frac{6}{16}$

$\frac{5}{16} = \frac{5}{16}$

Thus, $\frac{1}{4} = \frac{4}{16}, \frac{3}{8} = \frac{6}{16}$, and $\frac{5}{16} = \frac{5}{16}$

C. **10.** $\frac{3}{5}$ is larger than $\frac{2}{5}$

11. $\frac{2}{5} = \frac{2 \times 4}{5 \times 4} = \frac{8}{20}$

$\frac{1}{4} = \frac{1 \times 5}{4 \times 5} = \frac{5}{20}$

$\frac{8}{20}$ is larger than $\frac{5}{20}$

Therefore, $\frac{2}{5}$ is larger than $\frac{1}{4}$

12. $\frac{3}{7} = \frac{3 \times 8}{7 \times 8} = \frac{24}{56}$

$\frac{3}{8} = \frac{3 \times 7}{8 \times 7} = \frac{21}{56}$

$\frac{24}{56}$ is larger than $\frac{21}{56}$

Therefore, $\frac{3}{7}$ is larger than $\frac{3}{8}$

13. LCD = 36 (see problem 6)

$\frac{1}{9} = \frac{1 \times 4}{9 \times 4} = \frac{4}{36}$

$\frac{1}{3} = \frac{1 \times 12}{3 \times 12} = \frac{12}{36}$

$\frac{5}{36} = \frac{5}{36}$

$\frac{12}{36}$ is larger than either $\frac{5}{36}$ or $\frac{4}{36}$

Therefore, $\frac{1}{3}$ is the largest fraction.

SUPPLEMENTARY PROBLEMS

A. Find the LCD for each of the following groups of fractions:

1. $\frac{1}{4}$ $\frac{3}{14}$ **2.** $\frac{2}{7}$ $\frac{3}{11}$ **3.** $\frac{7}{15}$ $\frac{1}{21}$ **4.** $\frac{4}{18}$ $\frac{7}{30}$

5. $\frac{1}{26}$ $\frac{3}{14}$ **6.** $\frac{1}{4}$ $\frac{7}{10}$ $\frac{5}{12}$ **7.** $\frac{5}{6}$ $\frac{1}{15}$ $\frac{5}{18}$ **8.** $\frac{1}{27}$ $\frac{1}{63}$ $\frac{1}{72}$

9. $\frac{4}{9}$ $\frac{5}{6}$ $\frac{7}{12}$ $\frac{11}{15}$ **10.** $\frac{11}{12}$ $\frac{25}{42}$

B. Change to equivalent fractions with the same denominator. Use the LCD:

11. $\frac{1}{4}$ $\frac{5}{6}$ **12.** $\frac{1}{6}$ $\frac{5}{9}$ **13.** $\frac{2}{3}$ $\frac{3}{5}$ $\frac{5}{6}$ **14.** $\frac{8}{21}$ $\frac{9}{35}$

15. $\frac{1}{3}$ $\frac{3}{4}$ $\frac{2}{5}$ **16.** $\frac{7}{12}$ $\frac{5}{8}$ **17.** $\frac{1}{10}$ $\frac{3}{4}$ **18.** $\frac{1}{2}$ $\frac{5}{9}$

19. $\frac{5}{6}$ $\frac{3}{8}$ **20.** $\frac{3}{4}$ $\frac{2}{9}$ **21.** $\frac{3}{8}$ $\frac{7}{10}$ **22.** $\frac{3}{5}$ $\frac{4}{7}$ $\frac{2}{3}$

C. Determine which fraction is largest:

23. $\frac{5}{8}$ or $\frac{2}{3}$? **24.** $\frac{4}{5}$ or $\frac{3}{4}$? **25.** $\frac{1}{8}$ or $\frac{1}{10}$?

26. $\frac{1}{4}$, $\frac{1}{2}$, or $\frac{1}{6}$?

D. Arrange in order of size (smallest first):

27. $\frac{3}{4}$ $\frac{2}{3}$ $\frac{3}{5}$ **18.** $\frac{1}{2}$ $\frac{1}{5}$ $\frac{1}{3}$ **29.** $\frac{1}{2}$ $\frac{3}{8}$ $\frac{5}{6}$ **30.** $\frac{4}{5}$ $\frac{5}{6}$ $\frac{3}{8}$

Solutions to Study Exercises (7A)

Study Exercise One (Frame 7)

A. 1. Since 3 and 8 have to common factor,
LCD = 3 × 8 = 24

2. Since 11 and 7 have no common factor,
LCD = 11 × 7 = 77

3. Since 5 and 9 have no common factor,
LCD = 5 × 9 = 45

4. Since 4 and 5 have no common factor,
LCD = 4 × 5 = 20

B. 5. Since 3 and 5 have no common factor,
LCD = 3 × 5 = 15

$$\frac{2}{3} = \frac{2 \times 5}{3 \times 5} = \frac{10}{15}$$

$$\frac{1}{5} = \frac{1 \times 3}{5 \times 3} = \frac{3}{15}$$

6. Since 4 and 3 have no common factor,
LCD = 4 × 3 = 12

$$\frac{1}{4} = \frac{1 \times 3}{4 \times 3} = \frac{3}{12}$$

$$\frac{1}{3} = \frac{1 \times 4}{3 \times 4} = \frac{4}{12}$$

7. Since 8 and 9 have no common factor, LCD = 8 × 9 = 72

$$\frac{3}{8} = \frac{3 \times 9}{8 \times 9} = \frac{27}{72}$$

$$\frac{4}{9} = \frac{4 \times 8}{9 \times 8} = \frac{32}{72}$$

Study Exercise Two (Frame 11) (11A)

1. Since 2, 3, and 5 have no common factor,
$\text{LCD} = 2 \times 3 \times 5 = 30$

2. $3 = 3^1, \quad 12 = 2^2 \times 3^1$
$\text{LCD} = 2^2 \times 3^1 = 12$

3. $20 = 2^2 \times 5^1, \quad 12 = 2^2 \times 3^1$
$\text{LCD} = 2^2 \times 5^1 \times 3^1 = 60$

4. $2 = 2^1, \quad 16 = 2^4$
$\text{LCD} = 2^4 = 16$

5. $24 = 2^3 \times 3^1, \quad 8 = 2^3$
$\text{LCD} = 2^3 \times 3^1 = 24$

6. $6 = 2^1 \times 3^1, \quad 18 = 2^1 \times 3^2$
$\text{LCD} = 2^1 \times 3^2 = 18$

7. $3 = 3^1, \quad 6 = 2^1 \times 3^1$
$\text{LCD} = 2^1 \times 3^1 = 6$

Study Exercise Three (Frame 13) (13A)

1. $4 = 2^2, \quad 12 = 2^2 \times 3^1, \quad 6 = 2^1 \times 3^1$
$\text{LCD} = 2^2 \times 3^1 = 12$

2. $15 = 3^1 \times 5^1, \quad 25 = 5^2, \quad 10 = 2^1 \times 5^1$
$\text{LCD} = 2^1 \times 3^1 \times 5^2 = 150$

3. $8 = 2^3, \quad 10 = 2^1 \times 5^1, \quad 4 = 2^2, \quad 2 = 2^1$
$\text{LCD} = 2^3 \times 5^1 = 40$

Study Exercise Four (Frame 16) (16A)

1. $16 = 2^4, \quad 12 = 2^2 \times 3^1$
$\text{LCD} = 2^4 \times 3^1 = 48$
$\frac{1}{16} = \frac{1 \times 3}{16 \times 3} = \frac{3}{48}, \quad \frac{7}{12} = \frac{7 \times 4}{12 \times 4} = \frac{28}{48}$

2. $4 = 2^2, \quad 6 = 2^1 \times 3^1$
$\text{LCD} = 2^2 \times 3^1 = 12$
$\frac{3}{4} = \frac{3 \times 3}{4 \times 3} = \frac{9}{12}, \quad \frac{5}{6} = \frac{5 \times 2}{6 \times 2} = \frac{10}{12}$

3. $3 = 3^1, \quad 15 = 3^1 \times 5^1, \quad 18 = 2^1 \times 3^2$
$\text{LCD} = 2^1 \times 3^2 \times 5^1 = 90$
$\frac{1}{3} = \frac{1 \times 30}{3 \times 30} = \frac{30}{90}, \quad \frac{2}{15} = \frac{2 \times 6}{15 \times 6} = \frac{12}{90}, \quad \frac{5}{18} = \frac{5 \times 5}{18 \times 5} = \frac{25}{90}$

Study Exercise Five (Frame 19) (19A)

1. Since 3 parts out of 4 is larger than 2 parts out of four, $\frac{3}{4}$ is larger than $\frac{2}{4}$

2. $\frac{6}{8}$ is the largest.

3. $\frac{7}{13}$ is the largest.

Study Exercise Six (Frame 22) (22A)

1. $\frac{1}{4} = \frac{1 \times 3}{4 \times 3} = \frac{3}{12}, \quad \frac{2}{3} = \frac{2 \times 4}{3 \times 4} = \frac{8}{12}$
Since $\frac{8}{12}$ is larger than $\frac{3}{12}$, $\frac{2}{3}$ is larger than $\frac{1}{4}$

2. $\frac{1}{6} = \frac{1 \times 3}{6 \times 3} = \frac{3}{18}, \quad \frac{2}{9} = \frac{2 \times 2}{9 \times 2} = \frac{4}{18}$
$\frac{2}{9}$ is larger than $\frac{1}{6}$

3. $\frac{3}{5} = \frac{3 \times 2}{5 \times 2} = \frac{6}{10}, \quad \frac{7}{10}$
$\frac{7}{10}$ is larger than $\frac{3}{5}$

4. $\frac{3}{4} = \frac{3 \times 3}{4 \times 3} = \frac{9}{12}, \quad \frac{5}{6} = \frac{5 \times 2}{6 \times 2} = \frac{10}{12}$
$\frac{5}{6}$ is larger than $\frac{3}{4}$

5. $\frac{3}{5} = \frac{3 \times 6}{5 \times 6} = \frac{18}{30}, \quad \frac{1}{10} = \frac{1 \times 3}{10 \times 3} = \frac{3}{30}, \quad \frac{1}{6} = \frac{1 \times 5}{6 \times 5} = \frac{5}{30}$
$\frac{3}{5}$ is the largest.

UNIT

12 Addition and Subtraction of Fractions

OBJECTIVES

(1)

By the end of this unit you should be able to:

1. ADD LIKE AND UNLIKE FRACTIONS.
2. SUBTRACT LIKE AND UNLIKE FRACTIONS.

---

(2)

Let us add $\frac{2}{5} + \frac{1}{5}$

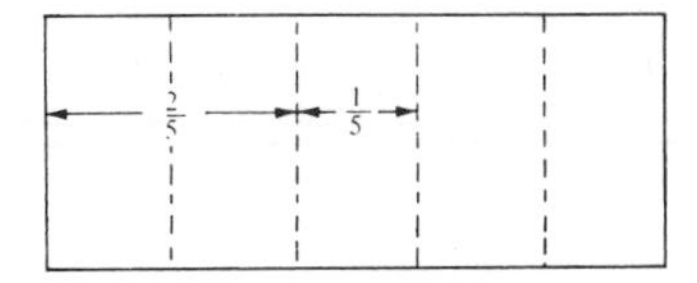

From the diagram we see the result is $\frac{2}{5} + \frac{1}{5} = \frac{3}{5}$

We are adding 2 of the 5 equal sub-units to 1 of the 5 equal sub-units.

Notice, when adding $\frac{2}{5} + \frac{1}{5} = \frac{3}{5}$, the numerators are added, but the denominator remains unchanged.

---

(3)

Let us add $\frac{2}{5} + \frac{4}{5}$

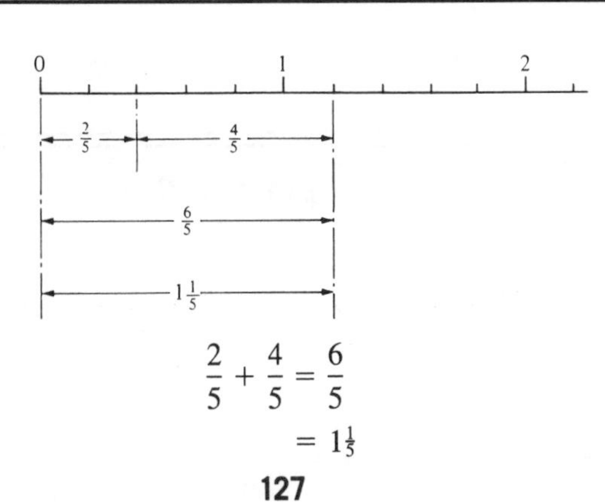

$$\frac{2}{5} + \frac{4}{5} = \frac{6}{5}$$
$$= 1\tfrac{1}{5}$$

Frame 3 (Continued)

Notice, when adding $\frac{2}{5} + \frac{4}{5}$, the numerators are added, but the denominator remains unchanged.

Remember to change improper fractions to mixed numerals.

Addition of Like Fractions

(4)

Rule: To add two or more fractions with like denominators, add the numerators and put the result over the denominator.

Example 1: $\frac{3}{7} + \frac{1}{7}$

Solution: $\frac{3}{7} + \frac{1}{7} = \frac{3 + 1}{7}$

$= \frac{4}{7}$

Example 2: $\frac{7}{8} + \frac{5}{8}$

(5)

Solution: $\frac{7}{8} + \frac{5}{8} = \frac{7 + 5}{8}$

$= \frac{12}{8}$

$= 1\frac{4}{8}$

$= 1\frac{1}{2}$

The answer must always be expressed in lowest terms. Remember to reduce all answers.

Two Formats of Addition

(6)

Add: $\frac{3}{7} + \frac{1}{7}$

Horizontal Format	**Vertical Format**
$\frac{3}{7} + \frac{1}{7} = \frac{4}{7}$	$\begin{array}{r} \frac{3}{7} \\ +\frac{1}{7} \\ \hline \frac{4}{7} \end{array}$

Study Exercise One

(7)

Add:

1. $\frac{3}{5} + \frac{1}{5}$ **2.** $\frac{1}{7} + \frac{3}{7}$ **3.** $\frac{5}{8} + \frac{3}{8}$

Study Exercise One **(Continued)**

4. $\frac{6}{7} + \frac{5}{7}$

5. $\begin{array}{r} \frac{1}{3} \\ +\frac{1}{3} \\ \hline \end{array}$

6. $\begin{array}{r} \frac{3}{10} \\ +\frac{9}{10} \\ \hline \end{array}$

Adding Mixed Numerals **(8)**

Mixed numerals may be added by combining the whole numbers and then by combining the fractions separately.

Example: Add: $5\frac{3}{7} + 2\frac{1}{7}$

Solution: **Horizontal Form** $5\frac{3}{7} + 2\frac{1}{7}$

Step (1): Add the whole numbers $(5 + 2)$

Step (2): Add the fractions $\left(\frac{3}{7} + \frac{1}{7}\right)$

Therefore,

$$5\tfrac{3}{7} + 2\tfrac{1}{7} = 7 + \frac{4}{7} = 7\tfrac{4}{7}$$

Vertical Form

$$\begin{array}{r} 5\frac{3}{7} \\ +2\frac{1}{7} \\ \hline 7\frac{4}{7} \end{array}$$

Example: $2\frac{7}{8} + 6\frac{5}{8}$ **(9)**

Solution:

Horizontal Form

$$\begin{aligned} 2\tfrac{7}{8} + 6\tfrac{5}{8} &= 8 + \tfrac{12}{8} \\ &= 8 + 1\tfrac{4}{8} \\ &= 8 + 1\tfrac{1}{2} \\ &= 9\tfrac{1}{2} \end{aligned}$$

Vertical Form

$$\begin{array}{r} 2\frac{7}{8} \\ +6\frac{5}{8} \\ \hline \end{array}$$

$$8\tfrac{12}{8} = 8 + \tfrac{12}{8} = 8 + 1\tfrac{4}{8} = 9\tfrac{4}{8} = 9\tfrac{1}{2}$$

Study Exercise Two **(10)**

Add using the horizontal form in the first three problems and the vertical form in the last three problems:

1. $2\frac{3}{11} + 5\frac{6}{11}$

2. $1\frac{15}{16} + 3\frac{13}{16}$

3. $4\frac{8}{12} + 2\frac{6}{12}$

4. $1\frac{13}{16} + 2\frac{4}{16} + \frac{3}{16}$

5. $2\frac{3}{6} + 4\frac{5}{6}$

6. $1\frac{3}{5} + \frac{4}{5}$

Changing Mixed Numerals to Improper Fractions **(11)**

A mixed numeral such as $3\frac{1}{2}$ can be expressed as an improper fraction. $3\frac{1}{2}$ means $3 + \frac{1}{2}$. However, $3 = \frac{6}{2}$. Thus,

$$\begin{aligned} 3 + \frac{1}{2} &= \frac{6}{2} + \frac{1}{2} \\ &= \frac{6 + 1}{2} \\ &= \frac{7}{2} \end{aligned}$$

Changing Mixed Numerals to Improper Fractions (Continued)

This process can be shortened as follows:

To change $3\frac{1}{2}$ to an improper fraction, multiply 3×2. Then add 1 to the result and divide your answer by 2. That is,

$$3\tfrac{1}{2} = \frac{(3 \times 2) + 1}{2} = \frac{7}{2}$$

Rule: $\dfrac{(\text{whole number} \times \text{denominator}) + \text{numerator}}{\text{denominator}}$

(12)

Example 1: Change $5\frac{1}{3}$ to an improper fraction:

Solution: $5\frac{1}{3} = \dfrac{(5 \times 3) + 1}{3} = \dfrac{15 + 1}{3} = \dfrac{16}{3}$

Example 2: Change $4\frac{2}{5}$ to an improper fraction:

Solution: $4\frac{2}{5} = \dfrac{(4 \times 5) + 2}{5} = \dfrac{20 + 2}{5} = \dfrac{22}{5}$

Adding Mixed Numerals by Changing to Improper Fractions (13)

Example: Add by changing to improper fractions: $2\frac{1}{3} + 1\frac{2}{3}$

Solution: We will change each mixed numeral to an improper fraction.

$$2\tfrac{1}{3} = \frac{(2 \times 3) + 1}{3} = \frac{7}{3}$$

$$1\tfrac{2}{3} = \frac{(1 \times 3) + 2}{3} = \frac{5}{3}$$

$$\frac{7}{3} + \frac{5}{3} = \frac{12}{3} = 4$$

Study Exercise Three (14)

1. Change $8\frac{2}{3}$ to an improper fraction.
2. Change $6\frac{3}{5}$ to an improper fraction.
3. Change $2\frac{1}{2}$ to an improper fraction.
4. Add $8\frac{2}{3} + 6$ by changing to improper fractions.
5. Add $5\frac{3}{5} + 6\frac{3}{5}$ by changing to improper fractions.
6. Add $1\frac{3}{4} + 2\frac{3}{4}$ by changing to improper fractions.

Addition of Unlike Fractions (15)

We have learned that fractions can be added when they have the same denominators. For example, $\frac{2}{5} + \frac{1}{5} = \frac{3}{5}$.

To find the sum of two or more fractions whose denominators are not the same, the fractions must be renamed so that they all have the same denominator.

To add fractions with unlike denominators, find the LCD and use the Fundamental Principle.

Example: Add: $\dfrac{3}{4} + \dfrac{1}{6}$

Solution:

Step (1): Find the LCD:

$4 = 2^2; \quad 6 = 2^1 \times 3^1$

$\text{LCD} = 2^2 \times 3^1 = 12$

Addition of Unlike Fractions (Continued)

Step (2): Expand each fraction so it has the LCD:

$$\frac{3}{4} = \frac{3 \times 3}{4 \times 3} = \frac{9}{12}$$

$$\frac{1}{6} = \frac{1 \times 2}{6 \times 2} = \frac{2}{12}$$

Step (3): Combine fractions:

$$\frac{3}{4} + \frac{1}{6} = \frac{9}{12} + \frac{2}{12}$$

$$= \frac{11}{12}$$

Geometric Meaning **(16)**

The sum, $\frac{3}{4} + \frac{1}{6}$, has the geometric meaning shown below.

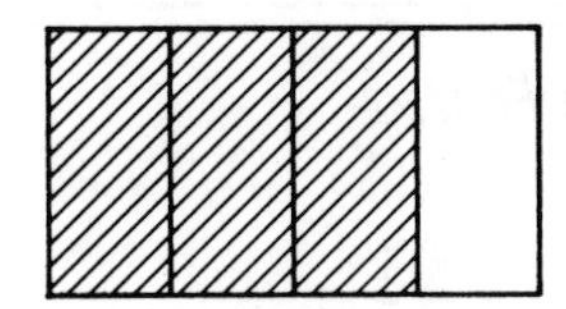 + 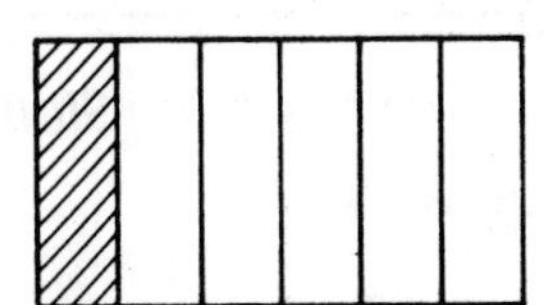 =

$\frac{3}{4}$ + 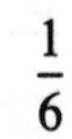

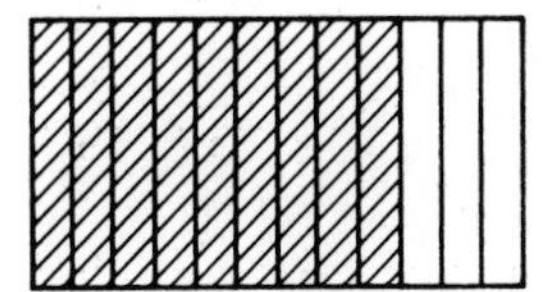 + 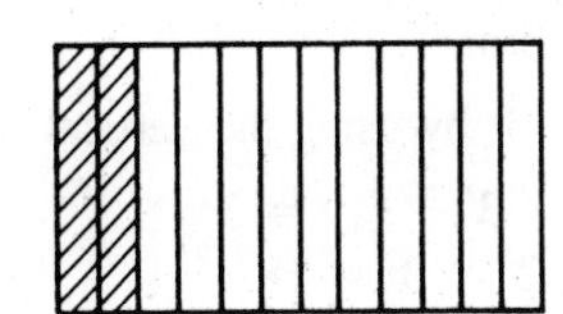 =

$\frac{9}{12}$ + $\frac{2}{12}$

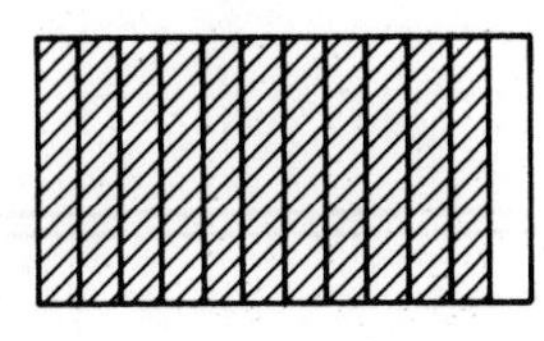

$\frac{11}{12}$

Another example: Add $\frac{1}{16} + \frac{5}{24}$: **(17)**

Solution:

Step (1): Find the LCD:

$16 = 2^4; \quad 24 = 2^3 \times 3^1$

$\text{LCD} = 2^4 \times 3^1 = 48$

Step (2): Expand each fraction so it has the LCD:

$$\frac{1}{16} = \frac{1 \times 3}{16 \times 3} = \frac{3}{48}$$

$$\frac{5}{24} = \frac{5 \times 2}{24 \times 2} = \frac{10}{48}$$

Step (3): Combine fractions:

$$\frac{1}{16} + \frac{5}{24} = \frac{3}{48} + \frac{10}{48}$$

$$= \frac{13}{48}$$

Study Exercise Four (18)

Add the following fractions:

1. $\frac{3}{5} + \frac{1}{4}$
2. $\frac{5}{8} + \frac{3}{4}$
3. $\frac{3}{8} + \frac{1}{6}$
4. $\frac{5}{14} + \frac{2}{7}$

Adding Mixed Numerals Containing Unlike Fractions (19)

When adding mixed numerals, add the whole numbers first and then add the fractions. You can use either the horizontal or vertical form.

Example: Add $1\frac{5}{6} + 3\frac{5}{9}$ by using the vertical form:

Solution: $6 = 2^1 \times 3^1, \quad 9 = 3^2$

$\text{LCD} = 2^1 \times 3^2 = 18$

$$1\tfrac{5}{6} = 1\frac{5 \times 3}{6 \times 3} = 1\tfrac{15}{18}$$

$$3\tfrac{5}{9} = 3\frac{5 \times 2}{9 \times 2} = 3\tfrac{10}{18}$$

$$1\tfrac{5}{6} + 3\tfrac{5}{9} = 4 + \tfrac{25}{18}$$

$$= 4 + 1\tfrac{7}{18}$$

$$= 5\tfrac{7}{18}$$

Horizontal Method (20)

The example in Frame 19 can be done as follows, using the horizontal method:

$$1\tfrac{5}{6} + 3\tfrac{5}{9} = 1\tfrac{15}{18} + 3\tfrac{10}{18}$$

$$= 4\tfrac{25}{18}$$

$$= 4 + \tfrac{25}{18}$$

$$= 4 + 1\tfrac{7}{18}$$

$$= 5\tfrac{7}{18}$$

Study Exercise Five (21)

Add as indicated, using either the horizontal or vertical format:

1. $5\frac{1}{5} + 2\frac{2}{5}$
2. $3\frac{2}{7} + \frac{4}{7}$
3. $1\frac{3}{4} + 3\frac{3}{8}$
4. $6\frac{7}{10} + 4\frac{1}{4}$
5. $4\frac{5}{6} + 2\frac{1}{8}$
6. $1\frac{1}{2} + 1\frac{1}{4} + 1\frac{1}{8}$

Simple Subtraction of Fractions (22)

We wish to subtract $\frac{5}{6} - \frac{3}{6}$

The diagram shows a rectangle divided in 6 parts. 5 parts are shaded to give $\frac{5}{6}$

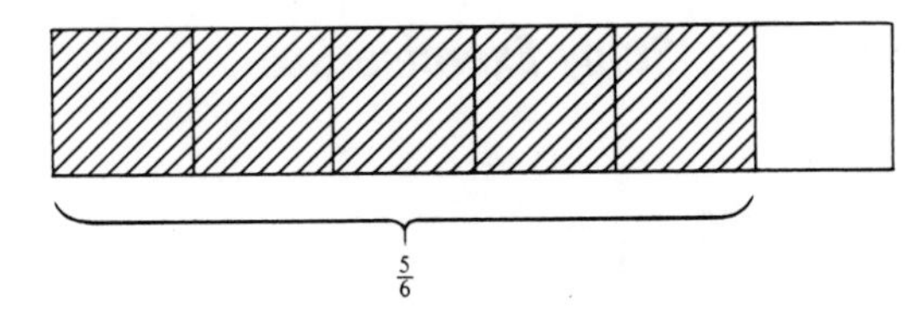

The next diagram shows the difference when $\frac{3}{6}$ are taken from $\frac{5}{6}$

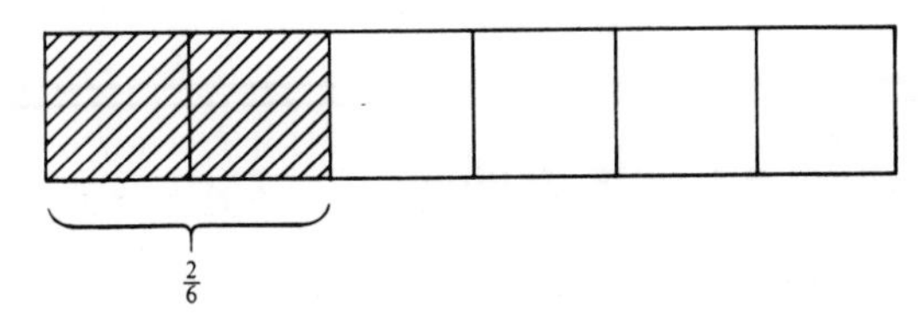

Observe that $\frac{2}{6}$ remain and $\frac{2}{6} = \frac{1}{3}$. Thus

$$\frac{5}{6} - \frac{3}{6} = \frac{2}{6} = \frac{1}{3}$$

Subtraction of Like Fractions (23)

To subtract like fractions, we subtract the numerators and keep the denominators the same.

Example 1: $\frac{5}{6} - \frac{4}{6} = \frac{5 - 4}{6} = \frac{1}{6}$

Example 2: $\frac{8}{11} - \frac{2}{11} = \frac{8 - 2}{11} = \frac{6}{11}$

Example 3: $\frac{5}{8} - \frac{3}{8} = \frac{5 - 3}{8} = \frac{2}{8} = \frac{1}{4}$

Subtraction of Unlike Fractions (24)

In subtracting fractions, if the fractions do not have the same denominator, we will use the LCD to write them with the same denominator.

Example: Subtract: $\frac{2}{3} - \frac{1}{12}$

Subtraction of Unlike Fractions (Continued)

Solution:

Step (1): Find the LCD:

$3 = 3^1, \qquad 12 = 2^2 \times 3^1$

$LCD = 2^2 \times 3^1 = 12$

Step (2): Expand each fraction so it has the LCD:

$$\frac{2}{3} = \frac{2 \times 4}{3 \times 4} = \frac{8}{12}$$

Step (3): Subtract:

$$\frac{8}{12} - \frac{1}{12} = \frac{7}{12}$$

Study Exercise Six (25)

Subtract as indicated:

1. $\frac{7}{10} - \frac{3}{10}$ **2.** $\frac{5}{8} - \frac{3}{8}$ **3.** $\frac{5}{6} - \frac{1}{4}$

4. $\frac{7}{8} - \frac{7}{10}$ **5.** $\frac{3}{4} - \frac{1}{2}$ **6.** $\frac{1}{4} - \frac{1}{5}$

7. $\frac{4}{5} - \frac{8}{25}$ **8.** $\frac{10}{9} - \frac{5}{6}$

Subtraction of a Fraction from a Whole Number (26)

Example: Subtract: $5 - \frac{3}{8}$

Solution: We will use the vertical method and borrowing.

Step (1): Write the problem vertically and draw a dotted vertical line to the right of the whole number

$$\begin{array}{r|l} 5 & \\ - & \frac{3}{8} \\ \hline \end{array}$$

Step (2): Borrow 1 from the 5 and write it on the other side of the dotted line

$$\begin{array}{r|l} {}^{4}\not{5} & 1 \\ - & \frac{3}{8} \\ \hline \end{array}$$

Step (3): Change the 1 to a fraction with denominator of 8

$$\begin{array}{r|l} {}^{4}\not{5} & \frac{8}{8} \\ - & \frac{3}{8} \\ \hline \end{array}$$

Step (4): Subtract

$$\begin{array}{r|l} {}^{4}\not{5} & \frac{8}{8} \\ - & \frac{3}{8} \\ \hline 4 & \frac{5}{8} \end{array}$$

Subtraction of Mixed Numerals With the Same Denominators (27)

Example: Subtract $3\frac{3}{7} - 2\frac{4}{7}$

Solution:

Step (1): Write the problem vertically and draw a dotted vertical line to the right of the whole number

$$\begin{array}{r|l} 3 & \frac{3}{7} \\ -\ 2 & \frac{4}{7} \\ \hline \end{array}$$

Step (2): Borrow 1 from the 3 and write it on the other side of the dotted line

$$\begin{array}{r|l} {}^{2}\cancel{3} & 1\frac{3}{7} \\ -\ 2 & \ \frac{4}{7} \\ \hline \end{array}$$

Step (3): Change the mixed numeral to an improper fraction

$$\begin{array}{r|l} {}^{2}\cancel{3} & \frac{10}{7} \\ -\ 2 & \frac{4}{7} \\ \hline \end{array}$$

Step (4): Subtract

$$\begin{array}{r|l} {}^{2}\cancel{3} & \frac{10}{7} \\ -\ 2 & \frac{4}{7} \\ \hline & \frac{6}{7} \end{array}$$

Study Exercise Seven (28)

Subtract as indicated:

1. $3 - \frac{1}{4}$
2. $7 - \frac{3}{5}$
3. $5 - 1\frac{2}{3}$
4. $4 - 3\frac{1}{8}$
5. $3\frac{1}{3} - 1\frac{2}{3}$
6. $6\frac{3}{10} - 3\frac{7}{10}$

Subtraction of Mixed Numerals with Different Denominators (29)

Example 1: Subtract: $5\frac{1}{4} - 3\frac{5}{8}$

Solution: The LCD is 8

$$\begin{array}{rcrcr} 5\frac{1}{4} & = & 5\frac{1\times2}{4\times2} & = & 5\frac{2}{8} \\ \underline{-3\frac{5}{8}} & = & \underline{-3\frac{5}{8}} & = & \underline{-3\frac{5}{8}} \end{array}$$

Next we use borrowing

$$\begin{array}{r|l} 5 & \frac{2}{8} \\ -3 & \frac{5}{8} \\ \hline \end{array} \longrightarrow \begin{array}{r|l} {}^{4}\cancel{5} & 1\frac{2}{8} \\ -\ 3 & \ \frac{5}{8} \\ \hline \end{array} \longrightarrow \begin{array}{r|l} {}^{4}\cancel{5} & \frac{10}{8} \\ -\ 3 & \frac{5}{8} \\ \hline \end{array} \longrightarrow \begin{array}{r|l} {}^{4}\cancel{5} & \frac{10}{8} \\ -\ 3 & \frac{5}{8} \\ \hline 1 & \frac{5}{8} \end{array}$$

Example 2: Subtract: $6\frac{1}{6} - 2\frac{3}{8}$

Solution:

Line (a): $6 = 2^1 \times 3^1, \quad 8 = 2^3$

Line (b): The LCD is $2^3 \times 3^1 = 24$

$$\begin{array}{rcrcr} 6\frac{1}{6} & = & 6\frac{1\times4}{6\times4} & = & 6\frac{4}{24} \\ \underline{-2\frac{3}{8}} & = & \underline{-2\frac{3\times3}{8\times3}} & = & \underline{-2\frac{9}{24}} \end{array}$$

Subtraction of Mixed Numerals with Different Denominators (Continued)

Line (c): Next we use borrowing

$$\begin{array}{r|l} 6 & \frac{4}{24} \\ -2 & \frac{9}{24} \\ \hline \end{array} \longrightarrow \begin{array}{r|l} {}^{5}\not{6} & 1\frac{4}{24} \\ -\ 2 & \ \ \frac{9}{24} \\ \hline \end{array} \longrightarrow \begin{array}{r|l} {}^{5}\not{6} & \frac{28}{24} \\ -\ 2 & \frac{9}{24} \\ \hline \end{array} \longrightarrow \begin{array}{r|l} {}^{5}\not{6} & \frac{28}{24} \\ -\ 2 & \frac{9}{24} \\ \hline 3 & \frac{19}{24} \end{array}$$

Study Exercise Eight (30)

Subtract as indicated. Reduce answers to lowest terms.

1. $4\frac{1}{6} - 2\frac{3}{4}$
2. $10\frac{1}{2} - 2\frac{5}{6}$
3. $11\frac{1}{10} - 3\frac{1}{12}$
4. $4\frac{1}{9} - 1\frac{5}{6}$

REVIEW EXERCISES (31)

Add or subtract as indicated. Reduce all answers to lowest terms:

1. $\frac{3}{8} + \frac{1}{8}$
2. $\frac{3}{4} + \frac{2}{4}$
3. $4\frac{4}{5} + \frac{2}{5}$
4. $1\frac{2}{7} + 3\frac{6}{7}$
5. $6\frac{3}{4} + 2$
6. $\frac{5}{9} + \frac{7}{12}$
7. $\frac{1}{8} + \frac{3}{16}$
8. $1\frac{5}{9} + 2\frac{1}{3}$
9. $\frac{5}{8} - \frac{3}{8}$
10. $\frac{5}{6} - \frac{7}{9}$
11. $4\frac{1}{4} - 3\frac{3}{8}$
12. $5 - 1\frac{2}{3}$
13. $\frac{1}{2} + 2\frac{2}{3} + \frac{5}{8}$
14. $1\frac{2}{3} - \frac{3}{4}$

Solutions to Review Exercises (32)

1. $\frac{3}{8} + \frac{1}{8} = \frac{3+1}{8} = \frac{4}{8} = \frac{1}{2}$

2. $\frac{3}{4} + \frac{2}{4} = \frac{3+2}{4} = \frac{5}{4} = 1\frac{1}{4}$

3. $4\frac{4}{5} + \frac{2}{5} = 4 + \frac{4+2}{5} = 4 + \frac{6}{5} = 4 + 1\frac{1}{5} = 5\frac{1}{5}$

4. $1\frac{2}{7} + 3\frac{6}{7} = 4 + \frac{2+6}{7} = 4 + \frac{8}{7} = 4 + 1\frac{1}{7} = 5\frac{1}{7}$

5. $6\frac{3}{4} + 2 = 8 + \frac{3}{4} = 8\frac{3}{4}$

6. $9 = 3^2, \quad 12 = 2^2 \times 3^1, \quad \text{LCD} = 2^2 \times 3^2 = 36$

   $\frac{5}{9} = \frac{5 \times 4}{9 \times 4} = \frac{20}{36}, \quad \frac{7}{12} = \frac{7 \times 3}{12 \times 3} = \frac{21}{36}$

   $\frac{5}{9} + \frac{7}{12} = \frac{20}{36} + \frac{21}{36} = \frac{41}{36} = 1\frac{5}{36}$

7. $8 = 2^3, \quad 16 = 2^4, \quad \text{LCD} = 2^4 = 16$

   $\frac{1}{8} + \frac{3}{16} = \frac{2}{16} + \frac{3}{16} = \frac{5}{16}$

8. $9 = 3^2, \quad 3 = 3^1, \quad \text{LCD} = 3^2 = 9$

   $1\frac{5}{9} + 2\frac{1}{3} = 1\frac{5}{9} + 2\frac{3}{9} = 3 + \frac{5+3}{9} = 3 + \frac{8}{9} = 3\frac{8}{9}$

9. $\frac{5}{8} - \frac{3}{8} = \frac{5-3}{8} = \frac{2}{8} = \frac{1}{4}$

10. $6 = 2^1 \times 3^1, \quad 9 = 3^2, \quad \text{LCD} = 2^1 \times 3^2 = 18$

    $\frac{5}{6} = \frac{5 \times 3}{6 \times 3} = \frac{15}{18}, \quad \frac{7}{9} = \frac{7 \times 2}{9 \times 2} = \frac{14}{18}$

    $\frac{5}{6} - \frac{7}{9} = \frac{15}{18} - \frac{14}{18} = \frac{1}{18}$

11. $4 = 2^2, \quad 8 = 2^3, \quad \text{LCD} = 2^3 = 8$

    $$\begin{array}{rcrcr} 4\frac{1}{4} & = & 4\frac{1\times 2}{4\times 2} & = & 4\frac{2}{8} \\ -3\frac{3}{8} & = & -3\frac{3}{8} & = & -3\frac{3}{8} \\ \hline \end{array}$$

    $$\begin{array}{r|l} 4 & \frac{2}{8} \\ -3 & \frac{3}{8} \\ \hline \end{array} \longrightarrow \begin{array}{r|l} {}^{3}\not{4} & 1\frac{2}{8} \\ -\ 3 & \ \ \frac{3}{8} \\ \hline \end{array} \longrightarrow \begin{array}{r|l} {}^{3}\not{4} & \frac{10}{8} \\ -\ 3 & \frac{3}{8} \\ \hline & \frac{7}{8} \end{array}$$

12. $$\begin{array}{r} 5 \\ -1\frac{2}{3} \\ \hline \end{array} \longrightarrow \begin{array}{r|l} {}^{4}\not{5} & 1 \\ -\ 1 & \frac{2}{3} \\ \hline \end{array} \longrightarrow \begin{array}{r|l} {}^{4}\not{5} & \frac{3}{3} \\ -\ 1 & \frac{2}{3} \\ \hline 3 & \frac{1}{3} \end{array}$$

Solutions to Review Exercises (Continued)

13. $2 = 2^1, \quad 3 = 3^1, \quad 8 = 2^3, \quad \text{LCD} = 2^3 \times 3^1 = 24$

$\frac{1}{2} = \frac{1\times12}{2\times12} = \frac{12}{24}, \quad 2\frac{2}{3} = 2\frac{2\times8}{3\times8} = 2\frac{16}{24}, \quad \frac{5}{8} = \frac{5\times3}{8\times3} = \frac{15}{24}$

$\frac{12}{24} + 2\frac{16}{24} + \frac{15}{24} = 2\frac{43}{24} = 3\frac{19}{24}$

14. LCD = 12

$$\begin{array}{rcrcr} 1\frac{2}{3} & = & 1\frac{2\times4}{3\times4} & = & 1\frac{8}{12} \\ -\ \frac{3}{4} & = & -\ \frac{3\times3}{4\times3} & = & -\ \frac{9}{12} \\ \hline \end{array} \longrightarrow \begin{array}{r|r} {}^{0}\not{1} & 1\frac{8}{12} \\ - & \frac{9}{12} \\ \hline \end{array} \longrightarrow \begin{array}{r|r} {}^{0}\not{1} & 1\frac{8}{12} \\ - & \frac{9}{12} \\ \hline \end{array} \longrightarrow \begin{array}{r|r} {}^{0}\not{1} & \frac{20}{12} \\ - & \frac{9}{12} \\ \hline & \frac{11}{12} \end{array}$$

SUPPLEMENTARY PROBLEMS

Add or subtract as indicated. Leave answers in reduced form:

1. $\frac{1}{8} + \frac{5}{8}$ **2.** $\frac{3}{12} + \frac{5}{12}$ **3.** $2\frac{1}{6} + \frac{5}{6}$ **4.** $3\frac{1}{12} + 1\frac{7}{12}$

5. $1\frac{7}{12} + \frac{11}{12}$ **6.** $3\frac{7}{9} + 1\frac{5}{9}$ **7.** $\frac{3}{10} + \frac{4}{5}$ **8.** $\frac{5}{16} + \frac{3}{4}$

9. $1\frac{1}{8} + \frac{3}{4}$ **10.** $\frac{3}{10} + 2\frac{1}{6}$ **11.** $1\frac{3}{10} + 3\frac{4}{5}$ **12.** $1\frac{5}{16} + \frac{3}{4}$

13. $3\frac{7}{10} + 2\frac{1}{2}$ **14.** $2\frac{7}{8} + 1\frac{3}{4}$ **15.** $\frac{7}{10} + 1\frac{3}{4} + \frac{1}{2}$ **16.** $\frac{2}{5} + \frac{1}{2} + \frac{9}{10}$

17. $3\frac{1}{3} + 1\frac{1}{4} + 4\frac{7}{8}$ **18.** $3\frac{2}{5} + \frac{3}{4} + 1\frac{7}{10}$ **19.** $\frac{5}{8} - \frac{3}{8}$ **20.** $4\frac{7}{12} - 1\frac{5}{12}$

21. $4\frac{1}{5} - 2\frac{3}{5}$ **22.** $5\frac{3}{8} - \frac{7}{8}$ **23.** $\frac{7}{12} - \frac{1}{3}$ **24.** $\frac{5}{6} - \frac{3}{4}$

25. $6\frac{7}{10} - \frac{1}{2}$ **26.** $2\frac{5}{6} - 1\frac{4}{9}$ **27.** $8\frac{1}{3} - 3\frac{5}{6}$ **28.** $2\frac{1}{4} - \frac{1}{2}$

29. $5\frac{1}{4} - 2\frac{5}{6}$ **30.** $1\frac{1}{2} - \frac{3}{4}$ **31.** $3\frac{1}{6} - 1\frac{5}{8}$ **32.** $2 - 1\frac{3}{8}$

33. Take $\frac{9}{10}$ from $1\frac{4}{5}$

34. Find the difference between $4\frac{5}{8}$ and $2\frac{9}{10}$

35. Find the difference between $9\frac{1}{3}$ and $2\frac{5}{6}$

36. Find the sum of $2 + \frac{1}{2} + \frac{1}{4} + \frac{7}{8}$

37. Find the sum of $\frac{7}{12} + \frac{3}{8} + \frac{5}{9} + 1\frac{1}{4}$

38. Take $\frac{5}{6}$ from $3\frac{5}{9}$

39. Take $2\frac{7}{10}$ from $4\frac{3}{5}$

40. Take $\frac{1}{5}$ from the sum of $\frac{1}{15}$ and $\frac{1}{3}$

Solutions to Study Exercises (7A)

Study Exercise One (Frame 7)

1. $\frac{3}{5} + \frac{1}{5} = \frac{3+1}{5} = \frac{4}{5}$ **2.** $\frac{1}{7} + \frac{3}{7} = \frac{1+3}{7} = \frac{4}{7}$ **3.** $\frac{5}{8} + \frac{3}{8} = \frac{5+3}{8} = \frac{8}{8} = 1$

Solutions to Study Exercises (Continued)

4. $\frac{6}{7} + \frac{5}{7} = \frac{6+5}{7} = \frac{11}{7} = 1\frac{4}{7}$

5. $\begin{array}{r} \frac{1}{3} \\ +\frac{1}{3} \\ \hline \frac{1+1}{3} = \frac{2}{3} \end{array}$

6. $\begin{array}{r} \frac{3}{10} \\ +\frac{9}{10} \\ \hline \frac{3+9}{10} = \frac{12}{10} = 1\frac{2}{10} = 1\frac{1}{5} \end{array}$

Study Exercise Two (Frame 10) **(10A)**

1. $2\frac{3}{11} + 5\frac{6}{11} = 7 + \frac{9}{11} = 7\frac{9}{11}$

2. $1\frac{15}{16} + 3\frac{13}{16} = 4 + \frac{28}{16} = 4 + 1\frac{12}{16} = 5\frac{12}{16} = 5\frac{3}{4}$

3. $4\frac{8}{12} + 2\frac{6}{12} = 6 + \frac{14}{12} = 6 + 1\frac{2}{12} = 7\frac{2}{12} = 7\frac{1}{6}$

4. $\begin{array}{l} 1\frac{13}{16} \\ 2\frac{4}{16} \\ \frac{3}{16} \\ \hline 3\frac{20}{16} = 3 + 1\frac{4}{16} = 4\frac{4}{16} = 4\frac{1}{4} \end{array}$

5. $\begin{array}{l} \quad 2\frac{3}{6} \\ +4\frac{5}{6} \\ \hline 6\frac{8}{6} = 6 + 1\frac{2}{6} \\ \quad = 7\frac{2}{6} \\ \quad = 7\frac{1}{3} \end{array}$

6. $\begin{array}{l} 1\frac{3}{5} \\ +\frac{4}{5} \\ \hline 1\frac{7}{5} = 1 + 1\frac{2}{5} \\ \quad = 2\frac{2}{5} \end{array}$

Study Exercise Three (Frame 14) **(14A)**

1. $8\frac{2}{3} = \frac{8 \times 3 + 2}{3} = \frac{24 + 2}{3} = \frac{26}{3}$

2. $6\frac{3}{5} = \frac{6 \times 5 + 3}{5} = \frac{30 + 3}{5} = \frac{33}{5}$

3. $2\frac{1}{2} = \frac{2 \times 2 + 1}{2} = \frac{4 + 1}{2} = \frac{5}{2}$

4. $8\frac{2}{3} + 6 = \frac{8 \times 3 + 2}{3} + \frac{6 \times 3}{3}$
$= \frac{26}{3} + \frac{18}{3}$
$= \frac{44}{3}$
$= 14\frac{2}{3}$

5. $5\frac{3}{5} + 6\frac{3}{5} = \frac{5 \times 5 + 3}{5} + \frac{6 \times 5 + 3}{5}$
$= \frac{28}{5} + \frac{33}{5}$
$= \frac{61}{5}$
$= 12\frac{1}{5}$

6. $1\frac{3}{4} + 2\frac{3}{4} = \frac{1 \times 4 + 3}{4} + \frac{2 \times 4 + 3}{4}$
$= \frac{7}{4} + \frac{11}{4}$
$= \frac{18}{4}$
$= 4\frac{2}{4}$
$= 4\frac{1}{2}$

Study Exercise Four (Frame 18) **(18A)**

1. $5 = 5^1, \quad 4 = 2^2, \quad \text{LCD} = 5^1 \times 2^2 = 20$
$\frac{3}{5} = \frac{3 \times 4}{5 \times 4} = \frac{12}{20}, \quad \frac{1}{4} = \frac{1 \times 5}{4 \times 5} = \frac{5}{20}$
$\frac{3}{5} + \frac{1}{4} = \frac{12}{20} + \frac{5}{20} = \frac{17}{20}$

2. $8 = 2^3, \quad 4 = 2^2, \quad \text{LCD} = 2^3 = 8$
$\frac{5}{8} + \frac{3}{4} = \frac{5}{8} + \frac{6}{8} = \frac{11}{8} = 1\frac{3}{8}$

3. $8 = 2^3, \quad 6 = 2^1 \times 3^1, \quad \text{LCD} = 2^3 \times 3^1 = 24$
$\frac{3}{8} = \frac{3 \times 3}{8 \times 3} = \frac{9}{24}, \quad \frac{1}{6} = \frac{1 \times 4}{6 \times 4} = \frac{4}{24}$
$\frac{3}{8} + \frac{1}{6} = \frac{9}{24} + \frac{4}{24} = \frac{13}{24}$

4. $14 = 2^1 \times 7^1, \quad 7 = 7^1, \quad \text{LCD} = 2^1 \times 7^1 = 14$
$\frac{5}{14} + \frac{2}{7} = \frac{5}{14} + \frac{4}{14} = \frac{9}{14}$

Study Exercise Five (Frame 21) (21A)

1. $5\frac{1}{5} + \frac{2}{5} = 7 + \frac{1+2}{5} = 7 + \frac{3}{5} = 7\frac{3}{5}$

2. $3\frac{2}{7} + \frac{4}{7} = 3 + \frac{2+4}{7} = 3 + \frac{6}{7} = 3\frac{6}{7}$

3. $4 = 2^2, \quad 8 = 2^3, \quad \text{LCD} = 8$

   $1\frac{3}{4} + 3\frac{3}{8} = 1\frac{6}{8} + 3\frac{3}{8} = 4 + \frac{9}{8} = 4 + 1\frac{1}{8} = 5\frac{1}{8}$

4. $10 = 2^1 \times 5^1, \quad 4 = 2^2, \quad \text{LCD} = 2^2 \times 5^1 = 20$

   $6\frac{7}{10} = 6\frac{7\times2}{10\times2} = 6\frac{14}{20}, \quad 4\frac{1}{4} = 4\frac{1\times5}{4\times5} = 4\frac{5}{20}$

   $6\frac{7}{10} + 4\frac{1}{4} = 6\frac{14}{20} + 4\frac{5}{20} = 10 + \frac{14+5}{20} = 10 + \frac{19}{20} = 10\frac{19}{20}$

5. $6 = 2^1 \times 3^1, \quad 8 = 2^3, \quad \text{LCD} = 2^3 \times 3^1 = 24$

   $4\frac{5}{6} = 4\frac{5\times4}{6\times4} = 4\frac{20}{24}, \quad 2\frac{1}{8} = 2\frac{1\times3}{8\times3} = 2\frac{3}{24}$

   $4\frac{5}{6} + 2\frac{1}{8} = 4\frac{20}{24} + 2\frac{3}{24} = 6 + \frac{20+3}{24} = 6 + \frac{23}{24} = 6\frac{23}{24}$

6. $2 = 2^1, \quad 4 = 2^2, \quad 8 = 2^3, \quad \text{LCD} = 2^3 = 8$

   $1\frac{1}{2} + 1\frac{1}{4} + 1\frac{1}{8} = 1\frac{4}{8} + 1\frac{2}{8} + 1\frac{1}{8} = 3 + \frac{4+2+1}{8} = 3 + \frac{7}{8} = 3\frac{7}{8}$

Study Exercise Six (Frame 25) (25A)

1. $\frac{7}{10} - \frac{3}{10} = \frac{7-3}{10} = \frac{4}{10} = \frac{2}{5}$

2. $\frac{5}{8} - \frac{3}{8} = \frac{5-3}{8} = \frac{2}{8} = \frac{1}{4}$

3. LCD is 12, $\quad \frac{5}{6} = \frac{5\times2}{6\times2} = \frac{10}{12}, \quad \frac{1}{4} = \frac{1\times3}{4\times3} = \frac{3}{12}$

   $\frac{5}{6} - \frac{1}{4} = \frac{10}{12} - \frac{3}{12} = \frac{7}{12}$

4. LCD = 40, $\quad \frac{7}{8} = \frac{7\times5}{8\times5} = \frac{35}{40}, \quad \frac{7}{10} = \frac{7\times4}{10\times4} = \frac{28}{40}$

   $\frac{7}{8} - \frac{7}{10} = \frac{35}{40} - \frac{28}{40} = \frac{7}{40}$

5. LCD = 4, $\quad \frac{1}{2} = \frac{2}{4}$

   $\frac{3}{4} - \frac{1}{2} = \frac{3}{4} - \frac{2}{4} = \frac{1}{4}$

6. LCD = 20, $\quad \frac{1}{4} = \frac{1\times5}{4\times5} = \frac{5}{20}, \quad \frac{1}{5} = \frac{1\times4}{5\times4} = \frac{4}{20}$

   $\frac{1}{4} - \frac{1}{5} = \frac{5}{20} - \frac{4}{20} = \frac{1}{20}$

7. LCD = 25, $\quad \frac{4}{5} = \frac{20}{25}$

   $\frac{4}{5} - \frac{8}{25} = \frac{20}{25} - \frac{8}{25} = \frac{12}{25}$

8. LCD = 18

   $\frac{10}{9} = \frac{10\times2}{9\times2} = \frac{20}{18}$

   $\frac{5}{6} = \frac{5\times3}{6\times3} = \frac{15}{18}$

   $\frac{10}{9} - \frac{5}{6} = \frac{20}{18} - \frac{15}{18} = \frac{5}{18}$

Study Exercise Seven (Frame 28) (28A)

1. $\begin{array}{r|l} 3 & \\ - & \frac{1}{4} \\ \hline \end{array} \longrightarrow \begin{array}{r|l} {}^{2}\cancel{3} & 1 \\ - & \frac{1}{4} \\ \hline \end{array} \longrightarrow \begin{array}{r|l} {}^{2}\cancel{3} & \frac{4}{4} \\ - & \frac{1}{4} \\ \hline 2 & \frac{3}{4} \end{array}$

2. $\begin{array}{r|l} 7 & \\ - & \frac{3}{5} \\ \hline \end{array} \longrightarrow \begin{array}{r|l} {}^{6}\cancel{7} & 1 \\ - & \frac{3}{5} \\ \hline \end{array} \longrightarrow \begin{array}{r|l} {}^{7}\cancel{7} & \frac{5}{5} \\ - & \frac{3}{5} \\ \hline 6 & \frac{2}{5} \end{array}$

3. $\begin{array}{r|l} 5 & \\ -1 & \frac{2}{3} \\ \hline \end{array} \longrightarrow \begin{array}{r|l} {}^{4}\cancel{5} & 1 \\ -1 & \frac{2}{3} \\ \hline \end{array} \longrightarrow \begin{array}{r|l} {}^{4}\cancel{5} & \frac{3}{3} \\ -1 & \frac{2}{3} \\ \hline 3 & \frac{1}{3} \end{array}$

4. $\begin{array}{r|l} 4 & \\ -3 & \frac{1}{8} \\ \hline \end{array} \longrightarrow \begin{array}{r|l} {}^{3}\cancel{4} & 1 \\ -3 & \frac{1}{8} \\ \hline \end{array} \longrightarrow \begin{array}{r|l} {}^{3}\cancel{4} & \frac{8}{8} \\ -3 & \frac{1}{8} \\ \hline & \frac{7}{8} \end{array}$

5. $\begin{array}{r|l} 3 & \frac{1}{3} \\ -1 & \frac{2}{3} \\ \hline \end{array} \longrightarrow \begin{array}{r|l} {}^{2}\cancel{3} & 1\frac{1}{3} \\ -1 & \frac{2}{3} \\ \hline \end{array} \longrightarrow \begin{array}{r|l} {}^{2}\cancel{3} & \frac{4}{3} \\ -1 & \frac{2}{3} \\ \hline 1 & \frac{2}{3} \end{array}$

6. $\begin{array}{r|l} 6 & \frac{3}{10} \\ -3 & \frac{7}{10} \\ \hline \end{array} \longrightarrow \begin{array}{r|l} {}^{5}\cancel{6} & 1\frac{3}{10} \\ -3 & \frac{7}{10} \\ \hline \end{array} \longrightarrow \begin{array}{r|l} {}^{5}\cancel{6} & \frac{13}{10} \\ -3 & \frac{7}{10} \\ \hline 2 & \frac{6}{10} \end{array} = 2\frac{3}{5}$

Study Exercise Eight (Frame 30) (30A)

1. $6 = 2^1 \times 3^1, \quad 4 = 2^2, \quad \text{LCD} = 2^2 \times 3^1 = 12$

   $4\frac{1}{6} = 4\frac{1\times2}{6\times2} = 4\frac{2}{12}$

   $-2\frac{3}{4} = -2\frac{3\times3}{4\times3} = -2\frac{9}{12}$

   $\begin{array}{r|l} {}^{3}\cancel{4} & 1\frac{2}{12} \\ -2 & \frac{9}{12} \\ \hline \end{array} \longrightarrow \begin{array}{r|l} {}^{3}\cancel{4} & \frac{14}{12} \\ -2 & \frac{9}{12} \\ \hline 1 & \frac{5}{12} \end{array}$

2. $2 = 2^1, \quad 6 = 2^1 \times 3^1, \quad \text{LCD} = 2^1 \times 3^1 = 6$

   $10\frac{1}{2} = 10\frac{1\times3}{2\times3} = 10\frac{3}{6}$

   $-2\frac{5}{6} = -2\frac{5}{6} = -2\frac{5}{6}$

   $\begin{array}{r|l} {}^{9}\cancel{10} & 1\frac{3}{6} \\ -2 & \frac{5}{6} \\ \hline \end{array} \longrightarrow \begin{array}{r|l} {}^{9}\cancel{10} & \frac{9}{6} \\ -2 & \frac{5}{6} \\ \hline 7 & \frac{4}{6} \end{array} = 7\frac{2}{3}$

Study Exercise Eight (Frame 30) (Continued)

3. $10=2^1\times5^1$, $12=2^2\times3^1$, $LCD=2^2\times3^1\times5^1=60$

$$\begin{array}{rcrcr} 11\frac{1}{10} & = & 11\frac{1\times6}{10\times6} & = & 11\frac{6}{60} \\ \underline{3\frac{1}{12}} & = & \underline{-\ 3\frac{1\times5}{12\times5}} & = & \underline{-\ 3\frac{5}{60}} \\ & & & & 8\frac{1}{60} \end{array}$$

4. $9 = 3^2$, $6 = 2^1 \times 3^1$, $LCD = 2^1 \times 3^2 = 18$

$$\begin{array}{rcrcr} 4\frac{1}{9} & = & 4\frac{1\times2}{9\times2} & = & 4\frac{2}{18} \\ \underline{-1\frac{5}{6}} & = & \underline{-1\frac{5\times3}{6\times3}} & = & \underline{-1\frac{15}{18}} \end{array}$$

$$\begin{array}{r|l} 4 & \frac{2}{18} \\ \underline{-1} & \underline{\frac{15}{18}} \end{array} \longrightarrow \begin{array}{r|l} {}^{3}\not{4} & 1\frac{2}{18} \\ \underline{-\ 1} & \underline{\ \ \frac{15}{18}} \end{array} \longrightarrow \begin{array}{r|l} {}^{3}\not{4} & \frac{20}{18} \\ \underline{-\ 1} & \underline{\frac{15}{18}} \\ 2 & \frac{5}{18} \end{array}$$

UNIT

13 Multiplication of Fractions

OBJECTIVES (1)

By the end of this unit you should be able to:

1. MULTIPLY FRACTIONS:
   (A) A WHOLE NUMBER BY A FRACTION.
   (B) TWO FRACTIONS.
   (C) A MIXED NUMERAL BY A WHOLE NUMBER, FRACTION, OR ANOTHER MIXED NUMERAL.
   (D) THREE FRACTIONS.
2. RAISE FRACTIONS TO POWERS.
3. TAKE SQUARE ROOTS OF CERTAIN FRACTIONS AND MIXED NUMERALS.
4. WORK VERBAL PROBLEMS WHERE WE CHANGE FROM A BASE OF ONE.

Review of Meaning of Multiplication (2)

In the unit on whole numbers we learned that multiplication is *repetitive addition*.

For example, 3×5 means 5 is added three times. That is, $5 + 5 + 5 = 15$. Also, 4×6 means $6 + 6 + 6 + 6$

(3)

Now let us multiply $4 \times \frac{1}{6}$ by the method of repetitive addition:

$$4 \times \frac{1}{6} \quad \text{means} \quad \frac{1}{6} + \frac{1}{6} + \frac{1}{6} + \frac{1}{6}$$

We know

$\frac{1}{6} + \frac{1}{6} + \frac{1}{6} + \frac{1}{6} = \frac{4}{6} = \frac{2}{3}$

Also, if we interchange the numbers, the product remains the same.

$$\frac{1}{6} \times 4 = \frac{1}{6} + \frac{1}{6} + \frac{1}{6} + \frac{1}{6} = \frac{4}{6} = \frac{2}{3}$$

Multiplication of a Whole Number by a Fraction

(4)

Multiply $3 \times \frac{5}{7}$:

$$3 \times \frac{5}{7} = \frac{5}{7} + \frac{5}{7} + \frac{5}{7}$$
$$= \frac{15}{7}$$
$$= 2\frac{1}{7}$$

Since we add 3 fives in the numerator, we can merely multiply 3×5

$$3 \times \frac{5}{7} = \frac{3 \times 5}{7} = \frac{15}{7} = 2\frac{1}{7}$$

Rule: To multiply a fraction by a whole number, multiply the numerator by the whole number and keep the denominator unchanged.

(5)

Example 1: $7 \times \frac{2}{3}$

Solution: $7 \times \frac{2}{3} = \frac{7 \times 2}{3} = \frac{14}{3} = 4\frac{2}{3}$

Example 2: $\frac{2}{5} \cdot 6$

Solution: $\frac{2}{5} \cdot 6 = \frac{2 \cdot 6}{5} = \frac{12}{5} = 2\frac{2}{5}$

Example 3: $\frac{2}{3}$ of 8

Solution: Remember that "of" means multiply.

$$\frac{2}{3} \text{ of } 8 = \frac{2}{3} \times 8 = \frac{2 \times 8}{3} = \frac{16}{3} = 5\frac{1}{3}$$

Study Exercise One

(6)

Find the products:

1. $\frac{2}{7} \times 3$

2. $6 \cdot \frac{3}{8}$

3. $2 \times \frac{1}{2}$

4. $\frac{2}{3}$ of 12

5. $\frac{3}{4}$ of 20

6. $\frac{2}{5}$ of 45

Multiplication Using Cancellation

(7)

By our rule, to multiply $8 \times \frac{5}{6}$, we write:

$$\frac{8 \times 5}{6} = \frac{40}{6} = \frac{20}{3} = 6\frac{2}{3}$$

Notice we must reduce our answer by dividing numerator and denominator by 2.

Multiplication Using Cancellation (Continued)

The work can be made easier by reducing before multiplying (that is, cancelling before multiplying).

$$\not{8}^4 \times \frac{5}{\not{6}_3} = \frac{20}{3}$$
$$= 6\frac{2}{3}$$

---

(8)

Example 1: Find $25 \times \frac{7}{15}$:

Solution: $\not{25}^5 \times \frac{7}{\not{15}_3} = \frac{35}{3}$

$= 11\frac{2}{3}$

Example 2: Find $\frac{5}{18} \cdot 20$:

Solution: $\frac{5}{\not{18}_9} \cdot \not{20}^{10} = \frac{50}{9}$

$= 5\frac{5}{9}$

Halves and Fourths of 100 (9)

Certain parts of 100 are used so often they should be memorized.

$\frac{1}{2}$ of 100 = 50 since $\frac{1}{\not{2}_1} \times \not{100}^{50} = 50$

$\frac{1}{4}$ of 100 = 25 since $\frac{1}{\not{4}_1} \times \not{100}^{25} = 25$

$\frac{3}{4}$ of 100 = 75 since $\frac{3}{\not{4}_1} \times \not{100}^{25} = 75$

The Thirds of 100 (10)

$\frac{1}{3}$ of 100 $= \frac{1}{3} \times 100 = \frac{100}{3} = 33\frac{1}{3}$

$\frac{2}{3}$ of 100 $= \frac{2}{3} \times 100 = \frac{200}{3} = 66\frac{2}{3}$

The thirds of 100 should also be memorized.

Study Exercise Two (11)

A. Find the products by first cancelling if possible and then multiplying:

1. $\frac{3}{4} \times 22$ **2.** $18 \times \frac{4}{9}$ **3.** $\frac{5}{8} \cdot 3$ **4.** $\frac{3}{4}$ of 6

B. Find the indicated parts of 100:

5. $\frac{1}{5}$ of 100 **6.** $\frac{3}{4}$ of 100 **7.** $\frac{1}{6}$ of 100 **8.** $\frac{1}{3}$ of 100

9. $\frac{3}{10}$ of 100 **10.** $\frac{2}{5}$ of 100 **11.** $\frac{7}{10}$ of 100 **12.** $\frac{4}{5}$ of 100

Multiplying Two Fractions (12)

Let us find $\frac{1}{2} \times \frac{3}{5}$

Line (a): We know $\frac{3}{5} = \frac{6}{10}$

Line (2): But $\frac{6}{10} = \frac{3}{10} + \frac{3}{10}$

Line (3): $\frac{1}{2}$ of $\frac{3}{5} = \frac{3}{10}$

Line (4): Since $\frac{1}{2}$ of $\frac{3}{5}$ means $\frac{1}{2} \times \frac{3}{5}$, $\frac{3}{10} = \frac{1}{2} \times \frac{3}{5}$

Line (5): Therefore, $\frac{1}{2} \times \frac{3}{5} = \frac{3}{10}$

Rule: To multiply fractions, multiply the numerators and then multiply the denominators.

(13)

Example 1: $\left(\frac{3}{4}\right)\left(\frac{6}{11}\right)$

Solution: $\left(\frac{3}{4}\right)\left(\frac{6}{11}\right) = \frac{(3)(6)}{(4)(11)} = \frac{18}{44} = \frac{9}{22}$

Example 2: $\frac{5}{8} \times \frac{4}{15}$

Solution: $\frac{5}{8} \times \frac{4}{15} = \frac{5 \times 4}{8 \times 15} = \frac{20}{120} = \frac{1}{6}$

To make the work easier, we cancel first, if we wish. We must cancel the same factor from both numerator and denominator.

$$\frac{\cancel{5}^{1}}{\cancel{8}_{2}} \times \frac{\cancel{4}^{1}}{\cancel{15}_{3}} = \frac{1}{6}$$

Multiplying Two Fractions (14)

Procedure:

Step (1): If possible, cancel the same factor from a numerator and a denominator.
Step (2): Multiply the remaining numerators and the remaining denominators.

Example: Multiply $\frac{15}{28} \times \frac{12}{25}$:

Solution:

Step (1): $\frac{\cancel{15}^{3}}{\cancel{28}_{7}} \times \frac{\cancel{12}^{3}}{\cancel{25}_{5}}$

Step (2): $\frac{\cancel{15}^{3}}{\cancel{28}_{7}} \times \frac{\cancel{12}^{3}}{\cancel{28}_{5}} = \frac{9}{35}$

Study Exercise Three (15)

Multiply as indicated and reduce all answers:

1. $\frac{5}{6} \cdot \frac{2}{3}$
2. $\frac{4}{15} \times \frac{3}{14}$
3. $\left(\frac{2}{3}\right)\left(\frac{5}{7}\right)$
4. $\frac{2}{3} \times \frac{5}{12}$
5. $\frac{3}{5} \cdot \frac{25}{30}$
6. $\frac{1}{4} \times \frac{12}{16}$

Multiplying Mixed Numerals (16)

Procedure:

Step (1): Change mixed numerals to improper fractions.
Step (2): Then proceed in the same manner as multiplication of two fractions.

Example: $3\frac{1}{2} \times 1\frac{4}{5}$

Solution:

Step (1): Change mixed numerals to improper fractions:

$$3\frac{1}{2} \times 1\frac{4}{5} = \frac{7}{2} \times \frac{9}{5}$$

Step (2): Multiply the two fractions:

$$\frac{7}{2} \times \frac{9}{5} = \frac{63}{10} = 6\frac{3}{10}$$

(17)

Example 1: $2\frac{1}{2} \times 5\frac{2}{3}$

Solution:

Step (1): $2\frac{1}{2} \times 5\frac{2}{3} = \frac{5}{2} \times \frac{17}{3}$

Step (2): $\frac{5}{2} \times \frac{17}{3} = \frac{85}{6} = 14\frac{1}{6}$

Example 2: $3\frac{1}{3} \times 6\frac{1}{2}$

Step (1): $3\frac{1}{3} \times 6\frac{1}{2} = \frac{10}{3} \times \frac{13}{2}$

Step (2): $\frac{\cancel{10}^{5}}{3} \times \frac{13}{\cancel{2}_{1}} = \frac{65}{3} = 21\frac{2}{3}$

Cancelling should be used when possible to make the work easier.

Study Exercise Four (18)

Multiply and reduce all answers:

1. $3\frac{1}{4} \times \frac{2}{5}$
2. $4\frac{1}{4} \cdot \frac{8}{9}$
3. $3\frac{2}{3} \times 2\frac{1}{4}$
4. $1\frac{4}{5} \times 3\frac{1}{3}$
5. $2\frac{6}{7} \times 1\frac{1}{4}$
6. $4\frac{3}{5} \times 8$
7. $\frac{5}{8} \cdot 2\frac{2}{5}$
8. $\frac{3}{4}$ of $2\frac{1}{3}$

Multiplication of Three Fractions (19)

Procedure:

Step (1): Cancel common factors which appear in numerators and denominators.
Step (2): Multiply remaining numerators and remaining denominators.

Multiplication of Three Fractions (Continued)

Example 1: Multiply $\frac{2}{3} \times \frac{4}{5} \times \frac{6}{7}$:

Solution:

Step (1): Cancel common factors which appear in numerators and denominators

$$\frac{2}{\cancel{3}_1} \times \frac{4}{5} \times \frac{\cancel{6}^2}{7}$$

Step (2): Multiply remaining numerators and remaining denominators

$$\frac{2}{\cancel{3}_1} \times \frac{4}{5} \times \frac{\cancel{6}^2}{7} = \frac{16}{35}$$

Example 2: Multiply $2\frac{1}{3} \times 1\frac{1}{5} \times 3$: **(20)**

Solution: First change to improper fractions, then cancel, and finally multiply.

Step (1): $2\frac{1}{3} \times 1\frac{1}{5} \times 3 = \frac{7}{\cancel{3}_1} \times \frac{6}{5} \times \frac{\cancel{3}^1}{1}$

Step (2): $\frac{7}{\cancel{3}_1} \times \frac{6}{5} \times \frac{\cancel{3}^1}{1} = \frac{7 \times 6 \times 1}{1 \times 5 \times 1} = \frac{42}{5} = 8\frac{2}{5}$

Example 3: Multiply $\frac{4}{9} \times \frac{5}{10} \times \frac{3}{14}$: **(21)**

Solution:

Step (1): *Line (a):* $\frac{4}{\cancel{9}_3} \times \frac{5}{10} \times \frac{\cancel{3}^1}{14}$

Line (b): $\frac{\cancel{4}^2}{\cancel{9}_3} \times \frac{5}{10} \times \frac{\cancel{3}^1}{\cancel{14}_7}$

Line (c): $\frac{\cancel{4}^2}{\cancel{9}_3} \times \frac{\cancel{5}^1}{\cancel{10}_2} \times \frac{\cancel{3}^1}{\cancel{14}_7}$

Line (d): $\frac{\cancel{4}^{\cancel{2}^1}}{\cancel{9}_3} \times \frac{\cancel{5}^1}{\cancel{10}_{\cancel{2}_1}} \times \frac{\cancel{3}^1}{\cancel{14}_7}$

Step (2): $\frac{\cancel{4}^{\cancel{2}^1}}{\cancel{9}_3} \times \frac{\cancel{5}^1}{\cancel{10}_{\cancel{2}_1}} \times \frac{\cancel{3}^1}{\cancel{14}_7} = \frac{1}{21}$

Example 4: Multiply $\frac{2}{7} \times \frac{5}{6} \times \frac{8}{25}$: **(22)**

Solution: $\frac{\cancel{2}^1}{7} \times \frac{\cancel{5}^1}{\cancel{6}_3} \times \frac{8}{\cancel{25}_5} = \frac{1 \times 1 \times 8}{7 \times 3 \times 5} = \frac{8}{105}$

Study Exercise Five (23)

Multiply as indicated and reduce all answers:

1. $\frac{1}{2} \times \frac{8}{15} \times \frac{5}{6}$
2. $\frac{2}{5} \times \frac{3}{4} \times \frac{15}{16}$
3. $1\frac{1}{2} \times \frac{4}{5} \times 3\frac{1}{6}$
4. $2\frac{3}{4} \times 2\frac{2}{3} \times 3\frac{1}{7}$
5. $2\frac{1}{2} \times \frac{4}{5} \times 2\frac{1}{6}$
6. $4\frac{1}{6} \times 1\frac{1}{4} \times \frac{3}{20}$

Raising Fractions to Powers (24)

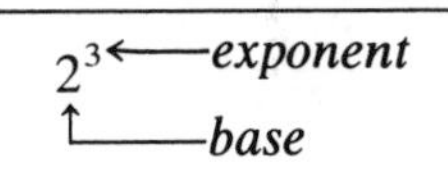

$$2^3 = 2 \cdot 2 \cdot 2 = 8$$

$$\left(\frac{3}{4}\right)^2 \leftarrow \textit{exponent}, \quad \frac{3}{4} \leftarrow \textit{base}$$

$$\left(\frac{3}{4}\right)^2 = \frac{3}{4} \cdot \frac{3}{4} = \frac{9}{16}$$

(25)

Example 1: Find $\left(\frac{2}{5}\right)^2$:

Solution: $\left(\frac{2}{5}\right)^2 = \frac{2}{5} \times \frac{2}{5} = \frac{4}{25}$

Example 2: Find $\left(1\frac{1}{3}\right)^2$:

Solution: Change the mixed numeral to an improper fraction.

$$\left(1\frac{1}{3}\right)^2 = \left(\frac{4}{3}\right)^2 = \frac{4}{3} \times \frac{4}{3} = \frac{16}{9} = 1\frac{7}{9}$$

Study Exercise Six (26)

Find:

1. $\left(\frac{1}{2}\right)^2$
2. $\left(\frac{7}{8}\right)^2$
3. $(2\frac{2}{3})^2$
4. $(1\frac{1}{5})^2$

Square Roots (27)

A *square root* of a number is one of its two equal factors.

$$3 \times 3 = 9$$

the two equal factors of 9

Thus, 3 is a square root of 9

Recall that the square root of 9 is written $\sqrt{9}$

$$\sqrt{9} = 3 \text{ since } 3 \times 3 = 9$$

Square Roots of Perfect Squares (28)

Study the following square roots until you know them:

$\sqrt{1} = 1$	$\sqrt{4} = 2$	$\sqrt{9} = 3$	$\sqrt{16} = 4$
$\sqrt{25} = 5$	$\sqrt{36} = 6$	$\sqrt{49} = 7$	$\sqrt{64} = 8$
$\sqrt{81} = 9$	$\sqrt{100} = 10$	$\sqrt{121} = 11$	$\sqrt{144} = 12$
$\sqrt{169} = 13$			

Square Roots of Fractions (29)

$$\sqrt{\frac{4}{9}} = \frac{2}{3} \quad \left(\textbf{Check: } \frac{2}{3} \times \frac{2}{3} = \frac{4}{9}\right)$$

$$\sqrt{\frac{25}{36}} = \frac{5}{6} \quad \left(\textbf{Check: } \frac{5}{6} \times \frac{5}{6} = \frac{25}{36}\right)$$

To find the square root of a fraction, find the square root of the numerator and denominator separately.

Square Roots of Mixed Numerals (30)

To find the square root of a mixed numeral, first change to an improper fraction. Then find the square root of the numerator and the square root of the denominator.

Example: $\sqrt{7\frac{1}{9}}$

Solution: $\sqrt{7\frac{1}{9}} = \sqrt{\frac{64}{9}}$

$= \frac{8}{3}$

$= 2\frac{2}{3}$

Check: $\frac{8}{3} \times \frac{8}{3} = \frac{64}{9} = 7\frac{1}{9}$

Study Exercise Seven (31)

Find the indicated square roots:

1. $\sqrt{\frac{4}{25}}$ **2.** $\sqrt{\frac{49}{100}}$ **3.** $\sqrt{\frac{1}{9}}$ **4.** $\sqrt{1\frac{7}{9}}$ **5.** $\sqrt{3\frac{1}{16}}$ **6.** $\sqrt{2\frac{7}{9}}$

Changing from a Base of One (32)

To change from a base of one, use the operation of multiplication.

Example 1: If one apple costs 25¢ find the cost of 3 apples:

Solution: Write down the known fact:
1 apple costs 25¢
Change from 1 to 3 by multiplying:
3 apples cost (3) (25¢) = 75¢

Example 2: If one pound of yellow onions costs 38¢, find the cost of $2\frac{1}{4}$ pounds of yellow onions.

Solution: 1 pound costs 38¢
Change from 1 to $2\frac{1}{4}$ by multiplying:

$$2\frac{1}{4} \text{ pound costs } 38¢ \times 2\frac{1}{4} = \frac{38}{1} \times \frac{9}{4}$$

$$= \frac{\cancel{38}^{19}}{1} \times \frac{9}{\cancel{4}_2}$$

$$= \frac{171}{2}$$

$$= 85\frac{1}{2}¢$$

The cost of $2\frac{1}{4}$ pounds of yellow onions is $85\frac{1}{2}$¢

Example 3: If 1 pound of candy costs 87¢, find the cost of $\frac{2}{3}$ pound of candy.

Solution: Change from 1 by multiplying by $\frac{2}{3}$:

$$\frac{2}{\cancel{3}_1} \times \frac{\cancel{87}^{29}}{1} = \frac{58}{1} = 58¢$$

The cost of $\frac{2}{3}$ pound of candy is 58¢

(Notice the similarity to the expansion problems found in Unit 5)

Study Exercise Eight (33)

1. If 1 pound of candy costs 49¢, find the cost of 2 pounds of candy.
2. If 1 pound of candy costs 99¢, find the cost of $\frac{2}{3}$ pound of candy.
3. If eggs cost 88¢ per dozen, what is the cost of $1\frac{1}{2}$ dozen?
4. A cookie recipe calls for $1\frac{1}{3}$ cups sugar. How much sugar is needed to make $2\frac{1}{2}$ times as many cookies?

REVIEW EXERCISES

(34)

Perform the indicated operations and reduce all answers to lowest terms:

1. $2 \times \frac{3}{5}$ **2.** $\frac{4}{5}$ of 20 **3.** $\frac{4}{21} \cdot \frac{7}{8}$ **4.** $\frac{10}{9} \cdot \frac{15}{8}$

5. $2\frac{1}{3} \times \frac{3}{16}$ **6.** $1\frac{7}{16} \times 1\frac{5}{9}$ **7.** $\left(\frac{3}{5}\right)^2$ **8.** $(3\frac{2}{3})^2$

9. $\frac{3}{4} \times \frac{4}{7} \times \frac{1}{2}$ **10.** $4\frac{1}{6} \times 3\frac{1}{5} \times 1\frac{1}{10}$ **11.** $\sqrt{\frac{81}{121}}$ **12.** $\sqrt{12\frac{1}{4}}$

13. $\frac{1}{4}$ of 100 **14.** $\frac{7}{10}$ of 100 **15.** $\frac{2}{3}$ of 100

16. If apples sell for 60¢ per pound, find the cost of $1\frac{1}{4}$ pounds.

Solutions to Review Exercises

(35)

1. $2 \times \frac{3}{5} = \frac{2 \times 3}{5} = \frac{6}{5} = 1\frac{1}{5}$

2. $\frac{4}{5}$ of $20 = \frac{4}{\cancel{5}_1} \times \frac{\cancel{20}^4}{1} = 16$

3. $\frac{\cancel{4}^1}{\cancel{21}_3} \cdot \frac{\cancel{7}^1}{\cancel{8}_2} = \frac{1 \cdot 1}{3 \cdot 2} = \frac{1}{6}$

4. $\frac{\cancel{10}^5}{\cancel{9}_3} \cdot \frac{\cancel{15}^5}{\cancel{8}_4} = \frac{5 \cdot 5}{3 \cdot 4} = \frac{25}{12} = 2\frac{1}{12}$

5. $2\frac{1}{3} \times \frac{3}{16} = \frac{7}{\cancel{3}_1} \times \frac{\cancel{3}^1}{16} = \frac{7 \times 1}{1 \times 16} = \frac{7}{16}$

6. $1\frac{7}{16} \times 1\frac{5}{9} = \frac{23}{\cancel{16}_8} \times \frac{\cancel{14}^7}{9} = \frac{23 \times 7}{8 \times 9} = \frac{161}{72} = 2\frac{17}{72}$

7. $\left(\frac{3}{5}\right)^2 = \frac{3}{5} \times \frac{3}{5} = \frac{9}{25}$

8. $\left(3\frac{2}{3}\right)^2 = \left(\frac{11}{3}\right)^2 = \frac{11}{3} \times \frac{11}{3} = \frac{11 \times 11}{3 \times 3} = \frac{121}{9} = 13\frac{4}{9}$

9. $\frac{3}{\cancel{4}_1} \times \frac{\cancel{4}^1}{7} \times \frac{1}{2} = \frac{3 \times 1 \times 1}{1 \times 7 \times 2} = \frac{3}{14}$

10. $4\frac{1}{6} \times 3\frac{1}{5} \times 1\frac{1}{10} = \frac{25}{6} \times \frac{\cancel{16}^8}{5} \times \frac{11}{\cancel{10}_5}$

$= \frac{\cancel{25}^5}{\cancel{6}_3} \times \frac{\cancel{16}^{\cancel{8}^4}}{5} \times \frac{11}{\cancel{10}_{\cancel{5}_1}}$

$= \frac{\cancel{25}^{\cancel{5}^1}}{\cancel{6}_3} \times \frac{\cancel{16}^{\cancel{8}^4}}{\cancel{5}_1} \times \frac{11}{\cancel{10}_{\cancel{5}_1}}$

$= \frac{1 \times 4 \times 11}{3 \times 1 \times 1}$

$= \frac{44}{3}$

$= 14\frac{2}{3}$

11. $\sqrt{\frac{81}{121}} = \frac{9}{11}$

12. $\sqrt{12\frac{1}{4}} = \sqrt{\frac{49}{4}} = \frac{7}{2} = 3\frac{1}{2}$

13. $\frac{1}{\cancel{4}_1} \times \frac{\cancel{100}^{25}}{1} = \frac{25}{1} = 25$

14. $\frac{7}{\cancel{10}_1} \times \frac{\cancel{100}^{10}}{1} = \frac{70}{1} = 70$

15. $\frac{2}{3} \times \frac{100}{1} = \frac{200}{3} = 66\frac{2}{3}$

16. 1 pound sells for 60¢

Change from 1 to $1\frac{1}{4}$ by multiplying

$60¢ \times 1\frac{1}{4} = \frac{\cancel{60}^{15}}{1} \times \frac{5}{\cancel{4}_1} = 75¢$

SUPPLEMENTARY PROBLEMS

Perform the indicated operations and reduce all fractions to lowest terms:

1. $4 \times \frac{5}{6}$

2. $12 \times \frac{3}{8}$

3. $6 \times 2\frac{1}{2}$

4. $\frac{5}{6}$ of 10

5. $\frac{3}{5} \cdot 5$

6. $\left(\frac{1}{4}\right)\left(\frac{2}{5}\right)$

7. $\frac{3}{8} \cdot \frac{4}{15}$

8. $\frac{1}{3} \times 6\frac{1}{5}$

9. $12 \times 2\frac{2}{3}$

10. $\frac{1}{3}$ of 100

11. $\frac{3}{5}$ of 100

12. $\frac{3}{8}$ of 100

13. $\frac{1}{8}$ of 100

14. $\frac{1}{4}$ of 100

15. $\frac{4}{5}$ of 100

16. $\frac{5}{8} \times \frac{5}{11}$

17. $\frac{10}{9} \cdot \frac{15}{8}$

18. $2\frac{1}{2} \times 1\frac{1}{5}$

19. $\frac{9}{16} \times 1\frac{1}{3}$

20. $2\frac{9}{16} \times 1\frac{1}{4}$

21. $\frac{4}{21} \cdot \frac{7}{8}$

22. $3\frac{3}{4} \cdot 3\frac{1}{5}$

23. $\frac{1}{2} \times \frac{2}{3} \times 8$

24. $\frac{2}{3} \times \frac{7}{8} \times \frac{3}{10}$

25. $\left(\frac{3}{4}\right)\left(\frac{3}{5}\right)\left(\frac{3}{6}\right)$

26. $1\frac{1}{8} \cdot 1\frac{1}{4} \cdot 1\frac{1}{3}$

27. $3\frac{3}{5} \times 1\frac{3}{8} \times \frac{10}{11}$

28. $\left(\frac{2}{5}\right)^2$

29. $\left(\frac{7}{8}\right)^2$

30. $\left(\frac{1}{6}\right)^2$

31. $(3\frac{1}{2})^2$

32. $(2\frac{3}{4})^2$

33. $(5\frac{1}{5})^2$

34. $(1\frac{1}{2})^2 \times \frac{6}{7}$

35. $\left(\frac{1}{3}\right)^2 \times 9$

36. $\left(\frac{2}{5}\right)^2 \times \frac{1}{2}$

37. $\frac{3}{4^2} \times \frac{2^2}{6}$

38. $\left(\frac{2}{3}\right)^3$

39. $(1\frac{1}{4})^3$

40. $\sqrt{\frac{64}{121}}$

41. $\sqrt{\frac{25}{36}}$

42. $\sqrt{\frac{9}{144}}$

43. $\sqrt{1\frac{15}{49}}$

44. $\sqrt{7\frac{1}{9}}$

45. $\sqrt{1\frac{11}{25}}$

46. $\sqrt{\frac{1}{100}}$

47. $2\frac{1}{3} \times \sqrt{\frac{36}{49}}$

48. $\frac{5}{6} \times \sqrt{1\frac{11}{25}}$

49. If 1 article costs $12\frac{2}{3}$¢, find the cost of 6 articles.

50. If a car travels $\frac{3}{4}$ mile in 1 minute, how far will it travel in $3\frac{1}{2}$ minutes?

51. If one certain item weighs $12\frac{1}{3}$ pounds, find the weight of 4 of these items.

52. If 1 cubic foot of water weighs $62\frac{1}{2}$ pounds, find the weight of $2\frac{1}{3}$ cubic feet of water.

53. Tom spends $\frac{3}{8}$ of his monthly income for food. If his take-home pay is $960 per month, how much does he spend for food?

54. A cookie recipe calls for $\frac{2}{3}$ cup butter. How much butter is needed to make $2\frac{1}{2}$ times as many cookies?

55. An automobile depreciates $\frac{1}{4}$ of its original value at the end of the first year. If the original value was $8000, find its value at the end of the first year.

Solutions to Study Exercises **(6A)**

Study Exercise One (Frame 6)

1. $\frac{2}{7} \times 3 = \frac{2 \times 3}{7} = \frac{6}{7}$

2. $6 \cdot \frac{3}{8} = \frac{6 \cdot 3}{8} = \frac{18}{8} = 2\frac{2}{8} = 2\frac{1}{4}$

3. $2 \times \frac{1}{2} = \frac{2 \times 1}{2} = \frac{2}{2} = 1$

Solutions to Study Exercises (Continued)

4. $\frac{2}{3}$ of $12 = \frac{2 \times 12}{3} = \frac{24}{3} = 8$

5. $\frac{3}{4}$ of $20 = \frac{3}{4} \times 20 = \frac{3 \times 20}{4} = \frac{60}{4} = 15$

6. $\frac{2}{5}$ of $45 = \frac{2}{5} \times 45 = \frac{2 \times 45}{5} = \frac{90}{5} = 18$

Study Exercise Two (Frame 11) **(11A)**

1. $\frac{3}{4} \times 22 = \frac{3}{\cancel{4}_2} \times \cancel{22}^{11} = \frac{33}{2} = 16\frac{1}{2}$
2. $18 \times \frac{4}{9} = \cancel{18}^2 \times \frac{4}{\cancel{9}_1} = \frac{8}{1} = 8$
3. $\frac{5}{8} \cdot 3 = \frac{5 \cdot 3}{8} = \frac{15}{8} = 1\frac{7}{8}$
4. $\frac{3}{4} \times 6 = \frac{3}{\cancel{4}_2} \times \cancel{6}^3 = \frac{9}{2} = 4\frac{1}{2}$
5. $\frac{1}{\cancel{5}_1} \times \cancel{100}^{20} = 20$
6. $\frac{3}{\cancel{4}_1} \times \cancel{100}^{25} = 75$
7. $\frac{1}{\cancel{6}_3} \times \cancel{100}^{50} = \frac{50}{3} = 16\frac{2}{3}$
8. $\frac{1}{3} \times 100 = \frac{100}{3} = 33\frac{1}{3}$
9. $\frac{3}{\cancel{10}_1} \times \cancel{100}^{10} = 30$
10. $\frac{2}{\cancel{5}_1} \times \cancel{100}^{20} = 40$
11. $\frac{7}{\cancel{10}_1} \times \cancel{100}^{10} = 70$
12. $\frac{4}{\cancel{5}_1} \times \cancel{100}^{20} = 80$

Study Exercise Three (Frame 15) **(15A)**

1. $\frac{5}{\cancel{6}_3} \cdot \frac{\cancel{2}^1}{3} = \frac{5}{9}$
2. $\frac{\cancel{4}^2}{\cancel{15}_5} \times \frac{\cancel{3}^1}{\cancel{14}_7} = \frac{2}{35}$
3. $\left(\frac{2}{3}\right)\left(\frac{5}{7}\right) = \frac{10}{21}$
4. $\frac{\cancel{2}^1}{3} \times \frac{5}{\cancel{12}_6} = \frac{5}{18}$
5. $\frac{\cancel{3}^1}{\cancel{5}_1} \cdot \frac{\cancel{25}^{\cancel{5}^1}}{\cancel{30}_{\cancel{10}_2}} = \frac{1}{2}$
6. $\frac{1}{\cancel{4}_1} \times \frac{\cancel{12}^3}{16} = \frac{3}{16}$

Study Exercise Four (Frame 18) **(18A)**

1. $3\frac{1}{4} \times \frac{2}{5} = \frac{13}{\cancel{4}_2} \times \frac{\cancel{2}^1}{5} = \frac{13}{10} = 1\frac{3}{10}$
2. $4\frac{1}{4} \cdot \frac{8}{9} = \frac{17}{\cancel{4}_1} \cdot \frac{\cancel{8}^2}{9} = \frac{34}{9} = 3\frac{7}{9}$
3. $3\frac{2}{3} \times 2\frac{1}{4} = \frac{11}{\cancel{3}_1} \times \frac{\cancel{9}^3}{4} = \frac{33}{4} = 8\frac{1}{4}$
4. $1\frac{4}{5} \times 3\frac{1}{3} = \frac{\cancel{9}^3}{\cancel{5}_1} \times \frac{\cancel{10}^2}{\cancel{3}_1} = \frac{6}{1} = 6$
5. $2\frac{6}{7} \times 1\frac{1}{4} = \frac{\cancel{20}^5}{7} \times \frac{5}{\cancel{4}_1} = \frac{25}{7} = 3\frac{4}{7}$
6. $4\frac{3}{5} \times 8 = \frac{23}{5} \times \frac{8}{1} = \frac{184}{5} = 36\frac{4}{5}$
7. $\frac{5}{8} \cdot 2\frac{2}{5} = \frac{\cancel{5}^1}{\cancel{8}_2} \cdot \frac{\cancel{12}^3}{\cancel{5}_1} = \frac{3}{2} = 1\frac{1}{2}$
8. $\frac{3}{4}$ of $2\frac{1}{3} = \frac{\cancel{3}^1}{4} \times \frac{7}{\cancel{3}_1} = \frac{7}{4} = 1\frac{3}{4}$

Study Exercise Five (Frame 23) **(23A)**

1. $\frac{1}{\cancel{2}_1} \times \frac{\cancel{8}^4}{\cancel{15}_3} \times \frac{\cancel{5}^1}{6} = \frac{1}{\cancel{2}_1} \times \frac{\cancel{8}^{\cancel{4}^2}}{\cancel{15}_3} \times \frac{\cancel{5}^1}{\cancel{6}_3} = \frac{1 \times 2 \times 1}{1 \times 3 \times 3} = \frac{2}{9}$
2. $\frac{\cancel{2}^1}{\cancel{5}_1} \times \frac{3}{\cancel{4}_2} \times \frac{\cancel{15}^3}{16} = \frac{1 \times 3 \times 3}{1 \times 2 \times 16} = \frac{9}{32}$
3. $1\frac{1}{2} \times \frac{4}{5} \times 3\frac{1}{6} = \frac{\cancel{3}^1}{\cancel{2}_1} \times \frac{\cancel{4}^{\cancel{2}^1}}{5} \times \frac{19}{\cancel{6}_{\cancel{2}_1}} = \frac{1 \times 1 \times 19}{1 \times 5 \times 1} = \frac{19}{5} = 3\frac{4}{5}$
4. $2\frac{3}{4} \times 2\frac{2}{3} \times 3\frac{1}{7} = \frac{11}{\cancel{4}_1} \times \frac{\cancel{8}^2}{3} \times \frac{22}{7} = \frac{11 \times 2 \times 22}{1 \times 3 \times 7} = \frac{484}{21} = 23\frac{1}{21}$
5. $2\frac{1}{2} \times \frac{4}{5} \times 2\frac{1}{6} = \frac{\cancel{5}^1}{\cancel{2}_1} \times \frac{\cancel{4}^{\cancel{2}^1}}{\cancel{5}_1} \times \frac{13}{\cancel{6}_3} = \frac{1 \times 1 \times 13}{1 \times 1 \times 3} = \frac{13}{3} = 4\frac{1}{3}$
6. $4\frac{1}{6} \times 1\frac{1}{4} \times \frac{3}{20} = \frac{25}{\cancel{6}_2} \times \frac{\cancel{5}^1}{4} \times \frac{\cancel{3}^1}{\cancel{20}_4} = \frac{25 \times 1 \times 1}{2 \times 4 \times 4} = \frac{25}{32}$

Study Exercise Six (Frame 26) **(26A)**

1. $\left(\frac{1}{2}\right)^2 = \frac{1}{2} \times \frac{1}{2} = \frac{1 \times 1}{2 \times 2} = \frac{1}{4}$

2. $\left(\frac{7}{8}\right)^2 = \frac{7}{8} \times \frac{7}{8} = \frac{7 \times 7}{8 \times 8} = \frac{49}{64}$

3. $\left(2\frac{2}{3}\right)^2 = \left(\frac{8}{3}\right)^2 = \frac{8}{3} \times \frac{8}{3} = \frac{8 \times 8}{3 \times 3} = \frac{64}{9} = 7\frac{1}{9}$

4. $\left(1\frac{1}{5}\right)^2 = \left(\frac{6}{5}\right)^2 = \frac{6}{5} \times \frac{6}{5} = \frac{6 \times 6}{5 \times 5} = \frac{36}{25} = 1\frac{11}{25}$

Study Exercise Seven (Frame 31) **(31A)**

1. $\sqrt{\frac{4}{25}} = \frac{2}{5}$

2. $\sqrt{\frac{49}{100}} = \frac{7}{10}$

3. $\sqrt{\frac{1}{9}} = \frac{1}{3}$

4. $\sqrt{1\frac{7}{9}} = \sqrt{\frac{16}{9}} = \frac{4}{3} = 1\frac{1}{3}$

5. $\sqrt{3\frac{1}{16}} = \sqrt{\frac{49}{16}} = \frac{7}{4} = 1\frac{3}{4}$

6. $\sqrt{2\frac{7}{9}} = \sqrt{\frac{25}{9}} = \frac{5}{3} = 1\frac{2}{3}$

Study Exercise Eight (Frame 33) **(33A)**

1. Change from 1 by multiplying by 2:
   $(2)(49¢) = 98¢$

2. Change from 1 by multiplying by $\frac{2}{3}$:

   $99¢ \times \frac{2}{3} = \frac{\cancel{99}^{33}}{1} \times \frac{2}{\cancel{3}_1} = \frac{66}{1} = 66¢$

3. Change from 1 by multiplying by $1\frac{1}{2}$:

   $88¢ \times 1\frac{1}{2} = \frac{\cancel{88}^{44}}{1} = \frac{3}{\cancel{2}_1} = \frac{132}{1} = 132¢$

4. Change from 1 by multiplying by $2\frac{1}{2}$:

   $1\frac{1}{3} \times 2\frac{1}{2} = \frac{\cancel{4}^2}{3} \times \frac{5}{\cancel{2}_1} = \frac{10}{3} = 3\frac{1}{3}$ cups sugar

UNIT

14 Division of Fractions

OBJECTIVES (1)

By the end of this unit you should be able to:

1. FIND THE RECIPROCAL OF A NUMBER.
2. SIMPLIFY COMPLEX FRACTIONS.
3. DIVIDE FRACTIONS.
4. WORK VERBAL PROBLEMS CHANGING BACK TO A BASE OF ONE.

Reciprocal (2)

The *reciprocal* of a fraction is found by inverting the fraction (that is, by switching the numerator and denominator).

	Fraction ↓	Reciprocal ↓
Example 1:	$\frac{2}{5}$	$\frac{5}{2}$ or $2\frac{1}{2}$
Example 2:	$\frac{1}{3}$	$\frac{3}{1}$ or 3

Reciprocals of Whole Numbers and Mixed Numerals (3)

Reciprocals of whole numbers and of mixed numerals are found by first changing to improper fractions and then inverting.

	Number ↓	Improper Fraction ↓	Reciprocal ↓
Example 1:	3	$\frac{3}{1}$	$\frac{1}{3}$
Example 2:	$2\frac{2}{5}$	$\frac{12}{5}$	$\frac{5}{12}$

Product of a Number by its Reciprocal (4)

	Number	Reciprocal	Number × Reciprocal
Example 1:	3	$\frac{1}{3}$	$3 \times \frac{1}{3} = 1$
Example 2:	$\frac{2}{5}$	$\frac{5}{2}$	$\frac{2}{5} \times \frac{5}{2} = 1$
Example 3:	$2\frac{2}{5}$ or $\frac{12}{5}$	$\frac{5}{12}$	$\frac{12}{5} \times \frac{5}{12} = 1$

Rule: The product of a number by its reciprocal is 1.

We can also state that whenever the product of two numbers is 1, each number is the reciprocal of the other. **(5)**

For example, in $\frac{2}{5} \times ? = 1$, the question mark must stand for the reciprocal of $\frac{2}{5}$. The reciprocal of $\frac{2}{5}$ is $\frac{5}{2}$ or $2\frac{1}{2}$.

Check: $\frac{\not{2}^1}{\not{5}_1} \times \frac{\not{5}^1}{\not{2}_1} = 1$

Instead of a question mark, the letter n can be used. If $\frac{2}{5} \cdot n = 1$, then $n = \frac{5}{2}$ or $2\frac{1}{2}$.

Study Exercise One (6)

Find the reciprocal of each of these numbers:

1. $\frac{3}{11}$ **2.** $\frac{4}{5}$ **3.** 7 **4.** $4\frac{2}{3}$

Find what number the question mark or n represents:

5. $\frac{2}{3} \times ? = 1$ **6.** $\frac{5}{3} \times ? = 1$ **7.** $\frac{3}{7} \cdot n = 1$ **8.** $\frac{12}{11} \times n = 1$

Division of Fractions (7)

Problem: $\frac{2}{3} \div \frac{5}{7}$

Solution:

Step (1): Recall that a division problem can be written in the form of a fraction:

$$\frac{2}{3} \div \frac{5}{7} = \frac{\frac{2}{3}}{\frac{5}{7}}$$

Division of Fractions (Continued)

Step (2): Make the denominator 1 by multiplying numerator and denominator by the reciprocal of the denominator:

$$\frac{\frac{2}{3} \times \frac{7}{5}}{\frac{5}{7} \times \frac{7}{5}} = \frac{\frac{2}{3} \times \frac{7}{5}}{1}$$

Step (3): Multiply fractions in the numerator:

$$\frac{\frac{2}{3} \times \frac{7}{5}}{1} = \frac{\frac{14}{15}}{1} = \frac{14}{15}$$

Shortcut: To divide one fraction by another, invert the fraction after the division sign and multiply.

Examples of Division of Fractions Using the Shortcut (8)

Example 1: $\frac{2}{3} \div \frac{4}{5}$

Solution:
$$\frac{2}{3} \div \frac{4}{5} = \frac{2}{3} \times \frac{5}{4}$$
$$= \frac{\not{2}^{1}}{3} \times \frac{5}{\not{4}_{2}}$$
$$= \frac{5}{6}$$

Example 2: $6 \div \frac{2}{3}$

Solution:
$$6 \div \frac{2}{3} = \frac{6}{1} \times \frac{3}{2}$$
$$= \frac{\not{6}^{3}}{1} \times \frac{3}{\not{2}_{1}}$$
$$= \frac{9}{1}$$
$$= 9$$

Example 3: $\frac{5}{6} \div \frac{3}{4}$

Solution:
$$\frac{5}{6} \div \frac{3}{4} = \frac{5}{6} \times \frac{4}{3}$$
$$= \frac{5}{\not{6}_{3}} \times \frac{\not{4}^{2}}{3}$$
$$= \frac{10}{9}$$
$$= 1\frac{1}{9}$$

Example 4: $\frac{3}{5} \div 4$

Solution:
$$\frac{3}{5} \div 4 = \frac{3}{5} \div \frac{4}{1}$$
$$= \frac{3}{5} \times \frac{1}{4}$$
$$= \frac{3}{20}$$

Study Exercise Two (9)

Divide as indicated and reduce all answers.

1. $\frac{4}{5} \div \frac{3}{10}$ **2.** $8 \div \frac{2}{3}$ **3.** $\frac{5}{9} \div \frac{3}{4}$ **4.** $\frac{6}{7} \div 3$

5. $\frac{3}{8} \div \frac{6}{7}$ **6.** $\frac{1}{6} \div \frac{1}{3}$ **7.** $4 \div \frac{2}{3}$ **8.** $\frac{5}{9} \div \frac{2}{5}$

9. $\frac{5}{9} \div \frac{4}{5}$ **10.** $4 \div \frac{1}{2}$ **11.** $\frac{3}{8} \div \frac{1}{4}$ **12.** $\frac{1}{2} \div 4$

Mixed Numerals in Division Problems (10)

Mixed numerals should first be changed to improper fractions. Then divide by inverting the divisor and multiplying.

Example: $2\frac{2}{3} \div \frac{2}{5} = \frac{8}{3} \div \frac{2}{5}$

$= \frac{\not{8}^{4}}{3} \times \frac{5}{\not{2}_{1}}$

$= \frac{20}{3}$

$= 6\frac{2}{3}$

Example: Find $5\frac{1}{3} \div 2\frac{3}{4}$: (11)

Solution: $5\frac{1}{3} \div 2\frac{3}{4} = \frac{16}{3} \div \frac{11}{4}$

$= \frac{16}{3} \times \frac{4}{11}$

$= \frac{64}{33}$

$= 1\frac{31}{33}$

Study Exercise Three (12)

Divide as indicated and reduce all answers to lowest terms:

1. $3\frac{1}{2} \div 2\frac{2}{3}$ **2.** $4\frac{2}{3} \div \frac{1}{3}$ **3.** $1\frac{2}{3} \div 2\frac{1}{2}$ **4.** $6\frac{1}{2} \div 2$ **5.** $1\frac{3}{4} \div 5\frac{1}{4}$ **6.** $1\frac{1}{3} \div 3\frac{5}{9}$

Complex Fractions (13)

A *complex fraction* is a fraction whose numerator or denominator or both are also fractions.

Example 1: $\dfrac{\frac{2}{3}}{\frac{1}{6}}$ **Example 2:** $\dfrac{8}{\frac{2}{5}}$ **Example 3:** $\dfrac{\frac{3}{8}}{4}$

It is important to draw a longer line between numerator and denominator.

Numerators and Denominators of Complex Fractions (14)

Example 1: $\dfrac{\frac{2}{3}}{\frac{1}{6}}$ $\frac{2}{3} \longleftarrow$ the numerator is $\frac{2}{3}$; $\frac{1}{6} \longleftarrow$ the denominator is $\frac{1}{6}$

Example 2: $\dfrac{\frac{3}{8}}{4}$ $\frac{3}{8} \longleftarrow$ the numerator is $\frac{3}{8}$; $4 \longleftarrow$ the denominator is 4

Example 3: $\dfrac{8}{\frac{2}{5}}$ $8 \longleftarrow$ the numerator is 8; $\frac{2}{5} \longleftarrow$ the denominator is $\frac{2}{5}$

Additional Examples (15)

Example 1: What is the denominator of $\dfrac{\frac{3}{5}}{6}$?

Solution: 6

Example 2: What is the numerator of $\dfrac{\frac{1}{3}}{\frac{3}{4}}$?

Solution: $\frac{1}{3}$

Example 3: Which of these fractions is a complex fraction?

$\dfrac{12}{5}$ $\dfrac{\frac{1}{3}}{6}$ $2\frac{1}{3}$

Solution: $\dfrac{\frac{1}{3}}{6}$

Meaning of a Complex Fraction (16)

Since one of the meanings of a fraction is an indicated division, a complex fraction will indicate division (numerator divided by denominator).

Example 1: $\dfrac{\frac{5}{6}}{\frac{3}{4}}$ means $\frac{5}{6} \div \frac{3}{4}$

Example 2: $\dfrac{7}{\frac{5}{6}}$ means $7 \div \frac{5}{6}$

Study Exercise Four (17)

1. What is the numerator of $\dfrac{8}{\frac{2}{3}}$?
2. What is the denominator of $\dfrac{\frac{1}{6}}{\frac{4}{7}}$?
3. Give the meaning of the complex fraction $\dfrac{\frac{3}{4}}{\frac{5}{6}}$
4. Give the meaning of the complex fraction $\dfrac{4}{\frac{5}{6}}$.

Simplifying a Complex Fraction (18)

To simplify a complex fraction:

Step (1): Write an indicated division.
Step (2): Invert the divisor and multiply.

Example: Simplify $\dfrac{\frac{5}{6}}{\frac{2}{3}}$:

Simplifying a Complex Fraction (Continued)

Solution:

Step (1): Write an indicated division:

$$\frac{\frac{5}{6}}{\frac{2}{3}} = \frac{5}{6} \div \frac{2}{3}$$

Step (2): Invert the divisor, cancel where possible, and multiply:

$$\frac{5}{6} \div \frac{2}{3} = \frac{5}{\cancel{6}_2} \times \frac{\cancel{3}^1}{2}$$

$$= \frac{5}{4}$$

$$= 1\frac{1}{4}$$

Study Exercise Five (19)

Simplify the following:

1. $\frac{\frac{3}{8}}{\frac{1}{4}}$
2. $\frac{1\frac{1}{2}}{2\frac{1}{3}}$
3. $\frac{\frac{2}{3}}{100}$
4. $\frac{1}{\frac{4}{5}}$

Changing Back to a Base of One (20)

To change back to a base of one, use the operation of division. Remember, you must divide by the quantity you are changing to one.

Example 1: If 4 pencils cost 76¢, find the cost of 1 pencil.

Solution: Write the known fact: 4 pencils cost 76¢
Change to one by dividing; find the cost of 1 pencil, so divide by pencils (4):

$$1 \text{ pencil costs } 76¢ \div 4 = 19¢$$

Example 2: If $2\frac{3}{4}$ pounds costs 48¢, find the cost of 1 pound:

Solution: $2\frac{3}{4}$ pounds costs 48¢
Change to one by dividing; find the cost of 1 pound, so divide by pounds ($2\frac{3}{4}$):

$$1 \text{ pound costs } 48¢ \div 2\frac{3}{4} = \frac{48}{1} \div \frac{11}{4}$$

$$= \frac{48}{1} \times \frac{4}{11}$$

$$= \frac{192}{11}$$

$$= 17\frac{5}{11}¢$$

Example 3: If $\frac{2}{3}$ pound of nuts costs 66¢, find the cost of 1 pound of nuts:

Changing Back to a Base of One (Continued)

Solution: $\frac{2}{3}$ pound costs 66¢

Change to one by dividing; find the cost of 1 pound, so divide by pounds ($\frac{2}{3}$):

$$1 \text{ pound costs } 66¢ \div \frac{2}{3} = \frac{66}{1} \times \frac{3}{2}$$
$$= \frac{\cancel{66}^{33}}{1} \times \frac{3}{\cancel{2}_1}$$
$$= \frac{99}{1}$$
$$= 99¢$$

(Notice the similarity to reduction problems found in Unit 5).

Study Exercise Six (21)

1. If $\frac{7}{16}$ pound of salad dressing costs 84¢, find the cost in cents of 1 pound.
2. A fuel oil tank $\frac{3}{8}$ full contains 540 gallons of oil. How many gallons would a full tank hold?
3. An airline offers a special fare for husband and wife at $1\frac{1}{2}$ times the cost of a full ticket. The cost of a special fare ticket is $111. What is the cost of one full fare ticket?
4. If $2\frac{1}{2}$ yards of cloth costs $5, what is the cost of 1 yard of the cloth?

REVIEW EXERCISES (22)

1. The reciprocal of $\frac{5}{7}$ is ________.
2. The reciprocal of $6\frac{3}{5}$ is ________.
3. The product of a number and its reciprocal is ________.
4. Find the number n if $\frac{5}{7} \cdot n = 1$.
5. Which of the following is a complex fraction?

$$\frac{2}{7} \qquad 1\frac{3}{5} \qquad \frac{5}{\frac{2}{3}} \qquad \frac{\frac{2}{3}}{\frac{1}{5}}$$

6. Give the meaning of $\dfrac{\frac{7}{10}}{\frac{2}{5}}$

7. Simplify $\dfrac{\frac{7}{8}}{1}$

8. Simplify $\dfrac{5}{\frac{3}{4}}$

REVIEW EXERCISES (Continued)

9. Simplify $\dfrac{\frac{2}{3}}{\frac{1}{4}}$

10. Find $\frac{3}{8} \div \frac{1}{6}$

11. Simplify $\dfrac{1\frac{2}{3}}{3\frac{5}{9}}$

12. Find $2\frac{5}{8} \div \frac{3}{16}$

13. If $1\frac{1}{4}$ pounds costs 60¢, find the cost of 1 pound.

14. If a car travels 5 miles in 9 minutes, how far does it travel in 1 minute?

Solutions to Review Exercises **(23)**

1. The reciprocal of $\frac{5}{7}$ is $\frac{7}{5}$ or $1\frac{2}{5}$

2. The reciprocal of $6\frac{3}{5}$ is $\frac{5}{33}$

3. The product of a number and its reciprocal is 1

4. $\frac{5}{7} \cdot n = 1$; n is the reciprocal of $\frac{5}{7}$; thus $n = \frac{7}{5}$ or $1\frac{2}{5}$

5. $\dfrac{5}{\frac{2}{3}}$ and $\dfrac{\frac{2}{3}}{\frac{1}{5}}$ are complex fractions.

6. The meaning of $\dfrac{\frac{7}{10}}{\frac{2}{5}}$ is $\frac{7}{10} \div \frac{2}{5}$

7. $\dfrac{\frac{7}{8}}{1} = \frac{7}{8}$

8. $\dfrac{5}{\frac{3}{4}} = 5 \div \frac{3}{4} = \frac{5}{1} \times \frac{4}{3} = \frac{20}{3} = 6\frac{2}{3}$

9. $\dfrac{\frac{2}{3}}{\frac{1}{4}} = \frac{2}{3} \div \frac{1}{4} = \frac{2}{3} \times \frac{4}{1} = \frac{8}{3} = 2\frac{2}{3}$

10. $\frac{3}{8} \div \frac{1}{6} = \frac{3}{\cancel{8}_4} \times \frac{\cancel{6}^3}{1} = \frac{9}{4} = 2\frac{1}{4}$

11. $\dfrac{1\frac{2}{3}}{3\frac{5}{9}} = 1\frac{2}{3} \div 3\frac{5}{9}$

$= \frac{5}{3} \div \frac{32}{9}$

$= \frac{5}{\cancel{3}_1} \times \frac{\cancel{9}^3}{32} = \frac{15}{32}$

12. $2\frac{5}{8} \div \frac{3}{16} = \frac{21}{8} \div \frac{3}{16}$

$= \frac{\cancel{21}^7}{\cancel{8}_1} \times \frac{\cancel{16}^2}{\cancel{3}_1}$

$= 14$

13. $1\frac{1}{4}$ pounds costs 60¢. To find the cost of 1 pound, divide by $1\frac{1}{4}$

$60¢ \div 1\frac{1}{4} = \frac{60}{1} \times \frac{4}{5}$

$= \frac{\cancel{60}^{12}}{1} \times \frac{4}{\cancel{5}_1}$

$= 48¢$

The cost of 1 pound is 48¢

14. Change to 1 minute by dividing by 9

5 miles $\div 9 = \frac{5}{1} \times \frac{1}{9}$

$= \frac{5}{9}$ miles

The car travels $\frac{5}{9}$ miles in 1 minute.

SUPPLEMENTARY PROBLEMS

A. Find the reciprocals of:

1. $\frac{8}{13}$ **2.** $\frac{1}{8}$ **3.** $\frac{18}{5}$ **4.** $2\frac{3}{4}$ **5.** 7 **6.** $1\frac{1}{8}$

B. 7. If the product of two numbers is 1, then each is the ________ of the other.

C. Find n such that each of the following will be true:

8. $\frac{1}{3} \times n = 1$ **9.** $\frac{9}{2} \times n = 1$ **10.** $\frac{1}{25} \times n = 1$ **11.** $\frac{3}{5} \cdot n = 1$ **12.** $1\frac{1}{2} \cdot n = 1$

D. Do as indicated and reduce final answers:

13. What is the numerator of $\dfrac{\frac{2}{3}}{5}$?

14. $\frac{5}{8} \div 15$ **15.** $\frac{2}{9} \div 6$ **16.** $5 \div \frac{3}{4}$ **17.** $8 \div \frac{2}{3}$

18. $\dfrac{1}{\frac{5}{12}}$ **19.** $\dfrac{\frac{3}{4}}{\frac{3}{8}}$ **20.** $\dfrac{2\frac{1}{4}}{\frac{5}{8}}$ **21.** $\dfrac{1\frac{1}{2}}{1\frac{1}{4}}$

22. $8\frac{3}{4} \div 2\frac{1}{6}$ **23.** $2\frac{5}{8} \div \frac{5}{8}$ **24.** $\frac{3}{20} \div \frac{4}{5}$ **25.** $1\frac{1}{6} \div 2\frac{2}{3}$

26. $6\frac{1}{4} \div 1\frac{3}{5}$ **27.** $1\frac{6}{7} \div 1\frac{6}{7}$ **28.** $4\frac{3}{8} \div 14$ **29.** $9 \div 1\frac{2}{3}$

30. $\dfrac{3\frac{3}{4}}{\frac{5}{8}}$ **31.** $\dfrac{3\frac{1}{8}}{2\frac{6}{7}}$

32. If $\frac{3}{4}$ pound costs $14\frac{1}{2}$¢, find the cost of 1 pound.

33. If $2\frac{1}{3}$ yards of cloth sells for \$7, what is the price of 1 yard of cloth?

34. A bus is scheduled to go a distance of $66\frac{1}{2}$ miles in $1\frac{1}{2}$ hours. How far must it travel in 1 hour to be on schedule?

Solutions to Study Exercises (6A)

Study Exercise One (Frame 6)

1. $\frac{11}{3}$ or $3\frac{2}{3}$ **2.** $\frac{5}{4}$ or $1\frac{1}{4}$ **3.** $\frac{1}{7}$ **4.** $\frac{3}{14}$

5. The question mark stands for the reciprocal of $\frac{2}{3}$, which is $\frac{3}{2}$ or $1\frac{1}{2}$

6. The question mark stands for the reciprocal of $\frac{5}{3}$, which is $\frac{3}{5}$

Solutions to Study Exercises (Continued)

7. n represents the reciprocal of $\frac{3}{7}$, which is $\frac{7}{3}$ or $2\frac{1}{3}$

8. n represents the reciprocal of $\frac{12}{11}$, which is $\frac{11}{12}$

Study Exercise Two (Frame 9) **(9A)**

1. $\frac{4}{5} \div \frac{3}{10} = \frac{4}{\cancel{5}_1} \times \frac{\cancel{10}^2}{3} = \frac{8}{3} = 2\frac{2}{3}$
2. $8 \div \frac{2}{3} = \frac{\cancel{8}^4}{1} \times \frac{3}{\cancel{2}_1} = \frac{12}{1} = 12$
3. $\frac{5}{9} \div \frac{3}{4} = \frac{5}{9} \times \frac{4}{3} = \frac{20}{27}$
4. $\frac{6}{7} \div 3 = \frac{\cancel{6}^2}{7} \times \frac{1}{\cancel{3}_1} = \frac{2}{7}$
5. $\frac{3}{8} \div \frac{6}{7} = \frac{\cancel{3}^1}{8} \times \frac{7}{\cancel{6}_2} = \frac{7}{16}$
6. $\frac{1}{6} \div \frac{1}{3} = \frac{1}{\cancel{6}_2} \times \frac{\cancel{3}^1}{1} = \frac{1}{2}$
7. $4 \div \frac{2}{3} = \frac{\cancel{4}^2}{1} \times \frac{3}{\cancel{2}_1} = \frac{6}{1} = 6$
8. $\frac{5}{9} \div \frac{2}{5} = \frac{5}{9} \times \frac{5}{2} = \frac{25}{18} = 1\frac{7}{18}$
9. $\frac{5}{9} \div \frac{4}{5} = \frac{5}{9} \times \frac{5}{4} = \frac{25}{36}$
10. $4 \div \frac{1}{2} = \frac{4}{1} \times \frac{2}{1} = \frac{8}{1} = 8$
11. $\frac{3}{8} \div \frac{1}{4} = \frac{3}{\cancel{8}_2} \times \frac{\cancel{4}^1}{1} = \frac{3}{2} = 1\frac{1}{2}$
12. $\frac{1}{2} \div 4 = \frac{1}{2} \div \frac{4}{1} = \frac{1}{2} \times \frac{1}{4} = \frac{1}{8}$

Study Exercise Three (Frame 12) **(12A)**

1. $3\frac{1}{2} \div 2\frac{2}{3} = \frac{7}{2} \div \frac{8}{3} = \frac{7}{2} \times \frac{3}{8} = \frac{21}{16} = 1\frac{5}{16}$
2. $4\frac{2}{3} \div \frac{1}{3} = \frac{14}{3} \div \frac{1}{3} = \frac{14}{\cancel{3}_1} \times \frac{\cancel{3}^1}{1} = \frac{14}{1} = 14$
3. $1\frac{2}{3} \div 2\frac{1}{2} = \frac{5}{3} \div \frac{5}{2} = \frac{\cancel{5}^1}{3} \times \frac{2}{\cancel{5}_1} = \frac{2}{3}$
4. $6\frac{1}{2} \div 2 = \frac{13}{2} \div \frac{2}{1} = \frac{13}{2} \times \frac{1}{2} = \frac{13}{4} = 3\frac{1}{4}$
5. $1\frac{3}{4} \div 5\frac{1}{4} = \frac{7}{4} \div \frac{21}{4} = \frac{\cancel{7}^1}{\cancel{4}_1} \times \frac{\cancel{4}^1}{\cancel{21}_3} = \frac{1}{3}$
6. $1\frac{1}{3} \div 3\frac{5}{9} = \frac{4}{3} \div \frac{32}{9} = \frac{\cancel{4}^1}{\cancel{3}_1} \times \frac{\cancel{9}^3}{\cancel{32}_8} = \frac{3}{8}$

Study Exercise Four (Frame 17) **(17A)**

1. The numerator of $\dfrac{8}{\frac{2}{3}}$ is 8
2. The denominator of $\dfrac{\frac{1}{6}}{\frac{4}{7}}$ is $\frac{4}{7}$
3. $\dfrac{\frac{3}{4}}{\frac{5}{6}}$ means $\frac{3}{4} \div \frac{5}{6}$
4. $\dfrac{4}{\frac{5}{6}}$ means $4 \div \frac{5}{6}$

Study Exercise Five (Frame 19) **(19A)**

1. $$\frac{\frac{3}{8}}{\frac{1}{4}} = \frac{3}{8} \div \frac{1}{4} = \frac{3}{\cancel{8}_2} \times \frac{\cancel{4}^1}{1} = \frac{3}{2} = 1\frac{1}{2}$$
2. $$\frac{1\frac{1}{2}}{2\frac{1}{3}} = 1\frac{1}{2} \div 2\frac{1}{3} = \frac{3}{2} \div \frac{7}{3} = \frac{3}{2} \times \frac{3}{7} = \frac{9}{14}$$
3. $$\frac{\frac{2}{3}}{100} = \frac{2}{3} \div 100 = \frac{\cancel{2}^1}{3} \times \frac{1}{\cancel{100}_{50}} = \frac{1}{150}$$
4. $$\frac{1}{\frac{4}{5}} = 1 \div \frac{4}{5} = 1 \times \frac{5}{4} = \frac{5}{4} = 1\frac{1}{4}$$

Study Exercise Six (Frame 21) (21A)

1. Change to 1 pound by dividing by $\frac{7}{16}$:

$$84¢ \div \frac{7}{16} = \frac{\not{84}^{12}}{1} \times \frac{16}{\not{7}_1} = \frac{192}{1}$$
$$= 192 \text{ cents}$$

2. Change to 1 full tank by dividing by $\frac{3}{8}$:

$$540 \text{ gallons} \div \frac{3}{8} = \frac{\not{540}^{180}}{1} \times \frac{8}{\not{3}_1} = \frac{1440}{1}$$
$$= 1440 \text{ gallons}$$

3. Change to one full ticket by dividing by $1\frac{1}{2}$:

$$\$111 \div 1\frac{1}{2} = \frac{111}{1} \div \frac{3}{2}$$
$$= \frac{\not{111}^{37}}{1} = \frac{2}{\not{3}_1}$$
$$= \frac{74}{1}$$
$$= \$74$$

4. Change to 1 yard by dividing by $2\frac{1}{2}$:

$$\$5 \div 2\frac{1}{2} = \frac{5}{1} \div \frac{5}{2}$$
$$= \frac{\not{5}^1}{1} \times \frac{2}{\not{5}_1}$$
$$= \frac{2}{1}$$
$$= \$2 \text{ a yard}$$

UNIT

15 Equations and Applied Problems Involving Fractions

OBJECTIVES (1)

By the end of this unit you should be able to:

1. READ AND SOLVE APPLIED PROBLEMS INVOLVING ADDITION, SUBTRACTION, MULTIPLICATION, OR DIVISION.
2. KNOW THE PARTS OF AN EQUATION AND BE ABLE TO SOLVE EQUATIONS SUCH AS $2\frac{1}{2} \cdot n = 1\frac{1}{3}$.
3. TRANSLATE VERBAL STATEMENTS INTO EQUATIONS AND SOLVE THE EQUATIONS.

Applied Problems (2)

The applied problems in this unit will be solved using one of the operations of addition, subtraction, multiplication, or division. The operation used will depend on the problem.

ADDITION

Example: Sue made a 2-piece dress requiring $1\frac{1}{4}$ yards for one part and $2\frac{5}{8}$ yards for the other. How much material did she use?

Solution:

Line (a): $1\frac{1}{4} + 2\frac{5}{8} = \frac{5}{4} + \frac{21}{8}$

Line (b): $= \frac{5 \times 2}{4 \times 2} + \frac{21}{8}$

Line (c): $= \frac{10}{8} + \frac{21}{8}$

Line (d): $= \frac{31}{8}$

Line (e): $= 3\frac{7}{8}$

Sue used $3\frac{7}{8}$ yards.

Applied Problems (3)

SUBTRACTION

Example: Two months ago, Joe weighed $143\frac{1}{4}$ pounds. Now he weighs $137\frac{3}{8}$ pounds. How much did he lose?

Solution:

$$\begin{array}{r} 143\frac{1}{4} = 143\frac{2}{8} \\ -137\frac{3}{8} = -137\frac{3}{8} \end{array} \longrightarrow \begin{array}{r|l} {}^{142}\cancel{143} & 1\frac{2}{8} \\ -\ 137 & \frac{3}{8} \end{array} \longrightarrow \begin{array}{r|l} {}^{142}\cancel{143} & \frac{10}{8} \\ -\ 137 & \frac{3}{8} \\ \hline 5 & \frac{7}{8} \end{array}$$

Joe lost $5\frac{7}{8}$ pounds.

Applied Problems (4)

MULTIPLICATION

Example 1: If 1 cubic foot of water weighs $62\frac{1}{2}$ pounds, find the weight of a column of water containing $2\frac{1}{3}$ cubic feet.

Solution:

Line (a): $62\frac{1}{2} \times 2\frac{1}{3} = \frac{125}{2} \times \frac{7}{3}$

Line (b): $= \frac{875}{6}$

Line (c): $= 145\frac{5}{6}$

The column of water weighs $145\frac{5}{6}$ pounds.

Example 2: A certain item regularly sells for 80¢. It is on sale for $\frac{2}{5}$ of its regular price. What is the sale price?

Solution:

Line (a): $\frac{2}{5}$ of 80¢

Line (b): $\frac{2}{5} \times \frac{80}{1}$

Line (c): $\frac{2}{\cancel{5}_1} \times \frac{\cancel{80}^{16}}{1}$

Line (d): $\frac{32}{1}$

Line (e): 32

The sale price is 32¢.

Applied Problems (5)

DIVISION

Example: A bus is scheduled to go a distance of $66\frac{1}{2}$ miles in $1\frac{1}{2}$ hours. What average speed must be maintained to arrive on schedule?

Applied Problems (Continued)

Solution:

Line (a): $66\frac{1}{2} \div 1\frac{1}{2} = \frac{133}{2} \div \frac{3}{2}$

Line (b): $= \frac{133}{\cancel{2}_1} \times \frac{\cancel{2}^1}{3}$

Line (c): $= \frac{133}{3}$

Line (d): $= 44\frac{1}{3}$

The average speed must be $44\frac{1}{3}$ miles per hour.

Study Exercise One (6)

1. How many feet of wood are needed to make 8 shelves each $6\frac{1}{3}$ feet long?
2. Find the net change in a stock if its high was $20\frac{3}{8}$ and its low $5\frac{1}{4}$.
3. Bob worked $3\frac{3}{4}$ hours on Monday, $2\frac{1}{8}$ hours on Tuesday, and $3\frac{1}{2}$ hours on Wednesday. How many hours did he work altogether?
4. Sam has a take-home pay of \$940 per month. His budget allows him to spend $\frac{2}{5}$ of his income on food. How much money does he spend per month on food?
5. How much does a pound of apples cost if $1\frac{5}{8}$ pounds costs 78¢?
6. From a $\frac{7}{8}$-yard remnant of fabric, a pillow was made with $\frac{5}{12}$ yard. How much fabric was left?
7. A house worth \$108,000 is assessed at $\frac{2}{3}$ of its value. Find the assessed value of the house.

Equation (7)

An *equation* is a statement of equality between two quantities.

Example 1: $4 + 3 = 7$ **Example 2:** $2 \cdot 3 = 6$

Parts of an Equation (8)

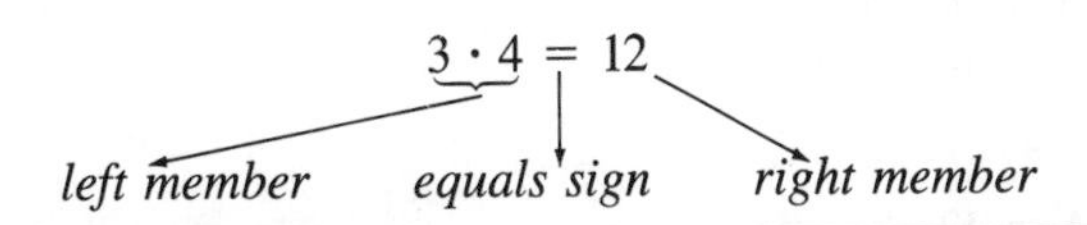

Changing Verbal Statements to Equations (9)

Verbal statement: Three times some number is 18

Equation statement: $3 \times ? = 18$

Instead of a question mark, let us use the letter n to represent the number.

$$3 \cdot n = 18$$

Remember, a dot may be used to represent multiplication.

Writing Equations (10)

Example 1: Change this verbal statement to an equation using n: $2\frac{1}{2}$ times some number is 40.

Solution: $2\frac{1}{2} \cdot n = 40$

Example 2: Change this verbal statement to an equation using n: $\frac{1}{3}$ of some number is 16.

Solution: $\frac{1}{3} \cdot n = 16$

Solving an Equation (11)

To *solve an equation* means to find the number that can be used for the letter n to make the equation a true statement.

Equation: $3 \cdot n = 18$
Solution: $n = 6$
since $3 \cdot 6 = 18$

Study Exercise Two (12)

1. Name the parts of the equation $5 \cdot n = 20$.
2. Change this verbal statement to an equation using the letter n: Four times some number is thirty-six.
3. Change this verbal statement to an equation using the letter n: Nine times some number is twenty-seven.
4. Guess the solution to $4 \cdot n = 28$ and check by multiplication.
5. Guess the solution to $7 \cdot n = 35$ and check by multiplication.

Rule for Solving an Equation (13)

Example 1: Solve $3 \cdot n = 12$

Solution: 3 times some number is 12
The number is $12 \div 3$
The number is 4
Check: $3 \cdot 4 = 12$

Example 2: **Solve:** $\frac{1}{2} \cdot n = 5$

Solution: $\frac{1}{2}$ of some number is 5

The number is $5 \div \frac{1}{2}$

$5 \div \frac{1}{2} = \frac{5}{1} \times \frac{2}{1} = 10$

The number is 10

Check: $\frac{1}{2} \cdot 10 = 5$

Rule for Solving an Equation (Continued)

Equations of the type $3 \cdot n = 12$ and $\frac{1}{2} \cdot n = 5$ are solved by division.

Rule: To solve an equation of this type, divide by the number in front of n.

Examples (14)

1. **Solve:** $5 \cdot n = 55$

   Solution: $n = 55 \div 5$

   $n = 11$

2. **Solve:** $\frac{1}{2} \cdot n = \frac{3}{2}$

   Solution: $n = \frac{3}{2} \div \frac{1}{2}$

   $n = \frac{3}{\cancel{2}_1} \times \frac{\cancel{2}^1}{1}$

   $n = 3$

3. **Solve:** $\frac{3}{4} \cdot n = \frac{5}{8}$

   Solution: $n = \frac{5}{8} \div \frac{3}{4}$

   $n = \frac{5}{\cancel{8}_2} \times \frac{\cancel{4}^1}{3}$

   $n = \frac{5}{6}$

Study Exercise Three (15)

Solve the following equations:

1. $2 \cdot n = 24$
2. $5 \cdot n = 60$
3. $9 \cdot n = 30$
4. $3 \cdot n = 4\frac{2}{3}$
5. $\frac{1}{4} \cdot n = 20$
6. $\frac{1}{2} \cdot n = \frac{4}{5}$
7. $2\frac{1}{2} \cdot n = 1\frac{1}{4}$
8. $3\frac{1}{3} \cdot n = 10$

To Find What Part One Number Is of Another (16)

To find what part of 6 is 5, recall that "of" means multiply and "is" means equal. We will write an equation for the statement: What part of 6 is 5?

What part	of	6	is	5?
n	$\cdot$	6	=	5

or $6 \cdot n = 5$

The equation is $6 \cdot n = 5$

The solution is $n = 5 \div 6$

$n = \frac{5}{6}$

To Find What Part One Number Is of Another (Continued)

Example: What part of 30 is 23?

Solution:
$$n \cdot 30 = 23$$
$$30 \cdot n = 23$$
$$n = 23 \div 30$$
$$n = \frac{23}{30}$$

---

(17)

Example 1: What part of 3 is $\frac{5}{6}$?

Solution:
$$n \cdot 3 = \frac{5}{6}$$
$$3 \cdot n = \frac{5}{6}$$
$$n = \frac{5}{6} \div 3$$
$$n = \frac{5}{6} \div \frac{3}{1}$$
$$n = \frac{5}{6} \times \frac{1}{3}$$
$$n = \frac{5}{18}$$

Example 2: What part of $3\frac{3}{4}$ is $5\frac{1}{2}$?

Solution:
$$n \cdot 3\frac{3}{4} = 5\frac{1}{2}$$
$$3\frac{3}{4} \cdot n = 5\frac{1}{2}$$
$$n = 5\frac{1}{2} \div 3\frac{3}{4}$$
$$n = \frac{11}{2} \div \frac{15}{4}$$
$$n = \frac{11}{2} \times \frac{4}{15}$$
$$n = \frac{11}{\not{2}_1} \times \frac{\not{4}^2}{15}$$
$$n = \frac{22}{15}$$
$$n = 1\frac{7}{15}$$

Study Exercise Four (18)

1. What part of 8 is $\frac{2}{3}$?
2. What part of $\frac{3}{4}$ is $\frac{2}{3}$?
3. What part of $3\frac{1}{4}$ is 5?
4. What part of $1\frac{5}{7}$ is $\frac{2}{3}$?
5. What part of $3\frac{1}{4}$ is $2\frac{1}{2}$?

Finding a Number When a Fractional Part Is Known (19)

Suppose we know that $\frac{1}{2}$ of a number is 7.

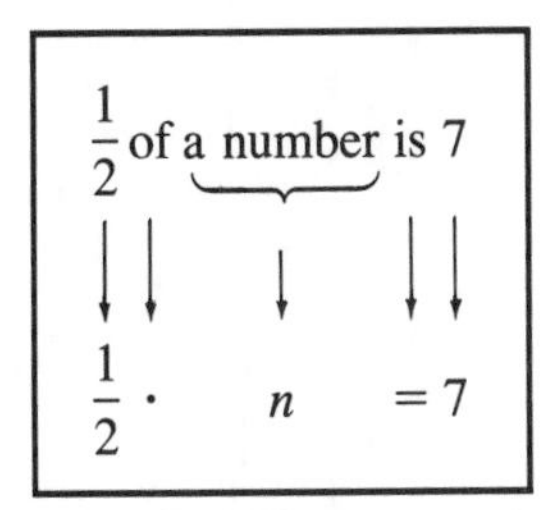

We can write the equation $\frac{1}{2} \cdot n = 7$

$$n = 7 \div \frac{1}{2}$$

$$n = \frac{7}{1} \times \frac{2}{1}$$

$$n = \frac{14}{1}$$

$$n = 14$$

Thus, if $\frac{1}{2}$ of the number is 7, the number is 14

(20)

Example 1: Find the number if $\frac{1}{8}$ of the number is 4:

Solution: $\frac{1}{8}$ of the number is 4

$$\frac{1}{8} \cdot n = 4$$

$$n = 4 \div \frac{1}{8}$$

$$n = \frac{4}{1} \times \frac{8}{1}$$

$$n = \frac{32}{1}$$

$$n = 32$$

The number is 32

Example 2: What is the number if $\frac{5}{8}$ of the number is $2\frac{1}{4}$? **(21)**

Solution: $\frac{5}{8}$ of the number is $2\frac{1}{4}$

$$\frac{5}{8} \cdot n = 2\frac{1}{4}$$

$$n = 2\frac{1}{4} \div \frac{5}{8}$$

$$n = 2\frac{1}{4} \times \frac{8}{5}$$

$$n = \frac{9}{\cancel{4}_1} \times \frac{\cancel{8}^2}{5}$$

$$n = \frac{18}{5}$$

$$n = 3\frac{3}{5}$$

The number is $3\frac{3}{5}$

Study Exercise Five **(22)**

1. Find the number if $\frac{1}{4}$ of the number is 15
2. Find the number if $\frac{2}{3}$ of the number is $\frac{4}{9}$
3. Find the number if $1\frac{1}{2}$ of the number is $1\frac{1}{4}$
4. Find the number if $1\frac{1}{2}$ of the number is $2\frac{1}{2}$.

REVIEW EXERCISES **(23)**

1. If a costume requires $3\frac{1}{3}$ yards of material, how many costumes can be made from a 30-yard bolt of material?
2. If $1\frac{1}{4}$ pounds costs 60¢, find the cost of 1 pound.
3. If 1 pound of nuts costs 90¢, find the cost of $\frac{3}{5}$ of a pound.
4. Solve: $6 \cdot n = 42$
5. Solve: $4 \cdot n = 1\frac{3}{4}$
6. Solve: $3\frac{2}{3} \cdot n = 4\frac{1}{3}$
7. Solve: $8 \cdot n = 5\frac{1}{3}$
8. What fractional part of 6 is $\frac{1}{3}$?
9. What fractional part of $1\frac{1}{2}$ is $\frac{2}{3}$?
10. Find the number if $\frac{2}{5}$ of the number is $\frac{1}{4}$

Solutions to Review Exercises (24)

1. $30 \div 3\frac{1}{3} = \frac{30}{1} \div \frac{10}{3}$

$= \frac{30}{1} \times \frac{3}{10}$

$= \frac{\not{30}^{3}}{1} \times \frac{3}{\not{10}_{1}}$

$= 9$

9 costumes can be made.

2. Change to 1 pound by dividing by $1\frac{1}{4}$:

$60¢ \div 1\frac{1}{4} = \frac{\not{60}^{12}}{1} \times \frac{4}{\not{5}_{1}}$

$= \frac{48}{1}$

$= 48¢$

The cost of 1 pound is 48¢.

3. 1 pound costs 90¢

To find the cost of $\frac{3}{5}$ pound, multiply by $\frac{3}{5}$:

$\frac{3}{5} \times 90¢ = \frac{3}{\not{5}_{1}} \times \frac{\not{90}^{18}}{1}$

$= 54¢.$

The cost of $\frac{3}{5}$ pound is 54¢

4. $6 \cdot n = 42$

$n = 42 \div 6$

$n = 7$

5. $4 \cdot n = 1\frac{3}{4}$

$n = 1\frac{3}{4} \div 4$

$n = \frac{7}{4} \times \frac{1}{4}$

$n = \frac{7}{16}$

6. $3\frac{2}{3} \cdot n = 4\frac{1}{3}$

$n = 4\frac{1}{3} \div 3\frac{2}{3}$

$n = \frac{13}{\not{3}_{1}} \times \frac{\not{3}^{1}}{11}$

$n = \frac{13}{11}$

$n = 1\frac{2}{11}$

7. $8 \cdot n = 5\frac{1}{3}$

$n = 5\frac{1}{3} \div 8$

$n = \frac{\not{16}^{2}}{3} \times \frac{1}{\not{8}_{1}}$

$n = \frac{2}{3}$

8. $6 \cdot n = \frac{1}{3}$

$n = \frac{1}{3} \div 6$

$n = \frac{1}{3} \times \frac{1}{6}$

$n = \frac{1}{18}$

9. $1\frac{1}{2} \cdot n = \frac{2}{3}$

$n = \frac{2}{3} \div 1\frac{1}{2}$

$n = \frac{2}{3} \div \frac{3}{2}$

$n = \frac{2}{3} \times \frac{2}{3}$

$n = \frac{4}{9}$

10. $\frac{2}{5} \cdot n = \frac{1}{4}$

$n = \frac{1}{4} \div \frac{2}{5}$

$n = \frac{1}{4} \times \frac{5}{2}$

$n = \frac{5}{8}$

The number is $\frac{5}{8}$.

SUPPLEMENTARY PROBLEMS

1. Bob has a take-home pay of \$1260 per month. His budget allows him to spend $\frac{2}{5}$ of his income on food and $\frac{1}{4}$ on housing. How much money does he spend each month on food and housing?
2. Joe bought 48 postage stamps and used $\frac{1}{3}$ of them the first day. How many did he use the first day?
3. Certain balls are on sale at $\frac{3}{4}$ of their former price of 60¢. What is the present price?
4. A 6-pound roast lost $\frac{1}{4}$ of its weight in cooking. How much did it weigh after cooking?
5. If one certain item weighs $12\frac{1}{3}$ pounds, what is the weight of 4 of these items?
6. At the end of the first year, a car loses $\frac{1}{4}$ its original value. At the end of the second year, the car loses $\frac{1}{5}$ the value at the end of the first year. If the original value of the car was \$8000, find its value at the end of the first year and at the end of the second year.
7. If $\frac{1}{4}$ of the height of the hull of a boat is above the water line and there is 3 feet of the hull above the water, what is the height of the hull?
8. If $\frac{3}{4}$ of a pound costs $14\frac{1}{2}$¢, find the cost of 1 pound.
9. How much heavier is a $5\frac{1}{3}$ pound package than a $2\frac{5}{6}$ pound package?
10. If $2\frac{1}{4}$ yards of material is needed for a suit, how many full suits can be cut from a roll of material 60 yards long?
11. A brass rod of certain length was cut into pieces $7\frac{5}{32}$ inches, $4\frac{3}{4}$ inches, and $2\frac{1}{8}$ inches long. How long was the rod if $\frac{1}{16}$ inch was wasted in each cut?
12. If a factory produces $8\frac{1}{3}$ articles in 5 days, how many will it produce in 1 day?
13. If a car travels $5\frac{1}{4}$ miles in 9 minutes, how long does it take to go 1 mile?
14. Solve: $3 \cdot n = 42$
15. Solve: $4 \cdot n = \frac{5}{6}$
16. Solve: $3\frac{1}{2} \cdot n = 6\frac{1}{4}$
17. Solve: $\frac{5}{16} \cdot n = 4\frac{1}{4}$
18. Solve: $1\frac{3}{14} \cdot n = 4\frac{1}{6}$
19. Solve: $3\frac{1}{9} \cdot n = \frac{5}{6}$
20. Solve: $\frac{3}{4} \cdot n = 12$
21. Solve: $9 \cdot n = 4\frac{3}{4}$
22. Solve: $10 \cdot n = \frac{1}{100}$
23. Solve: $1\frac{1}{2} \cdot n = 1\frac{1}{2}$
24. Solve: $\frac{1}{8} \cdot n = 12$
25. Solve: $2\frac{1}{3} \cdot n = 0$
26. What part of $\frac{1}{4}$ is $\frac{5}{16}$?
27. What part of $\frac{3}{4}$ is $\frac{2}{3}$?

SUPPLEMENTARY PROBLEMS (Continued)

28. What part of $3\frac{1}{2}$ is $2\frac{1}{3}$?

29. What part of 4 is $\frac{5}{6}$?

30. What part of $1\frac{3}{4}$ is $2\frac{1}{3}$?

31. What part of 100 is $3\frac{1}{8}$?

32. Find the number if $\frac{7}{10}$ of the number is 63

33. Find the number if $\frac{4}{5}$ of the number is 12

34. Find the number if $\frac{2}{3}$ of the number is $1\frac{1}{5}$

35. Find the number if $1\frac{1}{3}$ of the number is 9

Solutions to Study Exercises **(6A)**

Study Exercise One (Frame 6)

1. $8 \times 6\frac{1}{3} = \frac{8}{1} \times \frac{19}{3} = \frac{152}{3} = 50\frac{2}{3}$

$50\frac{2}{3}$ feet of wood are needed

2. $20\frac{3}{8} - 5\frac{1}{4} = 20\frac{3}{8} - 5\frac{2}{8} = 15\frac{1}{8}$

The net change is $15\frac{1}{8}$

3. $3\frac{3}{4} + 2\frac{1}{8} + 3\frac{1}{2} = 3\frac{6}{8} + 2\frac{1}{8} + 3\frac{4}{8}$

$= 8\frac{11}{8} = 8 + 1\frac{3}{8} = 9\frac{3}{8}$

Bob worked $9\frac{3}{8}$ hours

4. $\frac{2}{5} \times 940 = \frac{2}{\cancel{5}_1} \times \frac{\cancel{940}^{188}}{1}$

$= 376$

Sam spends \$376 per month on food

5. $78 \div 1\frac{5}{8} = \frac{78}{1} \div \frac{13}{8}$

$= \frac{\cancel{78}^6}{1} \times \frac{8}{\cancel{13}_1}$

$= 48$

A pound of apples costs 48¢

6. $\frac{7}{8} - \frac{5}{12} = \frac{21}{24} - \frac{10}{24}$

$= \frac{11}{24}$

$\frac{11}{24}$ yard was left

7. $\frac{2}{\cancel{3}_1} \times \frac{\cancel{108{,}000}^{36{,}000}}{1} = 72{,}000$

The assessed value is \$72,000

Study Exercise Two (Frame 12) **(12A)**

1. $\underbrace{5 \cdot n}$ = 20

left member ($5 \cdot n$), equal sign (=), right member (20)

2. $4 \cdot n = 36$

3. $9 \cdot n = 27$

4. $n = 7$ **Check:** $4 \cdot 7 = 28$

5. $n = 5$ **Check:** $7 \cdot 5 = 35$

Study Exercise Three (Frame 15) **(15A)**

1. $2 \cdot n = 24$

$n = 24 \div 2$

$n = 12$

2. $5 \cdot n = 60$

$n = 60 \div 5$

$n = 12$

3. $9 \cdot n = 30$

$n = 30 \div 9$

$n = 3\frac{1}{3}$

4. $3 \cdot n = 4\frac{2}{3}$

$n = 4\frac{2}{3} \div 3$

$n = \frac{14}{3} \times \frac{1}{3}$

$n = \frac{14}{9}$

$n = 1\frac{5}{9}$

Study Exercise Three (Frame 15) (Continued)

5. $\frac{1}{4} \cdot n = 20$

$n = 20 \div \frac{1}{4}$

$n = \frac{20}{1} \times \frac{4}{1}$

$n = \frac{80}{1}$

$n = 80$

6. $\frac{1}{2} \cdot n = \frac{4}{5}$

$n = \frac{4}{5} \div \frac{1}{2}$

$n = \frac{4}{5} \times \frac{2}{1}$

$n = \frac{8}{5}$

$n = 1\frac{3}{5}$

7. $2\frac{1}{2} \cdot n = 1\frac{1}{4}$

$n = 1\frac{1}{4} \div 2\frac{1}{2}$

$n = \frac{5}{4} \div \frac{5}{2}$

$n = \frac{\cancel{5}^1}{\cancel{4}_2} \times \frac{\cancel{2}^1}{\cancel{5}_1}$

$n = \frac{1}{2}$

8. $3\frac{1}{3} \cdot n = 10$

$n = 10 \div 3\frac{1}{3}$

$n = \frac{10}{1} \div \frac{10}{3}$

$n = \frac{\cancel{10}^1}{1} \times \frac{3}{\cancel{10}_1}$

$n = \frac{3}{1}$

$n = 3$

Study Exercise Four (Frame 18) (18A)

1. $8 \cdot n = \frac{2}{3}$

$n = \frac{2}{3} \div 8$

$n = \frac{\cancel{2}^1}{3} \times \frac{1}{\cancel{8}_4}$

$n = \frac{1}{12}$

2. $\frac{3}{4} \cdot n = \frac{2}{3}$

$n = \frac{2}{3} \div \frac{3}{4}$

$n = \frac{2}{3} \times \frac{4}{3}$

$n = \frac{8}{9}$

3. $3\frac{1}{4} \cdot n = 5$

$n = 5 \div 3\frac{1}{4}$

$n = \frac{5}{1} \times \frac{4}{13}$

$n = \frac{20}{13}$

$n = 1\frac{7}{13}$

4. $1\frac{5}{7} \cdot n = \frac{2}{3}$

$n = \frac{2}{3} \div 1\frac{5}{7}$

$n = \frac{\cancel{2}^1}{3} \times \frac{7}{\cancel{12}_6}$

$n = \frac{7}{18}$

5. $3\frac{1}{4} \cdot n = 2\frac{1}{2}$

$n = 2\frac{1}{2} \div 3\frac{1}{4}$

$n = \frac{5}{\cancel{2}_1} \times \frac{\cancel{4}^2}{13}$

$n = \frac{10}{13}$

Study Exercise Five (Frame 22) (22A)

1. $\frac{1}{4} \cdot n = 15$

$n = 15 \div \frac{1}{4}$

$n = \frac{15}{1} \times \frac{4}{1}$

$n = \frac{60}{1}$

$n = 60$

The number is 60

2. $\frac{2}{3} \cdot n = \frac{4}{9}$

$n = \frac{4}{9} \div \frac{2}{3}$

$n = \frac{\cancel{4}^2}{\cancel{9}_3} \times \frac{\cancel{3}^1}{\cancel{2}_1}$

$n = \frac{2}{3}$

The number is $\frac{2}{3}$

Study Exercise Five (Frame 22) **(Continued)**

3. $1\frac{1}{2} \cdot n = 1\frac{1}{4}$

$n = 1\frac{1}{4} \div 1\frac{1}{2}$

$n = \frac{5}{4} \div \frac{3}{2}$

$n = \frac{5}{\not{4}_2} \times \frac{\not{2}^1}{3}$

$n = \frac{5}{6}$

The number is $\frac{5}{6}$

4. $1\frac{1}{2} \cdot n = 2\frac{1}{2}$

$n = 2\frac{1}{2} \div 1\frac{1}{2}$

$n = \frac{5}{2} \div \frac{3}{2}$

$n = \frac{5}{\not{2}_1} \times \frac{\not{2}^1}{3}$

$n = \frac{5}{3}$

The number is $1\frac{2}{3}$

UNIT

16 Ratio and Proportion

OBJECTIVES (1)

By the end of this unit you should be able to:

1. GIVE THE MEANING OF RATIO AND PROPORTION.
2. DETERMINE WHEN RATIOS ARE PROPORTIONAL.
3. SOLVE PROPORTIONS.
4. USE PROPORTIONS IN SOLVING APPLIED PROBLEMS.

Ratio (2)

If you had to compare the number 60 with the number 30, you would say that 60 is twice as large as 30. When one number is twice as large as another, we say that the ratio of the first number to the second number is 2 to 1.

A *ratio* is a comparison of two quantities by division.

Example 1: The ratio of 8 to 4 is 2 to 1, since 8 divided by 4 is 2.

Example 2: The ratio of 20 to 10 is 2 to 1, since 20 divided by 10 is 2.

Different Ways of Writing a Ratio (3)

There are three different ways to indicate a ratio. The ratio of 2 to 1 can be written:

(1) 2 to 1

(2) $\frac{2}{1}$

(3) 2:1

The fractional numeral (method 2) is preferred.

Steps In Writing a Ratio (4)

To write a ratio:

Step (1): Write as a fraction.
Step (2): Reduce to lowest terms.

Example: Write a fractional numeral for the ratio of 10 to 8.

Solution:

Step (1): Write as a fraction:

$$\frac{10}{8}$$

Step (2): Reduce to lowest terms:

$$\frac{10}{8} = \frac{\not{10}^{5}}{\not{8}_{4}} = \frac{5}{4}$$

The ratio is $\frac{5}{4}$

Examples of Ratios (5)

Example 1: 3 of 6 squares are shaded. Write the ratio of the shaded squares to the total number of squares.

Solution: $\frac{3}{6} = \frac{1}{2}$

The ratio is $\frac{1}{2}$.

Example 2: Find the ratio in cents of a nickel to a dime.

Solution: A nickel is 5 cents and a dime is 10 cents. The ratio is $\frac{5}{10}$. But, $\frac{5}{10}$ reduces to $\frac{1}{2}$.

Thus, the ratio is $\frac{1}{2}$.

Notice that in forming ratios, the number following the word "to" is always the denominator.

Ratios of Like and Unlike Measurements (6)

When a ratio compares two numbers with the same unit of measurement, the unit of measurement does not need to be written.

Example 1: The ratio of 12 apples to 3 apples is

$$\frac{12 \text{ apples}}{3 \text{ apples}} = \frac{12}{3} = \frac{4}{1}$$

When a ratio compares two numbers with different units of measurement, the units of measurement must be written. This type of ratio is often called a *rate*.

Ratios of Like and Unlike Measurements (Continued)

Example 2: The ratio of 100 miles and 4 hours is

$$\frac{100 \text{ miles}}{4 \text{ hours}} = \frac{25 \text{ miles}}{1 \text{ hour}} \text{ or 25 miles per hour}$$

Examples of Ratios (7)

Example 1: Write a fractional numeral for the ratio of 10 to 8.

Solution: $\frac{10}{8} = \frac{5}{4}$

Example 2: If a student drives 125 miles on 5 gallons of gas, what is her rate in miles per gallon?

Solution: $\frac{125 \text{ miles}}{5 \text{ gallons}} = \frac{25 \text{ miles}}{1 \text{ gallon}} = 25 \text{ miles per gallon}$

Study Exercise One (8)

1. Write a fractional numeral for the ratio of 9 to 3.
2. Write a fractional numeral for the ratio of 5 books to 20 books.
3. Write a fractional numeral for the ratio of 200 miles to 10 gallons of gas.
4. If a student drives 126 miles on 7 gallons of gas, what is his rate in miles per gallon?
5. A baseball player got 112 hits in 336 times at bat. What is the ratio of the number of hits to the number of times at bat?
6. What is the ratio in cents of a dime to a dollar?

Proportion (9)

A *proportion* is an equation that states that two ratios are equal. The statement $\frac{2}{3} = \frac{4}{6}$ is a proportion, since the ratio $\frac{4}{6}$ is equal to the ratio $\frac{2}{3}$.

Cross Products Test (10)

In a proportion, the cross products are equal.

Example 1:

$\frac{2}{3} = \frac{4}{6}$

$\frac{2}{3} \times \frac{4}{6}$

$2 \times 6 = 3 \times 4$

$12 = 12$

Example 2:

$\frac{20}{8} = \frac{10}{4}$

$\frac{20}{8} \times \frac{10}{4}$

$20 \times 4 = 8 \times 10$

$80 = 80$

Determining When Ratios are Proportional (11)

To decide if two ratios are proportional, we can use the cross products test.

Example: Decide if $\frac{5}{6}$ and $\frac{6}{7}$ are proportional.

Solution: Use the cross products test.

$$\frac{5}{6} \times \frac{6}{7}$$

$5 \times 7 \quad 6 \times 6$

$35 \neq 36$

They are not proportional.

Study Exercise Two (12)

1. Determine if $\frac{4}{9}$ and $\frac{8}{16}$ are proportional.
2. Determine if $\frac{3}{10} = \frac{9}{30}$ is a proportion.
3. Determine if $\frac{12}{45}$ and $\frac{4}{15}$ are proportional.
4. True or False: $\frac{3}{8} = \frac{9}{24}$
5. True or False: $\frac{5}{7} = \frac{15}{21}$

Solving Proportions (13)

If three of the four numbers of a proportion are known, finding the fourth number is called *solving the proportion*.

To solve a proportion:

Step (1): Set the cross products equal.
Step (2): Divide by the multiplier of the unknown number.

Example: Solve $\frac{8}{20} = \frac{n}{30}$

Solution:

Step (1): Set the cross products equal.

$$\frac{8}{20} \times \frac{n}{30}$$

$20 \cdot n = 8(30)$

$20 \cdot n = 240$

Step (2): Divide by the multiplier of the unknown number.

$n = 240 \div 20$

$n = 12$

Examples (14)

Example 1: **Solve:** $\frac{4}{n} = \frac{8}{12}$

Solution: $\frac{4}{n} \times \frac{8}{12}$

$n \cdot 8 = 4(12)$

$8 \cdot n = 4(12)$

$8 \cdot n = 48$

$n = 48 \div 8$

$n = 6$

Example 2: **Solve:** $\frac{6}{22} = \frac{n}{88}$

Solution: $\frac{6}{22} \times \frac{n}{88}$

$22 \cdot n = 6(88)$

$22 \cdot n = 528$

$n = 528 \div 22$

$n = 24$

Study Exercise Three (15)

Solve each proportion.

1. $\frac{2}{3} = \frac{n}{9}$

2. $\frac{8}{10} = \frac{n}{5}$

3. $\frac{n}{3} = \frac{16}{12}$

4. $\frac{21}{12} = \frac{28}{n}$

5. $\frac{1}{n} = \frac{4}{12}$

6. $\frac{6}{20} = \frac{9}{n}$

Using Proportions in Applied Problems (16)

Many word problems can be solved using proportions.

Example: If 12 ounces of candy costs 240 cents, what would be the cost of 5 ounces of candy?

Solution: We will compare the ratio of ounces to cents. When writing a proportion, we will set up the ratios in the same way on each side of the equation. We let n be the cost.

$$\begin{matrix} \text{ounces} \longrightarrow \\ \text{cents} \longrightarrow \end{matrix} \frac{12}{240} = \frac{5}{n} \begin{matrix} \longleftarrow \text{ounces} \\ \longleftarrow \text{cents} \end{matrix}$$

$12 \cdot n = 5(240)$

$12 \cdot n = 1200$

$n = 1200 \div 12$

$n = 100$

The cost of 5 ounces of candy is 100 cents.

Additional Examples (17)

Example 1: A car uses 8 gallons of gasoline for a trip of 216 miles. How many miles can the car travel on 18 gallons of gas?

Solution: We will compare gallons to miles. Let n be the number of miles.

$$\begin{array}{r} \text{gallons} \longrightarrow \\ \text{miles} \longrightarrow \end{array} \frac{8}{216} = \frac{18}{n} \begin{array}{l} \longleftarrow \text{gallons} \\ \longleftarrow \text{miles} \end{array}$$

$$8 \cdot n = 216(18)$$
$$8 \cdot n = 3888$$
$$n = 3888 \div 8$$
$$n = 486$$

The car can travel 486 miles.

Example 2: The ratio of kilograms to pounds is 5 to 11. If a person weighs 176 pounds, how many kilograms is this?

Solution: Let n be the number of kilograms.

$$\begin{array}{r} \text{kilograms} \longrightarrow \\ \text{pounds} \longrightarrow \end{array} \frac{5}{11} = \frac{n}{176} \begin{array}{l} \longleftarrow \text{kilograms} \\ \longleftarrow \text{pounds} \end{array}$$

$$11 \cdot n = 5(176)$$
$$11 \cdot n = 880$$
$$n = 880 \div 11$$
$$n = 80$$

176 pounds is 80 kilograms.

Study Exercise Four (18)

1. If 2 ounces of a certain candy contain 160 calories, how many calories are in 12 ounces of the same candy?
2. A car uses 14 gallons of gas for a trip of 308 miles. How many miles could the car travel on 10 gallons of gas?
3. If it takes 3 gallons of paint to cover 1200 square feet, how many gallons would be needed to cover 2000 square feet?
4. The ratio of pounds to kilograms is 11 to 5. How many pounds is 60 kilograms?
5. The ratio of meters to yards is 9 to 10. The distance of a swimming meet is 360 meters. How many yards is this?
6. If an 18-foot pole casts a 15-foot shadow, what length of shadow does a 36-foot pole cast?

REVIEW EXERCISES (19)

1. What is a ratio?
2. Write a fractional numeral for the ratio of 14 pencils to 21 pencils.
3. Find the ratio of 80 miles to 4 hours.
4. Find the ratio in cents of a dime to a quarter.
5. Determine if $\frac{7}{9}$ and $\frac{12}{15}$ are proportional.

REVIEW EXERCISES (Continued)

6. True or False: $\frac{13}{17} = \frac{26}{68}$

7. Solve: $\frac{n}{3} = \frac{21}{9}$

8. Solve: $\frac{21}{18} = \frac{28}{n}$

9. Solve: $\frac{4}{n} = \frac{20}{35}$

10. A distance of 324 miles is traveled in 12 hours. At the same rate, how many miles are traveled in 8 hours?

11. The ratio of kilograms to pounds is 5 to 11. If a person weighs 165 pounds, how many kilograms is this?

Solutions to Review Exercises (20)

1. A ratio is a comparison of two quantities by division.

2. $\frac{14}{21} = \frac{2}{3}$

3. $\frac{80 \text{ miles}}{4 \text{ hours}} = \frac{20 \text{ miles}}{1 \text{ hour}}$ or 20 miles per hour

4. $\frac{10 \text{ cents}}{25 \text{ cents}} = \frac{2}{5}$

5. $\frac{7}{9} \times \frac{12}{15}$

   $(7)(15) \quad (9)(12)$

   $105 \neq 108$

   $\frac{7}{9}$ and $\frac{12}{15}$ are not proportional.

6. $\frac{13}{17} \times \frac{26}{68}$

   $(13)(68) \quad (17)(26)$

   $884 \neq 442$

   False.

7. $\frac{n}{3} = \frac{21}{9}$

   $n \cdot 9 = 3(21)$

   $9 \cdot n = 3(21)$

   $9 \cdot n = 63$

   $n = 63 \div 9$

   $n = 7$

8. $\frac{21}{18} = \frac{28}{n}$

   $21 \cdot n = 18(28)$

   $21 \cdot n = 504$

   $n = 504 \div 21$

   $n = 24$

9. $\frac{4}{n} = \frac{20}{35}$

   $n \cdot 20 = 4(35)$

   $20 \cdot n = 4(35)$

   $20 \cdot n = 140$

   $n = 140 \div 20$

   $n = 7$

10. Let n be the number of miles.

    $\text{miles} \longrightarrow \frac{324}{12} = \frac{n}{8} \longleftarrow \text{miles}$ (hours ⟶ 12, 8 ⟵ hours)

    $12 \cdot n = 324(8)$

    $12 \cdot n = 2592$

    $n = 2592 \div 12$

    $n = 216$

    216 miles

11. Let n be the number of kilograms.

    $\text{kilograms} \longrightarrow \frac{5}{11} = \frac{n}{165} \longleftarrow \text{kilograms}$ (pounds ⟶ 11, 165 ⟵ pounds)

    $11 \cdot n = 5(165)$

    $11 \cdot n = 825$

    $n = 825 \div 11$

    $n = 75$

    75 kilograms

SUPPLEMENTARY PROBLEMS

A. Determine the indicated ratios.

1. Find the ratio of 8 to 2.

2. Find the ratio of 342 to 9.

3. Find the ratio of 13 to 169.

4. Find the ratio in cents of a dime to a half dollar.

5. Find the ratio in minutes of 45 minutes to an hour.

6. Find the ratio in cents of a quarter to a dollar.

7. Find the ratio of 8 months to a year.

8. Express the ratio of the shaded region to the total region:

9. Express the ratio of the total region to the shaded region:

B. **10.** What is the definition of a proportion?

C. Determine if each is a true proportion. Use the cross products test.

11. $\frac{7}{42} = \frac{21}{14}$

12. $\frac{3}{8} = \frac{15}{40}$

13. $\frac{1\frac{1}{2}}{2} = \frac{3}{4}$

14. $\frac{3}{7} = \frac{28}{15}$

D. Solve each of the proportions.

15. $\frac{n}{12} = \frac{7}{12}$

16. $\frac{4}{6} = \frac{n}{3}$

17. $\frac{4}{17} = \frac{n}{34}$

18. $\frac{6}{20} = \frac{9}{n}$

19. $\frac{n}{40} = \frac{10}{16}$

20. $\frac{32}{n} = \frac{16}{3}$

21. $\frac{120}{18} = \frac{n}{36}$

22. $\frac{76}{9} = \frac{n}{36}$

23. $\frac{1\frac{1}{2}}{n} = \frac{3}{8}$

24. $\frac{1}{2\frac{3}{4}} = \frac{100}{n}$

SUPPLEMENTARY PROBLEMS (Continued)

E. Use proportions to solve the applied problems.

25. If a car uses 15 gallons of gas for a trip of 405 miles, how many gallons are needed for a trip of 324 miles?

26. If rain falls $\frac{1}{2}$ inch per hour, how many inches will fall in 10 minutes?

27. If 5 pieces of candy contain 11 grams of carbohydrates, how many grams of carbohydrates do 20 pieces contain?

28. A clock loses $1\frac{1}{2}$ seconds in $3\frac{3}{5}$ hours. How much time will it lose in 1 day (1 day equals 24 hours)?

29. A recipe calls for $7\frac{1}{2}$ cups of flour to make 2 loaves of bread. How much flour is needed to make 5 loaves?

30. The scale on a map reads "1 inch = 40 miles." How many inches are needed to show a distance of 100 miles?

31. Linda can type 11 pages in 2 hours. At this rate, how long will it take her to type an 88-page report?

32. On a map a 3-inch segment represents 51 miles. How many miles does a 5-inch segment represent?

33. The ratio of pounds to kilograms is 11 to 5. If a person weighs 70 kilograms, how many pounds is this?

34. Joe received \$21 interest on a savings account of \$900. If he had saved \$1200, how much interest would he have gotten?

35. At a rate of 2 defective TV sets in each 75 made, how many defective sets will there be in each 600?

36. If a car uses 7 gallons of gas to go 154 miles, how far can it go with 10 gallons of gas?

Solutions to Study Exercises (8A)

Study Exercise One (Frame 8)

1. $\frac{9}{3} = \frac{3}{1}$

2. $\frac{5}{20} = \frac{1}{4}$

3. $\frac{200 \text{ miles}}{10 \text{ gallons}} = \frac{20 \text{ miles}}{1 \text{ gallon}}$ or 20 miles per gallon

4. $\frac{126 \text{ miles}}{7 \text{ gallons}} = \frac{18 \text{ miles}}{1 \text{ gallon}}$ or 18 miles per gallon

5. $\frac{112 \text{ hits}}{336 \text{ times at bat}} = \frac{1 \text{ hit}}{3 \text{ times at bat}}$ or 1 hit per 3 times at bat

6. 1 dime is 10 cents.

1 dollar is 100 cents.

The ratio is $\frac{10}{100} = \frac{1}{10}$

Study Exercise Two (Frame 12) (12A)

1. $\frac{4}{9} \times \frac{8}{16}$

$4 \times 16 \quad 9 \times 8$

$64 \neq 72$

not proportional

2. $\frac{3}{10} \times \frac{9}{30}$

$30 \times 30 \quad 10 \times 9$

$90 = 90$

proportional

3. $\frac{12}{45} \times \frac{4}{15}$

$12 \times 15 \quad 45 \times 4$

$180 = 180$

proportional

4. $\frac{3}{8} \times \frac{9}{24}$

$3 \times 24 \quad 8 \times 9$

$72 = 72$

true

5. $\frac{5}{7} \times \frac{15}{21}$

$5 \times 21 \quad 7 \times 15$

$105 = 105$

true

Study Exercise Three (Frame 15) **(15A)**

1. $\frac{2}{3} = \frac{n}{9}$
   $3 \cdot n = 2(9)$
   $3 \cdot n = 18$
   $n = 18 \div 3$
   $n = 6$

2. $\frac{8}{10} = \frac{n}{5}$
   $10 \cdot n = 8(5)$
   $10 \cdot n = 40$
   $n = 40 \div 10$
   $n = 4$

3. $\frac{n}{3} = \frac{16}{12}$
   $n \cdot 12 = 3(16)$
   $12 \cdot n = 3(16)$
   $12 \cdot n = 48$
   $n = 48 \div 12$
   $n = 4$

4. $\frac{21}{12} = \frac{28}{n}$
   $21 \cdot n = 12(28)$
   $21 \cdot n = 336$
   $n = 336 \div 21$
   $n = 16$

5. $\frac{1}{n} = \frac{4}{12}$
   $n \cdot 4 = 1(12)$
   $4 \cdot n = 1(12)$
   $4 \cdot n = 12$
   $n = 12 \div 4$
   $n = 3$

6. $\frac{6}{20} = \frac{9}{n}$
   $6 \cdot n = 20(9)$
   $6 \cdot n = 180$
   $n = 180 \div 6$
   $n = 30$

Study Exercise Four (Frame 18) **(18A)**

1. $\frac{\text{ounces} \longrightarrow 2}{\text{calories} \longrightarrow 160} = \frac{12 \longleftarrow \text{ounces}}{n \longleftarrow \text{calories}}$
   $2 \cdot n = 160(2)$
   $2 \cdot n = 1920$
   $n = 1920 \div 2$
   $n = 960$
   There are 960 calories.

2. $\frac{\text{gallons} \longrightarrow 14}{\text{miles} \longrightarrow 308} = \frac{10 \longleftarrow \text{gallons}}{n \longleftarrow \text{miles}}$
   $14 \cdot n = 308(10)$
   $14 \cdot n = 3080$
   $n = 3080 \div 14$
   $n = 220$
   220 miles

3. $\frac{\text{gallons} \longrightarrow 3}{\text{square feet} \longrightarrow 1200} = \frac{n \longleftarrow \text{gallons}}{2000 \longleftarrow \text{square feet}}$
   $1200 \cdot n = 3(2000)$
   $1200 \cdot n = 6000$
   $n = 6000 \div 1200$
   $n = 5$
   5 gallons are needed.

4. $\frac{\text{pounds} \longrightarrow 11}{\text{kilograms} \longrightarrow 5} = \frac{n \longleftarrow \text{pounds}}{60 \longleftarrow \text{kilograms}}$
   $5 \cdot n = 11(60)$
   $5 \cdot n = 660$
   $n = 660 \div 5$
   $n = 132$
   60 kilograms equals 132 pounds.

5. $\frac{\text{meters} \longrightarrow 9}{\text{yards} \longrightarrow 10} = \frac{360 \longleftarrow \text{meters}}{n \longleftarrow \text{yards}}$
   $9 \cdot n = 10(360)$
   $9 \cdot n = 3600$
   $n = 3600 \div 9$
   $n = 400$
   The distance is 400 yards.

6. $\frac{\text{length} \longrightarrow 18}{\text{shadow} \longrightarrow 15} = \frac{36 \longleftarrow \text{length}}{n \longleftarrow \text{shadow}}$
   $18 \cdot n = 15(36)$
   $18 \cdot n = 540$
   $n = 540 \div 18$
   $n = 30$
   The shadow length is 30 feet.

Module 2 Practice Test

Units 9–16

1. State the Fundamental Principle of Fractions.

2. Change $\frac{4}{5}$ to an equivalent fraction by multiplying numerator and denominator by 5

3. Reduce to lowest terms: (a) $\frac{54}{60}$ (b) $\frac{35}{16}$

4. Which is largest: $\frac{3}{4}$, $\frac{5}{12}$ or $\frac{2}{3}$?

5. What is the LCD of $\frac{2}{3}$, $\frac{1}{4}$, $\frac{3}{8}$?

6. Add: $\frac{1}{8} + \frac{5}{12} + \frac{1}{6}$

7. Add: $2\frac{5}{9} + 3\frac{2}{3}$

8. Subtract: $\frac{5}{6} - \frac{3}{8}$

9. Subtract: $3\frac{3}{4} - 1\frac{7}{6}$

10. Multiply: $\frac{3}{5} \times \frac{20}{36}$

11. Multiply: $2\frac{2}{5} \times 1\frac{1}{10}$

12. Divide: $\frac{3}{8} \div \frac{5}{6}$

13. Divide: $6\frac{2}{3} \div 1\frac{3}{4}$

14. Simplify: $\dfrac{\frac{2}{9}}{\frac{4}{5}}$

15. What is the reciprocal of $1\frac{3}{7}$?

16. Find: $(2\frac{1}{3})^2$

17. Find: $\sqrt{12\frac{1}{4}}$

18. What fractional part of $2\frac{3}{4}$ is $1\frac{1}{3}$?

19. Find the number if $1\frac{5}{16}$ of the number is 7

20. From the sum of $4\frac{3}{4}$ and $1\frac{1}{6}$ subtract the sum of $2\frac{1}{3}$ and $1\frac{11}{12}$

21. Solve: $1\frac{2}{3} \cdot n = 6$

22. Solve: $\frac{21}{7} = \frac{24}{n}$

23. The ratio of yards to meters is 10 to 9. Find the number of yards in 900 meters.

24. Cathy typed 255 words in $4\frac{1}{4}$ minutes. How many words did she average per minute?

25. To change Fahrenheit temperature to Celsius temperature, first subtract 32 degrees from the Fahrenheit reading. Then take $\frac{5}{9}$ of the answer.

(a) Change 68°F to Celsius.

(b) Change 212°F to Celsius.

Answers to Module 2 Practice Test

1. The numerator and denominator of a fraction may be multiplied or divided by the same nonzero number without altering the value of the fraction.

2. $\frac{4}{5} = \frac{4 \times 5}{5 \times 5} = \frac{20}{25}$

3. (a) $\frac{9}{10}$ (b) $2\frac{3}{16}$

4. $\frac{3}{4}$

5. LCD is 24

6. $\frac{17}{24}$

7. $6\frac{2}{9}$

8. $\frac{11}{24}$

9. $1\frac{7}{12}$

10. $\frac{1}{3}$

11. $2\frac{16}{25}$

12. $\frac{9}{20}$

13. $3\frac{17}{21}$

14. $\frac{5}{18}$

15. $\frac{7}{10}$

16. $5\frac{4}{9}$

17. $3\frac{1}{2}$

18. $\frac{16}{33}$

19. $5\frac{1}{3}$

20. $1\frac{2}{3}$

21. $n = 3\frac{3}{5}$

22. $n = 8$

23. 1000 yards

24. 60 words per minute

25. (a) 20° Celsius.
(b) 100° Celsius.

MODULE

3

Operating with Decimals

UNIT

17 Decimal Numerals and Rounding

OBJECTIVES (1)

By the end of this unit you should be able to:

1. UNDERSTAND DECIMAL PLACE VALUE.
2. WRITE DECIMAL NUMERALS IN PLACE VALUE NOTATION.
3. READ DECIMAL NUMERALS.
4. ROUND DECIMAL NUMERALS TO A SPECIFIED PLACE VALUE OR A SPECIFIED NUMBER OF DECIMAL PLACES.

Review of the Place Value System (2)

In Unit 1 we discussed the place value of whole numbers. Study the place values below for the number 4,367,285

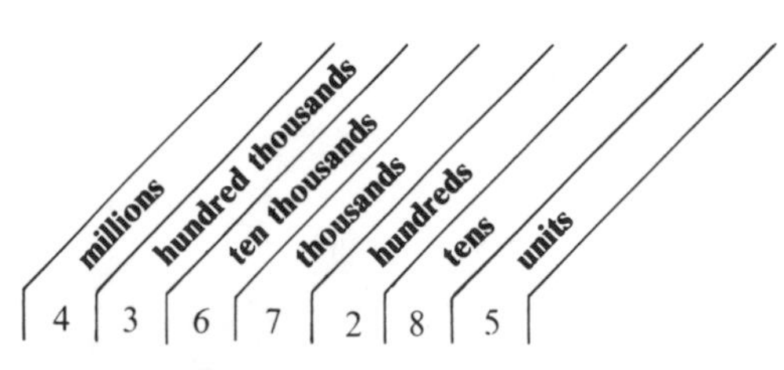

The place value of the digit 3 is hundred thousands.
The place value of the digit 2 is hundreds.
What is the place value of the digit 7?

---

(3)

thousands	hundreds	tens	units

Each place value is ten times the place value on its right.

1 thousand = 10 hundreds
1 hundred = 10 tens
1 ten = 10 units
1 unit = 10(?)

Decimal Point (4)

thousands	hundreds	tens	units	tenths

1 unit = 10 tenths

We will place a period between the units digit and the tenths digit. This period is called a *decimal point* and is read "and."

For example, 142.5 is read "one hundred forty-two and five tenths."

(5)

If we keep moving to the right, each place value is $\frac{1}{10}$ of the place value before it.

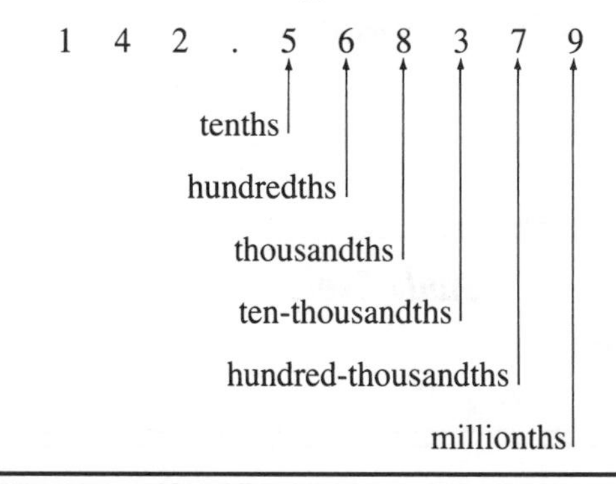

(6)

tenths	hundredths	thousandths	ten-thousandths	hundred-thousandths	millionths

Examples:

1. What is the place value of the digit 3 in 215.63804?
   Solution: hundredths.
2. What is the place value of the digit 8 in 215.63804?
   Solution: thousandths.
3. What is the place value of the digit 6 in 215.63804?
   Solution: tenths.
4. What is the place value of the digit 4 in 215.63804?
   Solution: hundred-thousandths.

Study Exercise One (7)

1. What is the place value of the digit 6 in 21.7065?
2. What is the place value of the digit 2 in 6.8752?
3. What is the place value of the digit 4 in 17.4?
4. What is the place value of each digit in 17.4?
5. What is the place value of the digit 8 in 23.041698?

Meaning of a Decimal (8)

Consider the decimal 8.365

The place value of 8 is units
3 is tenths
6 is hundredths
5 is thousandths

8.365 means 8 plus 3 tenths plus 6 hundredths plus 5 thousandths

(9)

Example 1: What is the meaning of .471?

Solution: .471 means 4 tenths plus 7 hundredths plus 1 thousandth.

Example 2: What is the meaning of 2.05?

Solution: 2.05 means 2 plus 0 tenths plus 5 hundredths.

Example 3: Write the decimal numeral that means 15 plus 2 tenths plus 0 hundredths plus 3 thousandths plus 7 ten-thousandths.

Solution: 15.2037

Study Exercise Two (10)

Give the meaning of the following decimals:

1. 2.32 **2.** .1005 **3.** 1.23134

Reading Decimals (11)

To read a decimal, read the numeral to the right of the decimal point as you would read a numeral for a whole number and use the name that applies to the place value of the last digit.

Example 1: Read .47

Solution: forty-seven hundredths.

Example 2: Read .731

Solution: seven hundred thirty-one thousandths.

Example 3: Read 5.047

Solution: five and forty-seven thousandths.

Decimal Point (12)

The decimal point plays an important role. For example, 1.33 is *not* equal to 13.3 since

1.33 means 1 plus 3 tenths plus 3 hundredths

and

13.3 means 13 plus 3 tenths

Omission of the Decimal Point (13)

When there are no digits after a decimal point, the decimal point is usually omitted.

Example 1: 35. = 35 **Example 2:** 2. = 2

Study Exercise Three (14)

1. Read .17 **2.** Read .049 **3.** Read 3.0724

4. Where is the decimal point in the number 64?

Writing Decimals (15)

Example 1: Write seventy-three thousandths as a decimal:

Solution: .073

Example 2: Write seven hundred ninety-three ten-thousandths:

Solution: .0793

Example 3: Write seven hundred and ninety-three ten-thousandths:

Solution: 700.0093

Study Exercise Four (16)

Write each of the following as a decimal numeral:

1. Two and fifteen thousandths.
2. Four and sixty-two hundredths.
3. Two hundred ten-thousandths.
4. Three hundred thirty-six thousandths.
5. Three hundred and thirty-six thousandths.
6. Fifty-one hundred-thousandths.

Converting Decimals to Fractions (17)

Decimals can be converted to fractions by using place value notation.

Example: Convert .43 to a fraction:

Solution: .43 means 4 tenths plus 3 hundredths

Line (a): $.43 = \frac{4}{10} + \frac{3}{100}$

Line (b): $= \frac{4 \times 10}{10 \times 10} + \frac{3}{100}$

Line (c): $= \frac{40}{100} + \frac{3}{100}$

Line (d): $= \frac{43}{100}$

Notice that the place value of the last digit determines the denominator.

$$.43 = \frac{43}{100}$$

place value of last digit is hundredths

Shortcut for Converting Decimals to Fractions (18)

Shortcut:

Step (1): Write a fraction and put in the numerator the digits that follow the decimal point.
Step (2): In the denominator write the place value of the last digit.

Example 1: $.3177 = \frac{3177}{10{,}000}$

Example 2: $.213 = \frac{213}{1000}$

Example 3: $.77 = \frac{77}{100}$

Example 4: $.039 = \frac{39}{1000}$

Example 5: $2.1703 = 2\frac{1703}{10{,}000}$

Sometimes when using the shortcut, fractions must be reduced to lowest terms. **(19)**

Example 1: Write .204 as a fraction in lowest terms:

Solution: $.204 = \frac{204}{1000}$

$= \frac{51}{250}$

Example 2: Write 10.002 as a mixed numeral in lowest terms:

Solution: $10.002 = 10\frac{2}{1000}$

$= 10\frac{1}{500}$

Study Exercise Five (20)

Convert the following decimals to fractions in lowest terms:

1. .893 **2.** .01121 **3.** 8.04 **4.** 2.2 **5.** .05 **6.** .75

Placing Zeros in a Decimal (21)

Rule: Zeros may be placed on either end of a decimal numeral without changing the number, if the decimal point remains in its original position.

Example 1: 12.2 = 12.20

Example 2: .302 = 0.30200

Now study the following examples: **(22)**

1. 123.1 equals 123.10
2. 203.04 equals 203.0400
3. 17.182 does not equal 17.0182
4. .091 equals 0.091
5. 91 equals 0091
6. 20.027 does not equal 20.27

Study Exercise Six (23)

True or false:

1. 2.007 = 2.700
2. 5.0 = 5
3. 1.700 = 1.7
4. 0.73 = .73
5. .1010 = .11
6. 3.018 = 3.18

Need for Rounding Decimals (24)

Some numbers when expressed as decimals have many decimal places. In order to make practical use of such numbers, we will cut them off and use only a limited number of decimal places.

To round a decimal to the nearest tenth, we will keep only the digits up to and including the tenths' place. But we want to keep the nearest tenth.

Rules for Rounding Decimals (25)

Step (1): Draw a box around the last digit to be kept.

Step (2): Look at the next digit after the box you drew.

(*a*) If that next digit is 0, 1, 2, 3, or 4, drop all the digits after the box.

(*b*) If that next digit is 5, 6, 7, 8, or 9, add one to the digit in the box and then drop all digits after the box.

Rules for Rounding Decimals (Continued)

Example: Round 12.8368 to the nearest hundredth:

Solution:

Step (1): Last digit to be kept is 3

12.8[3]68

Step (2): Next digit after 3 is 6
Therefore, add 1 to 3 and drop all digits after the box.

12.8[4]

Thus, 12.8368 rounds to 12.84 to the nearest hundredth.

More Examples (26)

Example 1: Round 4.2348 to the nearest tenth:

Solution:

Step (1): Last digit to be kept is 2

4.[2]348

Step (2): Next digit after 2 is 3. Drop all digits after 2

4.[2]

Thus, 4.2348 rounds to 4.2 to the nearest tenth.

Example 2: Round 31.4658 to the nearest thousandth:

Solution:

Step (1): Last digit to be kept is 5

31.46[5]8

Step (2): Next digit after 5 is 8
Therefore, add 1 to 5 and drop all digits after the box.

31.46[6]

Thus, 31.4658 rounds to 31.466 to the nearest thousandth.

Rounding to a Specified Number of Decimal Places (27)

Sometimes in rounding, the number of decimals is specified.

Example: Round 32.4178 to two decimal places:

Solution:

Step (1): Last digit to be kept is 1

32.4[1]78

Step (2): Next digit after 1 is 7
Therefore, add 1 to 1 and drop all digits after the box.

32.4[2]

Thus, 32.4178 rounds to 32.42 to two decimal places.

Study Exercise Seven (28)

1. Round .63148 to:
   (a) the nearest hundredth.
   (b) 4 decimal places.
2. Round .007891 to:
   (a) 3 decimal places.
   (b) the nearest thousandth.
3. Round .354506 to:
   (a) the nearest hundredth.
   (b) 3 decimal places.
4. Round 14.76 to:
   (a) the nearest tenth.
   (b) 1 decimal place.
5. Round 6.39918 to the nearest ten-thousandth.
6. Round 6.89 to one decimal place.
7. Round 19.507 to the nearest hundredth.

REVIEW EXERCISES (29)

A. Multiple choice—Select the letter of the correct answer:

1. What is the place value of the digit 8 in 61.283?
a. Tenths. **b.** Hundredths. **c.** Hundreds. **d.** Thousands.

2. What is the place value of the digit 4 in 2.004?
a. Tenths. **b.** Hundredths. **c.** Thousandths. **d.** Ten-thousandths.

3. Twenty-three and 23 hundredths is written:
a. 2323 **b.** 23.023 **c.** 23.23 **d.** 232.3

4. The number "107 ten-thousandths" is written:
a. .107 **b.** 107.0010 **c.** .0107 **d.** 100.007

5. Write .404 as a fraction in lowest terms:
a. $\frac{11}{25}$ **b.** $\frac{101}{250}$ **c.** $\frac{11}{250}$ **d.** $\frac{101}{125}$

6. .0081 as a fraction is:
a. $\frac{81}{10}$ **b.** $\frac{81}{100}$ **c.** $\frac{81}{1000}$ **d.** $\frac{81}{10,000}$

7. Which is correct?
a. $.09 = \frac{9}{100}$ **b.** $.009 = \frac{9}{100}$ **c.** $.0009 = \frac{9}{1000}$ **d.** $.00009 = \frac{9}{1000}$

B. True or false:

8. $.071 = .71$ **9.** $\frac{19}{100} = .19$ **10.** $.31 = .3100$

11. $.077 = 77$ **12.** $2.0 = 2$ **13.** $.20 = \frac{1}{5}$

C. **14.** What is the meaning of .234?
15. Round .5826 to the nearest hundredth.
16. Round .0784 to two decimal places.
17. Round 38.25 to the nearest tenth.
18. Round 1.866 to two decimal places.

Solutions to Review Exercises (30)

A. **1.** b **2.** c **3.** c **4.** c **5.** b **6.** d **7.** a

B. **8.** False. **9.** True. **10.** True. **11.** False. **12.** True. **13.** True.

C. **14.** .234 means 2 tenths plus 3 hundredths plus 4 thousandths.
15. .58 **16.** .08 **17.** 38.3 **18.** 1.87

SUPPLEMENTARY PROBLEMS

A. Give the meaning of each of the following:

1. .37 **2.** .034 **3.** 2.0007 **4.** .103

B. Write out the following decimals in words (that is, read each decimal):

5. .5 **6.** 9.25 **7.** 0.609 **8.** 3.217
9. 0.0013 **10.** 100.01 **11.** 708.080 **12.** 20.02
13. 0.00072

C. Find the place value of the underlined digit:

14. $17.6\underline{2}3$ **15.** $.117\underline{2}$ **16.** $.00\underline{1}17$ **17.** $107.062\underline{3}$
18. $0.00\underline{3}4$ **19.** $1.\underline{7}000$ **20.** $\underline{2}1.452$ **21.** $0.10170\underline{2}0$
22. $4.0\underline{9}9$

D. Write the decimal form of the following numbers:

23. forty-eight hundredths.
24. four thousandths.
25. five and ninety-three thousandths.
26. three and three tenths.
27. eighty-three ten-thousandths.
28. three thousand three ten-thousandths.
29. four hundred six hundred-thousandths.
30. five and seventy-two millionths.

E. Write the following as fractions reduced to lowest terms:

31. 0.25 **32.** .125 **33.** .050

F. **34.** Round 34.785:
(a) To nearest tenth.
(b) To two decimal places.

35. Round 6.39948:
(a) To nearest tenth.
(b) To one decimal place.

36. Round .0285:
(a) To two decimal places.
(b) To nearest thousandth.

37. Round .070707:
(a) To one decimal place.
(b) To nearest tenth.

38. Round 6.4079:
(a) To nearest hundredth.
(b) To one decimal place.

39. Round 1.2052:
(a) To three decimal places.
(b) To two decimal places.

40. Round 7.046:
(a) To one decimal place.
(b) To nearest hundredth.

41. Round 23.1646:
(a) To two decimal places.
(b) To nearest thousandth.

Solutions to Study Exercises (7A)

Study Exercise One (Frame 7)

1. Thousandths.
2. Ten-thousandths.
3. Tenths.
4. Place value of the digit 1 is tens.
Place value of the digit 7 is units.
Place value of the digit 4 is tenths.
5. Millionths.

Study Exercise Two (Frame 10) (10A)

1. 2 plus 3 tenths plus 2 hundredths.
2. 1 tenth plus 0 hundredths plus 0 thousandths plus 5 ten-thousandths.
3. 1 plus 2 tenths plus 3 hundredths plus 1 thousandth plus 3 ten-thousandths plus 4 hundred-thousandths.

Study Exercise Three (Frame 14) (14A)

1. Seventeen hundredths.
2. Forty-nine thousandths.
3. Three and seven hundred twenty-four ten-thousandths.
4. 64. (after the 4)

Study Exercise Four (Frame 16) (16A)

1. 2.015
2. 4.62
3. .0200
4. .336
5. 300.036
6. .00051

Study Exercise Five (Frame 20) (20A)

1. $\frac{893}{1000}$
2. $\frac{1,121}{100,000}$
3. $8\frac{4}{100} = 8\frac{1}{25}$
4. $2\frac{2}{10} = 2\frac{1}{5}$
5. $\frac{5}{100} = \frac{1}{20}$
6. $\frac{75}{100} = \frac{3}{4}$

Study Exercise Six (Frame 23) (23A)

1. False.
2. True.
3. True.
4. True.
5. False.
6. False.

Study Exercise Seven (Frame 28) (28A)

1. (a) .63
   (b) .6315
2. (a) .008
   (b) .008
3. (a) .35
   (b) .355
4. (a) 14.8
   (b) 14.8
5. 6.3992
6. 6.9
7. 19.51

UNIT

18 Changing Decimals to Fractions and Comparing Decimals

OBJECTIVES (1)

By the end of this unit you should be able to:

1. CHANGE DECIMALS TO FRACTIONS.
2. CHANGE DECIMALS ENDING IN A FRACTION TO COMMON FRACTIONS.
3. COMPARE DECIMALS.

Review of Changing Decimals to Fractions (2)

Recall the rule for changing decimals to fractions:

Write a fraction whose denominator is the place value of the last digit and whose numerator is composed of all the digits after the decimal point.

Example 1: $.357 = \frac{357}{1000}$

Example 2: $1.28 = 1\frac{28}{100} = 1\frac{7}{25}$

Example 3: $.07713 = \frac{7713}{100,000}$

Examples (3)

Change to fractions in lowest terms:

Example 1: .204

Solution: $.204 = \frac{204}{1000} = \frac{51}{250}$

Example 2: 3.14

Solution: $3.14 = 3\frac{14}{100} = 3\frac{7}{50}$

Example 3: .00701

Solution: $.00701 = \frac{701}{100,000}$

Study Exercise One (4)

Change to fractions or mixed numerals; then reduce to lowest terms.

1. .4 **2.** .0005 **3.** 2.12 **4.** 1.202

Decimals Ending with a Fraction (5)

Sometimes a decimal ends with a fraction.

Some examples include:

$$.12\tfrac{1}{4}, \quad .3\tfrac{1}{2}, \quad .007\tfrac{1}{8}$$

In converting this type of decimal to a fraction, the fraction at the end is not used to determine place value.

$$.12\tfrac{1}{4} = \frac{12\frac{1}{4}}{100}$$

$$.3\tfrac{1}{2} = \frac{3\frac{1}{2}}{10}$$

$$.007\tfrac{1}{8} = \frac{7\frac{1}{8}}{1000}$$

The Fraction $\frac{1}{2}$ At the End of a Decimal (6)

We will try to find a rule for replacing the fraction $\frac{1}{2}$ at the end of a decimal.

Example: Change $.13\frac{1}{2}$ to a decimal form without a fraction at the end:

Solution:

Line (a): $.13\frac{1}{2} = \dfrac{13\frac{1}{2}}{100}$

Line (b): $= \dfrac{13\frac{1}{2} \times 10}{100 \times 10}$

Line (c): $= \dfrac{\frac{27}{\cancel{2}_1} \times \frac{\cancel{10}^5}{1}}{1000}$

Line (d): $= \dfrac{135}{1000}$

Line (e): $= .135$

Thus, the $\frac{1}{2}$ in $.13\frac{1}{2}$ is replaced with a 5

Rules: (7)

If a decimal ends in $\frac{1}{2}$, replace the $\frac{1}{2}$ with 5

If a decimal ends in $\frac{1}{4}$, replace the $\frac{1}{4}$ with 25

If a decimal ends in $\frac{3}{4}$, replace the $\frac{3}{4}$ with 75

Example 1: $2.4\frac{1}{2} = 2.45$ **Example 2:** $.07\frac{1}{4} = .0725$ **Example 3:** $.347\frac{3}{4} = .34775$ **(8)**

Study Exercise Two (9)

Change to a decimal without a fraction at the end:

1. $.5\frac{1}{2}$ **2.** $1.07\frac{1}{4}$ **3.** $.403\frac{1}{2}$ **4.** $.62\frac{3}{4}$ **5.** $14.1\frac{1}{4}$ **6.** $.017\frac{3}{4}$

Changing a Decimal Ending in a Fraction to a Fraction (10)

If a decimal ends with the fraction $\frac{1}{3}$, it cannot be changed to a decimal without a fraction at the end. However, it can be changed to a common fraction.

In simplifying a decimal ending with a fraction to a common fraction, proceed as follows:

Example: Change $.12\frac{1}{3}$ to a common fraction:

Solution:

Line (a): $.12\frac{1}{3} = \frac{12\frac{1}{3}}{100}$

Line (b): $= \frac{\frac{37}{3}}{\frac{100}{1}}$

Line (c): $= \frac{37}{3} \div \frac{100}{1}$

Line (d): $= \frac{37}{3} \times \frac{1}{100}$

Line (e): $= \frac{37}{300}$

Thus, $.12\frac{1}{3} = \frac{37}{300}$

Example: Change $.13\frac{1}{3}$ to a fraction in lowest terms: **(11)**

Solution:

Line (a): $.13\frac{1}{3} = \frac{13\frac{1}{3}}{100}$

Line (b): $= \frac{\frac{40}{3}}{\frac{100}{1}}$

Line (c): $= \frac{40}{3} \div \frac{100}{1}$

Line (d): $= \frac{40}{3} \times \frac{1}{100}$

Line (e): $= \frac{\cancel{40}^{2}}{3} \times \frac{1}{\cancel{100}_{5}}$

Line (f): $= \dfrac{2}{15}$

Thus, $.13\frac{1}{3} = \dfrac{2}{15}$

Study Exercise Three **(12)**

Change to a fraction in lowest terms:

1. $.6\frac{2}{3}$ **2.** $.33\frac{1}{3}$ **3.** $.16\frac{2}{3}$ **4.** $.002\frac{1}{4}$ **5.** $.01\frac{1}{10}$

Example: Change $1.3\frac{1}{5}$ to a mixed numeral: **(13)**

Solution: Since $1.3\frac{1}{5} = 1 + .3\frac{1}{5}$, let us find the fractional equivalent of $.3\frac{1}{5}$ and then add it to 1.

Line (a): $.3\frac{1}{5} = \dfrac{3\frac{1}{5}}{10}$

Line (b): $= \dfrac{\frac{16}{5}}{\frac{10}{1}}$

Line (c): $= \dfrac{16}{5} \div \dfrac{10}{1}$

Line (d): $= \dfrac{\cancel{16}^{8}}{5} \times \dfrac{1}{\cancel{10}_{5}}$

Line (e): $= \dfrac{8}{25}$

Line (f): $1.3\frac{1}{5} = 1 + .3\frac{1}{5}$

Line (g): $= 1 + \frac{8}{25}$

Line (h): $= 1\frac{8}{25}$

Study Exercise Four **(14)**

Change to a mixed numeral:

1. $2.2\frac{1}{3}$ **2.** $1.04\frac{1}{5}$ **3.** $1.1\frac{1}{4}$

Comparing Decimals **(15)**

Which do you think is larger, .03 or .2?

To compare decimals:

> *Step (1):* Change all decimals to the same number of decimal places by putting zeros at the end.
> *Step (2):* Disregard decimal points and compare the numbers as if they were whole numbers.

Example: Which is larger, .03 or .2?

Solution:

Step (1): Change all decimals to the same number of decimal places by putting zeros at the end.

.03 = .03
.2 = .20

Comparing Decimals (Continued)

Step (2): Disregard decimal points and compare the numbers as if they were whole numbers.
20 is larger than 3

Thus, .2 is larger than .03

Additional Examples (16)

Example 1: Which is larger, .154 or .12?

Solution:

Step (1): .154 = .154
.12 = .120

Step (2): 154 is larger than 120

Thus, .154 is larger than .12

Example 2: Which is largest, .14, 1.02, or .6?

Solution:

Step (1): .14 = .14
1.02 = 1.02
.6 = .60

Step (2): 102 is larger than either 14 or 60

Thus, 1.02 is the largest.

Study Exercise Five (17)

Which is larger:

1. .47 or .278
2. .04 or .004
3. 2.9 or .89
4. .2 or .21
5. .0051 or .006
6. .4, .218, or .19

REVIEW EXERCISES (18)

A. Multiple choice—Select the letter of the correct response:

1. Change .002 to a fraction in lowest terms:
a. $\frac{1}{5}$ **b.** $\frac{1}{50}$ **c.** $\frac{1}{500}$ **d.** $\frac{1}{25}$

2. Which is correct?
a. $3.11 = 3\frac{11}{1000}$ **b.** $5.017 = 5\frac{17}{100}$ **c.** $2.013 = 2\frac{13}{1000}$ **d.** $.023 = \frac{23}{100}$

3. Find the decimal equivalent of $.42\frac{1}{2}$:
a. .425 **b.** 42.5 **c.** .005 **d.** .45

4. Which of the following is false?
a. $.2\frac{1}{2} = .25$ **b.** $.01\frac{1}{4} = .125$ **c.** $.7\frac{3}{4} = .775$ **d.** $1.01\frac{1}{2} = 1.015$

5. When $.08\frac{1}{3}$ is converted to a fraction, the result is:
a. $\frac{1}{7}$ **b.** $\frac{1}{8}$ **c.** $\frac{1}{12}$ **d.** $\frac{1}{9}$

6. When 7.12 is converted to a mixed numeral, the result is:
a. $7\frac{6}{10}$ **b.** $7\frac{3}{10}$ **c.** $7\frac{3}{25}$ **d.** $7\frac{3}{250}$

7. When $2.1\frac{2}{3}$ is converted to a mixed numeral, the result is:
a. $2\frac{1}{5}$ **b.** $2\frac{1}{6}$ **c.** $2\frac{1}{7}$ **d.** $2\frac{1}{8}$

B. **8.** Which is larger, .2 or .17?

9. Which is larger, .007 or .0012?

Solutions to Review Exercises **(19)**

A. **1.** c **2.** c **3.** a **4.** b **5.** c **6.** c **7.** b

B. **8.** *Step (1):* .2 = .20
.17 = .17
Step (2): 20 is larger than 17
Thus, .2 is larger than .17

9. *Step (1):* .007 = .0070
.0012 = .0012
Step (2): 70 is larger than 12
Thus, .007 is larger than .0012

SUPPLEMENTARY PROBLEMS

A. Write the following as fractions reduced to lowest terms:

1. 0.05 **2.** 0.500 **3.** .008
4. 0.600 **5.** .075 **6.** .2000
7. .45 **8.** .0002 **9.** .0012

B. Convert to mixed numerals:

10. 2.040 **11.** 12.001 **12.** 2.002
13. 004.02 **14.** 10.01 **15.** 1.001

C. Change to decimals without fractions at the end:

16. $.17\frac{1}{4}$ **17.** $1.08\frac{3}{4}$ **18.** $.017\frac{1}{2}$
19. $.0101\frac{1}{4}$ **20.** $.645\frac{3}{4}$ **21.** $1.00\frac{1}{2}$

D. Convert to fractions in lowest terms:

22. $.4\frac{2}{3}$ **23.** $.04\frac{2}{5}$ **24.** $2.3\frac{1}{3}$
25. $2.33\frac{1}{3}$ **26.** $1.16\frac{2}{3}$ **27.** $3.02\frac{1}{4}$

E. Determine which decimal is larger:

28. .7 or .088 **29.** .11 or .344 **30.** .007 or .4
31. 2.1 or 2.09 **32.** .154, .12, or .081 **33.** .088, .602, or .61

Solutions to Study Exercises **(4A)**

Study Exercise One (Frame 4)

1. $.4 = \frac{4}{10} = \frac{2}{5}$

2. $.0005 = \frac{5}{10,000} = \frac{1}{2000}$

3. $2.12 = 2\frac{12}{100} = 2\frac{3}{25}$

4. $1.202 = 1\frac{202}{1000} = 1\frac{101}{500}$

Study Exercise Two (Frame 9) **(9A)**

1. .55 **2.** 1.0725 **3.** .4035
4. .6275 **5.** 14.125 **6.** .01775

Study Exercise Three (Frame 12) **(12A)**

1. $.6\frac{2}{3} = \frac{6\frac{2}{3}}{10} = \frac{\frac{20}{3}}{\frac{10}{1}} = \frac{20}{3} \div \frac{10}{1} = \frac{\cancel{20}^{2}}{3} \times \frac{1}{\cancel{10}_{1}} = \frac{2}{3}$

2. $.33\frac{1}{3} = \frac{33\frac{1}{3}}{100} = \frac{\frac{100}{3}}{\frac{100}{1}} = \frac{100}{3} \div \frac{100}{1} = \frac{\cancel{100}^{1}}{3} \times \frac{1}{\cancel{100}_{1}} = \frac{1}{3}$

3. $.16\frac{2}{3} = \frac{16\frac{2}{3}}{100} = \frac{\frac{50}{3}}{\frac{100}{1}} = \frac{50}{3} \div \frac{100}{1} = \frac{\cancel{50}^{1}}{3} \times \frac{1}{\cancel{100}_{2}} = \frac{1}{6}$

4. $.002\frac{1}{4} = \frac{2\frac{1}{4}}{1000} = \frac{\frac{9}{4}}{\frac{1000}{1}} = \frac{9}{4} \div \frac{1000}{1} = \frac{9}{4} \times \frac{1}{1000} = \frac{9}{4000}$

5. $.01\frac{1}{10} = \frac{1\frac{1}{10}}{100} = \frac{\frac{11}{10}}{\frac{100}{1}} = \frac{11}{10} \div \frac{100}{1} = \frac{11}{10} \times \frac{1}{100} = \frac{11}{1000}$

Study Exercise Four (Frame 14) **(14A)**

1. $2.2\frac{1}{3} = 2 + .2\frac{1}{3}$

   $.2\frac{1}{3} = \frac{2\frac{1}{3}}{10} = \frac{\frac{7}{3}}{\frac{10}{1}} = \frac{7}{3} \div \frac{10}{1} = \frac{7}{3} \times \frac{1}{10} = \frac{7}{30}$

   $2.2\frac{1}{3} = 2 + \frac{7}{30} = 2\frac{7}{30}$

2. $1.04\frac{1}{5} = 1 + .04\frac{1}{5}$

   $.04\frac{1}{5} = \frac{4\frac{1}{5}}{100} = \frac{\frac{21}{5}}{\frac{100}{1}} = \frac{21}{5} \div \frac{100}{1} = \frac{21}{5} \times \frac{1}{100} = \frac{21}{500}$

   $1.04\frac{1}{5} = 1 + \frac{21}{500} = 1\frac{21}{500}$

3. $1.1\frac{1}{4} = 1 + .1\frac{1}{4}$

   $.1\frac{1}{4} = \frac{1\frac{1}{4}}{10} = \frac{\frac{5}{4}}{\frac{10}{1}} = \frac{5}{4} \div \frac{10}{1} = \frac{\cancel{5}^{1}}{4} \times \frac{1}{\cancel{10}_{2}} = \frac{1}{8}$

   $1.1\frac{1}{4} = 1 + \frac{1}{8} = 1\frac{1}{8}$

Study Exercise Five (Frame 17) (17A)

1. *Step (1):* .47 = .470
   .278 = .278
   Step (2): 470 is larger than 278
   Thus, .47 is larger than .278

2. *Step (1):* .04 = .040
   .004 = .004
   Step (2): 40 is larger than 4
   Thus, .04 is larger than .004

3. *Step (1):* 2.9 = 2.90
   .89 = .89
   Step (2): 290 is larger than 89
   Thus, 2.9 is larger than .89

4. *Step (1):* .2 = .20
   .21 = .21
   Step (2): 21 is larger than 20
   Thus, .21 is larger than .2

5. *Step (1):* .0051 = .0051
   .006 = .0060
   Step (2): 60 is larger than 51
   Thus, .006 is larger than .0051

6. *Step (1):* .4 = .400
   .218 = .218
   .19 = .190
   Step (2): 400 is larger than either 218 or 190
   Thus, .4 is the largest

UNIT

19 Addition and Subtraction of Decimals

OBJECTIVES (1)

By the end of this unit you should be able to:

1. ADD DECIMALS.
2. SUBTRACT DECIMALS.
3. SOLVE APPLIED PROBLEMS USING DECIMALS.

Decimals as Fractions (2)

Each digit of a decimal represents a fraction as indicated by its place value.

Line (a): $.23 = \frac{2}{10} + \frac{3}{100}$

Line (b): $= \frac{20}{100} + \frac{3}{100}$

Line (c): $= \frac{23}{100}$

Addition of Decimals (3)

We can add decimals by adding the fractions they represent.

Example: Let us add .13 + .21:

Solution:

Line (a): $.13 = \frac{1}{10} + \frac{3}{100}$

Line (b): $.21 = \frac{2}{10} + \frac{1}{100}$

Line (c): Add the tenths: $\frac{1}{10} + \frac{2}{10} = \frac{3}{10}$

Addition of Decimals (Continued)

Line (d): Add the hundredths: $\frac{3}{100} + \frac{1}{100} = \frac{4}{100}$

Line (e): $.13 + .21 = \frac{3}{10} + \frac{4}{100}$

$= .34$

Another Example (4)

Add .33 + .45:

Solution:

$$.33 = \frac{3}{10} + \frac{3}{100}$$

$$.45 = \frac{4}{10} + \frac{5}{100}$$

$$.33 + .45 = \frac{7}{10} + \frac{8}{100}$$

$$= .78$$

Study Exercise One (5)

Add the following decimals by using fractions:

1. .22 + .46

2. .03 + .21 + .24

3. .44 + .2 + .11

4. 1.2 + 2.3

A Shorter Method of Adding Decimals (6)

Let us add .23 + .14:

Line (a): $.23 + .14 = \left(\frac{2}{10} + \frac{3}{100}\right) + \left(\frac{1}{10} + \frac{4}{100}\right)$

Line (b): $= \frac{2}{10} + \frac{1}{10} + \frac{3}{100} + \frac{4}{100}$

Line (c): $= \frac{3}{10} + \frac{7}{100}$

Line (d): $= .37$ (3 tenths + 7 hundredths)

We will arrange the numbers in vertical columns so the tenths are in the first column and the hundredths are in the second column. This is done by lining up the decimal points one under the other.

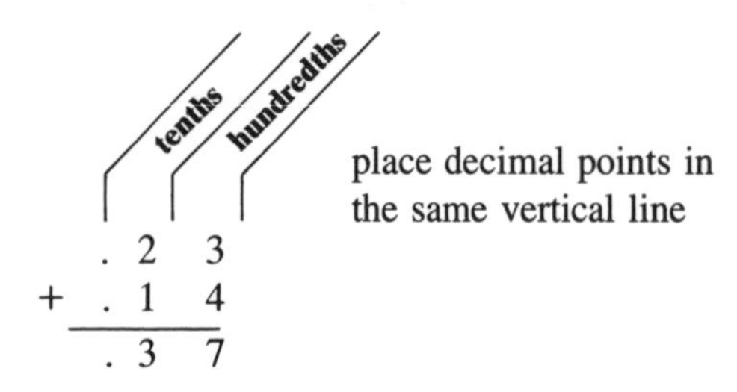

Examples (7)

1. Add .312 + .145:

Solution:

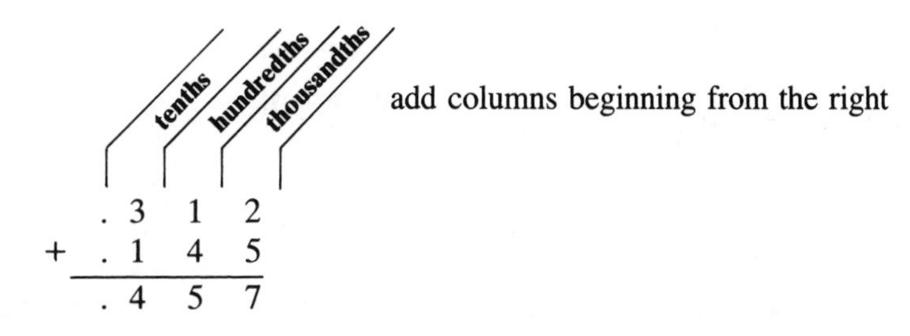

2. Add 1.02 + .143:

Solution:

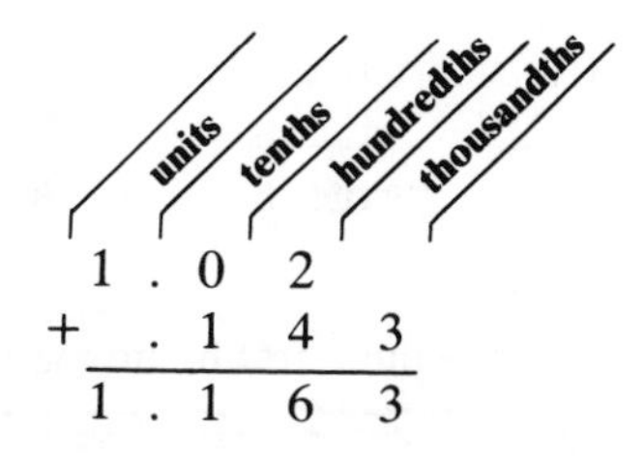

Carrying (8)

Remember that 10 hundredths is 1 tenth since $\frac{10}{100} = \frac{1}{10}$ and that 10 thousandths is 1 hundredth since $\frac{10}{1000} = \frac{1}{100}$

Example: Add .38 + .24:

Solution:

$$\begin{array}{r} .38 \\ +.24 \\ \hline \end{array}$$

When we add the hundredths column, we get 12 hundredths. Twelve hundredths is 10 hundredths plus 2 hundredths; but 10 hundredths is 1 tenth, so we simply carry 1 into the tenths' column.

$$\begin{array}{rl} {\scriptstyle 1} & \leftarrow \text{space for carrying} \\ .38 & \\ +.24 & \\ \hline .62 & \textit{answer} \end{array}$$

Addition Using Carrying (9)

We can use carrying in addition since each place value is ten times the one to its immediate right.

Example: Add 41.874 + 28.257:

Solution:

$$\begin{array}{rl} {\scriptstyle 1\,1\ \ 1\,1} & \leftarrow \text{space for carrying} \\ 41.874 & \\ +28.257 & \\ \hline 70.131 & \textit{answer} \end{array}$$

Carrying Mentally (10)

Most of the time, carrying is done mentally. In adding decimals, always remember to line up the decimal points.

Examples:

1. Add 32.85 + 1.46:

Solution:

```
  1 1
 32.85
 +1.46
 -----
 34.31
```

2. Add 2.04 + .762 + .3:

Solution:

```
 1 1
 2.04
  .762
 + .3
 -----
 3.102
```

(11)

In the previous example where we added 2.04 + .762 + .3, we could place zeros at each end of the decimals to make a more ordered arrangement.

```
 2.04    write as    2.040
  .762   write as    0.762
 +.3     write as   +0.300
 ----               ------
                       ↑
        remember to line up the decimal points
```

Study Exercise Two (12)

Find:

1. 1.34 + 2.41
2. 3.74 + 2.94
3. 6.273 + .198 + .223
4. 23.59 + 16.81 + 1.03
5. 1.07 + 31.2 + .0767
6. 2.6 + 5.983 + 11.75
7. 7.56 + 1.96 + .21
7. 31.8579 + 11.264 + 32.79
9. 7.456 + .923 + 1.04 + 7.3
10. 1.02 + .345 + 22.6 + .04

Subtraction of Decimals (13)

As an example of subtraction, let us subtract 2.78 − 1.42

```
  2.78    place decimal points in a vertical line
 −1.42
 -----
  1.36    arrange vertically and subtract from right to left
```

Checking a Subtraction Problem (14)

```
  2.856 ←— minuend
 −1.233 ←— subtrahend
 ------
  1.623 ←— difference
```

Our check is: difference + subtrahend = minuend

```
  1.623
 +1.233
 ------
  2.856    it checks!
```

Additional Examples (15)

Study the following examples of subtraction:

Example 1: Subtract 4.38 − 1.14:

Solution:

$$\begin{array}{r} 4.38 \\ -1.14 \\ \hline 3.24 \end{array}$$

Check:

$$\begin{array}{r} 3.24 \\ +1.14 \\ \hline 4.38 \end{array}$$

Example 2: Subtract 13.87 − 2.6:

Solution:

$$\begin{array}{r} 13.87 \\ -2.6 \\ \hline 11.27 \end{array}$$

Check:

$$\begin{array}{r} 11.27 \\ +2.6 \\ \hline 13.87 \end{array}$$

Example 3: Subtract 3.4897 − 1.1635:

Solution:

$$\begin{array}{r} 3.4897 \\ -1.1635 \\ \hline 2.3262 \end{array}$$

Check:

$$\begin{array}{r} 2.3262 \\ +1.1635 \\ \hline 3.4897 \end{array}$$

Study Exercise Three (16)

Perform the subtractions and check your answers by using the checking method shown in the previous frames.

1. 83.58 − 61.24

2. 17.77 − 6.5

3. 101.02 − 1.01

4. $\begin{array}{r} 2.7615 \\ -1.4102 \\ \hline \end{array}$

5. $\begin{array}{r} 8.899 \\ -4.685 \\ \hline \end{array}$

6. $\begin{array}{r} 6.0734 \\ -2.0112 \\ \hline \end{array}$

Borrowing (17)

Subtraction may require borrowing.

Example: Subtract .84 − .46:

Solution:

.84 ⟶ 8 tenths plus 4 hundredths ⟶ 7 tenths plus 14 hundredths
−.46 ⟶ 4 tenths plus 6 hundredths ⟶ 4 tenths plus 6 hundredths

Thus,

$$\begin{array}{r} .84 \\ -.46 \\ \hline \end{array} \quad \text{can be written} \quad \begin{array}{r} 7 \text{ tenths} + 14 \text{ hundredths} \\ 4 \text{ tenths} + 6 \text{ hundredths} \\ \hline 3 \text{ tenths} + 8 \text{ hundredths} \end{array} \text{ or } .38$$

In short form:

$$\begin{array}{r} .\overset{7}{\not{8}}{}^{1}4 \\ -.4\;6 \\ \hline .3\;8 \end{array} \quad \textit{answer}$$

Check:

$$\begin{array}{r} .38 \\ +.46 \\ \hline .84 \end{array}$$

(18)

In a similar fashion we may borrow from any column since our place value system is such that each place value is 10 times the value on its right.

Example: 4.347 − 1.478

Solution:

$$\begin{array}{r} 4.3\overset{3}{\not{4}}{}^{1}7 \\ -1.47\;8 \\ \hline ?\;??\;9 \end{array} \longrightarrow \begin{array}{r} 4.\overset{2}{\not{3}}\,{}^{1}\overset{3}{\not{4}}{}^{1}7 \\ -1.4\;7\;8 \\ \hline ?\;?\;6\;9 \end{array} \longrightarrow \begin{array}{r} \overset{3}{\not{4}}.{}^{1}\overset{2}{\not{3}}\,{}^{1}\overset{3}{\not{4}}{}^{1}7 \\ -1.4\;7\;8 \\ \hline 2.8\;6\;9 \end{array} \quad \textit{answer}$$

Check:

$$\begin{array}{r} 2.869 \\ +1.478 \\ \hline 4.347 \end{array} \quad \text{it checks}$$

Borrowing From Any Column (19)

Example: Let us subtract 4 − 1.86

Solution: First write 4 as 4.00:

$$\begin{array}{r} 4.00 \\ -1.86 \\ \hline \end{array} \longrightarrow \begin{array}{r} \overset{3}{\cancel{4}}.^{1}00 \\ -1.\ 86 \\ \hline \end{array} \longrightarrow \begin{array}{r} \overset{3}{\cancel{4}}.\overset{9}{\cancel{0}}{}^{1}0 \\ -1.8\ 6 \\ \hline 2.1\ 4 \end{array} \quad \textit{answer}$$

Check:

$$\begin{array}{r} 2.14 \\ +1.86 \\ \hline 4.00 \end{array} \quad \text{it checks}$$

Study Exercise Four (20)

Perform the indicated operations and check your answers:

1. 5.17 − 2.79
2. 4.681 − 3.022
3. 16.84 − 2.55
4. 4.2 − 1.892
5. 3 − 1.784
6. $\begin{array}{r} 4.8 \\ -1.9736 \\ \hline \end{array}$
7. $\begin{array}{r} 4.03 \\ -2.1856 \\ \hline \end{array}$
8. $\begin{array}{r} 3.004 \\ -\ \ .01 \\ \hline \end{array}$

Using Decimals to Solve Applied Problems—Addition (21)

Example: Joe went to the market for his mother and bought items which cost 69¢, 45¢, $1.93, 70¢, and 17¢. Excluding tax, how much did he spend?

Solution:

$$\begin{array}{r} {}^{2\ 2} \\ \$.69 \\ .45 \\ 1.93 \\ .70 \\ .17 \\ \hline \$3.94 \end{array}$$

Joe spent $3.94.

Using Decimals to Solve Applied Problems—Subtraction (22)

Example: Jim averaged 34.5 points in ten basketball games. Bob averaged 18.5 points. Jim's average is how much higher than Bob's?

Solution:

$$\begin{array}{r} \overset{2}{\cancel{3}}{}^{1}4.5 \\ -1\ 8.5 \\ \hline 1\ 6.0 \end{array}$$

Jim's average is 16 more than Bob's.

Study Exercise Five (23)

1. Shipments of 1,745.8 pounds, 2,348.72 pounds, and 1,694.2 pounds of coal are placed in a bin. Find the total number of pounds of coal in the bin.
2. Harry's deposit in the bank was $64.79 before depositing $62.34, $17.65, and $8.28. Find his balance after he made the deposits.
3. The odometer on Julie's car registered 3,567.3 miles before she left on a trip. Upon her return, it registered 4,020.9 miles. How many miles did she travel on her trip?

***Study Exercise Five* (Continued)**

4. A carpenter agrees to build a shelf for $75.00. The lumber and hardware cost him $20.48. What will be his profit?
5. Mrs. Jones had 106.4 yards of ribbon and used 78.8 yards. How many yards had she left?

REVIEW EXERCISES (24)

Perform the indicated operations:

1. .6 + 1.8 + 3.24
2. 24.16 + 90.75
3. $\begin{array}{r} 18.16 \\ 2.95 \\ +\ .91 \\ \hline \end{array}$
4. $\begin{array}{r} .83 \\ 2.9 \\ .7 \\ +1.0 \\ \hline \end{array}$
5. 35.18 − .72 (Check your answer)
6. 3.03 − .72 (Check your answer)
7. $\begin{array}{r} 3.191 \\ -1.063 \\ \hline \end{array}$ (Check your answer)
8. $\begin{array}{r} 1.2 \\ -.874 \\ \hline \end{array}$ (Check your answer)
9. Add by changing to fractions: .23 + .31
10. If a triangle has sides of 24.05 inches, 18.63 inches, and 9.54 inches, what is the sum of the three sides?
11. Art had $4.27 and spent $2.39 for gasoline. How much does he have remaining?

Solutions to Review Exercises (25)

1. $\begin{array}{r} .60 \\ 1.80 \\ +3.24 \\ \hline 5.64 \end{array}$
2. $\begin{array}{r} 24.16 \\ +90.75 \\ \hline 114.91 \end{array}$
3. $\begin{array}{r} 18.16 \\ 2.95 \\ +.91 \\ \hline 22.02 \end{array}$
4. $\begin{array}{r} .83 \\ 2.90 \\ .70 \\ +1.00 \\ \hline 5.43 \end{array}$
5. $\begin{array}{r} 35.18 \\ -.72 \\ \hline 34.46 \end{array}$ Check: $\begin{array}{r} 34.46 \\ +.72 \\ \hline 35.18 \end{array}$
6. $\begin{array}{r} 3.03 \\ -.72 \\ \hline 2.31 \end{array}$ Check: $\begin{array}{r} 2.31 \\ +.72 \\ \hline 3.03 \end{array}$
7. $\begin{array}{r} 3.191 \\ -1.063 \\ \hline 2.128 \end{array}$ Check: $\begin{array}{r} 2.128 \\ +1.063 \\ \hline 3.191 \end{array}$
8. $\begin{array}{r} 1.200 \\ -.874 \\ \hline .326 \end{array}$ Check: $\begin{array}{r} .326 \\ +.874 \\ \hline 1.200 \end{array}$
9. $$\begin{array}{rl} .23 &= \frac{2}{10} + \frac{3}{100} \\ +\ .31 &= \frac{3}{10} + \frac{1}{100} \\ \hline .23 + .31 &= \frac{5}{10} + \frac{4}{100} \\ &= .54 \end{array}$$
10. $\begin{array}{r} 24.05 \\ 18.63 \\ +9.54 \\ \hline 52.22 \end{array}$ The sum of the 3 sides is 52.22 inches.
11. $\begin{array}{r} \$4.27 \\ -2.39 \\ \hline \$1.88 \end{array}$ Art has $1.88 remaining.

SUPPLEMENTARY PROBLEMS

A. Add:

1. .92
.83

2. 5.01
2.99

3. 4.9
.13
2.648

4. 6.38 + 1.97
5. 4.79 + 6.829
6. 6.09 + 0.77
7. 7.56 + 3.2 + 8.579
8. 3.61 + 2.6 + .658
9. 3.071 + 2.0968 + .75
10. 74.382 + 9.76 + 11.5489 + 21.7
11. 32.071 + 1.009 + 23.6 + 1.5876
12. $3.41 + $.75 + $12.45 + $9.49 + $1.68

B. Subtract and check your answers:

13. 4.769 − 2.143
14. 6.529 − 3.417
15. 11.383 − 5.496
16. 5.0403 − 3.8767
17. 5 − 1.617
18. 6.2007 − 4.9568
19. 11.92 − 11.9
20. 32.214 − 17.839
21. 6.1
−2.005

22. 9
−.082

C. Applied problems:

23. Subtract 4.23 from the sum of 2.17 and 4.85.

24. Mike had $70. Then he spent $20.75 for shoes, $8.25 for a shirt, and $3.50 for a tie. How much money does he have left?

25. How much change should you receive from a $5 bill if you owe $3.62?

26. If you buy an article that regularly sells for $7.49 at a reduction of $1.98, how much will you pay?

27. The outside diameter of a copper pipe is 2.375 inches and its wall thickness is .083 inches. What is the inside diameter?

28. During a certain month, the weather bureau recorded rainfall of 1.02 inches, 2 inches, .58 inches, and 0.4 inches. What was the total rainfall for that month?

29. Study the following problems:

27.2
+1.3
28.5

2.04
−.11
1.93

Which of the above numbers is:
(a) A sum?
(b) An addend?
(c) A difference?
(d) A subtrahend?
(e) A minuend?
(f) A number less than one?

30. A car's odometer reading was 26,076.8 at the beginning of a trip and 27,423.9 at the end of the trip. How many miles did the car travel?

31. A car's gas tank holds 22.5 gallons. If 11.8 gallons is needed to fill the tank, how much was already in the tank?

32. Tim purchased a car for $7,865.20, including tax. If the tax was $445.20, what was the price of the car?

33. A pair of shoes that regularly sells for $72 is on sale for $59.99. How much is saved by buying at the sale price?

SUPPLEMENTARY PROBLEMS (Continued)

Find the total for the following restaurant bills. (Problems 34-35)

34.

	Item	Amount	
1	hamburger	2	50
1	french fries		95
1	fried chicken	3	95
1	coke		50
1	milk		60
	Tax		51
	Tip	1	35
	Total		

35.

	Item	Amount	
1	pancakes	2	85
1	bacon and eggs	4	25
2	coffee	1	20
1	orange juice		90
	Tax		55
	Tip	1	50
	Total		

Find the TOTAL and TOTAL DEPOSIT for the following bank deposit slips.

36.

Please List Each Check Separately by Financial Institution Number

	DOLLARS	CENTS
CURRENCY		
COIN		
Checks by Financial Institution No. 1	32	23
2	4	50
3	106	89
4	18	75
5	7	63
6		
7		
8		
9		
10		
TOTAL		
Less Cash Return	42	25
TOTAL DEPOSIT		

37.

Please List Each Check Separately by Financial Institution Number

	DOLLARS	CENTS
CURRENCY		
COIN		
Checks by Financial Institution No. 1	89	50
2	173	45
3	14	23
4	207	88
5	19	67
6		
7		
8		
9		
10		
TOTAL		
Less Cash Return	25	50
TOTAL DEPOSIT		

Solutions to Study Exercises

(5A)

Study Exercise One (Frame 5)

1.

$$.22 = \frac{2}{10} + \frac{2}{100}$$
$$.46 = \frac{4}{10} + \frac{6}{100}$$
$$.22 + .46 = \frac{6}{10} + \frac{8}{100} = .68$$

2.

$$.03 = \frac{0}{10} + \frac{3}{100}$$
$$.21 = \frac{2}{10} + \frac{1}{100}$$
$$.24 = \frac{2}{10} + \frac{4}{100}$$
$$.03 + .21 + .24 = \frac{4}{10} + \frac{8}{100} = .48$$

3.

$$.44 = \frac{4}{10} + \frac{4}{100}$$
$$.2 = \frac{2}{10}$$
$$.11 = \frac{1}{10} + \frac{1}{100}$$
$$.44 + .2 + .11 = \frac{7}{10} + \frac{5}{100} = .75$$

4.

$$1.2 = 1 + \frac{2}{10}$$
$$2.3 = 2 + \frac{3}{10}$$
$$1.2 + 2.3 = 3 + \frac{5}{10} = 3.5$$

Study Exercise Two (Frame 12)

(12A)

1. 3.75 **2.** 6.68 **3.** 6.694 **4.** 41.43
5. 32.3467 **6.** 20.333 **7.** 9.73 **8.** 75.9119
9. 16.719 **10.** 24.005

Study Exercise Three (Frame 16)

(16A)

1. $83.58 - 61.24 = 22.34$ Check: $22.34 + 61.24 = 83.58$

2. $17.77 - 6.5 = 11.27$ Check: $11.27 + 6.5 = 17.77$

3. $101.02 - 1.01 = 100.01$ Check: $100.01 + 1.01 = 101.02$

4. $2.7615 - 1.4102 = 1.3513$ Check: $1.3513 + 1.4102 = 2.7615$

5. $8.899 - 4.685 = 4.214$ Check: $4.214 + 4.685 = 8.899$

6. $6.0734 - 2.0112 = 4.0622$ Check: $4.0622 + 2.0112 = 6.0734$

Study Exercise Four (Frame 20)

(20A)

1. $5.17 - 2.79 = 2.38$ Check: $2.38 + 2.79 = 5.17$

2. $4.681 - 3.022 = 1.659$ Check: $1.659 + 3.022 = 4.681$

3. $16.84 - 2.55 = 14.29$ Check: $14.29 + 2.55 = 16.84$

4. $4.200 - 1.892 = 2.308$ Check: $2.308 + 1.892 = 4.200$

5. $3.000 - 1.784 = 1.216$ Check: $1.216 + 1.784 = 3.000$

6. $4.8000 - 1.9736 = 2.8264$ Check: $2.8264 + 1.9736 = 4.8000$

7. $4.0300 - 2.1856 = 1.8444$ Check: $1.8444 + 2.1856 = 4.0300$

8. $3.004 - .01 = 2.994$ Check: $2.994 + .01 = 3.004$

Study Exercise Five (Frame 23) **(23A)**

1.

```
   1,745.8
   2,348.72
 + 1,694.2
 ---------
   5,788.72
```

There are 5,788.72 pounds of coal.

2.

```
  $62.34
   17.65      $64.79
  + 8.28       88.27
  ------     -------
  $88.27     $153.06
```

Total balance is $153.06

3.

```
   4,020.9
 - 3,567.3
 ---------
     453.6
```

Julie traveled 453.6 miles.

4.

```
  $75.00
 - 20.48
 -------
  $54.52
```

His profit will be $54.52

5.

```
  106.4
 - 78.8
 ------
   27.6
```

There were 27.6 yards left.

UNIT

20 Multiplication of Decimals

OBJECTIVES (1)

By the end of this unit you should be able to:

1. MULTIPLY DECIMALS AND CORRECTLY PLACE THE DECIMAL POINT IN THE PRODUCT.
2. MULTIPLY BY POWERS OF TEN BY MOVING THE DECIMAL POINT.
3. RAISE DECIMALS TO POWERS.
4. WORK APPLIED PROBLEMS CHANGING FROM A BASE OF ONE.

Multiplication by Changing to Fractions (2)

We will multiply two decimals by first changing them to fractions. Then we will develop a shortcut for the multiplication process.

Let us multiply 2.3 by .7

Line (a): $2.3 \times .7 = 2\frac{3}{10} \times \frac{7}{10}$

Line (b): $= \frac{23}{10} \times \frac{7}{10}$

Line (c): $= \frac{161}{100}$

Line (d): $= 1.61$

Notice that each of the decimals to be multiplied consists of one decimal place, while the product consists of two decimal places.

Multiplication of Decimals—Shortcut (3)

Rule: When multiplying two decimals, the number of decimal places in the product is always the sum total of the decimal places of the original two numbers.

Multiplication of Decimals (Continued)

Example: Multiply 2.231 × .32:

Solution:

```
  2.231 ←— first factor has 3 decimal places
   ×.32 ←— second factor has 2 decimal places
   4462
  6693
 .71392 ←— product has 5 decimal places (count from right to left to set the decimal point)
```

Additional Examples (4)

Example 1: Multiply 213 × 1.28:

Solution:

```
   2 13 ←— 0 decimal places
  ×1.28 ←— 2 decimal places
  17 04
  42 6
 213
 272.64 ←— 2 decimal places
           in the product
           (count from right to left)
```

Example 2: Multiply 9 × .014:

Solution:

```
 .014 ←— 3 decimal places
   ×9 ←— 0 decimal places
 .126 ←— 3 decimal places
         in the product
         (count from right to left)
```

Example 3: Multiply 8 × .11 × 1.2:

Solution:

```
  .11 ←— 2 decimal places
  × 8 ←— 0 decimal places
  .88 ←— 2 decimal places in the product
         (count from right to left)

   .88 ←— 2 decimal places
  × 1.2 ←— 1 decimal place
   176
   88
 1.056 ←— 3 decimal places in the product
          (count from right to left)
```

Study Exercise One (5)

Multiply:

1. .612 × .14
2. 1.33 × 1.2
3. (1.97) · (1.8)
4. (2.48)(1.25)
5. 6 × 2.4 × .15

Inserting Zeros to Place the Decimal Point (6)

Example: Multiply .06 × .007:

Solution:

```
   .007 ←— 3 decimal places
   ×.06 ←— 2 decimal places
 .00042 ←— 5 decimal places in the product
```

count from the right 5 decimal places and attach zeros if necessary

Study Exercise Two (7)

Multiply:

1. $.002 \times 2.3$
2. $(.034) \cdot (.02)$
3. $(.104)(.003)$
4. $(.043)(.06)$

Multiplication by a Power of Ten (8)

$$\begin{array}{r} 3.318 \\ \times 10 \\ \hline 33.180 \end{array} \qquad \begin{array}{r} 3.318 \\ \times 100 \\ \hline 331.800 \end{array} \qquad \begin{array}{r} 3.318 \\ \times 1000 \\ \hline 3318.000 \end{array}$$

1. $3.318 \times 10 = 33.180$ or 33.18
2. $3.318 \times 100 = 331.800$ or 331.8
3. $3.318 \times 1000 = 3318.000$ or 3318

Notice the answer is simply found by moving the decimal point in the original number.

1. $3.318 \times 10 = 33.18$ (move decimal point 1 place to the right)
2. $3.318 \times 100 = 331.8$ (move decimal point 2 places to the right)
3. $3.318 \times 1000 = 3318$ (move decimal point 3 places to the right)

Multiplying Decimal Numerals by Powers of Ten (9)

Rule: To multiply by 10, move the decimal point 1 place to the right.
To multiply by 100, move the decimal point 2 places to the right.
To multiply by 1000, move the decimal point 3 places to the right.
To multiply by 10,000, move the decimal point 4 places to the right.

Inserting Zeros (10)

Remember, if there is no decimal point, it is assumed to be at the right end of the numeral. Thus 37 and 37. are equal. When multiplying by 10, 100, 1000, etc., it may be necessary to insert zeros.

For example, $100 \times 37 = 100 \times 37.00$
$= 3700$

Study these examples: (11)

1. $4.721 \times 10 \times 47.21$
2. $.0327 \times 100 = 03.27$ or 3.27
3. $2.23 \times 1000 = 2230$
4. $12.2 \times 1000 = 12,200$

Study Exercise Three (12)

Multiply as indicated:

1. $.00572 \times 100$
2. 3.51×1000
3. $.007 \times 10$
4. $5.072 \times 10,000$
5. $.0027 \times 10$
6. $100 \times .0453$
7. 1000×1.5
8. 100×1.4
9. 1.021×1000

Review of Exponents (13)

We will review exponents before proceeding to powers of decimals.

$4 \times 4 \times 4$ may be written 4^3

4^3 ← exponent (3), base (4)

Review of Exponents (Continued)

An *exponent* is a number that tells how many times the base appears as a factor in multiplication. Thus, 5^4 means $5 \times 5 \times 5 \times 5$

Powers of Decimals (14)

We will now find powers of decimals.

Example 1: Find $(.3)^2$:

Solution: $(.3)^2 = .3 \times .3$
$= .09$

Example 2: Find $(.02)^3$:

Solution: $(.02)^3 = .02 \times .02 \times .02$
$= .000008$

Example 3: Find $(1.06)^2$:

Solution: $(1.06)^2 = 1.06 \times 1.06$
$= 1.1236$

Study Exercise Four (15)

Find the following powers:

1. $(.4)^2$ **2.** $(.01)^3$ **3.** $(.12)^2$
4. $(1.2)^2$ **5.** $(.2)^4$ **6.** $(.03)^3$
7. $(12.1)^2$ **8.** $(.86)^2$ **9.** $(1.01)^2$

Applied Problems—Changing from a Base of One (16)

To change from a base of one, use the operation of multiplication.

Example: If Luis earns $3.75 per hour, what are his gross earnings for a 40-hour week?

Solution: Change from 1 hour to 40 hours by multiplication.

$40 \times \$3.75 = \$150$

Luis' earnings will be $150.

Computation

$$\begin{array}{r} 3.75 \\ \times 40 \\ \hline 150.00 \end{array}$$

Note the similarity of this type with expansion problems in Unit 6.

Additional Examples (17)

Example 1: The fuel consumption of a certain airplane is 32.5 gallons per hour. How many gallons will be consumed on a 3.25 hour trip?

Solution: Change from 1 hour to 3.25 hours by multiplication.

$32.5 \times 3.25 = 105.625$

Thus, 105.625 gallons will be used.

Computation

$$\begin{array}{r} 32.5 \\ \times 3.25 \\ \hline 1\ 625 \\ 6\ 50 \\ 97\ 5 \\ \hline 105.625 \end{array}$$

Additional Examples (Continued)

Example 2: A certain cut of meat sells for \$2.29 a pound. How much would .75 pounds of this meat cost?

Solution: Change from 1 pound to .75 pounds by multiplication.

$$.75 \times \$2.29 = \$1.7175$$

Round to nearest cent.
The cost is \$1.72.

Computation

```
  2.29
   .75
  1145
 1 603
1.7175
```

Study Exercise Five (18)

1. Frank has a small car which averages 27 miles to a gallon of gasoline. The car used 3.8 gallons of gas on a trip. What was the length of the trip?
2. How much will a person earn if she works 18 hours and is paid at the rate of \$2.75 an hour?
3. If it takes a factory 4.3 hours to produce a certain article, how long will it take to produce a dozen of these articles?
4. One inch is equivalent to 2.54 centimeters. How many centimeters are equivalent to 11.6 inches?

REVIEW EXERCISES (19)

A.
1. Multiply $.68 \times 1.4$
2. Multiply $2.432 \times .41$
3. Multiply 2.02 by 1.1
4. Multiply $(.003) \cdot (.21)$
5. Find $(.12)^3$

B. Multiply and select the one correct answer. Mark a, b, or c:

6. a. $100 \times 2.603 = 26.3$ b. $10 \times .814 = 81.4$ c. $1000 \times .026 = 26$
7. a. $.0041 \times 100 = .041$ b. $.0716 \times 1000 = 71.6$ c. $.325 \times 10 = 32.5$
8. a. $10 \times 1.34 = 13.4$ b. $100 \times .0217 = .217$ c. $36.5 \times 1000 = 3650$

C.
9. How much will a person earn if he is to be paid \$2.75 an hour and he works 35 hours?
10. If a factory produces one article in 3.2 hours, how long will it take to produce 7 articles?

Solutions to Review Exercises (20)

A. 1.
```
  .68
 ×1.4
  272
  68
 .952
```

2.
```
 2.432
  ×.41
  2432
 9728
.99712
```

3.
```
 2.02
 ×1.1
  202
 2 02
2.222
```

4.
```
   .003
   ×.21
    003
   006
 .00063
```

5. $(.12)^3 = (.12) \times (.12) \times (.12)$
```
  .12        .0144
 ×.12         ×.12
   24          288
  12          144
.0144      .001728
```
Thus, $(.12)^3 = .001728$

Solutions to Review Exercises (Continued)

B. **6.** c **7.** b **8.** a

C. **9.** Change from 1 hour to 35 hours by multiplication.

```
 $2.75
  ×35
 13 75
 82 5
$96.25   Earnings will be $96.25
```

10. Change from 1 article to 7 articles by multiplication.

```
 3.2
  ×7
22.4   It will take 22.4 hours to produce 7 articles.
```

SUPPLEMENTARY PROBLEMS

A. Multiply as indicated:

1. $6 \times .3$
2. $(.7) \cdot (.06)$
3. $.05 \times .12$
4. $.09 \times .09$
5. $.003 \times .2$
6. $.025 \times 8$
7. $(.016) \cdot (.205)$
8. 1.03×2.3
9. $(3.14)(3.5)$
10. 10×2.073
11. 100×33.4
12. $1{,}000 \times .036$
13. $.0816 \times 100$
14. $.0237 \times 10$
15. $.0273 \times 10{,}000$
16. 4.1427×100
17. $100 \times .5176$
18. $20.1 \times 1{,}000$
19. $(23.2)(1.3)(2.1)$
20. $12.22 \times 1.1 \times 3.2$
21. $(.31)^2$
22. $(1.21)^2$
23. $(.03)^3$
24. $(1.01)^3$
25. $(.3) \times (.3)^2$
26. $(.21)^2 \times (.02)$

B. **27.** How much will a person earn if he is paid $2.55 an hour and he works for 40 hours?

28. Certain meat sells for $2.75 a pound. How much would .75 pounds of this meat cost?

29. If a car averages 22 miles to a gallon and uses 4.8 gallons on a trip, what was the length of the trip?

30. A new rug costs $14.98 per square yard. How much will it cost to carpet a room with 22 square yards?

31. A mechanic is payed $15 per hour regular pay. He earns $18.50 per hour overtime pay. In a certain week he worked 40 regular hours and 5 overtime hours. Find his total pay for that week.

32. A person can buy a car by making 48 monthly payments of $184.50. How much will he have payed at the end of 48 months?

33. Bill went to the post office and bought twelve 22¢ stamps and thirteen 40¢ stamps. How much did he pay for all the stamps?

34. Find the total cost of the following grocery bill: 2 heads of lettuce at 49¢ each, 3 loaves of bread at 79¢ each, 3 bottles of cola at $1.29 a bottle, and 2 pounds of grapes at 89¢ a pound.

Solutions to Study Exercises (5A)

Study Exercise One (Frame 5)

1.

```
   .612 ←— 3 decimal places
   ×.14 ←— 2 decimal places
   2448
   612
 .08568 ←— 5 decimal places in the product
```

2.

```
  1.33 ←— 2 decimal places
  ×1.2 ←— 1 decimal place
   266
  1 33
 1.596 ←— 3 decimal places in the product
```

Solutions to Study Exercises (Continued)

3.
```
  1.97 <— 2 decimal places
  ×1.8 <— 1 decimal place
  1 576
  1 97
  3.546 <— 3 decimal places in the product
```

4.
```
  2.48 <— 2 decimal places
  ×1.25 <— 2 decimal places
  1240
  496
  2 48
  3.1000 or 3.1 <— 4 decimal places in the product
```

5.
```
  2.4 <— 1 decimal place
  ×6 <— 0 decimal places
  14.4 <— 1 decimal place in the product
  14.4 <— 1 decimal place
  ×.15 <— 2 decimal places
  720
  1 44
  2.160 or 2.16 <— 3 decimal places in the product
```

Study Exercise Two (Frame 7) (7A)

1.
```
  .002 <— 3 decimal places
  ×2.3 <— 1 decimal place
  006
  004
  .0046 <— 4 decimal places in the product
```

2.
```
  .034 <— 3 decimal places
  ×.02 <— 2 decimal places
  .00068 <— 5 decimal places in the product
```

3.
```
  .104 <— 3 decimal places
  ×.003 <— 3 decimal places
  .000312 <— 6 decimal places in the product
```

4.
```
  .043 <— 3 decimal places
  ×.06 <— 2 decimal places
  .00258 <— 5 decimal places in the product
```

Study Exercise Three (Frame 12) (12A)

1. .572 **2.** 3,510 **3.** .07 **4.** 50,720
5. .027 **6.** 4.53 **7.** 1,500 **8.** 140
9. 1,021

Study Exercise Four (Frame 15) (15A)

1. $(.4)^2 = .4 \times .4 = .16$

2. $(.01)^3 = .01 \times .01 \times .01 = .000001$

3. $(.12)^2 = .12 \times .12 = .0144$

4. $(1.2)^2 = 1.2 \times 1.2 = 1.44$

5. $(.2)^4 = .2 \times .2 \times .2 \times .2 = .0016$

6. $(.03)^3 = .03 \times .03 \times .03 = .000027$

7. $(12.1)^2 = 12.1 \times 12.1 = 146.41$

8. $(.86)^2 = .86 \times .86 = .7396$

9. $(1.01)^2 = (1.01) \times (1.01) = 1.0201$

Study Exercise Five (Frame 18) (18A)

1. Change from 1 gallon to 3.8 gallons by multiplication:
```
  27
  ×3.8
  21 6
  81
  102.6
```
The trip was 102.6 miles.

2. Change from 1 hour to 18 hours by multiplication:
```
  $2.75
  ×18
  22 00
  27 5
  $49.50
```
Earnings will be $49.50.

Study Exercise Five (Frame 18) **(Continued)**

3. Change from 1 article to 12 articles by multiplication:

```
  4.3
 ×1 2
 ----
  8 6
 43
 ----
 51.6
```

It will take 51.6 hours to produce a dozen articles.

4. Change from 1 inch to 11.6 inches by multiplication:

```
  2.54
  11.6
 ------
  1 524
  2 54
 25 4
 ------
 29.464
```

There are 29.464 centimeters in 11.6 inches.

UNIT

21 Division of Decimals

OBJECTIVES (1)

By the end of this unit you should be able to:

1. DIVIDE DECIMALS.
2. ROUND THE QUOTIENT IN A DIVISION PROBLEM.
3. CONVERT FRACTIONS TO DECIMALS.
4. DIVIDE BY MULTIPLES OF TEN BY THE SHORT METHOD.

Writing a Division Problem (2)

Division of decimals is similar to division of whole numbers.

A division problem may be written in three ways. For example, six divided by 3 equals 2 may be written:

1. $6 \div 3 = 2$
2. $\frac{6}{3} = 2$
3. $3\overline{)6}$ with quotient 2

Division of Decimals by Whole Numbers (3)

Problem: Divide .68 by 2:

Solution: Since .68 = 6 tenths plus 8 hundredths, .68 ÷ 2 gives 3 tenths plus 4 hundredths or .34:

$$\begin{array}{r} .34 \\ 2\overline{).68} \end{array}$$

.34 ← *quotient*

divisor — 2

dividend — .68 (decimal point in the quotient must line up with the decimal point in the dividend)

Examples (4)

Example 1: Find .84 ÷ 4:

Solution:

$$\begin{array}{r} .21 \\ 4\overline{).84} \\ \underline{8} \\ 4 \\ \underline{4} \\ 0 \end{array}$$

Example 2: Find 2.684 ÷ 4:

Solution:

$$\begin{array}{r} .671 \\ 4\overline{)2.684} \\ \underline{2\,4} \\ 28 \\ \underline{28} \\ 4 \\ \underline{4} \\ 0 \end{array}$$

Example 3: Find 6.003 ÷ 3:

Solution:

$$\begin{array}{r} 2.001 \\ 3\overline{)6.003} \\ \underline{6} \\ 0 \\ \underline{0} \\ 0 \\ \underline{0} \\ 3 \\ \underline{3} \\ 0 \end{array}$$

Study Exercise One (5)

1. Find 3.069 ÷ 3
2. Find 4.804 ÷ 4
3. Find 18.006 ÷ 2
4. Find 8.26402 ÷ 2

Division by a Decimal (6)

Division by a decimal will be changed to division by a whole number.

Example 1: Change .8614 ÷ .14 to division by a whole number:

Solution:

Line (a): $.8614 \div .14 = \dfrac{.8614}{.14}$

Line (b): $= \dfrac{.8614 \times 100}{.14 \times 100}$

Line (c): $= \dfrac{86.14}{14}$

Line (d): $= 86.14 \div 14$

Example 2: Change 23.1 ÷ .112 to division by a whole number:

Solution:

Line (a): $23.1 \div .112 = \dfrac{23.1}{.112}$

Line (b): $= \dfrac{23.1 \times 1000}{.112 \times 1000}$

Division by a Decimal (Continued)

Line (c): $= \dfrac{23100}{112}$

Line (d): $= 23100 \div 112$

Shortcut (7)

Notice in Example 1 of the preceding frame we could accomplish the same thing by moving both decimal points 2 places to the right. That is,

$$\frac{.8614}{.14} = \frac{86.14}{14.}$$

In Example 2, both decimal points are moved 3 places to the right. That is,

$$\frac{23.1}{.112} = \frac{23100.}{112.}$$

Important: Both decimal points must be moved the same number of places to the right.

---

(8)

The problem $.8614 \div .14$ was simplified in the preceding exercise to $86.14 \div 14$. As a long division problem $.8614 \div .14$ is written as follows:

divisor ⟶ $.14\overline{).8614}$ ⟵ *dividend*

or in simplified form as

divisor ⟶ $14.\overline{)86.14}$ ⟵ *dividend*

---

(9)

The usual practice is to show the position of the decimal points as follows: $.14\overline{).8614}$ is simplified by moving the decimal points 2 places to the right.

$$.14.\overline{).86.14}$$

Examples: **(10)**

1. $1.4\overline{)28.24}$ will be written $1.4.\overline{)28.2.4}$
2. $2.21\overline{).281}$ will be written $2.21.\overline{).28.1}$
3. $.131\overline{)42.3}$ will be written $.131.\overline{)42.300.}$

Study Exercise Two (11)

Rewrite so the divisor is a whole number:

1. $2.42 \div .21$ **2.** $1.7 \div .34$ **3.** $1.701\overline{)28.3456}$ **4.** $4.07\overline{)7.083}$

Steps in the Division by a Decimal (12)

Problem: Divide .048 by .2:

Solution:

Step (1): First arrange in long division form:

$.2\overline{).048}$

Steps in the Division by a Decimal (Continued)

Step (2): Next, the divisor must be converted to a whole number:

$.2.\overline{).0.48}$

Step (3): Then line up the decimal point in the quotient:

$.2.\overline{).0.48}$

Step (4): Finally, divide:

```
      .24
 .2.).0.48     The quotient is .24
      4
      ---
        8
        8
      ---
        0
```

---

(13)

Example 1: .216 ÷ .03

Solution:

$.03.\overline{).21.6}$

```
          7.2     The quotient is 7.2
 .03.).21.6
       21
       ---
          6
          6
       ---
          0
```

Example 2: 4.601 ÷ 4.3

Solution:

```
        1.07     The quotient is 1.07
 4.3.)4.6.01
      4 3
      ----
        3 0
          0
      ----
        3 01
        3 01
      ----
           0
```

Study Exercise Three

(14)

Divide as indicated:

1. .36 ÷ .6 **2.** .048 ÷ .2 **3.** .076 ÷ .04

4. 16.308 ÷ .36 **5.** .7881 ÷ 3.7 **6.** 9.128 ÷ .028

---

(15)

In a division problem like 9 ÷ 4, even though no decimal points are shown, it will be necessary to write 9 as 9.00.

Remember we can place as many zeros as we wish after the decimal point.

$4\overline{)9}$ will be written $4\overline{)9.00}$

```
    2.25
  4)9.00
    8
    ---
    1 0
      8
    ---
     20
     20
    ---
      0
```

Another Example

(16)

Divide .0006 by .012:

Solution: Move the decimal point three places to the right and insert the necessary zeros:

```
           .05
.012.).000.60
           60
            0
```

Study Exercise Four

(17)

Divide the following:

1. .00066 ÷ .022 **2.** 15 ÷ .625 **3.** 7 ÷ 1.75

Changing a Fraction to a Decimal

(18)

A fraction may be changed to a decimal by long division.

For example, $\frac{3}{4}$ is converted to a decimal as follows:

```
   .75
4)3.00
  2 8
    20
    20
     0
```

The decimal equivalent of $\frac{3}{4}$ is .75

Study Exercise Five

(19)

Find the decimal equivalent of the following fractions:

1. $\frac{1}{4}$ **2.** $\frac{7}{10}$ **3.** $\frac{3}{5}$ **4.** $\frac{4}{5}$

5. $\frac{2}{5}$ **6.** $\frac{1}{16}$ **7.** $\frac{5}{8}$ **8.** $\frac{3}{8}$

Repeating Decimals

(20)

When a division problem ends with a zero remainder, the division is said to *terminate*. Sometimes the division does not terminate.

Example:

```
  .1444...
9)1.3000...
   9
   40
   36
    40
    36
     40
     36
      4
```

.1444... is called a *repeating decimal*.

If we wish to terminate a quotient which is a repeating decimal, we will express the remainder as a fraction. **(21)**

Example: 1.3 ÷ 9

```
     .1            .14            .144
  9)1.3         9)1.30         9)1.300
    9              9              9
    4              40             40
                   36             36
                    4              40
                                   36
                                    4
```

to 1 decimal place	to 2 decimal places	to 3 decimal places
quotient is $.1\frac{4}{9}$	quotient is $.14\frac{4}{9}$	quotient is $.144\frac{4}{9}$

Study Exercise Six **(22)**

Divide as far as two decimal places and terminate the answer with a fraction:

1. 2 ÷ 3 **2.** 7 ÷ 16 **3.** .732 ÷ 3.4

Rounding **(23)**

A method of rounding a division problem to a specified place value or a specific number of decimal places is to carry the division one extra place and then round back by the rules of rounding.

Example: Divide and round to 1 decimal place .27163 ÷ .05:

Solution:

Step (1): Divide to 2 decimal places:

```
        5.43
 .05.).27.163
       25
       2 1
       2 0
         16
         15
          13
```

Step (2): Round to 1 decimal place:

5.[4]3 rounds to 5.4

The quotient is 5.4 to 1 decimal place.

Example: Divide .75 by .178 and round the answer to the nearest hundredth: **(24)**

Solution:

Step (1): Divide to one additional place value:

```
              4.213
     .178.).750.000
            712
            38 0
            35 6
             2 40
             1 78
              620
              534
               86
```

Step (2): Round to the nearest hundredth:

4.2[1]3 rounds to 4.21

The quotient is 4.21 to the nearest hundredth.

Study Exercise Seven **(25)**

1. Divide and round the result to two decimal places: .081 ÷ .14
2. Divide and round the result to the nearest tenth: .638 ÷ 2.6
3. Divide and round the result to the nearest hundredth: 1.37 ÷ .76
4. Divide and round the result to two decimal places: .06 ÷ .134
5. Divide and round the result to one decimal place: .134 ÷ .006

Division by Powers of Ten **(26)**

You may remember the short method we have for multiplying by 10, 100, 1000, etc. Now we want a short method of dividing by 10, 100, 1000, etc.

Study these examples:

```
14.3 ÷ 10             14.3 ÷ 100             14.3 ÷ 1000
     1.43                   .143                   .0143
10.)14.30            100)14.300            1000)14.3000
    10                   10 0                   10 00
    4 3                  4 30                   4 300
    4 0                  4 00                   4 000
      30                   300                   3000
      30                   300                   3000
       0                     0                      0
14.3 ÷ 10 = 1.43     14.3 ÷ 100 = .143     14.3 ÷ 1000 = .0143
```

Short Method of Division by Powers of Ten (27)

Rule: To divide by 10, move the decimal point one place to the left.
To divide by 100, move the decimal point two places to the left.
To divide by 1000, move the decimal point three places to the left.

Example 1: $142.3 \div 100 = 1.423$ **Example 2:** $17.21 \div 10 = 1.721$

Example 3: $.012 \div 1000 = .000012$

Study Exercise Eight (28)

Divide by the short method as indicated:

1. $7.56 \div 10$ **2.** $.304 \div 100$ **3.** $.701 \div 1000$ **4.** $3.11 \div 100$

5. $.051 \div 10$ **6.** $689 \div 1000$ **7.** $12 \div 100$ **8.** $.0101 \div 10$

REVIEW EXERCISES (29)

A. Change to division by a whole number:

1. $1.89 \div .3$ **2.** $.0417 \div .032$

B. Divide as indicated:

3. $.018 \div .02$ **4.** $.24 \div .6$ **5.** $16.308 \div .36$

C. Divide as indicated:

6. $3 \div .04$

D. Find the decimal equivalent:

7. $\frac{5}{4}$ **8.** $\frac{5}{8}$ **9.** $\frac{3}{4}$ **10.** $\frac{2}{5}$

E. Divide and round off as indicated:

11. $13 \div 17$ (to 2 decimals) **12.** $1 \div 6$ (to nearest tenth)

F. Divide by the short method:

13. $8.017 \div 100$ **14.** $.008 \div 10$ **15.** $2.15 \div 1000$ **16.** $17.24 \div 100$

Solutions to Review Exercises

A. **1.** $18.9 \div 3$ **2.** $41.7 \div 32$

B. **3.** $.02.\overline{)\,.01.8}$ = .9 **4.** $.6.\overline{)\,.2.4}$ = .4

5. $.36.\overline{)\,16.30.8}$ = 45.3

```
      45.3
.36.)16.30.8
     14 4
      1 90
      1 80
        10 8
        10 8
           0
```

Solutions to Review Exercises (Continued)

C. 6.

$$\begin{array}{r} 75. \\ .04.\overline{)3.00.} \\ \underline{2\,8} \\ 20 \\ \underline{20} \\ 0 \end{array}$$

D. 7.

$$\begin{array}{r} 1.25 \\ 4\overline{)5.00} \\ \underline{4} \\ 1\,0 \\ \underline{8} \\ 20 \\ \underline{20} \\ 0 \end{array}$$

8.

$$\begin{array}{r} .625 \\ 8\overline{)5.000} \\ \underline{4\,8} \\ 20 \\ \underline{16} \\ 40 \\ \underline{40} \\ 0 \end{array}$$

9.

$$\begin{array}{r} .75 \\ 4\overline{)3.00} \\ \underline{2\,8} \\ 20 \\ \underline{20} \\ 0 \end{array}$$

10.

$$\begin{array}{r} .4 \\ 5\overline{)2.0} \\ \underline{2\,0} \\ 0 \end{array}$$

E. 11.

$$\begin{array}{r} .764 \\ 17\overline{)13.00} \\ \underline{11\,9} \\ 1\,10 \\ \underline{1\,02} \\ 80 \\ \underline{68} \\ 12 \end{array}$$

quotient: .76 to 2 decimals

12.

$$\begin{array}{r} .16 \\ 6\overline{)1.00} \\ \underline{6} \\ 40 \\ \underline{36} \\ 4 \end{array}$$

quotient: .2 to nearest tenth

F. 13. 8.017 ÷ 100 = .08017

14. .008 ÷ 10 = .0008

15. 2.15 ÷ 1000 = .00215

16. 17.24 ÷ 100 = .1724

SUPPLEMENTARY PROBLEMS

A. Divide as indicated:

1. 35.6 ÷ .4 **2.** .0016 ÷ .8 **3.** 72 ÷ .4 **4.** 40 ÷ 12.8
5. .00012 ÷ .03 **6.** 18.7572 ÷ 5.39 **7.** 670.8 ÷ .78 **8.** 1.634 ÷ .043
9. 972 ÷ .108 **10.** .63 ÷ .0007

B. Find the decimal equivalent of the following fractions:

11. $\frac{6}{10}$ **12.** $\frac{3}{8}$ **13.** $\frac{9}{16}$ **14.** $\frac{1}{20}$

15. $\frac{5}{8}$ **16.** $\frac{4}{5}$ **17.** $\frac{7}{20}$

C. Divide as indicated:

18. .12 ÷ 3.2 **19.** 10.85 ÷ .0775 **20.** .1875 ÷ .25

SUPPLEMENTARY PROBLEMS (Continued)

D. Divide and round the answer as indicated:

21. 3.06 ÷ .11 to nearest tenth.
22. 24.1 ÷ 6.001 to 2 decimal places.
23. .527 ÷ 1.37 to nearest thousandth.
24. 731.8 ÷ 14.6 to 2 decimal places.
25. 51.7 ÷ .292 to 1 decimal place.
26. 12.2 ÷ .098 to 1 decimal place.

E. Divide by the short method:

27. 68.2 ÷ 100 **28.** 86.4 ÷ 100 **29.** .143 ÷ 10 **30.** 76.4 ÷ 1000
31. 2.824 ÷ 1000 **32.** 6.34 ÷ 10,000

Solutions to Study Exercises

(5A)

Study Exercise One (Frame 5)

1. 1.023 **2.** 1.201 **3.** 9.003 **4.** 4.13201

Study Exercise Two (Frame 11)

(11A)

1. $.21\overline{)2.42} = .21.\overline{)2.42.}$

2. $.34\overline{)1.7} = .34.\overline{)1.70.}$

3. $1.701\overline{)28.3456} = 1.701.\overline{)28.345.6}$

4. $4.07\overline{)7.083} = 4.07.\overline{)7.08.3}$

Study Exercise Three (Frame 14)

(14A)

1. $.6.\overline{).3.6}$ = .6
3 6
0

2. $.2.\overline{).0.48}$ = .24
4
8
8
0

3. $.04.\overline{).07.6}$ = 1.9
4
3 6
3 6
0

4. $.36.\overline{)16.30.8}$ = 45.3
14 4
1 90
1 80
10 8
10 8
0

5. $3.7.\overline{).7.881}$ = .213
7 4
48
37
111
111
0

6. $.028.\overline{)9.128.}$ = 326.
8 4
72
56
168
168
0

Study Exercise Four (Frame 17)

(17A)

1. $.022.\overline{).000.66}$ = .03
66
0

2. $.625.\overline{)15.000.}$ = 24.
12 50
2 500
2 500
0

3. $1.75.\overline{)7.00.}$ = 4.
7 00
0

Study Exercise Five (Frame 19)

(19A)

1. $4\overline{)1.00}$ = .25
8
20
20
0

2. .7

3. $5\overline{)3.0}$ = .6
3 0
0

***Study Exercise Five (Frame 19)* (Continued)**

4. $$\begin{array}{r} .8 \\ 5\overline{)4.0} \\ \underline{4\,0} \\ 0 \end{array}$$

5. $$\begin{array}{r} .4 \\ 5\overline{)2.0} \\ \underline{2\,0} \\ 0 \end{array}$$

6. $$\begin{array}{r} .0625 \\ 16\overline{)1.0000} \\ \underline{0} \\ 1\,00 \\ \underline{96} \\ 40 \\ \underline{32} \\ 80 \\ \underline{80} \\ 0 \end{array}$$

7. $$\begin{array}{r} .625 \\ 8\overline{)5.00} \\ \underline{4\,8} \\ 20 \\ \underline{16} \\ 40 \\ \underline{40} \\ 0 \end{array}$$

8. $$\begin{array}{r} .375 \\ 8\overline{)3.000} \\ \underline{2\,4} \\ 60 \\ \underline{56} \\ 40 \\ \underline{40} \\ 0 \end{array}$$

Study Exercise Six (Frame 22) (22A)

1. $$\begin{array}{r} .66 \\ 3\overline{)2.00} \\ \underline{1\,8} \\ 20 \\ \underline{18} \\ 2 \end{array}$$ quotient is $.66\frac{2}{3}$

2. $$\begin{array}{r} .43 \\ 16\overline{)7.00} \\ \underline{6\,4} \\ 60 \\ \underline{48} \\ 12 \end{array}$$ quotient is $.43\frac{3}{4}$

3. $$\begin{array}{r} .21 \\ 3.4.\overline{).7.32} \\ \underline{6\,8} \\ 52 \\ \underline{34} \\ 18 \end{array}$$ quotient is $.21\frac{9}{17}$

Study Exercise Seven (Frame 25) (25A)

1. $$\begin{array}{r} .578 \\ .14.\overline{).08.100} \\ \underline{7\,0} \\ 1\,10 \\ \underline{98} \\ 120 \\ \underline{112} \\ 8 \end{array}$$ quotient is .58 to 2 decimal places.

2. $$\begin{array}{r} .24 \\ 2.6.\overline{).6.38} \\ \underline{5\,2} \\ 1\,18 \\ \underline{1\,04} \\ 14 \end{array}$$ quotient is .2 to the nearest tenth.

3. $$\begin{array}{r} 1.802 \\ .76.\overline{)1.37.000} \\ \underline{76} \\ 61\,0 \\ \underline{60\,8} \\ 20 \\ \underline{0} \\ 200 \\ \underline{152} \\ 48 \end{array}$$ quotient is 1.80 to nearest hundredth.

4. $$\begin{array}{r} .447 \\ .134.\overline{).060.000} \\ \underline{53\,6} \\ 6\,40 \\ \underline{5\,36} \\ 1\,040 \\ \underline{938} \\ 102 \end{array}$$ quotient is .45 to 2 decimal places.

Study Exercise Seven (Frame 22) **(Continued)**

```
        22.33
5. .006).134.00      quotient is 22.3 to 1 decimal place.
        12
        14
        12
         2 0
         1 8
          20
          18
           2
```

Study Exercise Eight (Frame 28) **(28A)**

1. .756	**2.** .00304	**3.** .000701	**4.** .0311
5. .0051	**6.** .689	**7.** .12	**8.** .00101

UNIT

22 Applied Problems and Equations Involving Decimals

OBJECTIVES (1)

By the end of this unit you should be able to:

1. SOLVE APPLIED PROBLEMS CHANGING TO A BASE OF ONE.
2. FIND WHAT DECIMAL PART ONE NUMBER IS OF ANOTHER.
3. SOLVE EQUATIONS OF THE TYPE $(.3)n = .6$.

Changing to a Base of One (2)

To change to a base of one, use the operation of division.

In each example divide by the number with the same units as the base of one.

Example 1: If 1.5 pounds of potatoes cost 24¢, find the cost of one pound.

Solution: Change to a base of 1 pound by dividing.

Since the base of one is in pounds, divide by 1.5, which is the number of pounds.

$$24 \div 1.5 = 16$$

Computation

```
        1 6.
 1.5.)24.0.
      15
       9 0
       9 0
         0
```

One pound costs 16¢.

Example 2: If a car can travel 57.6 miles on 3.2 gallons of gas, how far can it travel on 1 gallon of gas?

Solution: Change to 1 by division:

Since the base of one is in gallons, divide by 3.2, which is the number of gallons.

Changing to a Base of One (Continued)

$$57.6 \div 3.2 = 18$$

The car travels 18 miles on 1 gallon of gas.

Computation

$$3.2\overline{)57.6} = 18.$$

```
      1 8.
3.2.)57.6.
     32
     25 6
     25 6
        0
```

Study Exercise One (3)

1. If a 100-pound bag of redi-mix concrete costs $1.78, find the cost per pound to the nearest cent.
2. A person made 36 equal payments on a car and paid out $4,337.28. What was the amount of each payment?
3. If a person works 32.5 hours and is paid $103.35, how much does he earn per hour?
4. A car travels 105.4 miles on 6.2 gallons of gas. How far can it travel on 1 gallon?

Solving an Equation (4)

Remember that an equation consists of a left member, an equal sign, and a right member.
Here are examples of equations:

1. $5 \cdot n = 10$ **2.** $.4 \cdot n = 1.6$

The following examples show how to solve these equations.

Example 1: Solve $5 \cdot n = 10$:

Solution: To solve the equation $5 \cdot n = 10$, we must divide 10 by 5

$$5 \cdot n = 10$$
$$n = 10 \div 5$$
$$n = 2$$

Example 2: Solve $.4 \cdot n = 1.6$

Solution: To solve the equation $.4 \cdot n = 1.6$, we must divide 1.6 by .4

$$.4 \cdot n = 1.6$$
$$n = 1.6 \div .4$$
$$n = 4$$

Computation

```
      4.
.4.)1.6.
    1 6
      0
```

Solving Equations Involving a Product (5)

Example 1: Find n if $3 \times n = 12$

Solution: $3 \times n = 12$

$$n = 12 \div 3$$
$$n = 4$$

Solving Equations Involving a Product (Continued)

Example 2: Find n if $n \cdot 6 = 24$

Solution:
$$n \times 6 = 24$$
$$n = 24 \div 6$$
$$n = 4$$

Rule: If a product and one of its factors are known, divide the product by the one factor to find the other factor.

Example 3: Find n if $.3 \times n = 84$

Solution:
$$.3 \times n = 84$$
$$n = 84 \div .3$$
$$n = 280$$

Example 4: Find n if $n \times 210 = 10.5$

Solution:
$$n \times 210 = 10.5$$
$$n = 10.5 \div 210$$
$$n = .05$$

Examples

(6)

1. **Solve:** $(.12)n = .72$

   Solution:
   $$(.12)n = .72$$
   $$n = .72 \div .12$$
   $$n = 6$$

   Computation

   $$.12.\overline{).72.}\quad \text{quotient } 6.$$
   72
   0

2. **Solve:** $(.4)n = 3.25$

   Solution:
   $$(.4)n = 3.25$$
   $$n = 3.25 \div .4$$
   $$n = 8.125$$

   Computation

   $$.4.\overline{)3.2.500}\quad \text{quotient } 8.125$$
   3 2
   5
   4
   10
   8
   20
   20
   0

Examples (Continued)

3. Solve: $(.003)n = 1.60101$

Solution:
$$(.003)n = 1.60101$$
$$n = 1.60101 \div .003$$
$$n = 533.67$$

Computation

```
         533.67
.003.)1.601.01
      1 5
        10
         9
         11
          9
          2 0
          1 8
            21
            21
             0
```

Study Exercise Two (7)

Solve the following equations:

1. $(.3)n = 1.5$
2. $n(.12) = 2.4$
3. $(.50)n = .75$
4. $(.025)n = 4.2$
5. $n(14.2) = 115.02$
6. $(.003)n = 17.01$

Finding What Decimal Part One Number Is of Another (8)

Many times the problem of finding what decimal part one number is of another can be done easily.

For example, what decimal part of 2 is 1? It is easy to see the answer is $\frac{1}{2}$ or .5. To check our answer we multiply .5 by 2. $(.5 \times 2 = 1)$

We want a method that would work with more complicated problems, such as what decimal part of .27 is .18?

Rule: To find what decimal part one number is of another, follow these three steps:
Step (1): Let $n =$ the decimal part
Step (2): Form an equation with 2 factors on the left side and their product on the right.
Step (3): Use the rule for finding an unknown factor.

Example 1: What decimal part of 30 is 6?

Solution: Let $n =$ the decimal part

What decimal part	of	30	is	6
↓	↓	↓	↓	↓
n	$\times$	30	=	6

$$n \times 30 = 6$$
$$n = 6 \div 30$$
$$n = .2$$

Computation

```
      .2
  30)6.0
     6 0
       0
```

Finding What Decimal Part One Number Is of Another (Continued)

Example 2: What decimal part of .27 is .18?

Solution: Let n = the decimal part

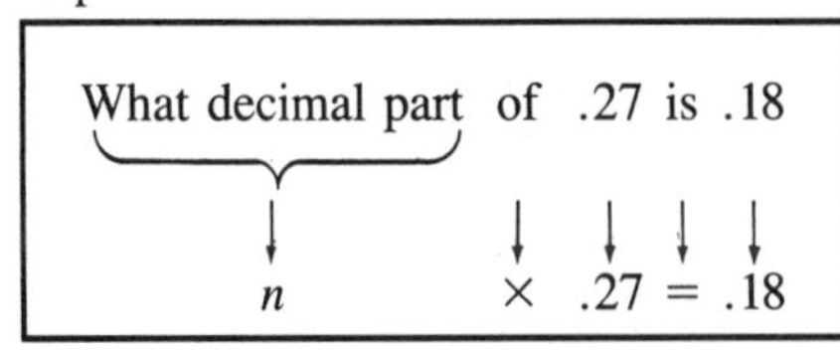

$n \times .27 = .18$

$n = .18 \div .27$

$n = .66\frac{2}{3}$

Computation

$$\begin{array}{r} .66\frac{18}{27} = .66\frac{2}{3} \\ .27.\overline{).18.0} \\ \underline{16\ 2} \\ 1\ 80 \\ \underline{1\ 62} \\ 18 \end{array}$$

Study Exercise Three (9)

1. What decimal part of 24 is 6?
2. What decimal part of .48 is .16?
3. What decimal part of 1.04 is .2496?
4. What decimal part of .12 is .4?
5. What decimal part of .51 is .34?
6. What decimal part of 2.8 is .08?

Finding a Number When a Decimal Part Is Known (10)

Example: If .3 of a number is 15, find the number:

Solution:

Step (1): Let n = the number

Step (2): Form an equation

$.3 \times n = 15$

Step (3): Solve the equation by dividing

$n = 15 \div .3$

$n = 50$

The number is 50.

Computation

$$\begin{array}{r} 5\ 0. \\ .3.\overline{)15.0.} \\ \underline{15} \\ 0 \\ \underline{0} \\ 0 \end{array}$$

Study Exercise Four (11)

1. Find the number if .25 of the number is 32.
2. Find the number if .4 of the number is 12.
3. If .02 of a number is .2, find the number.

Changing Verbal Statements to Equations (12)

Verbal statement: Four-tenths of some number is three and twenty-five hundredths.

Equation statement: We will let n represent the number.

$$(.4)n = 3.25$$

Steps in Forming and Solving Equations (13)

Example 1: If .25 of some number is 3.2, find the number.

Solution:

Step (1): Translate the verbal statement into an equation, letting n represent the unknown number.

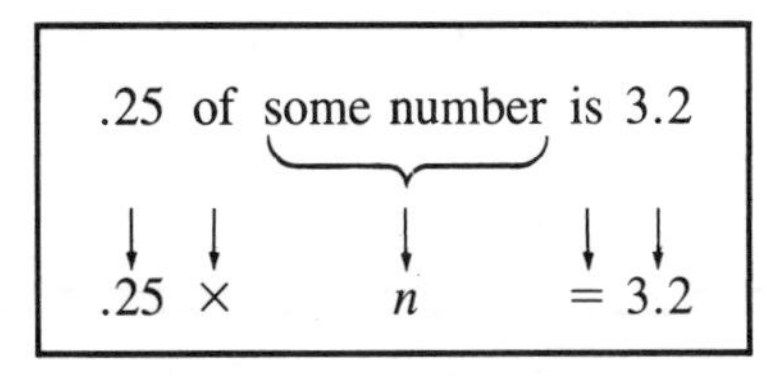

$$(.25)n = 3.2$$

Step (2): Solve the equation by division.

$$(.25)n = 3.2$$
$$n = 3.2 \div .25$$
$$n = 12.8$$

The number is 12.8

Computation

```
         12.8
.25.)3.20.
     2 5
       70
       50
       20 0
       20 0
```

Example 2: Find the number if .23 of the number is 6.1 (answer to nearest hundredth).

Solution:

Step (1): Translate the verbal statement into an equation, letting n represent the unknown number:

.23 of the number is 6.1

.23 · n = 6.1

$$(.23)n = 6.1$$

Steps in Forming and Solving Equations (Continued)

Step (2): Solve the equation by division:

$$(.23)n = 6.1$$
$$n = 6.1 \div .23$$

Step (3): Round as indicated:

$$n = 26.52 \text{ to nearest hundredth.}$$

Computation

```
       26.521
.23.)6.10.000
      4 6
      1 50
      1 38
        12 0
        11 5
           50
           46
            40
            23
```

Study Exercise Five (14)

Change the following statements to equations and solve:

1. If .8 of a number is 7.2, find the number.
2. What is the number if .25 of the number is 2.12?
3. If .3 of a number is .017, what is the number? (answer to nearest hundredth)
4. Find the number if .063 of the number is .2 (answer to nearest hundredth)

REVIEW EXERCISES (15)

1. If a 12.5-pound turkey sells for $11.00, find the cost per pound.
2. A jet travels 5,062.4 miles in 11.2 hours. What is its average rate per hour?
3. What decimal part of .48 is .8?
4. What decimal part of 2.6 is 1.3?
5. Find the number if .05 of the number is 12.1
6. Solve $(.6)n = .135$
7. Solve $(.25)n = 2.5$
8. Solve $(3.1)n = .0651$
9. If .06 of the number is .2, find the number to the nearest hundredth.

Solutions to Review Exercises (16)

1. Change back to 1 pound by division:
   $11.00 ÷ 12.5 = .88
   The cost per pound is $.88

```
            .88
12.5.)11.0.00
      10 0 0
       1 0 00
       1 0 00
            0
```

2. Change to 1 hour by division:
   5,062.4 ÷ 11.2 = 452
   The average rate is 452 miles per hour.

```
          45 2.
11.2.)5062.4.
      448
       582
       560
        22 4
        22 4
           0
```

Solutions to Review Exercises (Continued)

3. Let n = the decimal part

$n(.48) = .8$

$n = .8 \div .48$

$n = 1.66\frac{2}{3}$

$1.66\frac{32}{48} = 1.66\frac{2}{3}$

$.48.\overline{).80.00}$

48
32 0
28 8
3 20
2 88
32

4. Let n = the decimal part

$n \times 2.6 = 1.3$

$n = 1.3 \div 2.6$

$n = .5$

.5

$2.6.\overline{)1.3.0}$

1 3 0
0

5. Let n = the number

$.05 \times n = 12.1$

$n = 12.1 \div .05$

$n = 242$

2 42.

$.05.\overline{)12.10.}$

10
2 1
2 0
10
10
0

6. $(.6)n = .135$

$n = .135 \div .6$

$n = .225$

.225

$.6.\overline{).1.350}$

1 2
15
12
30
30
0

7. $(.25)n = 2.5$

$n = 2.5 \div .25$

$n = 10$

10.

$.25.\overline{)2.50.}$

2 5
0
0
0

8. $(3.1)n = .0651$

$n = .0651 \div 3.1$

$n = .021$

.021

$3.1.\overline{).0.651}$

62
31
31
0

9. .06 of the number is .2

$(.06)n = .2$

$n = .2 \div .06$

$n = 3.33$ (to nearest hundredth)

The number is 3.33

3.333

$.06.\overline{).20.000}$

18
2 0
1 8
20
18
2

SUPPLEMENTARY PROBLEMS

1. If a stack of 100 new dollar bills is 1.6 inches high, how thick is a single dollar bill?
2. If a plane travels 2,100 miles in 8.4 hours, what is its average rate per hour?
3. An automobile battery is guaranteed for 24 months and costs \$28.95. What is the cost per month?
4. A baseball player's batting average is found by dividing the number of hits by the total number of times at bat and expressing the result to 3 decimal places. A player has 182 hits out of 601 times at bat. What is his batting average?
5. A person receives a wage of \$148.75 for 42.5 hours of work. What is her wage per hour?
6. If a car travels 77.4 miles on 4.3 gallons of gas, how far can it travel on 1 gallon of gas?
7. What decimal part of .2 is .002?
8. What decimal part of .55 is 1.705?

SUPPLEMENTARY PROBLEMS (Continued)

9. What decimal part of 3.6 is 18?
10. What decimal part of .56 is .14?
11. What decimal part of .48 is .03?
12. What decimal part of .36 is .6? (answer to the nearest hundredth)
13. Solve $(.6)n = 25.65$
14. Solve $(3.2)n = 96$
15. Solve $(.02)n = 1.41$
16. Solve $(7.1)n = 16.543$
17. Solve $(.007)n = .1407$
18. Solve $(.25)n = .36$
19. Solve $(.035)n = 2.1$
20. Solve $(.54)n = .072$ (answer to nearest hundredth)
21. Solve $(17.1)n = 5$ (answer to nearest hundredth)
22. Write an equation and solve. If .34 of a number is .272, what is the number?
23. Write an equation and solve. If 3.01 of a number is 12.04, what is the number?
24. Write an equation and solve. If 5.4 of a number is 17.6, what is the number to the nearest hundredth?
25. Write an equation and solve. If 7.3 of a number is 18.4, what is the number to the nearest tenth?
26. If 75.7 liters of gas is the same as 20 gallons of gas, how many liters are in one gallon?
27. If a car travels 27.3 kilometers on 4.2 liters of gas, how far will it travel on 1 liter of gas?
28. A sign in an elevator states that its limits are 13 passengers or 2000 pounds. If these limits are equal, how much does an average passenger weigh? (Find answer to nearest tenth.)
29. Sam ran 6.25 miles in 48 minutes. How long did it take him to run an average mile?
30. Mary played 26 records in 57.2 minutes. For how long did an average record play?

Solutions to Study Exercises (3A)

Study Exercise One (Frame 3)

1. Change to 1 pound by division
   $1.78 ÷ 100 = .0178
   = .02 (to nearest cent)
   The cost per pound is $.02

2. $4337.28 ÷ 36 = $120.48
   Each payment is $120.48

```
       120.48
   36)4337.28
      36
       73
       72
        17
         0
        17 2
        14 4
         2 88
         2 88
            0
```

3. $103.35 ÷ 32.5 = $3.18
   The wage per hour is $3.18

```
            3.18
   32.5.)103.3.50
          97 5
           5 8 5
           3 2 5
           2 6 00
           2 6 00
                0
```

4. 105.4 ÷ 6.2 = 17
   The car travels 17 miles on 1 gallon of gas.

```
          1 7.
   6.2.)105.4.
         62
         43 4
         43 4
            0
```

Study Exercise Two (Frame 7) **(7A)**

1. $(.3)n = 1.5$
$n = 1.5 \div .3$
$n = 5$

$.3.\overline{)1.5.}$ quotient $5.$; 15; 0

2. $n(.12) = 2.4$
$n = 2.4 \div .12$
$n = 20$

$.12.\overline{)2.40.}$ quotient $20.$; 24; 0; 0; 0

3. $(.50)n = .75$
$n = .75 \div .50$
$n = 1.5$

$.50.\overline{).75.0}$ quotient 1.5; 50; $25\ 0$; $25\ 0$; 0

4. $(.025)n = 4.2$
$n = 4.2 \div .025$
$n = 168$

$.025.\overline{)4.200.}$ quotient $168.$; 25; $1\ 70$; $1\ 50$; 200; 200; 0

5. $n(14.2) = 115.02$
$n = 115.02 \div 14.2$
$n = 8.1$

$14.2.\overline{)115.0.2}$ quotient 8.1; $113\ 6$; $1\ 42$; $1\ 42$; 0

6. $(.003)n = 17.01$
$n = 17.01 \div .003$
$n = 5{,}670$

$.003.\overline{)17.010.}$ quotient $5\ 670.$; 15; $2\ 0$; $1\ 8$; 21; 21; 0; 0; 0

Study Exercise Three (Frame 9) **(9A)**

In each exercise, let n = the decimal part.

1. $n \times 24 = 6$
$n = 6 \div 24$
$n = .25$

2. $n \times .48 = .16$
$n = .16 \div .48$
$n = .33\frac{1}{3}$

$.48.\overline{).16.00}$ quotient $.33\frac{16}{48} = .33\frac{1}{3}$; $14\ 4$; $1\ 60$; $1\ 44$; 16

3. $n \times 1.04 = .2496$
$n = .2496 \div 1.04$
$n = .24$

$1.04.\overline{).24.96}$ quotient $.24$; $20\ 8$; $4\ 16$; $4\ 16$; 0

4. $n \times .12 = .4$
$n = .4 \div .12$
$n = 3.33\frac{1}{3}$

$.12.\overline{).40.00}$ quotient $3.33\frac{4}{12} = 3.33\frac{1}{3}$; 36; $4\ 0$; $3\ 6$; 40; 36; 4

5. $n \times .51 = .34$
$n = .34 \div .51$
$n = .66\frac{2}{3}$

$.51.\overline{).34.0}$ quotient $.66\frac{34}{51} = .66\frac{2}{3}$; $30\ 6$; $3\ 40$; $3\ 06$; 34

6. $n \times 2.8 = .08$
$n = .08 \div 2.8$
$n = .02\frac{6}{7}$

$2.8.\overline{).0.80}$ quotient $.02\frac{24}{28} = .02\frac{6}{7}$; 56; 24

Study Exercise Four (Frame 11) (11A)

1. Let n = the number

$.25 \times n = 32$

$n = 32 \div .25$

$n = 128$

The number is 128

```
       1 28.
.25.)32.00.
     25
      7 0
      5 0
      2 00
      2 00
         0
```

2. Let n = the number

$.4 \times n = 12$

$n = 12 \div .4$

$n = 30$

The number is 30.

```
     3 0.
.4.)12.0.
    12
      0
      0
      0
```

3. Let n = the number

$.02 \times n = .2$

$n = .2 \div .02$

$n = 10$

The number is 10

```
       10.
.02.).20.
      2
      0
      0
      0
```

Study Exercise Five (Frame 14) (14A)

1. .8 of a number is 7.2

$(.8)n = 7.2$

$n = 7.2 \div .8$

$n = 9$

The number is 9

```
      9.
.8.)7.2.
    7 2
      0
```

2. .25 of the number is 2.12

$(.25)n = 2.12$

$n = 2.12 \div .25$

$n = 8.48$

The number is 8.48

```
       8.48
.25.)2.12.00
     2 00
       12 0
       10 0
        2 00
        2 00
           0
```

3. .3 of a number is .017

$(.3)n = .017$

$n = .017 \div .3$

$n = .06$ (to nearest hundredth)

The number is .06

```
      .056
.3.).0.170
      15
       20
       18
        2
```

4. .063 of the number is .2

$(.063)n = .2$

$n = .2 \div .063$

$n = 3.17$ (to nearest hundredth)

The number is 3.17

```
        3.174
.063.).200.000
       189
        11 0
         6 3
         4 70
         4 41
          290
          252
           38
```

Module 3 Practice Test

Units 17–22

1. What is the place value of 3 in 1.203?
2. Add: $2.034 + .2 + 1.76 + .03$
3. Subtract: $7.86 - 2.9871$
4. Multiply: $2.107 \times .23$
5. Divide: $11.076 \div 1.2$
6. Arrange in order of size from the smallest to the largest:

   2.78, .0278, .278, 27.8
7. Divide and give the quotient correct to the nearest hundredth: $582 \div 39.37$
8. Change to a fraction or mixed numeral:

   (a) $.66\frac{2}{3}$ (b) 2.06
9. Change to decimals:

   (a) $\frac{5}{8}$ (b) $\frac{9}{2{,}000}$
10. Round to the nearest hundredth: 3.4892
11. Write as a decimal: six hundred and twenty-five thousandths.
12. Square 1.01
13. Solve: $(.1)n = .5$
14. Multiply by the short method: 1000×48.21
15. Divide by the short method: $37.2 \div 10$
16. Change $\frac{3}{7}$ to a decimal correct to 3 decimal places.
17. What decimal part of 3.2 is 15.12?
18. Convert $2.01\frac{1}{3}$ to a mixed numeral in lowest terms.
19. If 3.6 pounds cost 72¢, find the cost of 1 pound.
20. If a car can travel 148.4 miles on 11.2 gallons of gas, how far can it travel on 1 gallon of gas?
21. Solve: $(2.4)n = .078$
22. Simplify $\frac{.63}{.09}$
23. Solve: $.2n = 1.2$
24. Find $.03 \times 6 \times .4$
25. What is the number if $.03\frac{1}{2}$ of the number is 2.1?

Answers to Module 3 Practice Test

1. Thousandths.
2. 4.024
3. 4.8729
4. .48461
5. 9.23
6. .0278, .278, 2.78, 27.8
7. 14.78
8. (a) $\frac{2}{3}$ (b) $2\frac{3}{50}$
9. (a) .625 (b) .0045
10. 3.49
11. 600.025
12. 1.0201
13. $n = 5$
14. 48,210
15. 3.72
16. .429
17. 4.725
18. $2\frac{1}{75}$
19. 20¢
20. 13.25 miles.
21. $n = .0325$
22. 7
23. $n = 6$
24. .072
25. 60

MODULE

4

Introduction to Percent

UNIT

23 Changing Percents to Fractions and Decimals

OBJECTIVES (1)

By the end of this unit you should be able to:

1. EXPLAIN THE MEANING OF *PERCENT.*
2. CHANGE A PERCENT TO A FRACTION.
3. CHANGE A PERCENT TO A DECIMAL.
4. FIND A PERCENT OF A NUMBER.

Percent (2)

A *percent* is a fraction in which the numeral preceding the % symbol is the numerator and the % symbol represents a denominator of 100.

Example 1: $5\% = \frac{5}{100}$

Example 2: $286\% = \frac{286}{100}$

Example 3: $6\frac{1}{4}\% = \frac{6\frac{1}{4}}{100}$

Example 4: $4.25\% = \frac{4.25}{100}$

Example 5: $100\% = \frac{100}{100}$

Percent is not a new kind of number. It is merely a more convenient way of expressing hundredths. **(3)**

Example 1: "Our bank pays $5\frac{3}{4}\%$ interest on savings accounts."

Example 2: "The carrying charge on this refrigerator is 12%."

Example 3: "The salesman's commission amounts to $33\frac{1}{3}\%$ of the selling price."

The % symbol is convenient to write and to say, but we must be able to change percents to either fractions or decimals in order to solve problems.

Changing Percents to Fractions (4)

Step (1): Write the percent without the % symbol by placing the numeral over 100.

$$10\% = \frac{10}{100}$$

Step (2): Reduce the resulting fraction whenever possible.

$$\frac{10}{100} = \frac{10 \div 10}{100 \div 10} = \frac{1}{10}$$

Example 1: $25\% = \frac{25}{100} = \frac{25 \div 25}{100 \div 25} = \frac{1}{4}$

Example 2: $52\% = \frac{52}{100} = \frac{52 \div 4}{100 \div 4} = \frac{13}{25}$

Example 3: $19\% = \frac{19}{100}$ (can't be reduced)

Example 4: $500\% = \frac{500}{100} = \frac{5}{1} = 5$

Example 5: $425\% = \frac{425}{100} = 4\frac{25}{100} = 4\frac{1}{4}$

Study Exercise One (5)

Express each of the following percents as a numeral naming a fraction in lowest terms:

1. 50% **2.** 75% **3.** 20% **4.** 68% **5.** 59%
6. 100% **7.** 280% **8.** 150% **9.** 800%

Percents which Contain Fractions (6)

% means hundredths

$$\square\% = \frac{\square}{100}$$

Problem: Change $12\frac{1}{2}\%$ to a fraction in lowest terms.

Step (1): Write the percent without the % symbol by placing the numeral over 100.

$$12\frac{1}{2}\% = \frac{12\frac{1}{2}}{100}$$

Percents which Contain Fractions (Continued)

Step (2): If the numerator of the resulting complex fraction is a mixed numeral, change it to an improper fraction.

$$\frac{12\frac{1}{2}}{100} = \frac{\frac{25}{2}}{100}$$

Step (3): Divide by the denominator of 100.

$$\frac{\frac{25}{2}}{100} = \frac{25}{2} \div 100$$

$$= \frac{\cancel{25}^{1}}{2} \times \frac{1}{\cancel{100}_{4}}$$

$$= \frac{1}{8}$$

Examples of Percents which Contain Fractions (7)

Example 1: Change $33\frac{1}{3}\%$ to a fraction in lowest terms:

Solution:

Line (a): $33\frac{1}{3}\% = \frac{33\frac{1}{3}}{100}$

Line (b): $= \frac{\frac{100}{3}}{100}$

Line (c): $= \frac{100}{3} \div 100$

Line (d): $= \frac{\cancel{100}^{1}}{3} \times \frac{1}{\cancel{100}_{1}}$

Line (e): $= \frac{1}{3}$

Example 2: Change $116\frac{2}{3}\%$ to a mixed numeral in lowest terms:

Solution:

Line (a): $116\frac{2}{3}\% = \frac{116\frac{2}{3}}{100}$

Line (b): $= \frac{\frac{350}{3}}{100}$

Line (c): $= \frac{350}{3} \div 100$

Examples of Percents which Contain Fractions (Continued)

Line (d): $= \frac{\cancel{350}^{\cancel{35}^{7}}}{3} \times \frac{1}{\cancel{100}_{\cancel{10}_{2}}}$

Line (e): $= \frac{7}{6}$

Line (f): $= 1\frac{1}{6}$

Example 3: Change $\frac{1}{2}\%$ to a fraction in lowest terms:

Solution:

Line (a): $\frac{1}{2}\% = \frac{\frac{1}{2}}{100}$

Line (b): $= \frac{1}{2} \div 100$

Line (c): $= \frac{1}{2} \times \frac{1}{100}$

Line (d): $= \frac{1}{200}$

Example 4: Change $\frac{3}{4}\%$ to a fraction in lowest terms:

Solution:

Line (a): $\frac{3}{4}\% = \frac{\frac{3}{4}}{100}$

Line (b): $= \frac{3}{4} \div 100$

Line (c): $= \frac{3}{4} \times \frac{1}{100}$

Line (d): $= \frac{3}{400}$

Study Exercise Two (8)

Express each of the following percents as either a fraction or a mixed numeral in lowest terms:

1. $\frac{1}{4}\%$ **2.** $87\frac{1}{2}\%$ **3.** $16\frac{2}{3}\%$ **4.** $6\frac{1}{4}\%$ **5.** $112\frac{1}{2}\%$

Changing Percents to Decimals (9)

Problem: Change 14% to a decimal.

Step (1): Write the percent without the % symbol by placing the numeral over 100.

$$14\% = \frac{14}{100}$$

Changing Percents to Decimals (Continued)

Step (2): Divide by the denominator of 100.

$$\frac{14}{100} = 14 \div 100$$
$$= .14$$

Example 1: $115\% = \frac{115}{100}$
$= 115 \div 100$
$= 1.15$

Example 2: $215.8\% = \frac{215.8}{100}$
$= 215.8 \div 100$
$= 2.158$

A Short Cut (10)

The % symbol represents hundredths or 2 decimal places.

Therefore, to change a percent to a decimal: Move the decimal point 2 places to the left and remove the % symbol.

Example 1: $215.3\% = 2.153$

Example 2: $100\% = 1.00$
$= 1$

Example 3: $5\% = .05$

Example 4: $3\frac{1}{4}\% = 3.25\%$
$= .0325$

Example 5: $.6\% = .006$

Example 6: $2\frac{1}{2}\% = 2.5\%$
$= .025$

Study Exercise Three (11)

Express the following percents as decimals:

1. 24%
2. 4.7%
3. 148.63%
4. .3%
5. .475%
6. 102%
7. $4\frac{3}{4}\%$
8. $37\frac{1}{2}\%$

Finding a Percent of a Number (12)

To find a percent of a number, perform the following three steps:

Step (1): Change 'of' to 'times'.
Step (2): Change the percent to a decimal.
Step (3): Complete the multiplication.

Example 1: Find 25% of 84.

Step (1): $25\% \text{ of } 84 = 25\% \times 84$
Step (2): $25\% \times 84 = .25 \times 84$
Step (3):

```
   84
 ×.25
 ----
  4 20
 16 8
 -----
 21.00
```

Therefore, 25% of 84 = 21

Finding a Percent of a Number (Continued)

Example 2: Find 6% of 92

Step (1): 6% of $92 = 6\% \times 92$

Step (2): $6\% \times 92 = .06 \times 92$

Step (3):
$$\begin{array}{r} 92 \\ \times .06 \\ \hline 5.52 \end{array}$$

Therefore, 6% of 92 = 5.52

Example 3: Find 123% of 500

Step (1): 123% of $500 = 123\% \times 500$

Step (2): $123\% \times 500 = 1.23 \times 500$

Step (3):
$$\begin{array}{r} 1.23 \\ \times 500 \\ \hline 615.00 \end{array}$$

Therefore, 123% of 500 = 615

Study Exercise Four (13)

Find:

1. 50% of 36 **2.** 18% of 90 **3.** 2% of 48 **4.** 119% of 73

REVIEW EXERCISES (14)

A. **1.** Explain the meaning of *percent.*

B. **2.** The % symbol acts like a denominator of ________.

C. Change the following percents to either a fraction or a mixed numeral in lowest terms:

3. 30% **4.** $87\frac{1}{2}\%$ **5.** $\frac{1}{3}\%$ **6.** 80%

7. 100% **8.** 225% **9.** $166\frac{2}{3}\%$ **10.** 500%

D. Change the following percents to decimals:

11. 23.6% **12.** .7% **13.** $5\frac{3}{4}\%$ **14.** 25%

E. Find:

15. 40% of 65 **16.** 17% of 38 **17.** 3% of 8 **18.** 137% of 210

Solutions to Review Exercises (15)

A. **1.** See Frame 2

B. **2.** 100

C. **3.** $30\% = \dfrac{30 \div 10}{100 \div 10} = \dfrac{3}{10}$

4. $87\frac{1}{2}\% = \dfrac{87\frac{1}{2}}{100} = \dfrac{\frac{175}{2}}{100} = \dfrac{175}{2} \div 100$

$= \dfrac{\cancel{175}^{7}}{2} \times \dfrac{1}{\cancel{100}_{4}} = \dfrac{7}{8}$

Solutions to Review Exercises (Continued)

5. $\frac{1}{3}\% = \frac{\frac{1}{3}}{100} = \frac{1}{3} \div 100 = \frac{1}{3} \times \frac{1}{100} = \frac{1}{300}$

6. $80\% = \frac{80 \div 10}{100 \div 10} = \frac{8 \div 2}{10 \div 2} = \frac{4}{5}$

7. $100\% = \frac{100}{100} = 1$

8. $225\% = \frac{225}{100} = 2\frac{25}{100} = 2\frac{25 \div 25}{100 \div 25} = 2\frac{1}{4}$

9. $166\frac{2}{3}\% = \frac{166\frac{2}{3}}{100} = \frac{\frac{500}{3}}{100} = \frac{500}{3} \div 100$

$= \frac{\cancel{500}^5}{3} \times \frac{1}{\cancel{100}_1} = \frac{5}{3} = 1\frac{2}{3}$

10. $500\% = \frac{500}{100} = \frac{5}{1}$ or 5

D. 11. $23.6\% = .236$

12. $.7\% = .007$

13. $5\frac{3}{4}\% = 5.75\% = .0575$

14. $25\% = .25$

E. 15. 40% of $65 = 40\% \times 65$
$= .40 \times 65$
$= 26$

16. 17% of $38 = 17\% \times 38$
$= .17 \times 38$
$= 6.46$

17. 3% of $8 = 3\% \times 8$
$= .03 \times 8$
$= .24$

18. 137% of $210 = 137\% \times 210$
$= 1.37 \times 210$
$= 287.7$

SUPPLEMENTARY PROBLEMS

A. Change the following percents to either a fraction or a mixed numeral in lowest terms:

1. 6% **2.** 45% **3.** 13% **4.** $33\frac{1}{3}\%$ **5.** $5\frac{3}{4}\%$

6. 225% **7.** 120% **8.** $\frac{1}{5}\%$ **9.** $12\frac{1}{2}\%$ **10.** 175%

11. $6\frac{1}{4}\%$ **12.** 22% **13.** $16\frac{2}{3}\%$ **14.** 56% **15.** 700%

B. Change the following percents to decimals:

16. 23% **17.** 8% **18.** 15.3% **19.** 152.8% **20.** 108%

21. .2% **22.** $5\frac{3}{4}\%$ **23.** $37\frac{1}{2}\%$ **24.** .431% **25.** 3.6%

C. Find:

26. 36% of 42 **27.** 75% of 44 **28.** $33\frac{1}{3}\%$ of 96.

29. .3% of 420 **30.** 175% of 44 **31.** $6\frac{1}{4}\%$ of 144

D. Answer the following:

32. Meg says she will put 50% of her earnings into her savings account. If she earns $288, how much will she put into her savings account?

33. Reggie Jackson agreed to pay 8% of his earnings to his agent. If Reggie earns $946,000, how much will he pay his agent?

34. Henry says 75% of his class are females. If his class has 36 students, how many are females?

Solutions to Study Exercises

(5A)

Study Exercise One (Frame 5)

1. $50\% = \frac{50}{100} = \frac{50 \div 50}{100 \div 50} = \frac{1}{2}$

2. $75\% = \frac{75}{100} = \frac{75 \div 25}{100 \div 25} = \frac{3}{4}$

3. $20\% = \frac{20}{100} = \frac{20 \div 20}{100 \div 20} = \frac{1}{5}$

4. $68\% = \frac{68}{100} = \frac{68 \div 4}{100 \div 4} = \frac{17}{25}$

5. $59\% = \frac{59}{100}$

(can't be reduced)

6. $100\% = \frac{100}{100} = \frac{1}{1} = 1$

7. $280\% = \frac{280}{100} = 2\frac{80}{100} = 2\frac{4}{5}$

8. $150\% = \frac{150}{100} = 1\frac{50}{100} = 1\frac{1}{2}$

9. $800\% = \frac{800}{100} = \frac{8}{1} = 8$

Study Exercise Two (Frame 8)

(8A)

1. $\frac{1}{4}\% = \frac{\frac{1}{4}}{100} = \frac{1}{4} \div 100 = \frac{1}{4} \times \frac{1}{100} = \frac{1}{400}$

2. $87\frac{1}{2}\% = \frac{87\frac{1}{2}}{100} = \frac{\frac{175}{2}}{100} = \frac{175}{2} \div 100 = \frac{\cancel{175}^{\cancel{35}\,7}}{2} \times \frac{1}{\cancel{100}_{\cancel{20}\,4}} = \frac{7}{8}$

3. $16\frac{2}{3}\% = \frac{16\frac{2}{3}}{100} = \frac{\frac{50}{3}}{100} = \frac{50}{3} \div 100 = \frac{\cancel{50}^{1}}{3} \times \frac{1}{\cancel{100}_{2}} = \frac{1}{6}$

4. $6\frac{1}{4}\% = \frac{6\frac{1}{4}}{100} = \frac{\frac{25}{4}}{100} = \frac{25}{4} \div 100 = \frac{\cancel{25}^{1}}{4} \times \frac{1}{\cancel{100}_{4}} = \frac{1}{16}$

5. $112\frac{1}{2}\% = \frac{112\frac{1}{2}}{100} = \frac{\frac{225}{2}}{100} = \frac{225}{2} \div 100 = \frac{\cancel{225}^{\cancel{45}\,9}}{2} \times \frac{1}{\cancel{100}_{\cancel{20}\,4}} = \frac{9}{8} = 1\frac{1}{8}$

Study Exercise Three (Frame 11) (11A)

1. $24\% = .24$
2. $4.7\% = .047$
3. $148.63\% = 1.4863$
4. $.3\% = .003$
5. $.475\% = .00475$
6. $102\% = 1.02$
7. $4\frac{3}{4}\% = 4.75\% = .0475$
8. $37\frac{1}{2}\% = 37.5\% = .375$

Study Exercise Four (Frame 13) (13A)

1. 50% of $36 = 50\% \times 36$
   $= .50 \times 36$
   $= 18$
2. 18% of $90 = 18\% \times 90$
   $= .18 \times 90$
   $= 16.2$
3. 2% of $48 = 2\% \times 48$
   $= .02 \times 48$
   $= .96$
4. 119% of $73 = 119\% \times 73$
   $= 1.19 \times 73$
   $= 86.87$

UNIT

24 Changing Decimals and Fractions to Percents

OBJECTIVES (1)

By the end of this unit you should be able to:

1. CHANGE A DECIMAL TO A PERCENT.
2. CHANGE A FRACTION TO A PERCENT.
3. COMPLETE A CHART OF USEFUL EQUIVALENTS FROM MEMORY.
4. ROUND PERCENTS TO A GIVEN PLACE VALUE.

Changing Decimals to Percents (2)

Since we know the % symbol represents hundredths (two decimal places), we can use the % symbol instead of two decimal places. Therefore, to express a decimal as a percent do the following: Move the decimal point 2 places to the right and attach the % symbol.

Example 1: .63 = 63% **Example 2:** 4.2 = 420% **Example 3:** .0003 = .03%

Example 4: .2 = 20% **Example 5:** 8 = 800% **Example 6:** $.66\frac{2}{3} = 66\frac{2}{3}\%$

Study Exercise One (3)

Express each of the following decimals as a percent:

1. .35 **2.** .03 **3.** 5 **4.** $.33\frac{1}{3}$ **5.** .006 **6.** .3

Changing Fractions with Denominators of 100 to Percents (4)

The % symbol represents the denominator 100. Therefore, to change a fraction with a denominator of 100 to a percent do the following: Remove the denominator 100 and attach the % symbol.

Example 1: $\frac{25}{100} = 25\%$ **Example 2:** $\frac{42}{100} = 42\%$ **Example 3:** $\frac{387}{100} = 387\%$

Changing Fractions with Denominators of 100 to Percents (Continued)

Example 4: $\frac{100}{100} = 100\%$ **Example 5:** $\frac{2\frac{1}{2}}{100} = 2\frac{1}{2}\%$

Study Exercise Two (5)

Change the following fractions to percents:

1. $\frac{7}{100}$ 2. $\frac{82}{100}$ 3. $\frac{284}{100}$ 4. $\frac{100}{100}$ 5. $\frac{3\frac{3}{4}}{100}$ 6. $\frac{33\frac{1}{3}}{100}$

Fractions Not Having a Denominator of 100 (6)

Some denominators can easily be changed to 100.

Example 1: $\frac{3}{20} = \frac{3 \times 5}{20 \times 5} = \frac{15}{100} = 15\%$

Example 2: $\frac{7}{10} = \frac{7 \times 10}{10 \times 10} = \frac{70}{100} = 70\%$

Example 3: $\frac{8}{25} = \frac{8 \times 4}{25 \times 4} = \frac{32}{100} = 32\%$

Example 4: $\frac{15}{300} = \frac{15 \div 3}{300 \div 3} = \frac{5}{100} = 5\%$

Example 5: $\frac{3}{200} = \frac{3 \div 2}{200 \div 2} = \frac{1\frac{1}{2}}{100} = 1\frac{1}{2}\%$

Example 6: $\frac{300}{200} = \frac{300 \div 2}{200 \div 2} = \frac{150}{100} = 150\%$

Study Exercise Three (7)

Change the following fractions to percents:

1. $\frac{13}{20}$ 2. $\frac{9}{10}$ 3. $\frac{4}{25}$ 4. $\frac{3}{5}$

5. $\frac{124}{200}$ 6. $\frac{12}{400}$ 7. $\frac{17}{300}$ 8. $\frac{500}{200}$

If the denominator cannot be easily changed to 100, proceed as follows: **(8)**

Problem: Change $\frac{3}{8}$ to a percent.

Step (1): Change the fraction to a *two-place* decimal:

Line (a): $\frac{3}{8} = 3 \div 8$

$$.37\frac{4}{8} = .37\frac{1}{2}$$

Line (b):
$$\begin{array}{r} 8\overline{)3.00} \\ \underline{2\,4} \\ 60 \\ \underline{56} \\ 4 \end{array}$$

Line (c): $= .37\frac{1}{2}$

Step (2): Change the two-place decimal to a percent by moving the decimal point two places to the right and attaching the % symbol:

$$.37\frac{1}{2} = 37\frac{1}{2}\%$$

Examples

(9)

Example 1: $\frac{5}{6} = 5 \div 6$

$= .83\frac{1}{3}$

$= 83\frac{1}{3}\%$

Computation

$$.83\frac{2}{6} = .83\frac{1}{3}$$

$$\begin{array}{r} 6\overline{)5.00} \\ \underline{4\,8} \\ 20 \\ \underline{18} \\ 2 \end{array}$$

Example 2: $\frac{43}{7} = 43 \div 7$

$= 6.14\frac{2}{7}$

$= 614\frac{2}{7}\%$

Computation

$$6.14\frac{2}{7}$$

$$\begin{array}{r} 7\overline{)43.00} \\ \underline{42} \\ 1\,0 \\ \underline{7} \\ 30 \\ \underline{28} \\ 2 \end{array}$$

Example 3: $2\frac{1}{8} = \frac{17}{8}$

$= 17 \div 8$

$= 2.12\frac{1}{2}$

$= 212\frac{1}{2}\%$

Computation

$$2.12\frac{4}{8} = 2.12\frac{1}{2}$$

$$\begin{array}{r} 8\overline{)17.00} \\ \underline{16} \\ 1\,0 \\ \underline{8} \\ 20 \\ \underline{16} \\ 4 \end{array}$$

Study Exercise Four (10)

Convert the following to percents:

1. $\frac{3}{8}$ **2.** $\frac{1}{3}$ **3.** $\frac{4}{7}$ **4.** $5\frac{1}{6}$ **5.** $2\frac{3}{11}$

Useful Equivalents (11)

Memorize the following equivalents:

Percent	Fraction	Decimal
10%	$\frac{1}{10}$	.1
20%	$\frac{1}{5}$	.2
30%	$\frac{3}{10}$	.3
40%	$\frac{2}{5}$	.4
50%	$\frac{1}{2}$	.5
60%	$\frac{3}{5}$	.6
70%	$\frac{7}{10}$	.7
80%	$\frac{4}{5}$	.8
90%	$\frac{9}{10}$	.9
100%	1	1.0
25%	$\frac{1}{4}$	.25
75%	$\frac{3}{4}$	.75
$33\frac{1}{3}\%$	$\frac{1}{3}$	$.33\frac{1}{3}$
$66\frac{2}{3}\%$	$\frac{2}{3}$	$.66\frac{2}{3}$

Study Exercise Five (12)

Complete this chart from memory:

	Percent	Fraction	Decimal
1.	10%		
2.		$\frac{7}{10}$	
3.			.6
4.		$\frac{2}{5}$	
5.	50%		
6.			.3
7.		$\frac{1}{5}$	
8.	$66\frac{2}{3}\%$		
9.		$\frac{1}{4}$	
10.			.9
11.		$\frac{1}{3}$	
12.		1	
13.			.8
14.	75%		

Expressing a Percent as an Approximate Decimal by Rounding (13)

Example 1: Change $\frac{1}{7}$ to a percent:

Line (a): $\frac{1}{7} = 7\overline{)1}$

Expressing a Percent as an Approximate Decimal by Rounding (Continued)

Line (b):

$$\begin{array}{r} .142857\ldots \\ 7\overline{)1.000000} \\ \underline{7} \\ 30 \\ \underline{28} \\ 20 \\ \underline{14} \\ 60 \\ \underline{56} \\ 40 \\ \underline{35} \\ 50 \\ \underline{49} \\ 1 \end{array}$$

(where do we stop?)

Rounding Percents (14)

	Division	Percent	Nearest Whole Percent	Nearest Tenth Percent	Nearest Hundredth Percent
1.	$7\overline{)1.000}$ = .142	14.2%	14%		
2.	$7\overline{)1.0000}$ = .1428	14.28%		14.3%	
3.	$7\overline{)1.00000}$ = .14285	14.285%			14.29%

Note: The symbol ≈ means "is approximately equal to." On occasion we will use it to indicate that a number has been rounded.

Example: 14.2% ≈ 14%

Study Exercise Six (15)

Change each fraction to a percent and round as instructed:

A. Round to the nearest whole percent:

1. $\frac{1}{3}$ **2.** $\frac{4}{7}$

B. Round to the nearest tenth of a percent:

3. $\frac{4}{17}$ **4.** $\frac{1}{23}$

C. Round to the nearest hundredth of a percent:

5. $\frac{1}{23}$

D. Round to the nearest thousandth of a percent:

6. $\frac{1}{23}$

REVIEW EXERCISES (16)

A. Change each of the following decimals to a percent:

1. .57 **2.** .08 **3.** 6 **4.** $.16\frac{2}{3}$ **5.** .005

B. Change each of the following fractions to a percent:

6. $\frac{9}{100}$ **7.** $\frac{17}{25}$ **8.** $\frac{14}{300}$ **9.** $\frac{6}{7}$ **10.** $2\frac{1}{2}$

C. Complete the following chart from memory:

	Percent	Fraction	Decimal
11.	25%		
12.		$\frac{3}{4}$	
13.			$.33\frac{1}{3}$
14.		$\frac{2}{5}$	
15.	$66\frac{2}{3}\%$		
16.		$\frac{7}{10}$	
17.	30%		
18.		$\frac{1}{2}$	

D. Round to the nearest whole percent:

19. $\frac{3}{7}$ **20.** $\frac{9}{41}$

E. Round to the nearest tenth of a percent:

21. $\frac{4}{11}$ **22.** $\frac{2}{19}$

Solutions to Review Exercises (17)

A. **1.** .57 = 57% **2.** .08 = 8% **3.** 6 = 600%

4. $.16\frac{2}{3} = 16\frac{2}{3}\%$ **5.** .005 = .5%

Solutions to Review Exercises (Continued)

B. **6.** $\frac{9}{100} = 9\%$

7. $\frac{17}{25} = \frac{17 \times 4}{25 \times 4} = \frac{68}{100} = 68\%$

8. $\frac{14}{300} = \frac{14 \div 3}{300 \div 3} = \frac{4\frac{2}{3}}{100} = 4\frac{2}{3}\%$

9. $\frac{6}{7} = 6 \div 7 = .85\frac{5}{7} = 85\frac{5}{7}\%$

Computation

$$\begin{array}{r} .85\frac{5}{7} \\ 7\overline{)6.00} \\ \underline{5\ 6} \\ 40 \\ \underline{35} \\ 5 \end{array}$$

10. $2\frac{1}{2} = \frac{5}{2} = 5 \div 2 = 2.50 = 250\%$

Computation

$$\begin{array}{r} 2.50 \\ 2\overline{)5.00} \\ \underline{4} \\ 1\ 0 \\ \underline{1\ 0} \\ 0 \end{array}$$

C.

	Percent	Fraction	Decimal
11.	25%	$\frac{1}{4}$	.25
12.	75%	$\frac{3}{4}$	.75
13.	$33\frac{1}{3}\%$	$\frac{1}{3}$	$.33\frac{1}{3}$
14.	40%	$\frac{2}{5}$	.4
15.	$66\frac{2}{3}\%$	$\frac{2}{3}$	$.66\frac{2}{3}$
16.	70%	$\frac{7}{10}$	.7
17.	30%	$\frac{3}{10}$	.3
18.	50%	$\frac{1}{2}$	.5

D. **19.** $7\overline{)3.000}$ = .428 ≈ 43%

20. $41\overline{)9.000}$ = .219 ≈ 22%

E. **21.** $11\overline{)4.0000}$ = .3636 ≈ 36.4%

22. $19\overline{)2.0000}$ = .1052 ≈ 10.5%

SUPPLEMENTARY PROBLEMS

A. Change each of the following decimals to a percent:

1. .8327 **2.** 2.25 **3.** $.37\frac{1}{2}$ **4.** $.66\frac{2}{3}$

5. .008 **6.** 9 **7.** $.04\frac{3}{4}$

B. Change each of the following fractions to a percent:

8. $\frac{2}{5}$ **9.** $\frac{3}{16}$ **10.** $\frac{1}{8}$ **11.** $\frac{17}{400}$

12. $\frac{14}{20}$ **13.** $\frac{6}{25}$ **14.** $8\frac{1}{3}$ **15.** $\frac{3}{20}$

C. Complete the following chart from memory:

	Percent	Fraction	Decimal
16.	20%		
17.	25%		
18.	30%		
19.	$33\frac{1}{3}\%$		
20.	10%		
21.	75%		
22.	40%		
23.	$66\frac{2}{3}\%$		
24.	80%		
25.	50%		
26.	70%		
27.	90%		

D. Round to the nearest whole percent:

28. $\frac{5}{7}$ **29.** $\frac{8}{41}$

E. Round to the nearest tenth of a percent:

30. $\frac{3}{19}$ **31.** $\frac{5}{11}$

Solutions to Study Exercises

(3A)

Study Exercise One (Frame 3)

1. $.35 = 35\%$
2. $.03 = 3\%$
3. $5 = 500\%$
4. $.33\frac{1}{3} = 33\frac{1}{3}\%$
5. $.006 = .6\%$
6. $.3 = 30\%$

Study Exercise Two (Frame 5)

(5A)

1. $\frac{7}{100} = 7\%$
2. $\frac{82}{100} = 82\%$
3. $\frac{284}{100} = 284\%$
4. $\frac{100}{100} = 100\%$
5. $\frac{3\frac{3}{4}}{100} = 3\frac{3}{4}\%$
6. $\frac{33\frac{1}{3}}{100} = 33\frac{1}{3}\%$

Study Exercise Three (Frame 7)

(7A)

1. $\frac{13}{20} = \frac{13 \times 5}{20 \times 5} = \frac{65}{100} = 65\%$
2. $\frac{9}{10} = \frac{9 \times 10}{10 \times 10} = \frac{90}{100} = 90\%$
3. $\frac{4}{25} = \frac{4 \times 4}{25 \times 4} = \frac{16}{100} = 16\%$
4. $\frac{3}{5} = \frac{3 \times 20}{5 \times 20} = \frac{60}{100} = 60\%$
5. $\frac{124}{200} = \frac{124 \div 2}{200 \div 2} = \frac{62}{100} = 62\%$
6. $\frac{12}{400} = \frac{12 \div 4}{400 \div 4} = \frac{3}{100} = 3\%$
7. $\frac{17}{300} = \frac{17 \div 3}{300 \div 3} = \frac{5\frac{2}{3}}{100} = 5\frac{2}{3}\%$
8. $\frac{500}{200} = \frac{500 \div 2}{200 \div 2} = \frac{250}{100} = 250\%$

Study Exercise Four (Frame 10)

(10A)

1. $\frac{3}{8} = 3 \div 8 = .37\frac{1}{2} = 37\frac{1}{2}\%$

$$8\overline{)3.00} = .37\frac{4}{8} = .37\frac{1}{2};\quad 24,\ 60,\ 56,\ 4$$

2. $\frac{1}{3} = 1 \div 3 = .33\frac{1}{3} = 33\frac{1}{3}\%$

$$3\overline{)1.00} = .33\frac{1}{3};\quad 9,\ 10,\ 9,\ 1$$

3. $\frac{4}{7} = 4 \div 7 = .57\frac{1}{7} = 57\frac{1}{7}\%$

$$7\overline{)4.00} = .57\frac{1}{7};\quad 35,\ 50,\ 49,\ 1$$

4. $5\frac{1}{6} = \frac{31}{6} = 31 \div 6 = 5.16\frac{2}{3} = 516\frac{2}{3}\%$

$$6\overline{)31.00} = 5.16\frac{4}{6} = 5.16\frac{2}{3};\quad 30,\ 10,\ 6,\ 40,\ 36,\ 4$$

5. $2\frac{3}{11} = \frac{25}{11} = 25 \div 11 = 2.27\frac{3}{11} = 227\frac{3}{11}\%$

$$11\overline{)25.00} = 2.27\frac{3}{11};\quad 22,\ 30,\ 22,\ 80,\ 77,\ 3$$

Study Exercise Five (Frame 12) **(12A)**

	Percent	Fraction	Decimal
1.	10%	$\frac{1}{10}$	.1
2.	70%	$\frac{7}{10}$	.7
3.	60%	$\frac{3}{5}$	.6
4.	40%	$\frac{2}{5}$	.4
5.	50%	$\frac{1}{2}$	.5
6.	30%	$\frac{3}{10}$	.3
7.	20%	$\frac{1}{5}$	.2
8.	$66\frac{2}{3}\%$	$\frac{2}{3}$	$.66\frac{2}{3}$
9.	25%	$\frac{1}{4}$	.25
10.	90%	$\frac{9}{10}$	.9
11.	$33\frac{1}{3}\%$	$\frac{1}{3}$	$.33\frac{1}{3}$
12.	100%	1	1.0
13.	80%	$\frac{4}{5}$	.8
14.	75%	$\frac{3}{4}$	.75

Study Exercise Six (Frame 15) **(15A)**

A. 1. $3\overline{)1.000}$ → $.333 \approx 33\%$

2. $7\overline{)4.000}$ → $.571 \approx 57\%$

B. 3. $17\overline{)4.0000}$ → $.2352 \approx 23.5\%$

4. $23\overline{)1.0000}$ → $.0434 \approx 4.3\%$

C. 5. $23\overline{)1.00000}$ → $.04347 \approx 4.35\%$

D. 6. $23\overline{)1.000000}$ → $.043478 \approx 4.348\%$

UNIT

25 Three Types of Percent Problems

OBJECTIVES (1)

By the end of this unit you should be able to:

1. FIND A PERCENT OF A NUMBER.
2. FIND THE NUMBER WHEN A PERCENT OF IT IS KNOWN.
3. FIND WHAT PERCENT ONE NUMBER IS OF ANOTHER.
4. SOLVE APPLIED PROBLEMS INVOLVING PERCENT.

Finding a Percent of a Number (2)

Problem: Find 14% of 83:

Solution:

Step (1): Change "of" to "times."

$14\% \text{ of } 83 = 14\% \times 83$

Step (2): Change the percent to a decimal.

$14\% \times 83 = .14 \times 83$

Step (3): Complete the multiplication.

$$\begin{array}{r} 83 \\ \times .14 \\ \hline 3\ 32 \\ 8\ 3 \\ \hline 11.62 \end{array}$$

Therefore, 14% of 83 = 11.62

Examples

(3)

Example 1: Find 112% of 16:

Solution:

Step (1): 112% of $16 = 112\% \times 16$

Step (2): $112\% \times 16 = 1.12 \times 16$

Step (3):

$$\begin{array}{r} 1.12 \\ \times 16 \\ \hline 6\ 72 \\ 11\ 2 \\ \hline 17.92 \end{array}$$

Therefore, 112% of $16 = 17.92$

Example 2: Find $5\frac{3}{4}\%$ of 28:

Solution:

Step (1): $5\frac{3}{4}\%$ of $28 = 5\frac{3}{4}\% \times 28$

Step (2): $5\frac{3}{4}\% \times 28 = 5.75\% \times 28$

$= .0575 \times 28$

Step (3):

$$\begin{array}{r} .0575 \\ \times 28 \\ \hline 4600 \\ 1\ 150 \\ \hline 1.6100 \end{array}$$

Therefore, $5\frac{3}{4}\%$ of $28 = 1.61$

Study Exercise One

(4)

Find:

1. 42% of 57 **2.** 6.2% of 50 **3.** 120% of 16 **4.** $2\frac{3}{4}\%$ of 35

Using Percents Changed to Fractions

(5)

If we recognize the fractional equivalent of a given percent, we may find the problem easier to work by changing the percent to a fraction instead of a decimal.

Example 1: Find $33\frac{1}{3}\%$ of 12:

Solution: The fractional equivalent of $33\frac{1}{3}\%$ is $\frac{1}{3}$

Line (a): $33\frac{1}{3}\%$ of $12 = \frac{1}{3} \times 12$

Using Percents Changed to Fractions (Continued)

Line (b): $= \frac{1}{\not{3}_1} \times \not{12}^4$

Line (c): $= 4$

Example 2: Find 75% of 48:

Solution: The fractional equivalent of 75% is $\frac{3}{4}$

Line (a): $75\% \times 48 = \frac{3}{4} \times 48$

Line (b): $= \frac{3}{\not{4}_1} \times \not{48}^{12}$

Line (c): $= 36$

Study Exercise Two (6)

By changing percents to their fractional equivalents, find:

1. 50% of 86 **2.** 70% of 420 **3.** $33\frac{1}{3}\%$ of 96 **4.** 40% of 25

Finding a Percent of an Amount of Money (7)

Example 1: Find 3% of \$7.28, to the nearest cent:

Solution:

Line (a): 3% of \$7.28 = 3% × \$7.28
Line (b): = .03 × 7.28
Line (c): = \$.2184
Line (d): 3% of \$7.28 = \$.22, to the nearest cent.

Example 2: Find $33\frac{1}{3}\%$ of \$142, to the nearest cent:

Solution: The fractional equivalent of $33\frac{1}{3}\%$ is $\frac{1}{3}$

Line (a): $33\frac{1}{3}\%$ of \$142 = $33\frac{1}{3}\%$ × \$142

Line (b): $= \frac{1}{3} \times \$142$

Line (c): $\approx \$47.333$

Line (d): $33\frac{1}{3}\%$ of \$142 = \$47.33, to the nearest cent.

Computation

```
   47.333
3)142.000
  12
   22
   21
    1 0
      9
     10
      9
     10
      9
      1
```

Study Exercise Three (8)

Find each of the following to the nearest cent:

1. 36% of \$70
2. 6% of \$98.42
3. 125% of \$600
4. $66\frac{2}{3}\%$ of \$70

Applied Problems (9)

Example 1: 20% of the students in Lisa's class received a grade of A. If there are 30 students in Lisa's class, how many received grades of A?

Solution: Let n = the number of students receiving A's.

Line (a): $n = 20\%$ of 30
Line (b): $= 20\% \times 30$
Line (c): $= .20 \times 30$
Line (d): $= 6$

Therefore, 6 students received A's.

Example 2: 23% of the budget of Midtown City is spent for fire protection. If the budget of Midtown City is \$4,380,000, how much is spent for fire protection?

Solution: Let n = the amount spent for fire protection.

Line (a): $n = 23\%$ of \$4,380,000
Line (b): $= 23\% \times 4{,}380{,}000$
Line (c): $= .23 \times 4{,}380{,}000$
Line (d): $= 1{,}007{,}400$

Therefore, \$1,007,400 was spent for fire protection.

Study Exercise Four (10)

1. John leaves a tip equal to 15% of the bill at a restaurant. If the bill is \$19.20, what is the amount of the tip?
2. $33\frac{1}{3}\%$ of the calculators made by a company were defective. If 279 calculators were made, how many were defective?
3. In Fairtown, U.S.A., .7% of the annual budget is spent for trimming trees. If the annual budget is \$10,725,000, how much is spent for trimming trees?

Review of Rule for Finding a Factor (11)

Example 1: Find n if $3 \times n = 12$

Solution:
$3 \times n = 12$
$n = 12 \div 3$
$n = 4$

Example 2: Find n if $n \cdot 6 = 24$

Solution:
$n \times 6 = 24$
$n = 24 \div 6$
$n = 4$

Rule: If a product and one of its factors are known, divide the product by the one factor to find the other factor.

Review of Rule for Finding a Factor (Continued)

Example 3: Find n if $.3 \times n = 84$

Solution:

$$.3 \times n = 84$$
$$n = 84 \div .3$$
$$n = 280$$

Example 4: Find n if $n \times 210 = 10.5$

Solution:

$$n \times 210 = 10.5$$
$$n = 10.5 \div 210$$
$$n = .05$$

Study Exercise Five (12)

Find n:

1. $6 \times n = 144$
2. $.07 \times n = 21$
3. $n \times 8 = 224$
4. $n \times 250 = 30$
5. $n \times 123 = 17.22$
6. $.27 \times n = 9.99$

Finding the Number When a Percent of It Is Known (13)

Example 1: 12% of what number is 6?

Solution:

Step (1): Change the word statement to an equation:

12% of what number is 6

↓ ↓ ↓ ↓ ↓

$12\% \times n = 6$

Step (2): Solve the equation:

Line (a): $12\% \times n = 6$

Line (b): $(.12)n = 6$

Line (c): $n = 6 \div .12$

Line (d): $n = 50$

Computation

$$.12\overline{)6.00} = 50.$$

60
0
0
0

Example 2: 42% of what number is 14?

Solution:

Line (a): $42\% \times n = 14$

Line (b): $(.42)n = 14$

Line (c): $n = 14 \div .42$

Line (d): $n = 33\frac{1}{3}$

Computation

$$.42\overline{)14.00} = 33\frac{14}{42} = 33\frac{1}{3}$$

12 6
1 40
1 26
14

Finding the Number When a Percent of It Is Known (Continued)

Example 3: 75% of what number is 15?

Solution: The fractional equivalent of 75% is $\frac{3}{4}$

Line (a): $75\% \times n = 15$

Line (b): $\frac{3}{4} \cdot n = 15$

Line (c): $n = 15 \div \frac{3}{4}$

Line (d): $n = \frac{\cancel{15}^{5}}{1} \times \frac{4}{\cancel{3}_{1}}$

Line (e): $n = 20$

Example 4: 14% of what number is 2? (answer to nearest tenth)

Solution:

Line (a): $14\% \times n = 2$

Line (b): $(.14)n = 2$

Line (c): $n = 2 \div .14$

Line (d): $n = 14.3$ (to the nearest tenth)

Computation

```
        14.28
.14.)2.00.00
     1 4
       60
       56
        4 0
        2 8
        1 20
        1 12
           8
```

Example 5: 23% of what amount is \$5? (answer to nearest cent)

Solution:

Line (a): $23\% \times n = \$5$

Line (b): $(.23)n = \$5$

Line (c): $n = \$5 \div .23$

Line (d): $n = \$21.74$ (to the nearest cent)

Computation

```
         21.739
.23.)$5.00.000
      4 6
        40
        23
        17 0
        16 1
           90
           69
          210
          207
            3
```

Study Exercise Six (14)

Find each of the following:

1. $33\frac{1}{3}\%$ of what number is 6?
2. 32% of what number is 1.24 (to nearest tenth)?
3. 125% of what number is 15?
4. 70% of what number is 49?
5. $5\frac{3}{4}\%$ of what amount is $649 (to nearest dollar)?

Applied Problems (15)

Example 1: Wizard TV store sold 80% of their total stock of TV sets in a big weekend sale. If they sold 96 TV sets, what was their total stock?

Solution: Let n = number of total stock

Line (a): 80% of $n = 96$
Line (b): $.80 \times n = 96$
Line (c): $n = 96 \div 80$
Line (d): $n = 120$

Therefore, the number of their total stock was 120.

Example 2: A city spent 2% of the total budget for police protection. If the city spent $208,000 for police protection, what was the amount of the total budget?

Solution: Let n = amount of total budget

Line (a): 2% of n = $208,000
Line (b): $.02 \times n = 208{,}000$
Line (c): $n = 208{,}000 \div .02$
Line (d): $n = 10{,}400{,}000$

Therefore, the amount of the total budget was $10,400,000.

Study Exercise Seven (16)

1. At City College, 37% of the students wear glasses. If 7,400 students wear glasses, how many students are there at City College?
2. In Average Town, U.S.A., 9% of the total available workforce is unemployed. If 3,870 citizens are unemployed, how many citizens are there in the total available workforce?
3. In a study of traffic deaths in Florida, it was found that 1.5% of the victims were wearing seat belts. If 9 victims were wearing seat belts, how many deaths were involved in the total study?

Finding What Percent One Number is of Another (17)

Example 1: What percent of 24 is 18?

Solution: Let n = the percent

Finding What Percent One Number is of Another (Continued)

Step (1): Rewrite the statement into an equation:

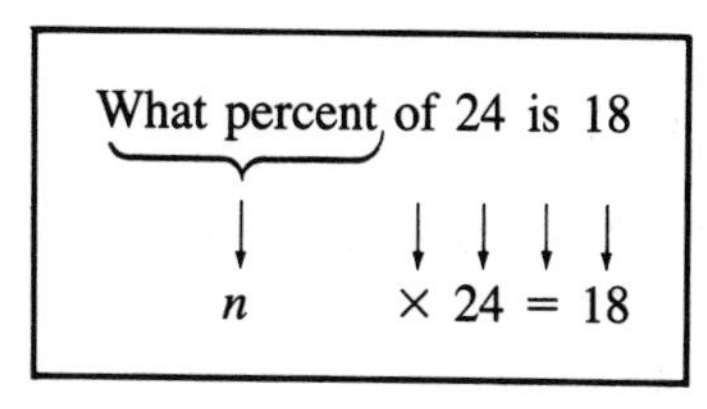

$$n \times 24 = 18$$

Step (2): Write the related division problem:

$n = 18 \div 24$

Step (3): Divide with 2 decimal places in the answer:

$$\begin{array}{r} .75 \\ 24\overline{)18.00} \\ \underline{16\ 8} \\ 1\ 20 \\ \underline{1\ 20} \end{array}$$

Step (4): Change the decimal answer to a percent:

$n = 75\%$

Therefore, 75% of 24 is 18.

Example 2: What percent of 15 is 7?

Solution: Let n = the percent

Step (1): Rewrite the statement into an equation:

$n \times 15 = 7$

Step (2): Write the related division problem:

$n = 7 \div 15$

Step (3): Divide with 2 decimal places in the answer:

$$\begin{array}{r} .46\frac{2}{3} \\ 15\overline{)7.00} \\ \underline{6\ 0} \\ 1\ 00 \\ \underline{90} \\ 10 \end{array}$$

Step (4): Change the decimal answer to a percent:

$$n = 46\frac{2}{3}\%$$

Therefore, $46\frac{2}{3}\%$ of 15 is 7.

Example 3: What percent of 18 is 2? (to nearest whole percent)

Solution: Let n = the percent

Line (a): $n \times 18 = 2$

Finding What Percent One Number is of Another (Continued)

Line (b): $n = 2 \div 18$

Line (c):
$$\begin{array}{r} .111 \\ 18\overline{)2.00} \\ \underline{1\ 8} \\ 20 \\ \underline{18} \\ 20 \\ \underline{18} \end{array}$$

.111 rounds off to .11 = 11%

Line (d): $n = 11\%$ (to nearest whole percent)

Example 4: What percent of 16 is 24?

Solution: Let n = the percent

Line (a): $n \times 16 = 24$

Line (b): $n = 24 \div 16$

Line (c):
$$\begin{array}{r} 1.50 \\ 16\overline{)24.00} \\ \underline{16} \\ 8\ 0 \\ \underline{8\ 0} \\ 0 \end{array}$$

Line (d): $n = 150\%$

Study Exercise Eight (18)

1. What percent of 6 is 2?
2. What percent of 14 is 30?
3. What percent of 8 is 3? (to nearest whole percent)
4. What percent of 7 is 3? (to nearest tenth of a percent)

Applied Problems (19)

Example 1: At a restaurant, the total bill came to $75 and a tip of $12 was left. What percent of the total bill was the tip?

Solution: Let n = the percent

Line (a): $n \times 75 = 12$

Line (b): $n = 12 \div 75$

Line (c):
$$\begin{array}{r} .16 \\ 75\overline{)12.00} \\ \underline{7\ 5} \\ 4\ 50 \\ \underline{4\ 50} \end{array}$$

Line (d): $n = 16\%$

Therefore, a tip of 16% was left.

Applied Problems (Continued)

Example 2: Jean earned \$24,000 and paid social security taxes of \$1,608. What percent of Jean's earnings were paid for social security taxes (find answer to nearest tenth of a percent)?

Solution: Let n = the percent

Line (a): $n \times 24{,}000 = 1608$

Line (b): $n = 1608 \div 24{,}000$

Line (c):

$$\begin{array}{r} .067 \\ 24000\overline{)1608.000} \\ \underline{1440\ 00} \\ 168\ 000 \\ \underline{168\ 000} \end{array}$$

Line (d): $n = 6.7\%$

Therefore, Jean paid 6.7% in social security taxes.

Study Exercise Nine (20)

1. Sam earned \$14,000 and paid \$420 in state income taxes. What percent of his earnings were paid in state income taxes?
2. A restaurant bill came to \$96 and a tip of \$15 was left. To the nearest tenth of a percent, what percent of the bill was left for a tip?

REVIEW EXERCISES (21)

A. Find:

1. 36% of 20 **2.** 75% of 128 **3.** 9% of \$384 **4.** 16% of 36

5. 92% of 1,842 **6.** 15% of \$34.25 (to nearest cent) **7.** 23% of 1.5 **8.** 20% of 35

B. Find the missing number:

9. $12\frac{1}{2}\%$ of what number is 32? **10.** 10% of what number is 430?

11. 25% of what number is 60? **12.** 35% of what number is 22? (to nearest tenth)

C. Find:

13. What percent of 6 is 1? **14.** What percent of 200 is 46?

15. What percent of 15 is 30? **16.** What percent of 60 is 20?

17. What percent of 41 is 11? (to nearest hundredth of a percent)

D. Answer the following:

18. Henry paid 8% of his total earnings in federal income taxes. If his total earnings were \$17,500, how much did he pay in federal income taxes?

19. Nancy spent 43% of her savings for a radio. If she spent \$109.65 for the radio, how much was in her savings before the purchase of the radio?

20. In a class with 40 students, 6 students failed the final exam. What percent of the students failed the final exam?

Solutions to Review Exercises (22)

A. **1.** 36% of $20 = 36\% \times 20$
$= .36 \times 20$
$= 7.2$

2. 75% of $128 = 75\% \times 128$
$= \frac{3}{\cancel{4}_1} \times \cancel{128}^{32}$
$= 96$

3. 9% of $\$384 = 9\% \times \$384$
$= .09 \times \$384$
$= \$34.56$

4. 16% of $36 = 16\% \times 36$
$= .16 \times 36$
$= 5.76$

5. 92% of $1842 = 92\% \times 1842$
$= .92 \times 1842$
$= 1694.64$

6. 15% of $\$34.25 = 15\% \times \$34.25$
$= .15 \times \$34.25$
$= \$5.1375$
$= \$5.14$ (to nearest cent)

7. 23% of $1.5 = 23\% \times 1.5$
$= .23 \times 1.5$
$= .345$

8. 20% of $35 = 20\% \times 35$
$= \frac{1}{\cancel{5}_1} \times \cancel{35}^{7}$
$= 7$

B. **9.** $12\frac{1}{2}\% \times n = 32$
$12.5\% \times n = 32$
$(.125)n = 32$
$n = 32 \div .125$
$n = 256$

10. $10\% \times n = 430$
$\frac{1}{10} \cdot n = 430$
$n = 430 \div \frac{1}{10}$
$n = 430 \times \frac{10}{1}$
$n = 4300$

11. $25\% \times n = 60$
$\frac{1}{4} \cdot n = 60$
$n = 60 \div \frac{1}{4}$
$n = 60 \times \frac{4}{1}$
$n = 240$

12. $35\% \times n = 22$
$(.35)n = 22$
$n = 22 \div .35$
$n \approx 62.85$
$n = 62.9$ (to nearest tenth)

C. **13.** $n \times 6 = 1$
$n = 1 \div 6$
$n = 16\frac{2}{3}\%$

14. $n \times 200 = 46$
$n = 46 \div 200$
$n = 23\%$

15. $n \times 15 = 30$
$n = 30 \div 15$
$n = 200\%$

16. $n \times 60 = 20$
$n = 20 \div 60$
$n = 33\frac{1}{3}\%$

17. $n \times 41 = 11$
$n = 11 \div 41$
$n = 26.83\%$

D. **18.** Let n = amount of federal income taxes
$n = 8\% \times 17{,}500$
$n = .08 \times 17{,}500$
$n = 1400$
Henry paid $1400 in federal income taxes

19. Let n = amount in savings
43% of $n = \$109.65$
$.43 \times n = 109.65$
$n = 109.65 \div .43$
$n = 255$
Nancy had $255 in savings

Solutions to Review Exercises (Continued)

20. Let n = the percent

$n \times 40 = 6$

$n = 6 \div 40$

$n = .15$

$n = 15\%$

15% of the students failed the final exam.

SUPPLEMENTARY PROBLEMS

A. Find:

1. 10% of 60 **2.** $66\frac{2}{3}\%$ of 27 **3.** 120% of 160 **4.** 13% of 16

5. $3\frac{3}{4}\%$ of 30 **6.** $12\frac{1}{2}\%$ of 48 **7.** 90% of 63 **8.** 150% of 1.8

9. 23% of 1.28 **10.** 4% of 30 **11.** 1% of 3,672 **12.** 7% of 6.3

13. $33\frac{1}{3}\%$ of 96 **14.** 1% of $6 **15.** 32% of $70

B. Find the missing number:

16. 8% of what number is 32?
17. 3.6% of what number is 40? (to the nearest whole number)
18. 300% of what number is 150?
19. 1% of what number is 68?
20. $33\frac{1}{3}\%$ of what number is 4.3? (Hint: Change $33\frac{1}{3}\%$ to its fractional equivalent.)

C. Find:

21. What percent of 12 is 8?
22. What percent of 144 is 96?
23. What percent of 50 is 60?
24. What percent of 49 is 13? (to the nearest whole percent)
25. What percent of 300 is 45?
26. What percent of 45 is 300?
27. What percent of 200 is 38?
28. What percent of 21 is 5? (to the nearest tenth of a percent)

D. Answer the following:

29. The bill in a restaurant comes to $35. What will be the amount of a tip of 15%?

30. John's bill in a restaurant is $8.50. If John leaves a tip equal to 16% of the bill, how much was the tip? Altogether how much did John spend at the restaurant?

31. Mr. Brown took his family to a restaurant and the bill for the food was $54. In addition, sales tax of 6% was added to the food bill. If Mr. Brown left a tip of 15% of the food portion of the bill, how much did he spend altogether at the restaurant?

32. Bill earned $27,000 last year and paid $1,809 in social security taxes. What percent of his earnings was paid in social security taxes?

33. Donna earned $37,000 last year and paid $8,880 in federal income taxes. What percent of her earnings was paid in federal income taxes?

SUPPLEMENTARY PROBLEMS (Continued)

34. Mr. Jones's taxable income last year was \$42,500. His federal income tax obligation was equal to \$6,274 plus 33% of the amount of taxable income above \$35,200. What was his total federal income tax obligation?

35. Dorothy paid \$5,238 in state income taxes last year. If she paid 9.7% of her earnings in state income taxes, how much were her earnings last year?

Solutions to Study Exercises (4A)

Study Exercise One (Frame 4)

1. 42% of $57 = 42\% \times 57 = .42 \times 57 = 23.94$

2. 6.2% of $50 = 6.2\% \times 50 = .062 \times 50 = 3.1$

3. 120% of $16 = 120\% \times 16 = 1.2 \times 16 = 19.2$

4. $2\frac{3}{4}\%$ of $35 = 2\frac{3}{4}\% \times 35 = 2.75\% \times 35 = .0275 \times 35 = .9625$

Study Exercise Two (Frame 6) (6A)

1. 50% of $86 = 50\% \times 86 = \frac{1}{\cancel{2}_1} \times \frac{\cancel{86}^{43}}{1} = 43$

2. 70% of $420 = 70\% \times 420 = \frac{7}{\cancel{10}_1} \times \frac{\cancel{420}^{42}}{1} = 294$

3. $33\frac{1}{3}\%$ of $96 = 33\frac{1}{3}\% \times 96 = \frac{1}{\cancel{3}_1} \times \frac{\cancel{96}^{32}}{1} = 32$

4. 40% of $25 = 40\% \times 25 = \frac{2}{\cancel{5}_1} \times \frac{\cancel{25}^{5}}{1} = 10$

Study Exercise Three (Frame 8) (8A)

1. 36% of $\$70 = 36\% \times \$70 = .36 \times \$70 = \$25.20$

2. 6% of $\$98.42 = 6\% \times \$98.42 = .06 \times \$98.42 = \$5.9052 = \$5.91$ (to nearest cent)

3. 125% of $\$600 = 125\% \times 600 = 1.25 \times \$600 = \$750$

4. $66\frac{2}{3}\%$ of $\$70 = 66\frac{2}{3}\% \times \$70 = \frac{2}{3} \times \$70 = \frac{\$140}{3} = \$46.67$ (to nearest cent)

Computation

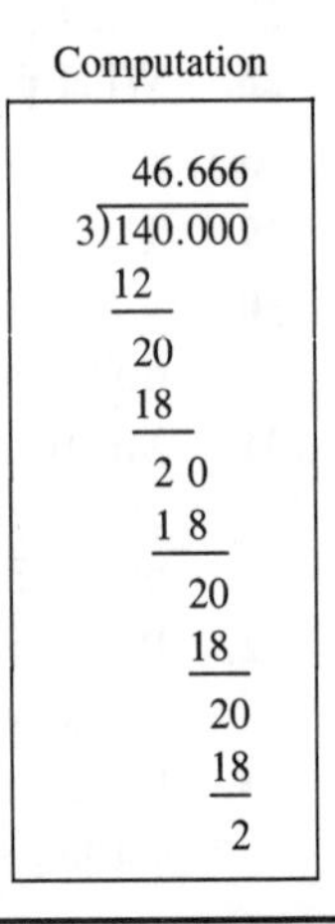

Study Exercise Four (Frame 10) (10A)

1. Let n = the amount of the tip

n = 15% of \$19.20
= 15% × 19.20
= .15 × 19.20
= 2.88

The tip is \$2.88

2. Let n = the number of defective calculators

$$n = 33\frac{1}{3}\% \text{ of } 279$$
$$= 33\frac{1}{3}\% \times 279$$
$$= \frac{1}{\not{3}_1} \times \frac{\not{279}^{93}}{1}$$
$$= 93$$

Therefore, 93 calculators were defective.

3. Let n = the amount spent for trimming trees

n = .7% of \$10,725,000
= .7% × 10,725,000
= .007 × 10,725,000
= 75,075

Therefore, \$75,075 is spent for trimming trees.

Study Exercise Five (Frame 12) (12A)

1. $6 \times n = 144$
$n = 144 \div 6$
$n = 24$

2. $.07 \times n = 21$
$n = 21 \div .07$
$n = 300$

3. $n \times 8 = 224$
$n = 224 \div 8$
$n = 28$

4. $n \times 250 = 30$
$n = 30 \div 250$
$n = .12$

5. $n \times 123 = 17.22$
$n = 17.22 \div 123$
$n = .14$

6. $.27 \times n = 9.99$
$n = 9.99 \div .27$
$n = 37$

Study Exercise Six (Frame 14) (14A)

1. $33\frac{1}{3}\% \times n = 6$

$$\frac{1}{3} \cdot n = 6$$
$$n = 6 \div \frac{1}{3}$$
$$n = 6 \times \frac{3}{1}$$
$$n = 18$$

2. $32\% \times n = 1.24$
$(.32)n = 1.24$
$n = 1.24 \div .32$
$n = 3.9$ (to the nearest tenth)

Computation

```
       3.87
.32.)1.24.00
      96
      28 0
      25 6
       2 40
       2 24
         16
```

3. $125\% \times n = 15$
$(1.25)n = 15$
$n = 15 \div 1.25$
$n = 12$

Computation

```
        12.
1.25.)15.00.
      12 5
       2 50
       2 50
          0
```

4. $70\% \times n = 49$

$$\frac{7}{10} \cdot n = 49$$
$$n = 49 \div \frac{7}{10}$$
$$n = \not{49}^{7} \times \frac{10}{\not{7}_1}$$
$$n = 70$$

Study Exercise Six (Frame 14) (Continued)

5. $5\frac{3}{4}\% \times n = \$649$

$5.75\% \times n = \$649$

$(.0575)n = \$649$

$n = \$649 \div .0575$

$n = \$11{,}287$ (to nearest dollar)

Computation

```
              1 1286.9
.0575.)649.0000.0
       575
        74 0
        57 5
        16 50
        11 50
         5 000
         4 600
           4000
           3450
            550 0
            517 5
             32 5
```

Study Exercise Seven (Frame 16) **(16A)**

1. Let n = number of students at City College

37% of $n = 7400$

$.37 \times n = 7400$

$n = 7400 \div .37$

$n = 20{,}000$

Therefore, there are 20,000 students at City College.

2. Let n = number of citizens in total workforce

9% of $n = 3870$

$.09 \times n = 3870$

$n = 3870 \div .09$

$n = 43{,}000$

Therefore, there are 43,000 citizens in the total workforce.

3. Let n = number of deaths in total study

1.5% of $n = 9$

$.015 \times n = 9$

$n = 9 \div .015$

$n = 600$

Therefore, the total study involved 600 deaths.

Study Exercise Eight (Frame 18) **(18A)**

1. Let n = the percent

$n \times 6 = 2$

$n = 2 \div 6$

$n = \frac{1}{3}$

$n = 33\frac{1}{3}\%$

2. Let n = the percent

$n \times 14 = 30$

$n = 30 \div 14$

$n = 2.14\frac{2}{7}$

$n = 214\frac{2}{7}\%$

Study Exercise Eight (Frame 18) (Continued)

3. Let n = the percent

   $n \times 8 = 3$

   $n = 3 \div 8$

   $n = .375$

   $n = 38\%$ (to nearest whole percent)

4. Let n = the percent

   $n \times 7 = 3$

   $n = 3 \div 7$

   $n = .4285$

   $n = 42.9\%$ (to nearest tenth of a percent)

Study Exercise Nine (Frame 20) **(20A)**

1. Let n = the percent

   $n \times 14000 = 420$

   $n = 420 \div 14000$

   $n = .03$

   $n = 3\%$

   Therefore, 3% of his earnings were paid in state income taxes

2. Let n = the percent

   $n \times 96 = 15$

   $n = 15 \div 96$

   $n = .1562$

   $n = 15.6\%$ (to nearest tenth of a percent)

   Therefore, 15.6% of the bill was left for a tip.

UNIT

26 Interest—Simple and Compound

OBJECTIVES (1)

Part I—Simple Interest

By the end of this part you should:

1. KNOW THE MEANING OF:

(a) INTEREST	(b) PRINCIPAL
(c) ANNUAL INTEREST RATE	(d) AMOUNT
(e) TRUTH-IN-LENDING ACT	(f) ANNUAL PERCENTAGE RATE

2. BE ABLE TO COMPUTE SIMPLE INTEREST FROM A FORMULA.
3. BE ABLE TO COMPUTE THE ANNUAL INTEREST RATE FROM A FORMULA.
4. BE ABLE TO DETERMINE THE MONTHLY INTEREST ON A CHARGE ACCOUNT BALANCE.

Part II—Compound Interest

By the end of this part you should:

1. KNOW THE MEANING OF THE FOLLOWING TERMS:
   (a) COMPOUND INTEREST (b) COMPOUND AMOUNT
2. BE ABLE TO FIND THE COMPOUND AMOUNT AND COMPOUND INTEREST FOR ONE YEAR WHEN INTEREST IS COMPOUNDED SEMIANNUALLY OR QUARTERLY.
3. BE ABLE TO FIND THE COMPOUND AMOUNT AND COMPOUND INTEREST BY MEANS OF A TABLE.

Part I—Simple Interest (2)

Key Terms

A man borrows $1,000 at a rate of 8% per year and agrees to pay back the $1,000 plus an additional $160 2 years later. At the end of the second year, he returns $1,160.

Line (a): Principal: $1,000
Principal is the sum of money borrowed or invested.

Line (b): Interest: $160
Interest is money paid for the use of borrowed money or money earned when funds are invested.

Part 1—Simple Interest (Continued)

Line (c): Amount: $1,160
Amount is the sum of the principal and the interest.

Line (d): Annual interest rate: 8%
Annual interest rate will be expressed as a "percent" number and is the rate of interest being charged for 1 year.

Line (e): Time: 2 years
Time is the interval over which the principal is kept.

Principal + Interest = Amount (3)

Example 1: A woman deposited $2,000 in a bank. At the end of 1 year she earned $113 interest. What was the amount of her account at the end of the year?

Line (a): Principal + Interest = Amount
↓ ↓ ↓
Line (b): $2,000 + $113 = $2,113

Example 2: A man borrowed $1,500 for 1 year. At the end of the year he paid back a total amount of $1,770. How much interest did he pay?

Line (a): Amount − Principal = Interest
↓ ↓ ↓
Line (b): $1,770 − $1,500 = $270

Example 3: The amount which must be paid back on a loan is $3,660. If $660 of this is interest, how much was borrowed?

Line (a): Amount − Interest = Principal
↓ ↓ ↓
Line (b): $3,660 − $660 = $3,000

Key Relationships

1. Principal + Interest = Amount
2. Amount − Principal = Interest
3. Amount − Interest = Principal

Study Exercise One (4)

Find the missing information:

	Amount	Principal	Interest
1.		$1,650	$103.13
2.	$964.80	$900	
3.		$722	$43.24
4.	$1,125.50	$1,000	
5.	$350		$15.75

Simple Interest Formula **(5)**

If p is the principal sum of money being borrowed or invested, r is the annual interest rate expressed as a percent, t is the time of the loan expressed in years, and i is the interest, then:

Interest = Principal × Rate × Time

or

$i = prt$

Note: When using letters to represent numbers, it is common practice to omit the multiplication signs.

Using the Formula to Find Simple Interest **(6)**

Example 1: A woman borrows $1,000 at a rate of 7% per year. She agrees to pay back $1,000 plus interest at the end of 2 years. How much interest must she pay? What is the total amount due at the end of the second year?

Method 1: Change the percent to a fraction and use cancellation:

Line (a): $i = prt$

Line (b): $= (\$1{,}000)(7\%)(2)$

Line (c): $= (\cancel{1000}^{10})\left(\frac{7}{\cancel{100}_1}\right)(2)$

Line (d): = $140 interest she must pay

Line (e): Principal + Interest = Amount

Line (f): $1,000 + $140 = $1,140 total due at the end of 2 years

Method 2: Change the percent to a decimal and multiply:

Line (g): $i = prt$

Line (h): $= (\$1{,}000)(.07)(2)$

Line (i): $= (\$1{,}000)(.14)$

Line (j): = $140 interest she must pay

Line (k): Principal + Interest = Amount

Line (l): $1,000 + $140 = $1,140 total due at the end of 2 years

Example 2: How much interest will be earned on $5,000 invested at an annual rate of 8% in 1 year, 9 months?

Solution:

Step (1): Convert the time to years.

$$1 \text{ year } 9 \text{ months} = 1\frac{9}{12} \text{ years} = 1\frac{3}{4} \text{ years}$$

Step (2): Substitute in the formula and solve.

Line (a): $i = prt$

Line (b): $= (\$5{,}000)(8\%)\left(1\frac{3}{4}\right)$

Line (c): $= \cancel{5{,}000}^{50} \cdot \frac{\cancel{8}^2}{\cancel{100}_1} \cdot \frac{7}{\cancel{4}_1}$

Line (d): = $700 interest

Example 3: What is the interest on $3,300 at 8% per year for 16 months?

Solution:

Step (1): Convert the time to years.

$$16 \text{ months} = \frac{16}{12} \text{ years} = \frac{4}{3} \text{ years}$$

Using the Formula to Find Simple Interest (Continued)

Step (2): Substitute in the formula and solve.

Line (a): $i = prt$

Line (b): $= (\$3{,}300)(8\%)\left(\frac{4}{3}\right)$

Line (c): $= \not{3{,}300}^{\not{33}\,11} \cdot \frac{8}{\not{100}_1} \cdot \frac{4}{\not{3}_1}$

Line (d): $= \$352$ interest

Study Exercise Two (7)

Find the missing information by using the simple interest formula. In problem 1, express the percent as a decimal and multiply. In the other problems, express the percent as a fraction and use cancellation.

	Principal	Annual Interest Rate	Time	Interest	Amount
1.	\$285	15%	3 years		
2.	\$4,500	6%	9 months		
3.	\$3,200	14%	1 year 6 months		
4.	\$6,000	8%	14 months		

A Formula for Finding the Annual Interest Rate (8)

If i is the interest, p is the principal sum of money being borrowed or invested, t is the time of the loan expressed in years, and r is the annual interest rate, then:

$$\text{Rate} = \frac{\text{Interest}}{\text{Principal} \times \text{Time}}$$

or

$$r = \frac{i}{pt}$$

To find the rate:

Step (1): Find the value of pt.
Step (2): Divide i by the value of pt and express as a percent.

Using the Formula to Find the Annual Interest Rate (9)

Example 1: A sum of \$4,000 was invested for 3 years. The interest earned was \$600. What was the annual interest rate?

Solution:

Step (1): Find the value of pt.

Line (a): $pt = (4{,}000)(3) = 12{,}000$

Using the Formula to Find the Annual Interest Rate (Continued)

Step (2): Substitute in the formula and solve, expressing your answer as a percent.

Line (b): $r = \frac{i}{pt}$

Line (c): $= \frac{600}{12{,}000}$

Line (d): $= \frac{6}{120}$

Line (e): $= \frac{1}{20}$

Line (f): $= .05$

Line (g): $= 5\%$

Computation

$$20\overline{)1.00} = .05$$
$$\begin{array}{r} .05 \\ 20\overline{)1.00} \\ \underline{1\ 00} \\ 0 \end{array}$$

Example 2: A sum of \$3,000 was invested for 15 months. If the interest earned was \$458, what was the annual interest rate to the nearest whole percent?

Solution:

Step (1): Find the value of *pt*.

Line (a): 15 months $= \frac{15}{12}$ years $= \frac{5}{4}$ years

Line (b): $pt = \cancel{3{,}000}^{750} \cdot \frac{5}{\cancel{4}_1} = 3{,}750$

Step (2): Substitute in the formula and solve, expressing your answer as a percent.

Line (c): $r = \frac{i}{pt}$

Line (d): $= \frac{458}{3{,}750}$

Line (e): $= .122$

Line (f): $= 12\%$ to the nearest whole percent

Computation

$$\begin{array}{r} .122 \\ 3{,}750\overline{)458.000} \\ \underline{375\ 0} \\ 83\ 00 \\ \underline{75\ 00} \\ 8\ 000 \\ \underline{7\ 500} \\ 500 \end{array}$$

Study Exercise Three (10)

In problems 1–3 find the annual interest rate to the nearest whole percent:

	Principal	Time	Interest	Annual Interest Rate
1.	\$600	3 years	\$144	
2.	\$4,000	1 year 6 months	\$510	
3.	\$870	4 months	\$18	

4. A man invested \$4,000 in a piece of land and sold it 2 years later for \$5,000. What annual interest rate did he earn on his money?

Computing Interest on Charge Accounts Using the Monthly Interest Rate (11)

Many department stores, gasoline service companies, banks, and other business institutions issue credit cards which permit the customer to "buy now and pay later." The loans which are created by using these cards require interest to be paid on the balance. The balance is the amount owed at any given time. For use in our formula, the balance becomes the principal.

A Formula for Finding Interest for One Month (12)

If P is the principal or balance for 1 month, R is the monthly interest rate, and I is the interest for 1 month, then:

Interest (for 1 month) = Principal (unpaid balance for 1 month) × Rate (monthly interest rate)

or

$$I = PR$$

Example: A customer's balance for April was \$300. If the monthly rate on this account is $1\frac{1}{2}\%$, how much interest must be paid?

Solution:

Line (a): $I = PR$

Line (b): $= (\$300)\left(1\frac{1}{2}\%\right)$

Line (c): $= (300)(1.5\%)$

Line (d): $= (300)(.015)$

Line (e): $= \$4.50$ interest for one month

Study Exercise Four (13)

Find the missing information to the nearest whole cent.

	Monthly Interest Rate	Balance	Interest for One Month
1.	$\frac{3}{4}\%$	\$280	
2.	$1\frac{1}{2}\%$	\$520	
3.	$1\frac{3}{4}\%$	\$328	
4.	2%	\$57.84	

Truth-in-Lending Act (14)

The federal Truth-in-Lending Act went into effect July 1, 1969. It requires conspicuous disclosure of the Annual Percentage Rate on the vast majority of credit contracts, private house first mortgages, and time-sale agreements. It also requires stores to indicate on their bills the method they use to compute their balance.

Look for the Annual Percentage Rate on any contract you may be asked to sign. This is the figure you should use to compare the cost of borrowing money.

Summary, Part I–Simple Interest (15)

A. Key Relationships

1. Principal + Interest = Amount
2. Amount − Principal = Interest
3. Amount − Interest = Principal

B. Simple Interest Formula

i: interest
p: principal
r: annual interest rate
t: time expressed in years

$$i = prt$$

C. Annual Interest Rate Formula

i: interest
p: principal
r: annual interest rate
t: time expressed in years

$$r = \frac{i}{pt}$$

D. Monthly Interest Formula

I: interest for 1 month
R: monthly interest rate
P: balance

$$I = PR$$

This Completes Part I–Simple Interest

If you are not doing Part II, proceed to frame 27 and work the Review Exercises for Part I; otherwise, proceed to frame 16.

Part II–Compound Interest (16)

Mr. Doe has $1,000 to invest for 1 year and finds that he has 2 choices.

Choice 1: He can invest the $1,000 for 1 year at an annual rate of 8%.

Choice 2: He can invest the $1,000 for the first 6 months of the year at an annual rate of 8% and then invest what he receives for the last six months of the year, also at an 8% annual rate.

What is the best course of action?

Analysis of Choice 1 (17)

Line (a): $i = prt$

Line (b): $= \$(1{,}000)(8\%)(1)$

Line (c): $= \cancel{1{,}000}^{10} \cdot \frac{8}{\cancel{100}_1} \cdot 1$

Analysis of Choice 1 (Continued)

Line (d): $= \$80$ interest

At the end of the year, Mr. Doe has $80 more than his original $1,000.

Analysis of Choice 2 (18)

First Six Months (First Period)

Line (a): $i = prt$

Line (b): $(\$1{,}000)(8\%)\left(\frac{1}{2}\right)$

Line (c): $= (\$1{,}000)(4\%)$

Line (d): $= \cancel{1{,}000}^{10} \cdot \frac{4}{\cancel{100}_1}$

Line (e): $= \$40$ interest for first 6 months

Line (f): Principal + Interest = Amount

Line (g): $1,000 + $40 = $1,040 at the end of the first 6 months

Second Six Months (Second Period)

Line (a): $i = prt$

Line (b): $= (\$1{,}040)(8\%)\left(\frac{1}{2}\right)$

Line (c): $= (\$1{,}040)(4\%)$

Line (d): $= (1{,}040)(.04)$

Line (e): $= \$41.60$ interest for the second 6 months

Results for the Year

Line (a):

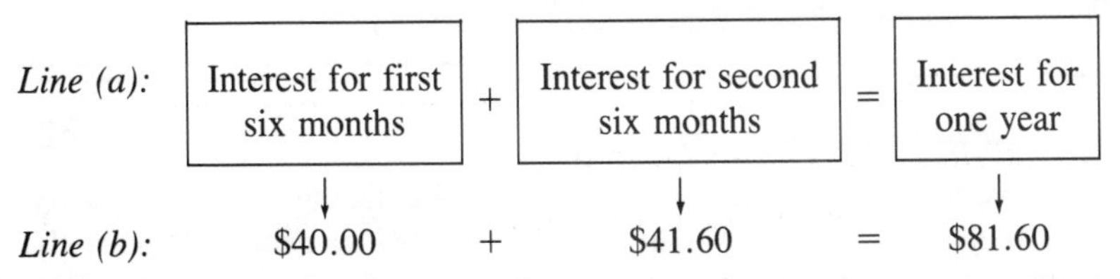

Line (b): $40.00 + $41.60 = $81.60

At the end of the year Mr. Doe has $81.60 more than his original thousand.

Choice 2 is an example of compound interest.

Compound Interest (19)

Compound interest is interest paid on both the principal and the previously earned interest.

Procedure for Calculating Compound Interest for 1 Year (20)

Example: Find the compound interest on $1,000 invested for 1 year at an annual rate of 12% compounded quarterly.

Solution:

Step (1): Find the number of periods per year the money is to be compounded. In this example, since quarterly means 4 times a year, the number of periods would be 4.

Compound Interest (Continued)

Step (2): Divide the annual rate by the number of periods per year to get the rate per period.

$12\% \div 4 = 3\%$ rate per period

Step (3): Use the rate found in *Step (2)* to find the interest for each period, accumulating the interest as you proceed. Simplify calculations by rounding to the nearest cent.

Period (1): *Line (a):* $i = (\$1{,}000)(.03) = \$30$
Line (b): $\$1{,}000 + \$30 = \$1{,}030$

Period (2): *Line (c):* $i = (\$1{,}030)(.03) = \$30.90$
Line (d): $\$1{,}030 + \$30.90 = \$1{,}060.90$

Period (3): *Line (e):* $i = (\$1{,}060.90)(.03) = \$31.827 \approx \$31.83$
Line (f): $\$1{,}060.90 + \$31.83 = \$1{,}092.73$

Period (4): *Line (g):* $i = (\$1{,}092.73)(.03) = \$32.7819 \approx \$32.78$
Line (h): $\$1{,}092.73 + \$32.78 = \$1{,}125.51$ (this is called the compound amount)

Step (4): Subtract to find the compound interest.

Line (i): Compound Amount − Principal = Compound Interest
↓ ↓ ↓
Line (j): $\$1{,}125.51 - \$1{,}000.00 = \$125.51$

Compound Amount (21)

Compound amount is the total amount accumulated at the end of a given interval of time when interest has been compounded.

Key Relationship

Compound Amount − Principal = Compound Interest

Study Exercise Five (22)

1. Find the compound interest on $500 invested for 1 year at an annual rate of 6% compounded semiannually.
2. Find the compound interest on $1,000 invested for 1 year at an annual rate of 8% compounded quarterly.

A Short Cut (23)

Finding compound interest and compound amount can be very difficult and tedious without the aid of a calculator or table. Even banks which have computers available still find it convenient to refer to tables on occasion.

Using a Compound Interest Table (24)

Example 1: Find the compound interest on $400 invested for 2 years at an annual rate of 6% compounded quarterly.

Solution:

Step (1): Find the number of periods per year.
Line (a): Quarterly means 4 times a year

Step (2): Find the interest rate per period by dividing the annual rate by the number of periods per year.
Line (b): $6\% \div 4 = 1\frac{1}{2}\%$ interest rate per period

Using a Compound Interest Table (Continued)

Step (3): Find the total number of periods by multiplying the periods per year times the number of years.

Line (c): 4 periods per year × 2 years = 8 total periods

Step (4): Find how much $1 will amount to by using the interest-rate-per-period column and the number-of-periods row. See Table II in the appendix.

Line (d): Total periods = 8. Interest rate per period = $1\frac{1}{2}\%$

Line (e): Here is a portion of Table II.

Number of Periods	Interest Rate Per Period ···	1%	$1\frac{1}{2}\%$	2% ···
1		1.0100	1.0150	1.0200
⋮		⋮	⋮	⋮
7		1.0721	1.1098	1.1487
8		1.0829	1.1265	1.1717
9		1.0937	1.1434	1.1051

Step (5): Multiply the number obtained in *Step (4)* by the principal, and you will obtain the compound amount.

Line (f): (1.1265) ($400) = $450.60 compound amount

Step (6): Subtract the principal from the compound amount and you will have the compound interest.

Line (g): Compound Amount − Principal = Compound Interest

Line (h): $450.60 − $400.00 = $50.60 compound interest

Example 2: Find the compound interest on $850 invested for 7 years at an annual rate of 12% compounded semiannually (to nearest cent).

Solution:

Step (1): Semiannually = 2 periods per year

Step (2): 12% ÷ 2 = 6% interest rate per period

Step (3): 2 periods per year × 7 years = 14 total periods

Step (4): Table II (14th row; 6% column): 2.2609

Step (5): (2.2609) ($850) = $1,921.765 ≈ $1,921.77

Step (6): $1,921.77 − $850 = $1,071.77 compound interest

Summary, Part II—Compound Interest (25)

A. Compound Amount − Principal = Compound Interest

B. Procedure for Using a Compound Interest Table:

Step (1): Find the number of periods per year.

Step (2): Find the interest rate per period by dividing the annual rate by the number of periods per year.

Step (3): Find the total number of periods by multiplying the periods per year times the number of years.

Step (4): Find out how much $1 will amount to by using the interest-rate-per-period column and the number-of-periods row.

Step (5): Find the compound amount by multiplying the number found in *Step (4)* by the principal.

Step (6): Find the compound interest by subtracting the principal from the compound amount.

Study Exercise Six (26)

Using Table II, find the missing information to the nearest cent. Refer to Frame 25 for the procedure.

	Amount Invested	Time Invested	Annual Interest Rate	Compounded	Compound Amount	Compound Interest
1.	$200	4 years	6%	annually		
2.	$500	5 years	8%	semiannually		
3.	$700	$\frac{1}{2}$ year	12%	monthly		
4.	$1,000	25 years	10%	semiannually		

This Completes Part II—Compound Interest

REVIEW EXERCISES (27)

Part I—Simple Interest

A. Answer the following:

1. Principal + ________ = Amount.
2. State the simple interest formula.

B. Find the interest and the amount, given the following information:

	Annual Interest Rate	Principal	Time	Interest	Amount
3.	5%	$200	2 years		
4.	10%	$1,800	8 months		
5.	7%	$900	1 year 4 months		
6.	9%	$720	3 months		

C. Answer the following:

7. State the formula for finding the annual interest rate.

REVIEW EXERCISES (Continued)

8. A sum of $1,500 was invested for 2 years. The interest earned was $210. What was the annual interest rate?
9. An investor bought a home for $30,000 and sold it 4 years later for $40,000. What annual interest rate was earned on this investment? (Answer to the nearest tenth of a percent.)
10. A credit card account had a balance of $600 for the month of January. If the monthly rate is $1\frac{1}{2}\%$, how much interest will be charged for this month?
11. What figure should you use to compare the costs of borrowing money?

Part II–Compound Interest

D. Answer the following:

12. What is compound interest?
13. What is compound amount?
14. Find the compound interest on $4,000 invested for 1 year at an annual rate of 10% compounded semiannually.

E. Using a table, determine the compound amount and compound interest on the following:

15. $600 invested for 3 years at a 14% annual interest rate compounded semiannually.
16. $800 invested for 2 years at a 6% annual interest rate compounded quarterly.

Solutions to Review Exercises (28)

Part I–Simple Interest

A. 1. See Frame 3

2. See Frame 5

B. 3.
$$i = prt$$
$$= (\$200)(5\%)(2)$$
$$= \cancel{200}^{2} \cdot \frac{5}{\cancel{100}_{1}} \cdot 2$$
$$= \$20 \text{ interest}$$
$$\$200 + \$20 = \$220 \text{ amount}$$

4. $8 \text{ months} = \frac{8}{12} \text{ or } \frac{2}{3} \text{ year}$
$$i = prt$$
$$= (\$1{,}800)(10\%)\left(\frac{2}{3}\right)$$
$$= \cancel{1{,}800}^{\cancel{180}\,60} \cdot \frac{1}{\cancel{10}_{1}} \cdot \frac{2}{\cancel{3}_{1}}$$
$$= \$120 \text{ interest}$$
$$\$1{,}800 + \$120 = \$1{,}920 \text{ amount}$$

5. $1 \text{ year } 4 \text{ months} = 1\frac{4}{12} \text{ or } 1\frac{1}{3} \text{ years}$
$$i = prt$$
$$= (\$900)(7\%)\left(1\frac{1}{3}\right)$$
$$= \cancel{900}^{\cancel{9}\,3} \cdot \frac{7}{\cancel{100}_{1}} \cdot \frac{4}{\cancel{3}_{1}}$$
$$= \$84 \text{ interest}$$
$$\$900 + \$84 = \$984 \text{ amount}$$

6. $3 \text{ months} = \frac{3}{12} \text{ or } \frac{1}{4} \text{ year}$
$$i = prt$$
$$= (\$720)(9\%)\left(\frac{1}{4}\right)$$
$$= (\cancel{720}^{180})(.09)\left(\frac{1}{\cancel{4}_{1}}\right)$$
$$= (180)(.09)$$
$$= \$16.20 \text{ interest}$$
$$\$720 + \$16.20 = \$736.20 \text{ amount}$$

Solutions to Review Exercises (Continued)

C. **7.** See Frame 8

8. $pt = (1{,}500)(2)$
$= 3{,}000$

$r = \frac{i}{pt}$
$= \frac{210}{3{,}000}$
$= \frac{21}{300}$
$= \frac{7}{100}$
$= 7\%$ annual interest rate

9. $\$40{,}000 - \$30{,}000 = \$10{,}000$ gain

$pt = (30{,}000)(4)$
$= 120{,}000$

$r = \frac{i}{pt}$
$= \frac{10{,}000}{120{,}000}$
$= \frac{10}{120}$
$= .0833$
$= 8.3\%$ annual interest rate to the nearest tenth of a percent

10. $I = PR$
$= (\$600)\left(1\frac{1}{2}\%\right)$
$= (600)(1.5\%)$
$= (600)(.015)$
$= \$9$ interest for month of January

11. See Frame 14

Part II—Compound Interest

D. **12.** See Frame 19

13. See Frame 21

14. *Step (1):* 2 periods per year
Step (2): $10\% \div 2 = 5\%$ or .05 per period
Step (3): *Period (1):*
$i = (\$4{,}000)(.05) = \$200$
$\$4{,}000 + \$200 = \$4.200$
Period (2):
$i = (\$4{,}200)(.05) = \$210$
$\$4{,}200 + \$210 = \$4{,}410$ compound amount
Step (4): Compound Amount − Principal = Compound Interest
↓ ↓ ↓
\$4,410 − \$4,000 = \$410

E. **15.** *Step (1):* 2 periods per year
Step (2): $14\% \div 2 = 7\%$ per period
Step (3): $2 \times 3 = 6$ total periods
Step (4): Table II (6th row; 7% column): 1.5007
Step (5): $(1.5007)(\$600) = \$900.42$ compound amount
Step (6): $\$900.42 - \$600.00 = \$300.42$ compound interest

16. *Step (1):* 4 periods per year
Step (2): $6\% \div 4 = 1\frac{1}{2}\%$ per period
Step (3): $4 \times 2 = 8$ total periods
Step (4): Table II (8th row; $1\frac{1}{2}\%$ column): 1.1265
Step (5): $(1.1265)(\$800) = \$901.20$ compound amount
Step (6): $\$901.20 - \$800.00 = \$101.20$ compound interest

SUPPLEMENTARY PROBLEMS

Part I—Simple Interest

A. Find the interest and the amount given the following information (answers to nearest cent):

	Annual Interest Rate	Principal	Time	Interest	Amount
1.	$6\frac{1}{2}\%$	\$1,000	2 years		
2.	9%	\$480	3 months		
3.	8%	\$600	4 months		
4.	13%	\$2,400	4 years		
5.	7%	\$625	6 months		
6.	9.5%	\$2,000	3 years		

B. Find the annual interest rate given the following information (answers to nearest whole percent):

	Principal	Time	Interest	Annual Interest Rate
7.	\$1,000	2 years	\$140	
8.	\$600	3 months	\$12	
9.	\$4,200	5 years	\$840	
10.	\$3,000	4 months	\$58	
11.	\$2,500	$4\frac{1}{2}$ years	\$1,012.50	
12.	\$900	3 years	\$240	

C. Answer the following:

13. A sum of \$2,000 was invested for 3 years. The interest earned was \$480. What was the annual interest rate?

14. A piece of land was purchased for \$50,000 and sold $4\frac{1}{2}$ years later for \$75,000. What annual interest rate was earned on this investment? (answer to nearest whole percent)

15. What figure should be used to determine the cost of borrowing money?

SUPPLEMENTARY PROBLEMS (Continued)

D. Find the missing information to the nearest whole cent:

	Monthly Interest Rate	Balance	Interest for One Month
16.	2%	$370	
17.	$1\frac{1}{2}\%$	$42	
18.	$\frac{3}{4}\%$	$560	
19.	$1\frac{3}{4}\%$	$490	
20.	$1\frac{1}{2}\%$	$652	
21.	1%	$32.48	

Part II—Compound Interest

E. Answer the following:

22. What is compound interest?

23. What is compound amount?

24. Find the compound interest on $2,000 invested for 1 year at an annual rate of 20% compounded semiannually.

25. Find the compound interest on $2,000 invested for 1 year at an annual rate of 4% compounded quarterly.

F. If $1,000 were invested the day you were born at an annual rate of 6% compounded annually, how much would it amount to on your following birthdays:

26. 10th birthday. **27.** 15th birthday. **28.** 20th birthday. **29.** 25th birthday.

G. Find the missing information to the nearest whole cent:

	Amount Invested	Time Invested	Annual Interest Rate	Compounded	Compound Amount	Compound Interest
30.	$1,000	3 years	10%	quarterly		
31.	$800	5 years	6%	semiannually		
32.	$500	20 years	11%	annually		
33.	$200	2 years	6%	quarterly		
34.	$900	4 years	8%	quarterly		
35.	$500	7 months	12%	monthly		

Solutions to Study Exercises

(4A)

Study Exercise One (Frame 4)

1. Principal + Interest = Amount
   $1,650 + $103.13 = $1,753.13

2. Amount − Principal = Interest
   $964.80 − $900 = $64.80

3. Principal + Interest = Amount
   $722 + $43.24 = $765.24

4. Amount − Principal = Interest
   $1,125.50 − $1,000 = $125.50

5. Amount − Interest = Principal
   $350 − $15.75 = $334.25

Study Exercise Two (Frame 7)

(7A)

1. $i = prt$
   $= (285)(15\%)(3)$
   $= (285)(.15)(3)$
   $= (285)(.45)$
   $= \$128.25$ interest
   $285 + $128.25 = $413.25 amount

2. 9 months $= \frac{9}{12}$ year
   $i = prt$
   $= (\$4,500)(6\%)\left(\frac{9}{12}\right)$
   $= \cancel{4,500}^{45} \cdot \frac{\cancel{6}^{3}}{\cancel{100}_{1}} \cdot \frac{3}{\cancel{4}_{2}}$
   $= \frac{405}{2}$
   $= \$202.50$ interest
   $4,500 + $202.50 = $4,702.50 amount

3. 1 year 6 months $= 1\frac{1}{2}$ years
   $i = prt$
   $= (\$3,200)(14\%)\left(1\frac{1}{2}\right)$
   $= \cancel{3,200}^{32} \cdot \frac{\cancel{14}^{7}}{\cancel{100}_{1}} \cdot \frac{3}{\cancel{2}_{1}}$
   $= \$672$ interest
   $3,200 + $672 = $3,872 amount

4. 14 months $= \frac{14}{12}$ years
   $i = prt$
   $= (\$6,000)(8\%)\left(\frac{14}{12}\right)$
   $= \cancel{6,000}^{\cancel{60}10} \cdot \frac{8}{\cancel{100}_{1}} \cdot \frac{7}{\cancel{6}_{1}}$
   $= \$560$ interest
   $6,000 + $560 = $6,560 amount

Study Exercise Three (Frame 10)

(10A)

1. $pt = (600)(3)$
   $= 1,800$
   $r = \frac{i}{pt}$
   $= \frac{144}{1,800}$
   $= 144 \div 1,800$
   $= .08$
   $= 8\%$ annual interest rate

2. 1 year 6 months $= 1\frac{1}{2}$ years
   $pt = (4,000)\left(1\frac{1}{2}\right)$
   $= (\cancel{4,000}^{2,000})\left(\frac{3}{\cancel{2}_{1}}\right)$
   $= 6,000$
   $r = \frac{i}{pt}$
   $= \frac{510}{6,000}$
   $= 510 \div 6,000$
   $= .085$
   $\approx 9\%$ annual interest rate

***Study Exercise Three (Frame 10)* (Continued)**

3. 4 months $= \frac{4}{12}$ year

$$pt = (870)\left(\frac{4}{12}\right) = (870)\left(\frac{1}{3}\right) = 290$$

$$r = \frac{i}{pt} = \frac{18}{290} = 18 \div 290 \approx .062$$

$\approx$ 6% annual interest rate

4. $pt = (4{,}000)(2) = 8{,}000$

$i = 5{,}000 - 4{,}000 = 1{,}000$

$$r = \frac{i}{pt} = \frac{1{,}000}{8{,}000} = .125$$

= 12.5% annual interest rate

Study Exercise Four (Frame 13) (13A)

1. $I = PR$

$= (280)\left(\frac{3}{4}\%\right)$

$= (280)(.75\%)$

$= (280)(.0075)$

= \$2.10 interest for one month

2. $I = PR$

$= (520)\left(1\frac{1}{2}\%\right)$

$= (520)(1.5\%)$

$= (520)(.015)$

= \$7.80 interest for one month

3. $I = PR$

$= (328)\left(1\frac{3}{4}\%\right)$

$= (328)(1.75\%)$

$= (328)(.0175)$

= \$5.74 interest for one month

4. $I = PR$

$= (57.84)(2\%)$

$= (57.84)(.02)$

$= 1.1568$

= \$1.16 interest for one month to nearest cent

Study Exercise Five (Frame 22) (22A)

1. *Step (1):* 2 periods per year
Step (2): 6% ÷ 2 = 3% or .03 per period
Step (3): *Period (1):*
i = (\$500)(.03) = \$15
\$500 + \$15 = \$515
Period (2):
i = (\$515)(.03) = \$15.45
\$515 + \$15.45 = \$530.45 compound amount
Step (4): Compound Amount − Principal = Compound Interest
\$530.45 − \$500 = \$30.45

2. *Step (1):* 4 periods per year
Step (2): 8% ÷ 4 = 2% or .02 per period
Step (3): *Period (1):*
i = (1,000)(.02) = \$20.00
\$1,000 + \$20 = \$1,020
Period (2):
i = (\$1,020)(.02) = \$20.40
\$1,020 + \$20.40 = \$1,040.40
Period (3):
i = (\$1,040.40)(.02) = \$20.808 ≈ \$20.81
\$1,040.40 + \$20.81 = \$1,061.21
Period (4):
i = (\$1,061.21)(.02) = \$21.2242 ≈ \$21.22
\$1,061.21 + \$21.22 = \$1,082.43 compound amount

Study Exercise Five (Frame 22) (Continued)

Step (4): Compound Amount − Principal = Compound Interest

$1,082.43 − $1,000 = $82.43

Study Exercise Six (Frame 26) (26A)

1. *Step (1):* 1 period per year
   Step (2): 6% ÷ 1 = 6% per period
   Step (3): 1 × 4 = 4 total periods
   Step (4): Table II (4th row; 6% column): 1.2625
   Step (5): (1.2625) ($200) = $252.50 compound amount
   Step (6): $252.50 − $200.00 = $52.50 compound interest

2. *Step (1):* 2 periods per year
   Step (2): 8% ÷ 2 = 4% per period
   Step (3): 2 × 5 = 10 total periods
   Step (4): Table II (10th row; 4% column): 1.4802
   Step (5): (1.4802) ($500) = $740.10 compound amount
   Step (6): $740.10 − $500.00 = $240.10 compound interest

3. *Step (1):* 12 periods per year
   Step (2): 12% ÷ 12 = 1% per period
   Step (3): $12 \times \frac{1}{2} = 6$ total periods
   Step (4): Table II (6th row; 1% column): 1.0615
   Step (5): (1.0615) ($700) = $743.05 compound amount
   Step (6): $743.05 − $700.00 = $43.05 compound interest

4. *Step (1):* 2 periods per year
   Step (2): 10% ÷ 2 = 5% per period
   Step (3): 2 × 25 = 50 total periods
   Step (4): Table II (50th row; 5% column): 11.4674
   Step (5): (11.4674) ($1,000) = $11,467.40 compound amount
   Step (6): $11,467.40 − $1,000.00 = $10,467.40 compound interest

Module 4 Practice Test

Units 23–26

Change these percents to fractions:

1. 25% **2.** 43% **3.** 600% **4.** $62\frac{1}{2}\%$

Change these percents to decimals:

5. 33% **6.** 123% **7.** $7\frac{1}{2}\%$

Change these decimals to percents:

8. .58 **9.** .004 **10.** $.66\frac{2}{3}$

Change these fractions to percents:

11. $\frac{58}{100}$ **12.** $\frac{9}{20}$ **13.** $\frac{3}{8}$

14. Round $\frac{2}{21}$ to the nearest tenth of a percent.

15. Find 6.2% of 80 **16.** Find $1\frac{1}{2}\%$ of $500 **17.** 17% of what number is 85?

18. 30% of what number is 4?(answer to nearest tenth)

19. What percent of 28 is 17? (answer to nearest whole percent)

20. What percent of 85 is 17?

21. The amount is $298.50 and the principal is $252.40. How much is the interest?

22. Find the simple interest on $800 invested for 3 years at an annual interest rate of $7\frac{1}{2}\%$.

23. $400 was invested for 3 years and earned $144 interest. What was the annual interest rate?

24. The balance on a charge account for the month of June was $640. What will the interest charge for this month be if a $1\frac{1}{2}\%$ monthly interest rate is used?

25. Find the compound interest on $700 invested for 5 years at an 8% annual interest rate compounded semiannually.

Answers to Module 4 Practice Test

1. $\frac{1}{4}$
2. $\frac{43}{100}$
3. $\frac{6}{1}$ or 6
4. $\frac{5}{8}$
5. .33
6. 1.23
7. .075
8. 58%
9. .4%
10. $66\frac{2}{3}\%$
11. 58%
12. 45%
13. $37\frac{1}{2}\%$
14. 9.5%
15. 4.96
16. $7.50
17. 500
18. 13.3
19. 61%
20. 20%
21. $46.10
22. $180
23. 12%
24. $9.60
25. $336.14

MODULE

5

Optional Units Involving Applications of Percent

UNIT

27 Percent Decrease

OBJECTIVES (1)

By the end of this unit you should be able to:

1. IDENTIFY THE ORIGINAL AMOUNT, THE NEW AMOUNT, AND THE AMOUNT OF DECREASE IN A PROBLEM.
2. FIND THE RATE OF DECREASE.
3. SOLVE PROBLEMS INVOLVING RATE OF DECREASE.

Original Amount–New Amount–Amount of Decrease (2)

Example 1: Yesterday a bicycle sold for $110. Today it sells for $100. What is the amount of decrease?

Solution:

Line (a): Original Amount − New Amount = Amount of Decrease

↓ ↓ ↓

Line (b): $110 − $100 = $10

Example 2: A club that previously had 500 members lost 40 of its members this year. What is the current membership?

Solution:

Line (a): Original Amount − Amount of Decrease = New Amount

↓ ↓ ↓

Line (b): 500 − 40 = 460

Example 3: A factory now employs 30 people. This is a decrease of 15 people from the previous year. How many people did it employ last year?

Solution:

Line (a): Amount of Decrease + New Amount = Original Amount

↓ ↓ ↓

Line (b): 15 + 30 = 45

Key Relationships (3)

1. Original Amount − New Amount = Amount of Decrease
2. Original Amount − Amount of Decrease = New Amount
3. Amount of Decrease + New Amount = Original Amount

Study Exercise One (4)

Find the missing information:

	Original Amount	New Amount	Amount of Decrease
1.		18	2
2.	55	32	
3.	$105		$16
4.	5,672	63	
5.		$48	$26

Determining the Rate of Decrease (5)

The rate of decrease is established by the fraction whose numerator is the amount of decrease and whose denominator is the original amount.

$$\text{Rate of Decrease} = \frac{\text{Amount of Decrease}}{\text{Original Amount}}$$

We will convert this fraction to a percent.

Example 1: Yesterday a bicycle sold for $110. Today it sells for $100. What is the rate of decrease?

Solution:

Step (1): Find the amount of decrease and the original amount:
Line (a): $110 − $100 = $10 (the amount of decrease)
Line (b): $110 (the original amount)

Step (2): Arrange the information in fraction form:

$$\frac{\text{Amount of Decrease}}{\text{Original Amount}} = \frac{\$10}{\$110}$$

Step (3): Change the fraction to a percent:

$$\frac{\$10}{\$110} = \frac{1}{11} = 9\frac{1}{11}\%\text{ rate of decrease}$$

Computation

$$\begin{array}{r} .09\frac{1}{11} \\ 11\overline{)1.00} \\ \underline{99} \\ 1 \end{array}$$

Determining the Rate of Decrease (Continued)

Example 2: A club that previously had 500 members lost 40 of its members this year. What rate of decrease did it experience?

Solution:

Step (1): 40 is the amount of decrease
500 is the original amount

Step (2): $\frac{\text{Amount of Decrease}}{\text{Original Amount}} = \frac{40}{500}$

Step (3): $\frac{40}{500} = \frac{40 \div 5}{500 \div 5} = \frac{8}{100} = 8\%$ rate of decrease

Example 3: A factory now employs 30 people. This is a decrease of 15 people from the previous year. What is the rate of decrease to the nearest whole percent?

Solution:

Step (1): 15 is the amount of decrease
15 + 30 = 45 is the original amount

Step (2): $\frac{\text{Amount of Decrease}}{\text{Original Amount}} = \frac{15}{45} = \frac{1}{3}$

Step (3): $\frac{1}{3} = 33\frac{1}{3}\%$
= 33% rate of decrease (to the nearest whole percent)

Remember (6)

$$\text{Rate of Decrease} = \frac{\text{Amount of Decrease}}{\text{Original Amount}}$$

Study Exercise Two (7)

Find the missing information to the nearest whole dollar or nearest whole percent:

	Original Amount	New Amount	Amount of Decrease	Rate of Decrease
1.	$75		$50	
2.	$328	$311		
3.		$142	$44	
4.	$1,879	$1,800		

Using the Rate of Decrease to Find the Amount of Decrease (8)

Example: A school with an enrollment of 1,200 students expects a rate of decrease of 20% next year. What will be the amount of this decrease?

Solution:

Step (1): Write the word formula:

Rate of Decrease × Original Amount = Amount of Decrease

Step (2): Replace the words with numerals where possible:

Rate of Decrease × Original Amount = Amount of Decrease

$$20\% \times 1{,}200 = \text{Amount of Decrease}$$

Step (3): Perform the computation:

$$20\% \times 1{,}200 = \frac{1}{\cancel{5}_1} \times \cancel{1{,}200}^{240}$$

$$= 240 \text{ students}$$

Using the Rate of Decrease to Find the Original Amount (9)

Example: A factory reduced the number of articles it produced in one week by 20 articles. This represented a 16% decrease from the previous week. How many articles did it produce in the previous week?

Solution:

Step (1): Write the word formula:

Rate of Decrease × Original Amount = Amount of Decrease

Step (2): Replace the words with numerals where possible and then form an equation:

Rate of Decrease × Original Amount = Amount of Decrease

Line (a): 16% × Original Amount = 20

Line (b): $(.16) \times n = 20$

Step (3): Solve the equation:

$$n = 20 \div .16$$
$$= 125 \text{ articles}$$

Computation

```
        1 25.
.16.)20.00.
      16
       4 0
       3 2
         80
         80
          0
```

Formula for Finding the Original Amount or the Amount of Decrease (10)

This formula is used when you are given the rate of decrease and either the original amount or the amount of decrease.

Rate of Decrease × Original Amount = Amount of Decrease

Study Exercise Three (11)

1. An article originally cost \$21.60. If the price was decreased by $33\frac{1}{3}\%$, what was the amount of decrease?
2. A factory which formerly produced 728 articles per day reduced its output by 23%. This represents a reduction of how many articles? (Give your answer to the nearest whole article.)
3. A person who was making \$5.40 an hour had a pay cut of 8%. To the nearest cent, how much less is this person now making?
4. A store reduced the number of its employees by 72 people. This amounted to a decrease of 25%. How many people were employed before the reduction?
5. A family decided to spend 30% of its income per year on housing. This amount was \$4,800. What was the original yearly income of this family?

Using the Rate of Decrease to Find the New Amount (12)

Example: A furniture store decided to decrease the price of a \$1,200 bedroom suite by a rate of 20%. What will be the new price?

Solution:

Step (1): Determine the amount of decrease:

Line (a): Rate of Decrease × Original Amount = Amount of Decrease

Line (b): 20% × \$1,200 = Amount of Decrease

Line (c): $\frac{1}{\cancel{5}_1} \times \cancel{1,200}^{240} = \$240$ (amount of decrease)

Step (2): Subtract to find the new amount:

Line (d): Original Amount − Amount of Decrease = New Amount

Line (e): \$1,200 − \$240 = \$960

A Short Cut (13)

Gold Bar Representing Original Price (\$1,200)

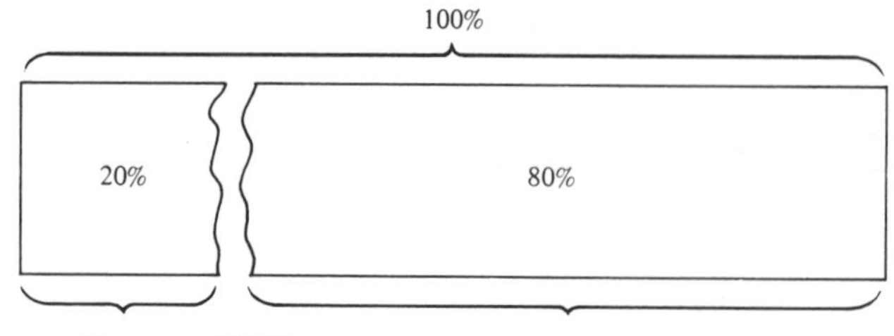

A Short Cut (Continued)

Since we are only interested in the new amount, which represents 80% of the original amount, we can get our information as follows:

Line (a): (100% − Rate of Decrease) × Original Amount = New Amount

Line (b): (100% − 20%) × $1,200 = New Amount

Line (c): 80% × $1,200 = New Amount

Line (d):

$$\begin{array}{r} \$1,200 \\ \underline{\times .80} \\ \$960.00 \end{array} = \text{New Amount}$$

Using 100% Minus the Rate of Decrease to Find the Original Amount (14)

Example: An automobile dealer sold a car for $4,200. He claimed that this was a reduction of 30% from the original price. What was the original price?

Solution:

Gold Bar Representing Original Price 100%

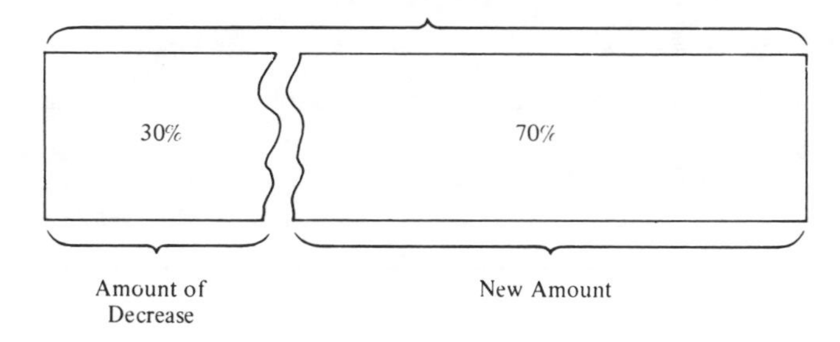

Line (a): (100% − Rate of Decrease) × Original Amount = New Amount

Line (b): (100% − 30%) × Original Amount = $4,200

Line (c): 70% × n = $4,200

Line (d): $(.7)n$ = $4,200

Line (e): n = $4,200 ÷ 7

Line (f): n = $6,000

Formula for Finding the Original Amount or the New Amount (15)

(100% minus the rate of decrease)

This formula is used when you are given the rate of decrease and either the original amount or the new amount.

(100% − Rate of Decrease) × Original Amount = New Amount

Study Exercise Four (16)

1. A factory which regularly produced 820 articles per day decreased production by a rate of 35%. How many articles is it now producing?
2. An automobile dealer sold 40 cars last month. This month sales decreased by a rate of 15%. How many cars were sold this month?

***Study Exercise Four* (Continued)**

3. An employee who made $10.20 an hour had a pay cut of 7%. To the nearest cent, what will be the new hourly rate?
4. A merchant claimed that the new price of a stereo set is $196.80 and that their price represents a rate of decrease of 18%. What was the original price?

Summary (17)

A. Addition and subtraction relationships.
   1. Original Amount − New Amount = Amount of Decrease
   2. Original Amount − Amount of Decrease = New Amount
   3. Amount of Decrease + New Amount = Original Amount

B. This fraction, which we will change to a percent, produces the rate of decrease.

$$\frac{\text{Amount of Decrease}}{\text{Original Amount}} = \text{Rate of Decrease}$$

C. This formula is used when you are given the rate of decrease and either the original amount or the amount of decrease.

$$\text{Rate of Decrease} \times \text{Original Amount} = \text{Amount of Decrease}$$

D. This formula is used when you are given the rate of decrease and either the original amount or the new amount.

$$(100\% - \text{Rate of Decrease}) \times \text{Original Amount} = \text{New Amount}$$

REVIEW EXERCISES (18)

A. Answer the following:
   1. The rate of decrease is produced from a fraction whose numerator is _________ and whose denominator is _________.
   2. Original Amount − _________ = New Amount.
   3. _________ + New Amount = Original Amount.
   4. Original Amount − _________ = Amount of Decrease.
   5. Rate of Decrease × Original Amount = _________.
   6. (100% − Rate of Decrease) × Original Amount = _________.
   7. A man bought a house for $70,500 and sold it for $60,000. What was the rate of decrease to the nearest whole percent?
   8. A coat which once sold for $94 is now priced at $87. Find the rate of decrease to the nearest whole percent.
   9. A dress marked at $32 is to be decreased by 12%. What will be the new sale price?
   10. A file cabinet now sells for $78.96. The clerk claims this was a reduction of 16% from the original price. What was the original price?
   11. A school which had an enrollment of 1,500 students this year expects to have a decrease of 17% next year. What will be the amount of this decrease?

REVIEW EXERCISES (Continued)

B. Find the missing information. Round rates to nearest whole percent:

	Original Price	New Price	Amount of Decrease	Rate of Decrease
12.	\$110	\$100		
13.	\$54	\$50		
14.	\$300	\$213		

Solutions to Review Exercises (19)

A. **1.** See Frames 5 and 17 **2.** See Frames 3 and 17 **3.** See Frames 3 and 17
4. See Frames 3 and 17 **5.** See Frames 10 and 17 **6.** See Frames 15 and 17

7. \$70,500 − \$60,000 = \$10,500 amount of decrease.

$$\frac{\text{Amount of Decrease}}{\text{Original Amount}} = \frac{\$10,500}{\$70,500} \approx 15\%$$

8. \$94 − \$87 = \$7 amount of decrease.

$$\frac{\text{Amount of Decrease}}{\text{Original Amount}} = \frac{\$7}{\$94} \approx 7\%$$

9. (100% − Rate of Decrease) × Original Amount = New Amount

$$(100\% - 12\%) \times \$32 = \text{New Amount}$$
$$88\% \times \$32 = \text{New Amount}$$
$$.88 \times \$32 = \$28.16 \text{ new price}$$

10. (100% − Rate of Decrease) × Original Amount = New Amount

$$(100\% - 16\%) \times \text{Original Amount} = \$78.96$$
$$84\% \times n = \$78.96$$
$$(.84) \times n = \$78.96$$
$$n = \$78.96 \div 84$$
$$n = \$94 \text{ original price}$$

11. Rate of Decrease × Original Amount = Amount of Decrease

$$17\% \times 1,500 = \text{Amount of Decrease}$$
$$.17 \times 1,500 = 255 \text{ students}$$

B. **12.** \$110 − \$100 = \$10 amount of decrease.

$$\frac{\text{Amount of Decrease}}{\text{Original Amount}} = \frac{\$10}{\$110} \approx 9\%$$

13. \$54 − \$50 = \$4 amount of decrease.

$$\frac{\text{Amount of Decrease}}{\text{Original Amount}} = \frac{\$4}{\$54} \approx 7\%$$

14. \$300 − \$213 = \$87

$$\frac{\text{Amount of Decrease}}{\text{Original Amount}} = \frac{\$87}{\$300} = \frac{87 \div 3}{300 \div 3} = \frac{29}{100} = 29\%$$

SUPPLEMENTARY PROBLEMS

A. Find the missing information. Round rates to nearest whole percent:

	Original Price	New Price	Amount of Decrease	Rate of Decrease
1.	$84	$72		
2.	$720	$680		
3.	$20	$17		
4.	$500	$475		

B. Answer the following:

5. (Rate of Decrease) × Original Amount = _________ .

6. (100% − Rate of Decrease) × Original Amount = _________ .

7. A factory which produced 8,200 articles last month experienced a 23% decrease this month. What is the amount of this decrease?

8. Yesterday a bicycle sold for $250. Today it sells for 17% less. What is the amount of decrease?

9. A school with an enrollment of 1,600 students expects a decrease of 21% next year. What will be the amount of this decrease?

10. In a class of 40 students, 15% were absent. How many students were absent?

11. A stove which sells for $228 may be purchased for a 25% down payment. How much money is needed for the down payment?

12. A company reduced the number of its employees by 30 people. This was a decrease of 8%. How many people were originally employed by this company?

13. A coat originally priced at $240 is to be reduced by 18%. What will be the new price?

14. In a city of 43,000 voters, it is estimated that 35% will not vote. How many people will vote?

15. A desk now sells for $284.70. The seller claims that this price was a reduction of 22% from the original price. What was the original price?

16. The population of Sometown, U.S.A., is presently 84,000. The population has decreased $12\frac{1}{2}$% in the last 5 years. What was the population of Sometown, U.S.A., 5 years ago?

17. The number of Datsun trucks sold in the United States decreased from 104,000 last year to 95,000 this year. To the nearest tenth of a percent, find the percent of decrease.

18. The price of gasoline dropped from $1.35 per gallon last year to a current price of $1.19. To the nearest tenth of a percent, find the percent of decrease.

19. Last week the index used to measure the status of the stock market was equal to 943. This week the index is equal to 928. To the nearest hundredth of a percent, find the percent of decrease.

20. After being reduced by 15%, a TV sold for $601. To the nearest cent, what was the original price?

Solutions to Study Exercises

(4A)

Study Exercise One (Frame 4)

1. Amount of Decrease + New Amount = Original Amount

   ↓ ↓ ↓

   2 + 18 = 20

2. Original Amount − New Amount = Amount of Decrease

   ↓ ↓ ↓

   55 − 32 = 23

3. Original Amount − Amount of Decrease = New Amount

   ↓ ↓ ↓

   $105 − $16 = $89

4. Original Amount − New Amount = Amount of Decrease

   ↓ ↓ ↓

   5,672 − 63 = 5,609

5. Amount of Decrease + New Amount = Original Amount

   ↓ ↓ ↓

   $26 + $48 = $74

(7A)

Study Exercise Two (Frame 7)

1. $75 − $50 = $25 new amount.

   $$\frac{\text{Amount of Decrease}}{\text{Original Amount}} = \frac{\$50}{\$75} = \frac{50 \div 25}{75 \div 25} = \frac{2}{3} \approx 67\%$$

2. $328 − $311 = $17 amount of decrease.

   $$\frac{\text{Amount of Decrease}}{\text{Original Amount}} = \frac{\$17}{\$328} \approx 5\%$$

3. $142 + $44 = $186 original amount.

   $$\frac{\text{Amount of Decrease}}{\text{Original Amount}} = \frac{\$44}{\$186} \approx 24\%$$

4. $1,879 − $1,800 = $79

   $$\frac{\text{Amount of Decrease}}{\text{Original Amount}} = \frac{\$79}{\$1{,}879} \approx 4\%$$

(11A)

Study Exercise Three (Frame 11)

1. Rate of Decrease × Original Amount = Amount of Decrease

   ↓ ↓ ↓

   $33\frac{1}{3}\%$ × $21.60 = Amount of Decrease

   $\frac{1}{\not{3}_1} \times \not{21.60}^{\$7.20} = \$7.20$

2. Rate of Decrease × Original Amount = Amount of Decrease

   ↓ ↓ ↓

   23% × 728 = Amount of Decrease

   .23 × 728 = 167.44 ≈ 167 articles

3. Rate of Decrease × Original Amount = Amount of Decrease

   ↓ ↓ ↓

   8% × $5.40 = Amount of Decrease

   .08 × $5.40 = $.432 ≈ $.43 or 43¢

4. Rate of Decrease × Original Amount = Amount of Decrease

   ↓ ↓ ↓

   25% × Original Amount = 72

   $\frac{1}{4} \times n = 72$

   $$n = 72 \div \frac{1}{4} = 72 \times \frac{4}{1} = 288 \text{ people}$$

5. Rate of Decrease × Original Amount = Amount of Decrease

   ↓ ↓ ↓

   30% × Original Amount = $4,800

   $(.30)n = \$4{,}800$

   $n = \$4{,}800 \div .30$

   $n = \$16{,}000$

Study Exercise Four (Frame 16) (16A)

1. (100% − Rate of Decrease) × Original Amount = New Amount

(100% − 35%)	×	820	= New Amount
65%	×	820	= New Amount
.65	×	820	= 533 articles

2. (100% − Rate of Decrease) × Original Amount = New Amount

(100% − 15%)	×	40	= New Amount
85%	×	40	= New Amount
.85	×	40	= 34 cars

3. (100% − Rate of Decrease) × Original Amount = New Amount

(100% − 7%)	×	$10.20	= New Amount
93%	×	$10.20	= New Amount
.93	×	$10.20	= $9.486 ≈ $9.49

4. (100% − Rate of Decrease) × Original Amount = New Amount

(100% − 18%)	×	Original Amount	= $196.80
82%	×	n	= $196.80
(.82)	×	n	= $196.80
		n	= $196.80 ÷ .82
		n	= $240 original price

UNIT

28 Commission

OBJECTIVES

(1)

By the end of this unit you should:

1. KNOW THE MEANING OF *RATE OF COMMISSION*, *AMOUNT OF COMMISSION*, AND *NET PROCEEDS*.
2. BE ABLE TO COMPUTE A RATE OF COMMISSION FROM A BASIC FRACTION.
3. BE ABLE TO SOLVE PROBLEMS INVOLVING RATE OF COMMISSION.
4. UNDERSTAND HOW RATE OF COMMISSION PROBLEMS MAY BE SOLVED IN EXACTLY THE SAME MANNER AS RATE OF DECREASE PROBLEMS.

Sales Amount–Net Proceeds–Amount of Commission

(2)

Example 1: A boat sold for $5,000 (this is the *sales amount*). The manufacturer of the boat kept $4,000 (this is called *net proceeds*). The difference was paid as a commission to the person who made the sale. What was the amount of commission?

Solution:

Line (a): Sales Amount − Net Proceeds = Amount of Commission
(original amount) − (new amount) = (amount of decrease)
↓ ↓ ↓
Line (b): $5,000 − $4,000 = $1,000

Example 2: A used car was sold for $1,000. The person who made the sale received a commission in the amount of $200. How much was left to the company after the commission was paid?

Solution:

Line (a): Sales Amount − Amount of Commission = Net Proceeds
↓ ↓ ↓
Line (b): $1,000 − $200 = $800

Example 3: An accountant discovered that the net proceeds for one sale was $600 and that the amount of commission paid was $200. What was the sales amount?

Sales Amount–Net Proceeds–Amount of Commission (Continued)

Solution:

Line (a): Amount of Commission + Net Proceeds = Sales Amount

Line (b): $200 + $600 = $800

Key Relationships (3)

1. Sales Amount − Net Proceeds = Amount of Commission
2. Sales Amount − Amount of Commission = Net Proceeds
3. Amount of Commission + Net Proceeds = Sales Amount

Study Exercise One (4)

Find the missing information:

	Sales Amount	Net Proceeds	Amount of Commission
1.	$72		$6
2.	$55	$32	
3.	$115		$26
4.		$48.35	$16.25
5.	$22.80		$3.50

Determining the Rate of Commission (5)

The rate of commission is established by the fraction whose numerator is the commission (amount of decrease) and whose denominator is the sales amount (original amount).

$$\underset{\text{(rate of decrease)}}{\text{Rate of Commission}} = \frac{\text{Amount of Commission}}{\text{Sales Amount}} \quad \frac{\text{(amount of decrease)}}{\text{(original amount)}}$$

We will convert this fraction to a percent.

Example 1: A boat sold for $5,000. The manufacturer of the boat kept $4,000. The difference was paid as a commission. What was the rate of commission?

Solution:

Step (1): Find the amount of commission and the sales amount:

Line (a): $5,000 − $4,000 = $1,000 is the amount of commission.

Line (b): $5,000 is the sales amount.

Step (2): Arrange the information in fraction form:

$$\frac{\text{Amount of Commission}}{\text{Sales Amount}} = \frac{\$1,000}{\$5,000}$$

Determining the Rate of Commission (Continued)

Step (3): Change the fraction to a percent:

$$\frac{\$1{,}000}{\$5{,}000} = \frac{1}{5}$$
$$= 20\% \text{ rate of commission}$$

Example 2: A used car was sold for $1,000. If a commission in the amount of $200 was paid on this sale, what was the rate of commission?

Solution:

Step (1): $200 is the amount of commission.
$1,000 is the sales amount.

Step (2): $$\frac{\text{Amount of Commission}}{\text{Sales Amount}} = \frac{\$200}{\$1{,}000}$$

Step (3): $$\frac{\$200}{\$1{,}000} = \frac{20}{100}$$
$$= 20\% \text{ rate of commission}$$

Example 3: An accountant discovered that the net proceeds for one sale was $600 and that the amount of commission paid was $200. What was the rate of commission?

Solution:

Step (1): $200 is the amount of commission.
$200 + $600 = $800 is the sales amount.

Step (2): $$\frac{\text{Amount of Commission}}{\text{Sales Amount}} = \frac{\$200}{\$800}$$

Step (3): $$\frac{\$200}{\$800} = \frac{2}{8}$$
$$= \frac{1}{4}$$
$$= 25\% \text{ rate of commission}$$

Remember

(6)

$$\text{Rate of Commission} = \frac{\text{Amount of Commission}}{\text{Sales Amount}}$$

Study Exercise Two

(7)

Find the missing information to the nearest whole dollar and nearest whole percent:

	Sales Amount	Net Proceeds	Amount of Commission	Rate of Commission
1.	$75		$25	
2.	$580	$464		
3.		$45.50	$19.50	
4.	$640	$449		

Using the Rate of Commission to Find the Amount of Commission (8)

Example: A diamond ring was sold for $2,000 and the clerk who made the sale was paid a rate of commission of 15%. What was the amount of the clerk's commission?

Solution:

Step (1): Write the word formula:

Rate of Commission × Sales Amount = Amount of Commission

Step (2): Replace the words with numerals where possible:

Rate of Commission × Sales Amount = Amount of Commission

15% × $2,000 = Amount of Commission

Step (3): Perform the computation:

15% × $2,000 =

(.15) × $2,000 = $300 amount of commission

Using the Rate of Commission to Find the Sales Amount (9)

Example: An employee whose rate of commission is 25% wishes to make $300 per week. What sales amount must be achieved in order that the amount of commission will be $300?

Solution:

Step (1): Write the word formula:

Rate of Commission × Sales Amount = Amount of Commission

Step (2): Replace the words with numerals where possible and then form an equation:

Rate of Commission × Sales Amount = Amount of Commission

Line (a): 25% × Sales Amount = $300

Line (b): (.25) × n = $300

Step (3): Solve the equation:

n = $300 ÷ (.25)

= $1,200 sales amount

Computation

```
        12 00.
 .25.)300.00.
      25
       50
       50
```

Formula for Finding the Sales Amount or Amount of Commission (10)

This formula is used when you are given the rate of commission and either the sales amount or the amount of commission:

Rate of Commission × Sales Amount = Amount of Commission

Study Exercise Three (11)

1. What will be the amount of commission on a $45,000 house if the rate of commission is 6%?
2. A used car was sold for $2,800. What is the amount of commission if the rate of commission was 8%?
3. A person whose rate of commission was 30% wanted to make $240 per week. What must the sales amount be to achieve this?
4. A sales clerk whose rate of commission was 22% wanted to make $198 per week. What sales amount is necessary to achieve this?

Using the Rate of Commission to Find the Net Proceeds (12)

Example: A furniture store sold a bedroom suite for $1,200 (original amount). If a 20% rate of commission (rate of decrease) were paid to the person making the sale, how much did the store receive (new amount)?

Solution:

Step (1): Determine the Amount of Commission (amount of decrease):

Line (a): Rate of Commission × Sales Amount = Amount of Commission

Line (b): 20% × $1,200 = Amount of Commission

Line (c): .2 × $1,200 = $240 (amount of commission)

Step (2): Subtract to find the net proceeds:

Line (a): Sales Amount − Amount of Commission = Net Proceeds

Line (b): $1,200 − $240 = $960

A Short Cut (13)

Gold Bar Representing 100% of Sales Amount

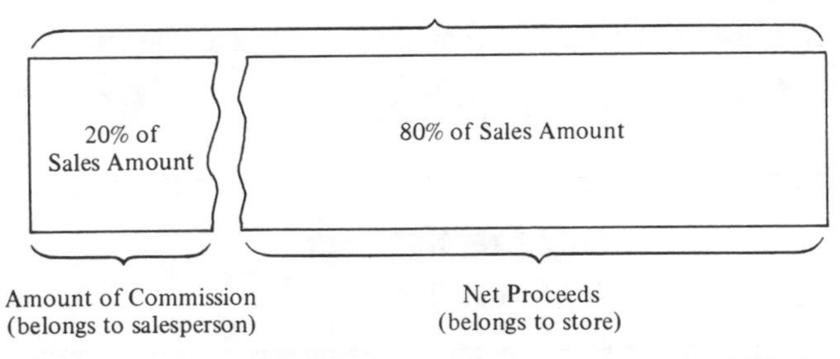

Since we are only interested in the net proceeds (new amount), which represents 80% of the sales amount (original amount), we can get our information as follows:

Line (a): (100% − Rate of Commission) × Sales Amount = Net Proceeds

Line (b): (100% − 20%) × $1,200 = Net Proceeds

Line (c): 80% × $1,200 = Net Proceeds

Line (d):

$$\begin{array}{r} \$1,200 \\ \underline{\times .80} \\ \$960.00 \end{array}$$ Net Proceeds

Using 100% Minus the Rate of Commission to Find the Sales Amount (14)

Example: In order to stay in business a manufacturer determines that his net proceeds (new amount) must be a minimum of $210,000 per week. He pays his sales personnel a 30% rate of commission. What is the minimum sales amount (original amount) required?

Solution:

Gold Bar Representing Sales Amount 100%

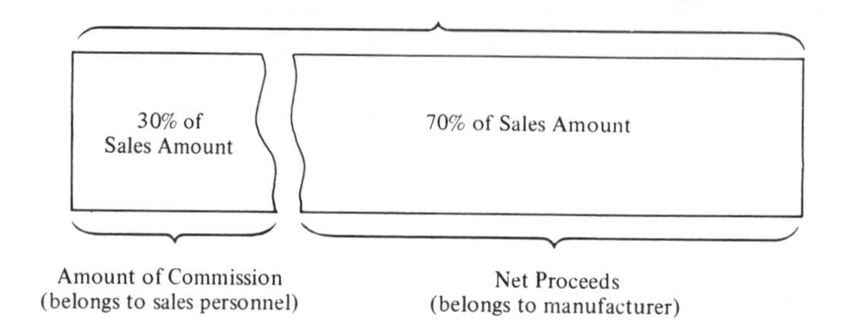

Line (a): (100% − Rate of Commission) × Sales Amount = Net Proceeds

Line (b): (100% − 30%) × Sales Amount = $210,000

Line (c): 70% × n = $210,000

Line (d): (.7) × n = $210,000

Line (e): n = $210,000 ÷ (.7)

Line (f): n = $300,000 sales amount

Formula for Finding the Sales Amount or Net Proceeds (15)

(100% minus the rate of commission)

This formula is used when you are given the rate of commission and either the sales amount or the net proceeds.

(100% − Rate of Commission) × Sales Amount = Net Proceeds

Study Exercise Four (16)

Find the missing information to the nearest whole dollar:

	Rate of Commission	Sales Amount	Net Proceeds
1.	25%	$288	
2.	16%		$2,870
3.	34%		$260
4.	12%	$362	

Summary (17)

A. Addition and subtraction relationships:

1. Sales Amount − Net Proceeds = Amount of Commission
2. Sales Amount − Amount of Commission = Net Proceeds
3. Amount of Commission + Net Proceeds = Sales Amount

B. This fraction, which we will change to a percent, produces the rate of commission:

$$\frac{\text{Amount of Commission}}{\text{Sales Amount}} = \text{Rate of Commission}$$

C. This formula is used when you are given the rate of commission and either the sales amount or the amount of commission:

$$\text{Rate of Commission} \times \text{Sales Amount} = \text{Amount of Commission}$$

D. This formula is used when you are given the rate of commission and either the sales amount or the net proceeds:

$$(100\% - \text{Rate of Commission}) \times \text{Sales Amount} = \text{Net Proceeds}$$

REVIEW EXERCISES (18)

A. A car was sold for $2,000. The salesperson received $200 and the dealer received $1,800.

1. The $200 is the ________.
2. The $2,000 is the ________.
3. The $1,800 is the ________.
4. What was the rate of commission on this sale?

B.

5. The rate of commission is established by the fraction whose numerator is the ________ and whose denominator is the ________.
6. If you know the amount of commission and sales amount, you can find the net proceeds by ________ the amount of commission from the ________.
7. If the commission on a sale of $840 is $48, find the rate of commission to the nearest whole percent.
8. A salesclerk receives $60 per week plus a 4% rate of commission on the sales amount. What are the total weekly earnings if the sales for one week amounted to $1,562?
9. An employee whose rate of commission is 27% wishes to make at least $200 per week. What is the minimum sales amount necessary? (to nearest whole dollar)
10. A used car was sold for $1,700. If the person who made the sale was paid a rate of commission of 16%, how much did the dealer receive?

Solutions to Review Exercises (19)

A. 1. Amount of commission. 2. Sales amount. 3. Net proceeds.

4. $\frac{\$200}{\$2{,}000} = \frac{\$200 \div 20}{\$2{,}000 \div 20} = \frac{10}{100} = 10\%$

Solutions to Review Exercises (Continued)

B. **5.** Amount of commission, sales amount. **6.** Subtracting, sales amount.

7. $\frac{\$48}{\$840} = 48 \div 840 = 840\overline{)48{,}000}$ (quotient .057) $= 6\%$ to nearest whole percent.

8. Rate of Commission × Sales Amount = Amount of Commission

4%	×	\$1,562	= Amount of Commission
.04	×	\$1,562	= \$62.48 amount of commission
\$62.48	+	\$60.00	= \$122.48 weekly earnings

9. Rate of Commission × Sales Amount = Amount of Commission

27%	× Sales Amount	=	\$200
(.27)	× n	=	\$200
	n	=	\$200 ÷ (.27)
	n	=	\$741 to nearest dollar

Computation

.27.)\$200.00.0 = \$7 40.7

10. (100% − Rate of Commission) × Sales Amount = Net Proceeds

(100% − 16%)	×	\$1,700	= Net Proceeds
84%	×	\$1,700	= Net Proceeds
(.84)	×	(\$1,700)	= \$1,428 net proceeds

SUPPLEMENTARY PROBLEMS

A. Fill in the blanks:

1. The rate of commission is produced from a fraction whose numerator is the __________ and whose denominator is the __________.
2. The Sales Amount − __________ = Net Proceeds.
3. The Sales Amount − __________ = Amount of Commission.
4. __________ + Net Proceeds = Sales Amount.
5. (100% − Rate of Commission) × Sales Amount = __________.

B. Find the missing information to the nearest whole percent or to the nearest dollar:

	Amount of Commission	Net Proceeds	Sales Amount	Rate of Commission
6.	\$48	\$62		
7.	\$32	\$98		
8.	\$5	\$20		
9.	\$72	\$360		

SUPPLEMENTARY PROBLEMS (Continued)

C. Answer the following:

10. What will be the amount of commission on a $36,000 house if the realtor charges a 6% rate of commission?

11. What will be the amount of commission on a $2,400 diamond ring if the rate of commission is 25%?

12. A salesperson who is paid a rate of commission of 16% wishes to make $200 per week. What must the sales amount be to achieve this goal?

13. A person whose rate of commission was 35% wanted to make at least $150 per week. To the nearest dollar, what minimum sales amount must be reached?

14. What will be the net proceeds on a car costing $3,400 if the rate of commission is 18%?

15. In order to stay in business, a manufacturer determines that his net proceeds must be a minimum of $150,000 per week. He pays his sales personnel a 20% rate of commission. What is the minimum sales amount required?

D. Find the missing information to the nearest dollar:

	Rate of Commission	Sales Amount	Amount of Commission	Net Proceeds
16.	10%	$1,287		
17.	32%			$64
18.	15%	$287		
19.	19%		$42	
20.	40%	$162		

E. Answer the following:

21. A realtor receives a 7% rate of commission on her sales. If her amount of commission last year was $76,650, what was her sales amount?

22. A salesman, who received a rate of commission of 5.5%, sold two cars for the same price. If the total commission was $1,287, what was the price of each car?

23. The Brown's house was sold for $126,000. The rate of commission for the realtor was 7%. What was the net proceeds the Brown's received?

24. Jennifer receives a rate of commission for each house she sells equal to 5% of the first $100,000 and 3% of any amount of the selling price above $100,000. Find the amount of commission if she sells a house for $138,000.

25. The Smiths paid a realtor a 6% rate of commission to sell their house. If the net proceeds for the Smiths was $134,420, what was the sales amount of the house?

Solutions to Study Exercises (4A)

Study Exercise One (Frame 4)

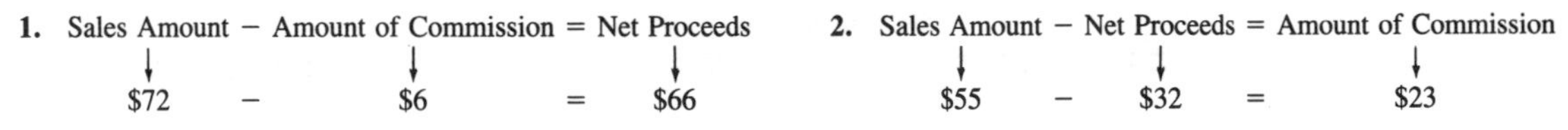

Solutions to Study Exercises (Continued)

3. Sales Amount − Amount of Commission = Net Proceeds
   $115 − $26 = $89

4. Amount of Commission + Net Proceeds = Sales Amount
   $16.25 + $48.35 = $64.60

5. Sales Amount − Amount of Commission = Net Proceeds
   $22.80 − $3.50 = $19.30

Study Exercise Two (Frame 7) (7A)

1. $75 − $25 = $50 net proceeds.

$$\frac{\text{Amount of Commission}}{\text{Sales Amount}} = \frac{\$25}{\$75} = \frac{1}{3} = 33\% \text{ rate of commission to nearest whole percent.}$$

2. $580 − $464 = $116 amount of commission.

$$\frac{\text{Amount of Commission}}{\text{Sales Amount}} = \frac{\$116}{\$580} = 20\% \text{ rate of commission.}$$

3. $45.50 + $19.50 = $65 sales amount.

$$\frac{\text{Amount of Commission}}{\text{Sales Amount}} = \frac{\$19.50}{\$65} = 30\% \text{ rate of commission.}$$

4. $640 − $449 = $191 amount of commission.

$$\frac{\text{Amount of Commission}}{\text{Sales Amount}} = \frac{\$191}{\$640} = 30\% \text{ rate of commission to nearest whole percent.}$$

Study Exercise Three (Frame 11) (11A)

1. Rate of Commission × Sales Amount = Amount of Commission
   6% × $45,000 = Amount of Commission
   (.06) × $45,000 = $2,700 amount of commission

2. Rate of Commission × Sales Amount = Amount of Commission
   8% × $2,800 = Amount of Commission
   (.08) × $2,800 = $224 amount of commission

3. Rate of Commission × Sales Amount = Amount of Commission
   30% × Sales Amount = $240
   (.3) × n = $240
   n = $240 ÷ .3
   n = $800 sales amount

4. Rate of Commission × Sales Amount = Amount of Commission
   22% × Sales Amount = $198
   (.22) × n = $198
   n = $198 ÷ .22
   n = $900 sales amount

Study Exercise Four (Frame 16) **(16A)**

1. (100% − Rate of Commission) × Sales Amount = Net Proceeds

(100% − 25%) × \$288 = Net Proceeds

75% × \$288 = Net Proceeds

.75 × \$288 = \$216 net proceeds

2. (100% − Rate of Commission) × Sales Amount = Net Proceeds

(100% − 16%) × Sales Amount = \$2,870

84% × n = \$2,870

(.84) × n = \$2,870

n = \$2,870 ÷ .84

n = \$3,417 sales amount to nearest whole dollar

3. (100% − Rate of Commission) × Sales Amount = Net Proceeds

(100% − 34%) × Sales Amount = \$260

66% × n = \$260

(.66) × n = \$260

n = \$260 ÷ .66

n = \$394 sales amount to nearest whole dollar

4. (100% − Rate of Commission) × Sales Amount = Net Proceeds

(100% − 12%) × \$362 = Net Proceeds

88% × \$362 = Net Proceeds

(.88) × \$362 = \$318.56

= \$319 net proceeds to nearest whole dollar

UNIT

29 Discount

OBJECTIVES (1)

By the end of this unit you should:

1. KNOW THE MEANING OF *RATE OF DISCOUNT, AMOUNT OF DISCOUNT, LIST PRICE,* AND *NET PRICE*.
2. BE ABLE TO COMPUTE A RATE OF DISCOUNT FROM A BASIC FRACTION.
3. BE ABLE TO SOLVE PROBLEMS INVOLVING RATE OF DISCOUNT.
4. UNDERSTAND HOW RATE OF DISCOUNT PROBLEMS MAY BE SOLVED IN EXACTLY THE SAME MANNER AS RATE OF DECREASE PROBLEMS.

List Price–Net Price–Amount of Discount (2)

Example 1: A desk originally sold for $100 (this is called the *list price*). It now is on sale for $80 (this is called the *net price*). The difference between $100 and $80 is called the *amount of discount*. What is the amount of discount?

Solution:

Line (a): List Price − Net Price = Amount of Discount
(original amount) − (new amount) = (amount of decrease)
↓ ↓ ↓
Line (b): $100 − $80 = $20

Example 2: A discount in the amount of $30 was offered on a bicycle which had a list price of $120. What is the net price of this bicycle?

Solution:

Line (a): List Price − Amount of Discount = Net Price
↓ ↓ ↓
Line (b): $120 − $30 = $90

Example 3: A discount in the amount of $60 was given on a refrigerator which had a net price of $340. What was the list price?

List Price–Net Price–Amount of Discount (Continued)

Solution:

Line (a): Amount of Discount + Net Price = List Price

Line (b): $60 + $340 = $400

Key Relationships (3)

1. List Price − Net Price = Amount of Discount
2. List Price − Amount of Discount = Net Price
3. Amount of Discount + Net Price = List Price

Study Exercise One (4)

Fill in the blanks:

	List Price	Net Price	Amount of Discount
1.	$65.00		$12.00
2.	$43.80	$39.30	
3.	$287.00		$14.35
4.		$62.98	$12.23
5.	$37.50		$6.90

Determining the Rate of Discount (5)

The rate of discount is established by the fraction whose numerator is the amount of discount (amount of decrease) and whose denominator is the list price (original amount).

$$\text{Rate of Discount (rate of decrease)} = \frac{\text{Amount of Discount}}{\text{List Price}} \quad \frac{\text{(amount of decrease)}}{\text{(original amount)}}$$

We will convert this fraction to a percent.

Example 1: A desk that originally sold for $100 now sells for $80. What is the rate of discount?

Solution:

Step (1): Find the amount of discount and the list price:
Line (a): $100 − $80 = $20 is the amount of discount.
Line (b): $100 is the list price.

Step (2): Arrange the information in fraction form:

$$\frac{\text{Amount of Discount}}{\text{List Price}} = \frac{\$20}{\$100}$$

Step (3): Change the fraction to a percent:

$$\frac{\$20}{\$100} = \frac{20}{100}$$

$$= 20\% \text{ rate of discount}$$

Determining the Rate of Discount (Continued)

Example 2: A discount in the amount of $30 was offered on a bicycle which had a list price of $120. What is the rate of discount?

Solution:

Step (1): $30 is the amount of discount.
$120 is the list price.

Step (2): $\dfrac{\text{Amount of Discount}}{\text{List Price}} = \dfrac{\$30}{\$120}$

Step (3): $\dfrac{\$30}{\$120} = \dfrac{3}{12}$

$= \dfrac{1}{4}$

$= 25\%$ rate of discount.

Example 3: A discount in the amount of $60 was offered on a refrigerator which now has a net price of $340. What was the rate of discount?

Solution:

Step (1): $60 is the amount of discount.
$60 + $340 = $400 is the list price.

Step (2): $\dfrac{\text{Amount of Discount}}{\text{List Price}} = \dfrac{\$60}{\$400}$

Step (3): $\dfrac{\$60}{\$400} = \dfrac{60 \div 4}{400 \div 4}$

$= \dfrac{15}{100}$

$= 15\%$ rate of discount.

Remember (6)

$$\text{Rate of Discount} = \frac{\text{Amount of Discount}}{\text{List Price}}$$

Study Exercise Two (7)

Find the missing information to the nearest cent or nearest whole percent:

	List Price	Net Price	Amount of Discount	Rate of Discount
1.	$90		$35.00	
2.	$1,160	$928.00		
3.		$22.75	$9.75	
4.	$52	$48.00		

Using the Rate of Discount to Find the Amount of Discount (8)

Example: An automobile had a list price of $5,200. The sales manager decided to sell the car for less and used a 12% rate of discount. What will be the amount of this discount?

Solution:

Step (1): Write the word formula:

Rate of Discount × List Price = Amount of Discount

Step (2): Replace the words with numerals where possible:

Rate of Discount × List Price = Amount of Discount

↓ ↓ ↓

12% × $5,200 = Amount of Discount

Step (3): Perform the computation:

12% × $5,200 =
(.12) × $5,200 = $624 amount of discount

Using the Rate of Discount to Find the List Price (9)

Example: Your neighbors tell you that they saved $795 on a new automobile and that this was "15% off the sticker price." Determine the sticker price from this information.

Solution:

Step (1): Write the word formula:

Rate of Discount × List Price (sticker price) = Amount of Discount

Step (2): Replace the words with numerals where possible and then form an equation:

Rate of Discount × List Price = Amount of Discount

↓ ↓ ↓

Line (a): 15% × List Price = $795

↓ ↓ ↓

Line (b): (.15) n = $795

Step (3): Solve the equation:

n = $795 ÷ (.15)

n = $5,300 List Price (sticker price)

Computation

```
      53 00.
.15.)795.00.
     75
      45
      45
        00
```

Formula for Finding the List Price or the Amount of Discount (10)

This formula is used when you are given the rate of discount and either the list price or the amount of discount.

Rate of Discount × List Price = Amount of Discount

Study Exercise Three (11)

Find the missing information to the nearest whole dollar:

	Rate of Discount	List Price	Amount of Discount
1.	22%	$1,285	
2.	8%	$72	
3.	14%		$399.00
4.	6%		$43.50

Using the Rate of Discount to Find the Net Price (12)

Example: A merchant wishes to give his customer a 14% rate of discount on a $500 television set. What should be the new price?

Solution:

Step (1): Determine the amount of discount:

Line (a): Rate of Discount × List Price = Amount of Discount

Line (b): 14% × $500 = Amount of Discount

Line (c): (.14) × $500 = $70 (amount of discount)

Step (2): Subtract to find the net price:

Line (a): List Price − Amount of Discount = Net Price

Line (b): $500 − $70 = $430

A Short Cut (13)

Gold Bar Representing 100% of List Price

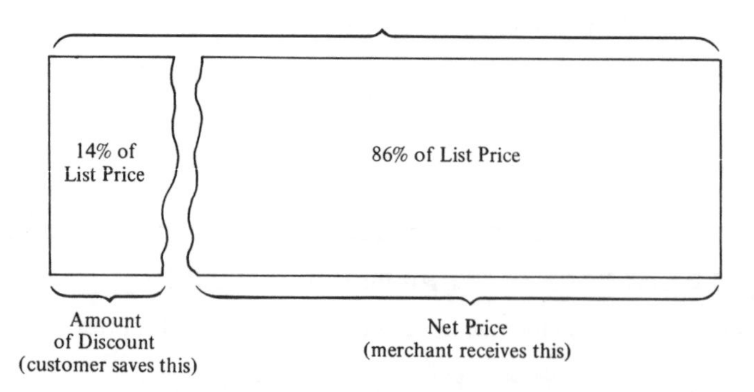

A Short Cut (Continued)

Since we are only interested in the net price (new amount) and this represents 86% of the list price (original amount), we can get our information as follows:

Line (a): (100% − Rate of Commission) × List Price = Net Price

Line (b): (100% − 14%) × $500 = Net Price

Line (c): 86% × $500 = Net Price

Line (d):

```
  $500
  ×.86
 -----
 30 00
 400 0
 -----
$430.00  Net Price
```

Using 100% Minus the Rate of Discount to Find the List Price (14)

Example: A washing machine is on sale for $216 and is advertised as "20% off our regular price." What was its regular price?

Solution:

Gold Bar Representing 100% of List Price ("Our regular price")

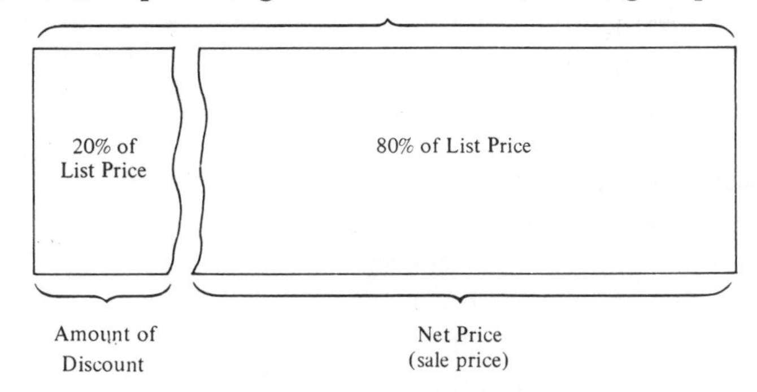

Line (a): (100% − Rate of Discount) × List Price = Net Price

Line (b): (100% − 20%) × List Price = $216

Line (c): 80% × n = $216

Line (d): (.8) × n = $216

Line (e): n = $216 ÷ .8

Line (f): n = $270 List Price ("our regular price")

Computation

```
      $27 0.
  .8.)$216.0.
       16
       --
        56
        56
        --
         0
```

Formula for Finding the List Price or the Net Price (15)

(100% minus the rate of discount)

This formula is used when you are given the rate of discount and either the list price or the net price.

(100% − Rate of Discount) × List Price = Net Price

Study Exercise Four (16)

Find the missing information:

	Rate of Discount	List Price	Net Price
1.	20%	$500	
2.	16%	$3,250	
3.	12%		$2,332
4.	32%		$291.72

Summary (17)

A. Addition and subtraction relationships:

1. List Price − Net Price = Amount of Discount
2. List Price − Amount of Discount = Net Price
3. Amount of Discount + Net Price = List Price

B. This fraction, which we will change to a percent, produces the rate of discount:

$$\frac{\text{Amount of Discount}}{\text{List Price}} = \text{Rate of Discount}$$

C. This formula is used when you are given the rate of discount and either the list price or the amount of discount:

Rate of Discount × List Price = Amount of Discount

D. This formula is used when you are given the rate of discount and either the list price or the net price:

(100% − Rate of Discount) × List Price = Net Price

REVIEW EXERCISES (18)

A. Fill in the blanks:

1. If we multiply the rate of discount times the list price, we obtain the __________.
2. If we subtract the rate of discount from 100% and then multiply the list price by the result, we obtain the __________.
3. The amount of discount + __________ = __________.

REVIEW EXERCISES (Continued)

4. The rate of discount is established by the fraction which has for its numerator the __________ and which has for its denominator the __________.
5. A merchant wishes to give a 4% discount on every item in his store. To get the net price of each article without finding the amount of discount he should multiply each list price by __________.

B. Answer the following:

6. An article which originally sold for $22 now sells for $18. To the nearest whole percent, what is the rate of discount?
7. What will be the amount of discount on a $450 television set if the rate of discount is 15%?
8. A discount of $42 was offered on a washing machine which had a net price of $258. What was the rate of discount?
9. A merchant wishes to give her customer a 12% rate of discount on some window drapes which have a list price of $380. How much should the customer pay?
10. George claimed that he saved $3.60 on a camera because the merchant gave him "8% off the list price." What was the list price?

C. Find the missing information to the nearest dollar or whole percent:

	List Price	Net Price	Amount of Discount	Rate of Discount
11.	$42		$15	
12.	$82	$78		
13.	$150			$22%
14.		$30	$10	
15.		$279		38%
16.			$420	12%

Solutions to Review Exercises (19)

A. **1.** Amount of Discount. **2.** Net Price. **3.** Amount of Discount + Net Price = List Price
4. Amount of discount, list price. **5.** 96%

B. **6.** $22 − $18 = $4 amount of discount.

$$\frac{\text{Amount of Discount}}{\text{List Price}} = \frac{\$4}{\$22} = 22\overline{)4.000}^{\;.181} = 18\% \text{ rate of discount (to nearest whole percent)}$$

7. Rate of Discount × List Price = Amount of Discount

↓		↓		↓
15%	×	$450	=	Amount of Discount
(.15)	×	($450)	=	$67.50 Amount of Discount

8. $42 + $258 = $300 list price.

$$\frac{\text{Amount of Discount}}{\text{List Price}} = \frac{\$42}{\$300} = \frac{42 \div 3}{300 \div 3} = \frac{14}{100} = 14\% \text{ rate of discount}$$

Solutions to Review Exercises (Continued)

9. (100% − Rate of Discount) × List Price = Net Price

(100% − 12%) × $380 = Net Price

88% × $380 = Net Price

(.88) × ($380) = $334.40 Net Price

10. Rate of Discount × List Price = Amount of Discount

8% × List Price = $3.60

(.08) × n = $3.60

n = $3.60 ÷ (.08)

n = $45 List Price

Computation

```
      45.
.08.)3.60.
     3 2
       40
       40
        0
```

C. 11. $42 − $15 = $27 net price.

$$\frac{\text{Amount of Discount}}{\text{List Price}} = \frac{\$15}{\$42} \approx 36\% \text{ Rate of Discount}$$

12. $82 − $78 = $4 amount of discount.

$$\frac{\text{Amount of Discount}}{\text{List Price}} = \frac{\$4}{\$82} \approx 5\% \text{ Rate of Discount}$$

13. Rate of Discount × List Price = Amount of Discount

22% × $150 = Amount of Discount

(.22) × ($150) = $33 Amount of Discount

$150 − $33 = $117 Net Price

14. $30 + $10 = $40 list price.

$$\frac{\text{Amount of Discount}}{\text{List Price}} = \frac{\$10}{\$40} = \frac{1}{4} = 25\% \text{ Rate of Discount}$$

15. (100% − Rate of Discount) × List Price = Net Price

(100% − 38%) × List Price = $279

62% × List Price = $279

(.62) × n = $279

n = $279 ÷ (.62)

n = $450 List Price

$450 − $279 = $171 Amount of Discount

Computation

```
      4 50.
.62.)279.00.
     248
      31 0
      31 0
         0
```

16. Rate of Discount × List Price = Amount of Discount

12% × List Price = $420

(.12) × n = $420

n = $420 ÷ (.12)

n = $3,500 List Price

$3,500 − $420 = $3,080 Net Price

Computation

```
      35 00.
.12.)420.00.
     36
      60
      60
       0
```

SUPPLEMENTARY PROBLEMS

A. Fill in the blanks:

1. The _________ + the net price = list price.

2. The rate of discount times the _________ yields the amount of discount.

3. The amount of discount subtracted from the list price yields _________.

4. The rate of discount is established by a fraction which has _________ for its numerator and _________ for the denominator.

5. If you know the rate of discount and list price it is possible to find the net price without finding the amount of discount by multiplying the list price by _________.

B. Answer the following:

6. A dress formerly marked at $79 was offered at $62. To the nearest whole percent, what is the rate of discount?

7. The net price on a chair is $122 and the amount of discount is $28. What is the rate of discount to the nearest whole percent?

8. A sand and gravel company offers a 12% rate of discount if a bill is paid within 10 days. The bill is $1,628. How much can be saved if the bill is promptly paid?

9. What will be the amount of discount on an automobile which has a list price of $6,258, if the rate of discount is 12%?

10. Your neighbors tell you that they saved $5,175 on the asking price of their home because the builder gave them a "4% allowance on the asking price for carpeting since they intend to carpet themselves." What was the asking price of the house?

11. A lawn mower which had a list price of $62 was offered at a 12% rate of discount. What was the net price?

12. One automobile dealer lists a car at $9,738 with a "15% end-of-the-model-year discount." A second dealer lists the same vehicle at $8,250. Which is the better buy and by how much?

13. A calculator is on sale for $84.15 and is advertised at "15% off our regular price." What was its regular price?

C. Complete the following chart to the nearest whole percent or dollar:

	Rate of Discount	List Price	Net Price	Amount of Discount
14.	14%	$700		
15.	6%		$423	
16.	28%			$238
17.		$98		$22
18.			$635	$55
19.		$358	$342	
20.	10%	$990		
21.	10%		$990	
22.	10%			$990

SUPPLEMENTARY PROBLEMS (Continued)

D. Answer the following:

23. The net price of a TV is \$484. If the amount of discount is \$78, what is the rate of discount to the nearest tenth of a percent?

24. The net price of a microcomputer is \$537. If the rate of discount is 13%, what is the amount of discount?

25. The net price of a sewing machine is \$217. If the list price is \$275, what is the rate of discount to the nearest tenth of a percent?

26. A vacuum cleaner outlet offers a $12\frac{1}{2}$% rate of discount on a vacuum with a list price of \$168. Find the net price.

27. The list price on a microwave oven is \$423 and the amount of discount is \$29. What is the rate of discount to the nearest hundredth of a percent?

Solutions to Study Exercises (4A)

Study Exercise One (Frame 4)

1. List Price − Amount of Discount = Net Price
↓ ↓ ↓
\$65 − \$12 = \$53

2. List Price − Net Price = Amount of Discount
↓ ↓ ↓
\$43.80 − \$39.30 = \$4.50

3. List Price − Amount of Discount = Net Price
↓ ↓ ↓
\$287 − \$14.35 = \$272.65

4. Amount of Discount + Net Price = List Price
↓ ↓ ↓
\$12.23 + \$62.98 = \$75.21

5. List Price − Amount of Discount = Net Price
↓ ↓ ↓
\$37.50 − \$6.90 = \$30.60

Study Exercise Two (Frame 7) (7A)

1. \$90 − \$35 = \$55 net price.

$$\frac{\text{Amount of Discount}}{\text{List Price}} = \frac{\$35}{\$90} = 39\% \text{ rate of discount (to nearest whole percent)}$$

2. \$1,160 − \$928 = \$232 amount of discount.

$$\frac{\text{Amount of Discount}}{\text{List Price}} = \frac{\$232}{\$1,160} = 20\% \text{ rate of discount}$$

3. \$22.75 + \$9.75 = \$32.50 list price.

$$\frac{\text{Amount of Discount}}{\text{List Price}} = \frac{\$9.75}{\$32.50} = 30\% \text{ rate of discount}$$

4. \$52 − \$48 = \$4 amount of discount.

$$\frac{\text{Amount of Discount}}{\text{List Price}} = \frac{\$4}{\$52} = 8\% \text{ rate of discount (to nearest whole percent)}$$

Study Exercise Three (Frame 11) (11A)

1. Rate of Discount × List Price = Amount of Discount
↓ ↓ ↓
22% × \$1,285 = Amount of Discount
(.22) × \$1,285 = \$282.70 Amount of Discount

2. Rate of Discount × List Price = Amount of Discount
↓ ↓ ↓
8% × \$72 = Amount of Discount
(.08) × \$72 = \$5.76 Amount of Discount

Study Exercise Three (Frame 11) (Continued)

3. Rate of Discount × List Price = Amount of Discount

14% × List Price = $399

(.14) × n = $399

n = $399 ÷ (.14)

n = $2,850 List Price

Computation

```
      28 50.
.14.)399.00.
     28
     119
     112
       7 0
       7 0
         0
```

4. Rate of Discount × List Price = Amount of Discount

6% × List Price = $43.50

(.06) × n = $43.50

n = $43.50 ÷ (.06)

n = $725 List Price

Computation

```
       7 25.
.06.)43.50.
     42
      1 5
      1 2
        30
        30
         0
```

Study Exercise Four (Frame 16) **(16A)**

1. (100% − Rate of Discount) × List Price = Net Price

(100% − 20%) × $500 = Net Price

80% × $500 = Net Price

(.8) × $500 = $400 Net Price

2. (100% − Rate of Discount) × List Price = Net Price

(100% − 16%) × $3,250 = Net Price

84% × $3,250 = Net Price

(.84) × $3,250 = $2,730 Net Price

3. (100% − Rate of Discount) × List Price = Net Price

(100% − 12%) × List Price = $2,332

88% × n = $2,332

(.88) × n = $2,332

n = $2,332 ÷ (.88)

n = $2,650

Computation

```
       $26 50.
.88.)$2332.00.
      176
      572
      528
       44 0
       44 0
          0
```

4. (100% − Rate of Discount) × List Price = Net Price

(100% − 32%) × List Price = $291.72

68% × n = $291.72

(.68) × n = $291.72

n = $291.72 ÷ (.68)

n = $429

Computation

```
       $4 29.
.68.)$291.72.
      272
       19 7
       13 6
        6 12
        6 12
           0
```

UNIT

30 Percent Increase

OBJECTIVES (1)

By the end of this unit you should be able to:

1. IDENTIFY THE *ORIGINAL AMOUNT,* THE *NEW AMOUNT,* AND THE *AMOUNT OF INCREASE* IN A PROBLEM.
2. FIND THE RATE OF INCREASE.
3. SOLVE PROBLEMS INVOLVING RATE OF INCREASE.

New Amount–Original Amount–Amount of Increase (2)

Example 1: Yesterday a bicycle sold for \$100. Today it sells for \$110. What is the amount of increase?

Solution:

Line (a): New Amount − Original Amount = Amount of Increase

Line (b): \$110 − \$100 = \$10

Example 2: A factory now employs 45 people. This is an increase of 15 people from the previous year. How many people did it employ last year?

Solution:

Line (a): New Amount − Amount of Increase = Original Amount

Line (b): 45 − 15 = 30

Example 3: A club that had 500 members last year gained 40 new members this year. What is the current membership?

Solution:

Line (a): Original Amount + Amount of Increase = New Amount

Line (b): 500 + 40 = 540

Key Relationships (3)

1. New Amount − Original Amount = Amount of Increase
2. New Amount − Amount of Increase = Original Amount
3. Original Amount + Amount of Increase = New Amount

Study Exercise One (4)

Find the missing information:

	Original Amount	New Amount	Amount of Increase
1.		14	2
2.	67	135	
3.	\$208.90		\$17.50
4.	6,785	8,492	
5.		\$59.36	\$12.42

Determining the Rate of Increase (5)

The rate of increase is established by the fraction whose numerator is the amount of increase and whose denominator is the original amount.

$$\text{Rate of Increase} = \frac{\text{Amount of Increase}}{\text{Original Amount}}$$

We will convert this fraction to a percent.

Example 1: Yesterday a bicycle sold for \$100. Today it sells for \$110. What is the rate of increase?

Solution:

Step (1): Find the amount of increase and the original amount:
Line (a): \$110 − \$100 = \$10 the amount of increase
Line (b): \$100 is the original amount

Step (2): Arrange the information in fraction form:

$$\frac{\text{Amount of Increase}}{\text{Original Amount}} = \frac{\$10}{\$100}$$

Step (3): Change the fraction to a percent:

$$\frac{\$10}{\$100} = 10\% \text{ rate of increase}$$

Determining the Rate of Increase (Continued)

Example 2: A factory now employs 45 people. This is an increase of 15 people from the previous year. What is the rate of increase?

Solution:

Step (1): 15 is the amount of increase.
45 − 15 = 30 is the original amount

Step (2): $\frac{\text{Amount Of Increase}}{\text{Original Amount}} = \frac{15}{30}$

Step (3): $\frac{15}{30} = \frac{15 \div 15}{30 \div 15}$

$= \frac{1}{2}$

= 50% rate of increase

Example 3: A club that had 500 members last year gained 40 new members this year. What is the rate of increase?

Solution:

Step (1): 40 is the amount of increase.
500 is the original amount.

Step (2): $\frac{\text{Amount of Increase}}{\text{Original Amount}} = \frac{40}{500}$

Step (3): $\frac{40}{500} = \frac{40 \div 5}{500 \div 5}$

$= \frac{8}{100}$

= 8% rate of increase

Remember (6)

$$\text{Rate of Increase} = \frac{\text{Amount of Increase}}{\text{Original Amount}}$$

Study Exercise Two (7)

Find the missing information to the nearest whole dollar or nearest whole percent:

	Original Amount	New Amount	Amount of Increase	Rate of Increase
1.	$75		$25	
2.	$720	$936		
3.		$19	$5	
4.	$38	$41		

Using the Rate of Increase to Find the Amount of Increase (8)

Example: A school with an enrollment of 1,200 students expects a 20% rate of increase next year. What will be the amount of this increase?

Solution:

Step (1): Write the word formula:

Rate of Increase × Original Amount = Amount of Increase

Step (2): Replace the words with numerals where possible:

Rate of Increase × Original Amount = Amount of Increase
↓ ↓ ↓
20% × 1,200 = Amount of Increase

Step (3): Perform the computation:

$$20\% \times 1{,}200 = \frac{1}{\cancel{5}_1} \times \cancel{1{,}200}^{240}$$
$$= 240 \text{ students}$$

Using the Rate of Increase to Find the Original Amount (9)

Example: A factory increased the number of articles it produced in one week by 42 articles. This represented a 15% increase from the previous week. How many articles did it produce in the previous week?

Solution:

Step (1): Write the word formula:

Rate of Increase × Original Amount = Amount of Increase

Step (2): Replace the words with numerals where possible and then form an equation:

Rate of Increase × Original Amount = Amount of Increase
↓ ↓ ↓
Line (a): 15% × Original Amount = 42
↓ ↓ ↓
Line (b): (.15) × n = 42

Step (3): Solve the equation:

$n = 42 \div .15$
$n = 280$ articles

Computation

```
      2 80.
.15.)42.00.
     30
     12 0
     12 0
        0
```

Formula for Finding the Original Amount or the Amount of Increase (10)

This formula is used when you are given the rate of increase and either the original amount or the amount of increase.

Rate of Increase × Original Amount = Amount of Increase

Study Exercise Three (11)

1. A television set which formerly sold for $480 was increased by a rate of 12%. What was the amount of this increase?
2. A factory which formerly produced 800 articles per day purchased new machinery which increased its output by a rate of 23%. How many additional articles will it now be able to produce per day?
3. An automobile sells for $6,250. However, this price will be increased by a rate of 6% for sales tax. How much sales tax must be paid?
4. A town increased its population by 528 people this year. If this was an 11% rate of increase over last year, what was the population last year?

Using the Rate of Increase to Find the New Amount (12)

Example: A factory produces 240 television sets per day. If production is increased by a rate of 20%, what will be the total number of sets produced in one day?

Solution:

Step (1): Determine the amount of increase:

Line (a): Rate of Increase × Original Amount = Amount of Increase

Line (b): 20% × 240 = Amount of Increase

Line (c) : $\frac{1}{\not{5}_1} \times \not{240}^{48}$ = 48 (amount of increase)

Step (2): Add to find the new amount:

Line (d): Original Amount + Amount of Increase = New Amount

Line (e): 240 + 48 = 288 sets

A Short Cut (13)

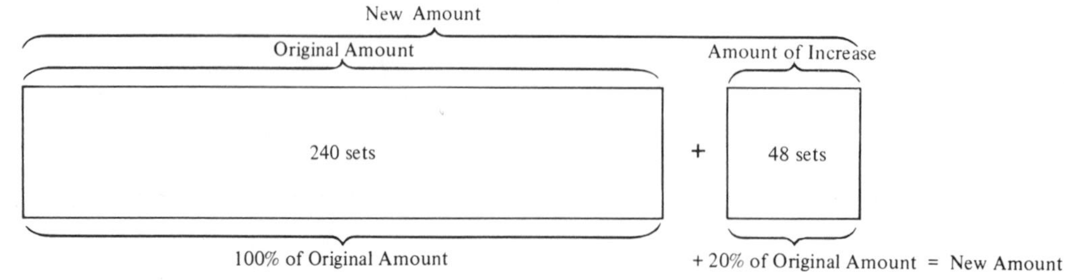

Since we are only interested in the new amount, which represents 20% more than the original amount, we can get our information as follows:

Line (a): (100% + Rate of Increase) × Original Amount = New Amount

Line (b): (100% + 20%) × 240 = New Amount

Line (c): 120% × 240 = New Amount

Line (d):

$$\begin{array}{r} 24\,0 \\ \times 1.2 \\ \hline 48\,0 \\ 240 \\ \hline 288.0 \end{array}$$ sets (new amount)

Using 100% Plus the Rate of Increase to Find the Original Amount (14)

Example: An automobile dealer purchases his cars from the factory for a certain amount and then increases this amount by a rate of 20% to obtain his gross profit. How much did the dealer pay for a car which has a selling price of $6,000?

Solution:

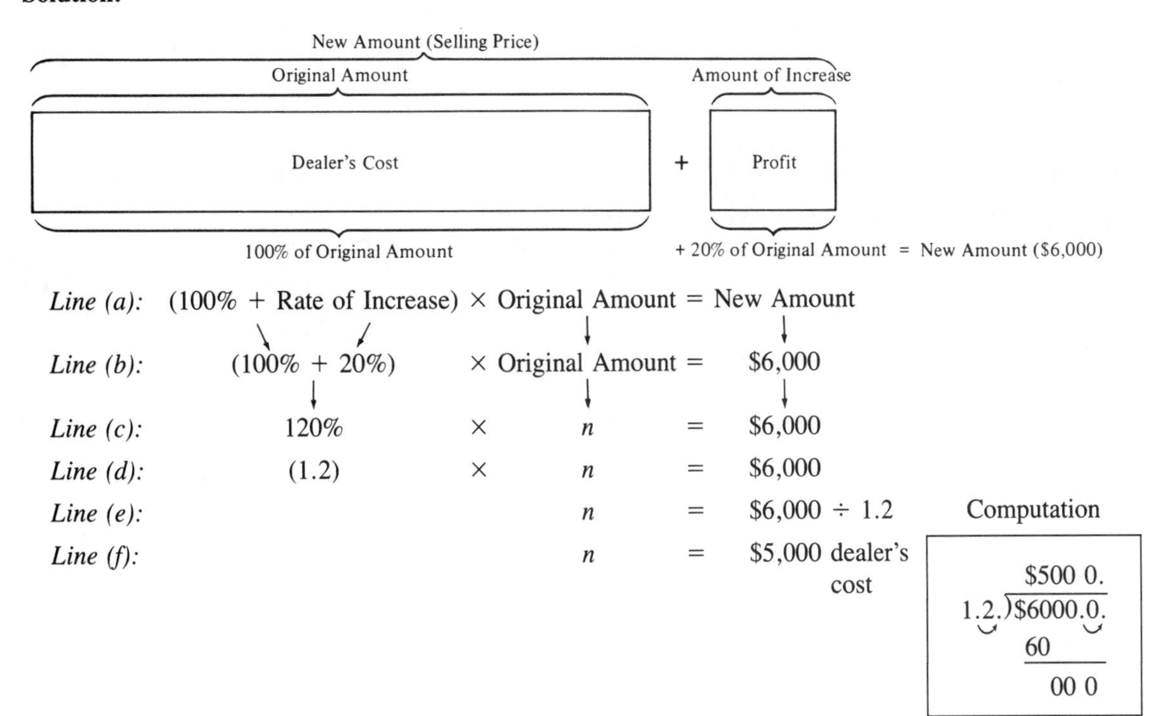

Formula for Finding Original Amount or New Amount (15)

(100% plus the rate of increase)

This formula is used when you are given the rate of increase and either the original amount or the new amount.

(100% + Rate of Increase) × Original Amount = New Amount

Study Exercise Four (16)

1. If the sales tax rate is 6%, what would be the total cost of a sweater that sells for $28.50?
2. The price of a typewriter which formerly sold for $225 was increased by a rate of 15%. What is the new price?
3. A company which employs 300 people decides to increase the number of its employees by 16%. If this is done, how many people will be employed by this company?
4. The "stricker price" on a certain automobile is $5,664. A consumers' magazine informs you that this price has been obtained by an 18% rate of increase. How much did the dealer pay for the car?

Summary (17)

A. Addition and subtraction relationships:

1. New Amount − Original Amount = Amount of Increase
2. New Amount − Amount of Increase = Original Amount
3. Original Amount + Amount of Increase = New Amount

B. This fraction, which we will change to a percent, produces the rate of increase:

$$\frac{\text{Amount of Increase}}{\text{Original Amount}} = \text{Rate of Increase}$$

C. This formula is used when you are given the rate of increase and either the original amount or the amount of increase:

$$\text{Rate of Increase} \times \text{Original Amount} = \text{Amount of Increase}$$

D. This formula is used when you are given the rate of increase and either the original amount or the new amount:

$$(100\% + \text{Rate of Increase}) \times \text{Original Amount} = \text{New Amount}$$

REVIEW EXERCISES (18)

A. Answer the following:

1. The rate of increase is produced by a fraction whose numerator is __________ and whose denominator is __________
2. Original Amount + __________ = New Amount.
3. New Amount − __________ = Original Amount.
4. Rate of Increase × Original Amount = __________
5. (100% + Rate of Increase) × Original Amount = __________
6. A man bought some stocks for $3,000 and sold them for $4,000. What was his rate of increase?
7. A coat formerly sold for $79. Its price will be increased by 17%. What will be the new price?
8. What is the amount of sales tax that must be paid on a $7,480 automobile if the rate is 6%?

B. Find the missing information; round rates to the nearest whole percent:

	Original Amount	New Amount	Amount of Increase	Rate of Increase
9.	$73		$5	
10.		$16	$7	
11.	$700			15%
12.	$51	$63		
13.			$69	23%
14.		$445.20		6%

Solutions to Review Exercises

(19)

A. **1.** See Frame 5 **2.** See Frame 3 **3.** See Frame 3

4. See Frame 10 **5.** See Frame 15

6. \$4,000 − \$3,000 = \$1,000 amount of increase.

$$\frac{\text{Amount of Increase}}{\text{Original Amount}} = \frac{\$1{,}000}{\$3{,}000} = \frac{1}{3} = 33\tfrac{1}{3}\% \text{ rate of increase}$$

7. (100% + Rate of Increase) × Original Amount = New Amount

(100% + 17%) × \$79 = New Amount

117% × \$79 = New Amount

1.17 × \$79 = \$92.43

8. Rate of Increase × Original Amount = Amount of Increase

6% × \$7,480 = Amount of Increase

(.06) × \$7,480 = \$448.80 sales tax

B. **9.** Original Amount + Amount of Increase = New Amount

\$73 + \$5 = \$78

$$\frac{\text{Amount of Increase}}{\text{Original Amount}} = \frac{\$5}{\$73} \approx 7\% \text{ rate of increase}$$

10. New Amount − Amount of Increase = Original Amount

\$16 − \$7 = \$9

$$\frac{\text{Amount of Increase}}{\text{Original Amount}} = \frac{\$7}{\$9} \approx 78\% \text{ rate of increase}$$

11. Rate of Increase × Original Amount = Amount of Increase

15% × \$700 = Amount of Increase

(.15) × \$700 = \$105 amount of increase

Original Amount + Amount of Increase = New Amount

\$700 + \$105 = \$805

12. New Amount − Original Amount = Amount of Increase

\$63 − \$51 = \$12

$$\frac{\text{Amount of Increase}}{\text{Original Amount}} = \frac{\$12}{\$51} \approx 24\% \text{ rate of increase}$$

13. Rate of Increase × Original Amount = Amount of Increase

23% × Original Amount = \$69

(.23) × n = \$69

n = \$69 ÷ (.23)

n = \$300 original amount

Original Amount + Amount of Increase = New Amount

\$300 + \$69 = \$369

Solutions to Review Exercises (Continued)

14. (100% + Rate of Increase) × Original Amount = New Amount

(100% + 6%) × Original amount = $445.20

106% × n = $445.20

(1.06) × n = $445.20

n = $445.20 ÷ (1.06)

n = $420 original amount

New Amount − Original Amount = Amount of Increase

$445.20 − $420 = $25.20

SUPPLEMENTARY PROBLEMS

A. Answer the following:

1. Rate of Increase × ________ = ________
2. (100% + Rate of Increase) × ________ = ________
3. ________ + ________ = New Amount.
4. The price of eggs increased from 82¢ a dozen to 89¢ a dozen. What was the rate of increase to the nearest whole percent?
5. The price of a child's toy increased from $1.98 to $2.40. What was the rate of increase to the nearest whole percent?
6. The price of a can of pet food increased from 48¢ to 62¢. What was the rate of increase to the nearest whole percent?
7. A bicycle which formerly sold for $130 now sells for $160. To the nearest whole percent, what rate of increase is this?

B. Using a rate of 6% compute the tax on the following:

8. A $5,280 automobile.
9. A $79 coat.
10. A $225 power saw.
11. An $11.95 textbook (round to nearest cent).

C. Answer the following:

12. What will be the total price on a lawn mower costing $78 if the sales tax rate is 5%?
13. Production in a factory increased by a 12% rate. It formerly produced 628 articles per week. To the nearest whole article, how many articles does it now produce per week?
14. An employee formerly made $8.72 per hour. This was increased by an 8% rate. To the nearest cent, what is the new hourly wage?
15. It was reported that the population of a city increased 17% over its last census. If at that time the population was 879,000, what is its new population?
16. An automobile has a "sticker price" of $9,243. This price is arrived at by using a 17% rate of increase over the dealer's cost. How much did the dealer pay for the car?
17. The "sticker price" on a truck is $9,512. This price is arrived at by using a 16% rate of increase over the dealer's cost. How much did the dealer pay for the truck?

SUPPLEMENTARY PROBLEMS (Continued)

D. Find the missing information; round rates to the nearest whole percent:

	Original Amount	New Amount	Amount of Increase	Rate of Increase
18.	\$72	\$84		
19.	\$20		\$2	
20.	\$120			15%
21.		\$55	\$6	
22.		\$208		4%
23.			\$182	13%

E. Answer the following:

24. An index used to measure the status of the stock market increased from 968 to 994. To the nearest tenth of a percent, find the rate of increase.

25. In a two-year period the number of trucks sold by General Motors increased from 78,000 to 206,000. To the nearest whole percent, find the rate of increase.

26. The price of an automobile was increased to \$9,430. If the rate of increase was 15%, what was the original price?

27. In a 3-hour period the temperature increased by 4°. If the rate of increase was 5%, find the new temperature.

Solutions to Study Exercises (4A)

Study Exercise One (Frame 4)

1. New Amount − Amount of Increase = Original Amount
   14 − 2 = 12

2. New Amount − Original Amount = Amount of Increase
   135 − 67 = 68

3. Original Amount + Amount of Increase = New Amount
   \$208.90 + \$17.50 = \$226.40

4. New Amount − Original Amount = Amount of Increase
   8,492 − 6,785 = 1,707

5. New Amount − Amount of Increase = Original Amount
   \$59.36 − \$12.42 = \$46.94

Study Exercise Two (Frame 7) (7A)

1. $75 + $25 = $100 new amount.

$$\frac{\text{Amount of Increase}}{\text{Original Amount}} = \frac{\$25}{\$75} = \frac{1}{3} = 33\% \text{ rate of increase to nearest whole percent}$$

2. $936 − $720 = $216 amount of increase.

$$\frac{\text{Amount of Increase}}{\text{Original Amount}} = \frac{\$216}{\$720} = 30\% \text{ rate of increase to nearest whole percent}$$

3. $19 − $5 = $14 original amount.

$$\frac{\text{Amount of Increase}}{\text{Original Amount}} = \frac{\$5}{\$14} = 36\% \text{ rate of increase to nearest whole percent}$$

4. $41 − $38 = $3 amount of increase.

$$\frac{\text{Amount of Increase}}{\text{Original Amount}} = \frac{\$3}{\$38} = 8\% \text{ rate of increase to nearest whole percent}$$

Study Exercise Three (Frame 11) (11A)

1. Rate of Increase × Original Amount = Amount of Increase

↓		↓		↓
12%	×	$480	=	Amount of Increase
(.12)	×	$480	=	$57.60 amount of increase

2. Rate of Increase × Original Amount = Amount of Increase

↓		↓		↓
23%	×	800	=	Amount of Increase
(.23)	×	800	=	184 additional articles per day

3. Rate of Increase × Original Amount = Amount of Increase

↓		↓		↓
6%	×	$6,250	=	Amount of Increase
(.06)	×	$6,250	=	$375 sales tax

4. Rate of Increase × Original Amount = Amount of Increase

↓		↓		↓
11%	×	Original Amount	=	528
(.11)	×	n	=	528
		n	=	528 ÷ (.11)
		n	=	4,800 people

Study Exercise Four (Frame 16) (16A)

1. (100% + Rate of Increase) × Original Amount = New Amount

↘ ↙		↓		↓
(100% + 6%)	×	$28.50	=	New Amount
106%	×	$28.50	=	New Amount
(1.06)	×	$28.50	=	$30.21 total cost

2. (100% + Rate of Increase) × Original Amount = New Amount

↘ ↙		↓		↓
(100% + 15%)	×	$225	=	New Amount
115%	×	$225	=	New Amount
(1.15)	×	$225	=	$258.75 new price

Study Exercise Four (Frame 16) **(Continued)**

3. (100% + Rate of Increase) × Original Amount = New Amount

(100% + 16%)	×	300	=	New Amount
116%	×	300	=	New Amount
(1.16)	×	300	=	348 people

4. (100% + Rate of Increase) × Original Amount = New Amount

(100% + 18%)	×	Original Amount	=	\$5,664
118%	×	n	=	\$5,664
(1.18)	×	n	=	\$5,664
		n	=	\$5,664 ÷ 1.18
		n	=	\$4,800 dealer cost

UNIT

31 Profit Based on Cost

OBJECTIVES (1)

By the end of this unit you should be able to:

1. IDENTIFY THE *SELLING PRICE, COST PRICE,* AND *PROFIT* IN A PROBLEM.
2. FIND THE RATE OF PROFIT BASED ON THE COST.
3. SOLVE PROBLEMS INVOLVING THE RATE OF PROFIT.
4. UNDERSTAND HOW RATE OF PROFIT PROBLEMS MAY BE SOLVED IN EXACTLY THE SAME MANNER AS RATE OF INCREASE PROBLEMS.

Selling Price–Cost Price–Profit (2)

Example 1: A merchant bought a chair for $100 (cost price) and sold it for $125 (selling price). What was his profit (amount of increase)?

Solution:

Line (a): Selling Price − Cost Price = Profit*

(new amount) − (original amount) = (amount of increase)

↓ ↓ ↓

Line (b): $125 − $100 = $25

**Note:* We will use the words *profit, mark-up,* and *amount of profit* to mean the same thing. In each case these words refer to a "dollar and cents" number.

Example 2: A dress was sold for $90. It provided the merchant with a $30 mark-up. How much did the dress cost the merchant?

Solution:

Line (a): Selling Price − Profit = Cost Price

↓ ↓ ↓

Line (b): $90 − $30 = $60

Selling Price–Cost Price–Profit (Continued)

Example 3: A profit of $1,000 was made on an automobile which cost the dealer $5,000. What was the selling price of this automobile?

Solution:

Line (a): Cost Price + Profit = Selling Price

Line (b): $5,000 + $1,000 = $6,000

Key Relationships (3)

1. Selling Price − Cost Price = Profit
2. Selling Price − Profit = Cost Price
3. Cost Price + Profit = Selling Price

Study Exercise One (4)

Find the missing information:

	Selling Price	Cost Price	Profit
1.	$80	$60	
2.	$49.50		$17.20
3.		$325.40	$48.50
4.	$179		$48
5.	$90.80	$53.50	

Determining the Rate of Profit Based on Cost (5)

The rate of profit based on cost is established by the fraction whose numerator is the amount of profit and whose denominator is the cost price.

$$\underset{\text{(rate of increase)}}{\text{Rate of Profit}} = \frac{\text{Amount of Profit (amount of increase)}}{\text{Cost Price (original price)}}$$

We will change this fraction to a percent.

Example 1: A merchant bought a chair for $100 and sold it for $125. What was his rate of profit?

Solution:

Step (1): Determine the amount of profit and the cost price:
Line (a): $125 − $100 = $25 profit
Line (b): $100 is the cost price

Step (2): Arrange the information in fraction form:

$$\frac{\text{Amount of Profit}}{\text{Cost Price}} = \frac{\$25}{\$100}$$

Determining the Rate of Profit Based on Cost (Continued)

Step (3): Change the fraction to a percent:

$$\frac{\$25}{\$100} = 25\% \text{ rate of profit}$$

Example 2: A dress which sold for \$90 provided a \$30 profit to the seller. What was the rate of profit?

Solution:

Step (1): \$90 − \$30 = \$60 cost price.
\$30 is the profit

Step (2): $$\frac{\text{Amount of Profit}}{\text{Cost Price}} = \frac{\$30}{\$60}$$

Step (3): $$\frac{\$30}{\$60} = \frac{1}{2} = 50\% \text{ rate of profit}$$

Example 3: A profit of \$1,000 was made on an automobile which cost the dealer \$5,000. What was her rate of profit?

Solution:

Step (1): \$1,000 is the profit
\$5,000 is the cost price

Step (2): $$\frac{\text{Amount of Profit}}{\text{Cost Price}} = \frac{\$1,000}{\$5,000}$$

Step (3): $$\frac{\$1,000}{\$5,000} = \frac{1}{5} = 20\% \text{ rate of profit}$$

Remember (6)

$$\text{Rate of Profit} = \frac{\text{Amount of Profit}}{\text{Cost Price}}$$

Study Exercise Two (7)

Find the missing information; round rates to nearest whole percent:

	Cost Price	Selling Price	Profit	Rate of Profit
1.		\$70	\$20	
2.	\$300	\$348		
3.	\$62		\$14	
4.		\$190	\$32	

Using the Rate of Profit to Find the Amount of Profit (8)

Example: A toy store determines that it needs to make a 40% rate of profit on its merchandise. What will be the profit on a toy which cost the store $25?

Solution:

Step (1): Write the word formula:

Rate of Profit × Cost Price = Profit
(rate of increase) × (original amount) = (amount of increase)

Step (2): Replace the words with numerals where possible:

Rate of Profit × Cost Price = Profit
↓ ↓ ↓
40% × $25 = Profit

Step (3): Perform the computation:

$$40\% \times \$25 = \frac{2}{\cancel{5}_1} \times \$\cancel{25}^5$$
$$= \$10 \text{ profit}$$

Using the Rate of Profit to Find the Cost Price (9)

Example: A dressmaker determines that she must have a mark-up of $12 on each dress she sells. If she uses a 30% rate of profit, what should each dress cost her to make?

Solution:

Step (1): Write the word formula:

Rate of Profit × Cost Price = Profit

Step (2): Replace the words with numerals where possible and then form an equation:

Rate of Profit × Cost Price = Profit
↓ ↓ ↓
Line (a): 30% × Cost Price = $12
↓ ↓ ↓
Line (b): (.3) × n = $12

Step (3): Solve the equation:

n = $12 ÷ .3
n = $40 cost price

Computation

$$.3\overline{)12.0}$$ quotient 40.
12
00

Formula for Finding Cost Price or Profit (10)

This formula is used when you are given the rate of profit and either the cost price or the profit.

Rate of Profit × Cost Price = Profit

Study Exercise Three (11)

1. A desk costs a merchant $180. What will be the profit if a 40% rate of profit is used?
2. An automobile costs the dealer $4,250. What will be the profit if the dealer uses a 17% rate of profit?
3. A typewriter which cost the merchant $120 is to be sold at a 35% rate of profit. What is the amount of this profit?
4. A pottery maker determines that she must make $6 on each pot sold. If she uses a 40% rate of profit, what should each pot cost her to make?

Using the Rate of Profit to Find the Selling Price (12)

Example: What should the selling price be on a file cabinet which costs $80 if the rate of profit is 20%?

Solution:

Step (1): Determine the profit:

Line (a): Rate of Profit × Cost Price = Profit

Line (b): 20% × $80 = Profit

Line (c): (.2) × $80 = $16 (profit)

Step (2): Add to find the selling price:

Line (d): Cost Price + Profit = Selling Price

Line (e): $80 + $16 = $96

A Short Cut (13)

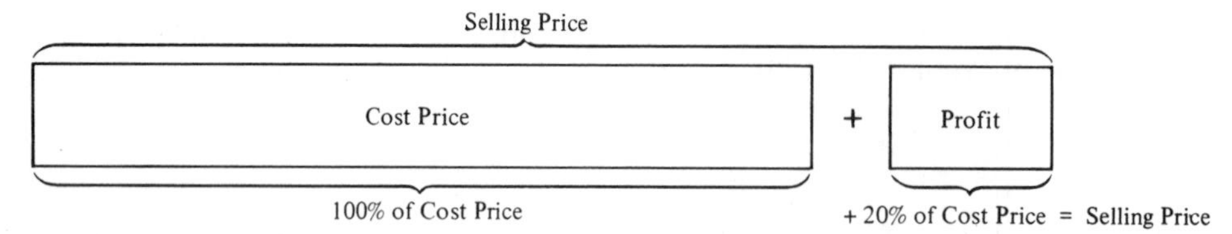

Since we are only interested in the selling price, which represents 20% more than the cost price, we can get our information as follows:

Line (a): (100% + Rate of Profit) × Cost Price = Selling Price

Line (b): (100% + 20%) × $80 = Selling Price

Line (c): 120% × $80 = Selling Price

Line (d):

```
  $80
 ×1.2
 ----
 16 0
 80
 ----
$96.0 selling price
```

Using 100% plus the Rate of Profit to Find the Cost Price (14)

Example: The sticker price of an automobile is $4,720. This is 18% more than the dealer's cost. How much did the dealer pay for this automobile?

Solution:

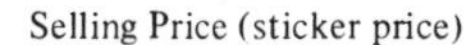

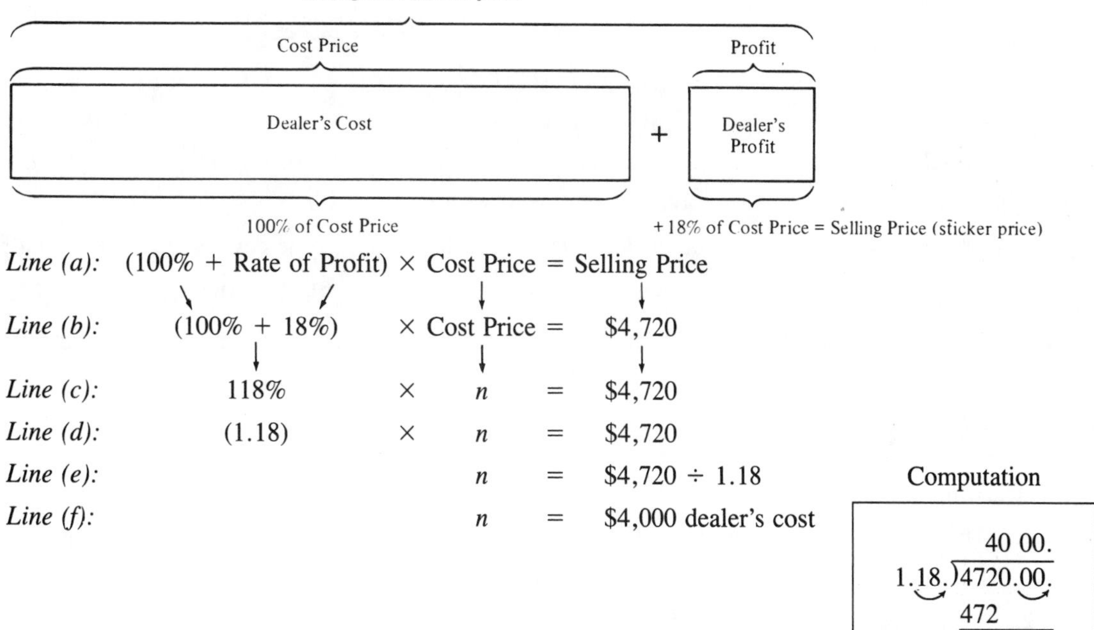

Line (a): (100% + Rate of Profit) × Cost Price = Selling Price

Line (b): (100% + 18%) × Cost Price = $4,720

Line (c): 118% × n = $4,720

Line (d): (1.18) × n = $4,720

Line (e): n = $4,720 ÷ 1.18

Line (f): n = $4,000 dealer's cost

Computation

$$\begin{array}{r} 40\,00. \\ 1.18.\overline{)4720.00.} \\ 472 \\ \hline 00\,00 \end{array}$$

Formula for Finding the Cost Price or the Selling Price (15)

(100% plus the rate of profit)

This formula is used when you are given the rate of profit and either the cost price or the selling price.

(100% + Rate of Profit) × Cost Price = Selling Price

Study Exercise Four (16)

Find the missing information to the nearest cent:

	Cost Price	Selling Price	Rate of Profit
1.	$126		42%
2.	$196		23%
3.	$48		10%
4.		$566.40	18%
5.		$1,247	45%

Summary (17)

A. Addition and subtraction relationships:
  1. Selling Price − Cost Price = Profit
  2. Selling Price − Profit = Cost Price
  3. Cost Price + Profit = Selling Price

B. This fraction, which we will change to a percent, produces the rate of profit based on cost:

$$\frac{\text{Amount of Profit}}{\text{Cost Price}} = \text{Rate of Profit}$$

C. This formula is used when you are given the rate of profit and either the cost price or the profit:

$$\text{Rate of Profit} \times \text{Cost Price} = \text{Profit}$$

D. This formula is used when you are given the rate of profit and either the cost price or the selling price:

$$(100\% + \text{Rate of Profit}) \times \text{Cost Price} = \text{Selling Price}$$

REVIEW EXERCISES (18)

A. Answer the following:
  1. The rate of profit is produced by a fraction whose numerator is _________ and whose denominator is _________.
  2. Cost Price + _________ = Selling Price.
  3. Selling Price − _________ = Cost Price.
  4. (100% + Rate of Profit) × Cost Price = _________
  5. Rate of Profit × Cost Price = _________
  6. A table was purchased for $90 and sold for $130. To the nearest whole percent, what is the rate of profit?
  7. What will be the amount of profit on a refrigerator which cost the seller $220, if a 40% rate of profit is used?
  8. A bookstore which uses a 25% rate of profit sells a book for $10. How much did the store pay for this book?

B. Find the missing information; round rates to nearest whole percent:

	Cost Price	Selling Price	Profit	Rate of Profit
9.		$32	$8	
10.	$39		$6	
11.	$500			46%
12.	$74	$98		
13.			$230.51	37%
14.		$1,024		28%

Solutions to Review Exercises (19)

A. 1. See Frame 5
2. See Frame 3
3. See Frame 3
4. See Frame 15
5. See Frame 10
6. $130 − $90 = $40 profit

$\frac{\text{Profit}}{\text{Cost Price}} = \frac{\$40}{\$90} \approx 44\%$ rate of profit

7. Rate of Profit × Cost Price = Profit

$40\% \times \$220 = \text{Profit}$

$(.4) \times \$220 = \$88$ profit

8. (100% + Rate of Profit) × Cost Price = Selling Price

$(100\% + 25\%) \times \text{Cost Price} = \$10$

$125\% \times n = \$10$

$(1.25) \times n = \$10$

$n = \$10 \div 1.25$

$n = \$8$ cost price

B. 9. Selling Price − Profit = Cost Price

$\$32 - \$8 = \$24$

$\frac{\text{Amount of Profit}}{\text{Cost Price}} = \frac{\$8}{\$24} = \frac{1}{3} = 33\%$ rate of profit to nearest whole percent

10. Cost Price + Profit = Selling Price

$\$39 + \$6 = \$45$

$\frac{\text{Amount of Profit}}{\text{Cost Price}} = \frac{\$6}{\$39} = 15\%$ rate of profit to nearest whole percent

11. Rate of Profit × Cost Price = Profit

$46\% \times \$500 = \text{Profit}$

$(.46) \times \$500 = \$230$ profit

Cost Price + Profit = Selling Price

$\$500 + \$230 = \$730$

12. Selling Price − Cost Price = Profit

$\$98 - \$74 = \$24$

$\frac{\text{Amount of Profit}}{\text{Cost Price}} = \frac{\$24}{\$74} = 32\%$ rate of profit to nearest whole percent

13. Rate of Profit × Cost Price = Profit

$37\% \times \text{Cost Price} = \$230.51$

$(.37) \times n = \$230.51$

$= \$230.51 \div .37$

$= \$623$ cost price

Cost Price + Profit = Selling Price

$\$623 + \$230.51 = \$853.51$

14. (100% + Rate of Profit) × Cost Price = Selling Price

$(100\% + 28\%) \times \text{Cost Price} = \$1,024$

$128\% \times n = \$1,024$

$(1.28) \times n = \$1,024$

$n = \$1,024 \div 1.28$

$n = \$800$ cost price

Selling Price − Cost Price = Profit

$\$1,024 - \$800 = \$224$

SUPPLEMENTARY PROBLEMS

A. Answer the following:

1. Selling Price − _________ = Profit.
2. Profit + _________ = Selling Price.

SUPPLEMENTARY PROBLEMS (Continued)

3. ________ × Cost Price = Profit.
4. ________ × Cost Price = Selling Price.
5. What is the rate of profit on a refrigerator which cost $320 and sold for $410? (to nearest whole percent)
6. What is the rate of profit on a motorcycle which cost $920 and sold for $1,100? (to nearest whole percent)
7. What is the amount of profit on a calculator which cost $70 if a 32% rate of profit is used?
8. What is the amount of profit on a lamp which cost $120 if the rate of profit is 42%?
9. A manufacturer needs to have a $1.20 mark-up on each wallet he sells. If his rate of profit is 30%, what should each wallet cost to produce?
10. A nursery needs to make a $2 profit on each tree it sells. If a 40% rate of profit is used, how much should each tree cost the nursery?
11. What will be the selling price of a set of dishes which cost $160 if the rate of profit is 35%?
12. What will be the selling price of a food freezer which costs $325 if the rate of profit is 43%?
13. An automobile has a sticker price of $9,794. If this is 18% over the dealer's cost, how much did the dealer pay for this car?
14. A truck has a sticker price of $8,352. If this is 16% over the dealer's cost, how much did the dealer pay for the truck?

B. Find the missing information; round rates to nearest whole percent:

	Cost Price	Selling Price	Profit	Rate of Profit
15.	$49			52%
16.		$67	$17	
17.	$480	$516		
18.			$713	31%
19.	$1,560		$380	
20.		$639.60		23%

C. Find the following:

21. A 20% rate of profit is included in the price of every book sold in a bookstore. If a book sells for $18, how much did it cost the bookstore?
22. What is the amount of profit on a TV which cost the dealer $320 if a 40% rate of profit is used?
23. What is the rate of profit on a watch that costs $316 and sells for $720 (to nearest whole percent)?
24. What is the cost to the store of a set of golf clubs with a selling price of $375, if the rate of profit is 25%?
25. What is the cost to the store of a ring with a selling price of $840, if the rate of profit is 250%?

Solutions to Study Exercises

(4A)

Study Exercise One (Frame 4)

1. Selling Price − Cost Price = Profit
   $80 − $60 = $20
2. Selling Price − Profit = Cost Price
   $49.50 − $17.20 = $32.30
3. Cost Price + Profit = Selling Price
   $325.40 + $48.50 = $373.90
4. Selling Price − Profit = Cost Price
   $179 − $48 = $131
5. Selling Price − Cost Price = Profit
   $90.80 − $53.50 = $37.30

Study Exercise Two (Frame 7)

(7A)

1. $70 − $20 = $50 cost price

   $\frac{\text{Amount of Profit}}{\text{Cost Price}} = \frac{\$20}{\$50} = \frac{2}{5} = 40\%$ rate of profit

2. $348 − $300 = $48 profit

   $\frac{\text{Amount of Profit}}{\text{Cost Price}} = \frac{\$48}{\$300} = \frac{48 \div 3}{300 \div 3} = \frac{16}{100} = 16\%$ rate of profit

3. $62 + $14 = $76 selling price

   $\frac{\text{Amount of Profit}}{\text{Cost Price}} = \frac{\$14}{\$62} = 23\%$ rate of profit to nearest whole percent

4. $190 − $32 = $158 cost price

   $\frac{\text{Amount of Profit}}{\text{Cost Price}} = \frac{\$32}{\$158} = 20\%$ rate of profit to nearest whole percent

Study Exercise Three (Frame 11)

(11A)

1. Rate of Profit × Cost Price = Profit
   40% × $180 = Profit
   (.4) × $180 = $72 profit
2. Rate of Profit × Cost Price = Profit
   17% × $4,250 = Profit
   (.17) × $4,250 = $722.50 profit
3. Rate of Profit × Cost Price = Profit
   35% × $120 = Profit
   (.35) × $120 = $42 profit
4. Rate of Profit × Cost Price = Profit
   40% × Cost Price = $6
   (.4) × n = $6
   n = $6 ÷ 4
   n = $15

   Computation

   $$\begin{array}{r} 15 \\ .4\overline{)6.0.} \\ \underline{4} \\ 20 \\ \underline{20} \\ 0 \end{array}$$

Study Exercise Four (Frame 16)

(16A)

1. (100% + Rate of Profit) × Cost Price = Selling Price
   (100% + 42%) × $126 = Selling Price
   142% × $126 = Selling Price
   (1.42) × $126 = $178.92 selling price

Study Exercise Four (Frame 16) **(Continued)**

2. (100% + Rate of Profit) × Cost Price = Selling Price

(100% + 23%)	×	\$196	=	Selling Price
123%	×	\$196	=	Selling Price
(1.23)	×	\$196	=	\$241.08 selling price

3. (100% + Rate of Profit) × Cost Price = Selling Price

(100% + 10%)	×	\$48	=	Selling Price
110%	×	\$48	=	Selling Price
(1.1)	×	\$48	=	\$52.80 selling price

4. (100% + Rate of Profit) × Cost Price = Selling Price

(100% + 18%)	×	Cost Price	=	\$566.40
118%	×	n	=	\$566.40
(1.18)	×	n	=	\$566.40
		n	=	\$566.40 ÷ 1.18
		n	=	\$480 cost price

5. (100% + Rate of Profit) × Cost Price = Selling Price

(100% + 45%)	×	Cost Price	=	\$1,247
145%	×	n	=	\$1,247
(1.45)	×	n	=	\$1,247
		n	=	\$1,247 ÷ 1.45
		n	=	\$860 cost price

Module 5 Practice Test

Units 27–31

1. Original Amount − _________ = New Amount
2. Find the missing information to the nearest whole percent:

Original Amount	New Amount	Amount of Decrease	Rate of Decrease
\$120		\$50	

3. A school with an enrollment of 1,200 students expects a 28% decrease next year. What will be the amount of this decrease?
4. A family decided to spend 30% of its income per year on rent. This amount is \$9,000. What is the yearly income of this family?
5. A factory which regularly produced 1420 articles per day decreased production by 35%. How many articles is it now producing?
6. A merchant claimed that the "new low price" on a television set represents a 22% decrease from the former price. If the set is marked at \$764.40, what was the former price?
7. Sales Amount − Net Proceeds = _________
8. Find the missing information:

Sales Amount	Net Proceeds	Amount of Commission	Rate of Commission
\$900		\$81	

9. A used car sold for \$5,200. What is the amount of commission if a 9% rate of commission was paid to the person who made the sale?
10. A sales clerk whose rate of commission is 30% wishes to make \$480 per week. What weekly sales amount is necessary to achieve this?
11. A real estate agent whose rate of commission is 6% sells a lot for \$21,000. How much does the owner of the lot receive?
12. In order to stay in business, a manufacturer determines that his net proceeds must be a minimum of \$27,000 per month. He pays his sales personnel a 20% rate of commission. What is the minimum sales amount required?

Module 5 Practice Test (Continued)

13. List Price − Amount of Discount = ________

Find the missing information to the nearest whole percent:

	List Price	Net Price	Amount of Discount	Rate of Discount
14.	\$150		\$24	
15.	\$840			15%

16. A refrigerator is on sale for \$516 and is advertised as "20% off our regular price." How much was the "regular price"?

17. New Amount − Original Amount = ________

18. A factory which regularly produced 1420 articles per day increased production by 35%. How many articles is it now producing?

19. A bicycle which formerly sold for \$120 now sells for \$150. What rate of increase is this?

20. How much is the sales tax on a \$7,920 automobile if the rate is 6%?

21. Selling Price − ________ = Profit

22. What is the rate of profit based on cost of a dress which cost \$80 and was sold for \$90?

23. A manufacturer determines that he must have a mark-up of \$18 on each article he sells. If a 30% rate of profit is standard in his industry, how much should each article cost to manufacture?

24. Using a rate of profit of 22%, determine the selling price of an article which costs \$72.

25. (100% + Rate of Profit) × Cost Price = ________

Answers to Module 5 Practice Test

1. Amount of Decrease.
2. New Amount = \$70, Rate of Decrease = 42%
3. 336 students.
4. \$30,000
5. 923 articles.
6. \$980
7. Amount of Commission.
8. Net Proceeds = \$819, Rate of Commission = 9%
9. \$468
10. \$1600
11. \$19,740
12. \$33,750
13. Net Price.
14. Net Price = \$126, Rate of Discount = 16%
15. Amount of Discount = \$126, Net Price = \$714
16. \$645
17. Amount of Increase.
18. 1,917 articles.
19. 25%
20. \$475.20
21. Cost Price.
22. $12\frac{1}{2}$%
23. \$60
24. \$87.84
25. Selling Price.

MODULE

6

Measurement

UNIT

32 Denominate Numerals and the English System of Measurement

OBJECTIVES (1)

By the end of this unit you should be able to:

1. USE DENOMINATE NUMERALS TO DESCRIBE MEASUREMENTS.
2. USE A TABLE OF MEASURES FOR THE ENGLISH SYSTEM OF MEASUREMENT.
3. CONVERT TO SMALLER UNITS OF MEASURE BY USING MULTIPLICATION.
4. CONVERT TO LARGER UNITS OF MEASURE BY USING DIVISION.
5. SIMPLIFY MEASUREMENTS.
6. OPERATE WITH DENOMINATE NUMERALS USING ADDITION, SUBTRACTION, MULTIPLICATION, AND DIVISION.

Denominate Numerals (2)

A denominate numeral is a symbol representing a number together with a unit of measure.

Examples:

1. 1 inch **2.** 3 quarts **3.** 5 pounds **4.** 2 hours

Measure of a Denominate Numeral (3)

The measure of a denominate numeral is the numeral without the label of measurement.

Denominate Numerals	Measure
2 feet	2
3.5 gallons	3.5
$4\frac{1}{2}$ tons	$4\frac{1}{2}$
6 days	6

Equivalent Denominate Numerals (4)

Equivalent denominate numerals are different measures of the same object. A different unit of measurement is used.

Examples:

1. 12 inches = 1 foot
2. 4 quarts = 1 gallon
3. 2 pounds = 32 ounces
4. $1\frac{1}{2}$ hours = 90 minutes
5. 3 feet = 1 yard
6. 2 days = 48 hours

Units of Measurement (5)

The fundamental quantities of measurement are length, volume, weight, time, and temperature. The two most basic systems of measurement are the English system and the metric system.

The English System of Measurement (6)

Length	Volume	Weight	Time	Temperature
inch (in) foot (ft) yard (yd) rod (rd) mile (mi)	fluid ounce (fl oz) pint (pt) quart (qt) gallon (gal)	ounce (oz) pound (lb) ton	second (sec) minute (min) hour (hr) day (da) week (wk)	degree (°F)

Using a Table of Measure (7)

Measure of Length

1 foot (ft) = 12 inches (in)
1 yard (yd) = 36 inches (in)
= 3 feet (ft)
1 rod (rd) = 16.5 feet (ft)
= 5.5 yards (yd)
1 mile (mi) = 5,280 feet (ft)
= 1,760 yards (yd)
= 320 rods (rd)

Measure of Volume

1 pint (pt) = 16 fluid ounces (fl oz)
1 quart (qt) = 2 pints (pt)
1 gallon (gal) = 4 quarts (qt)

Measure of Weight

1 pound (lb) = 16 ounces (oz)
1 short ton = 2,000 pounds (lb)
1 long ton = 2,240 pounds (lb)

Measure of Time

1 minute (min) = 60 seconds (sec)
1 hour (hr) = 60 minutes (min)
1 day (da) = 24 hours (hr)
1 week (wk) = 7 days (da)
1 year (yr) = 365 days (da)
= 52 weeks (wk)
= 12 months (mo)

Conversions (8)

Using equivalent denominate numerals, we can convert from one unit of measure to another.

Example 1: Convert 2 lb to ounces.

Solution: We know that 16 oz is equivalent to 1 lb. Thus, 16 is said to be the *conversion factor.* We now multiply 2 by 16 to obtain 32 oz. Notice that we changed from a larger unit of measure (pounds) to a smaller unit of measure (ounces), and we multiplied by the conversion factor.

Conversions (Continued)

Example 2: Convert 6 ft to yards.

Solution: We know that 3 ft is equivalent to 1 yd. Thus, 3 is said to be the conversion factor. We now divide 6 by 3 to obtain 2 yd. Notice that we changed from a smaller unit of measure (feet) to a larger unit of measure (yards), and we divided by the conversion factor.

Summary (1): When converting from a larger unit of measure to a smaller unit of measure, we multiply by the conversion factor.
(2): When converting from a smaller unit of measure to a larger unit of measure, we divide by the conversion factor.

Converting to Smaller Units of Measure (9)

When converting to a smaller unit of measure, follow these two steps:

Step (1): Determine the conversion factor by finding the number of units of the smaller measurement which is equivalent to one unit of the larger measurement.
Step (2): Multiply by this conversion factor.

Example: Convert 7 ft to inches:

Solution: 12 in = 1 ft

Step (1): The conversion factor is 12
Step (2): Multiply by 12:

$$\begin{aligned} 7 \text{ ft} &= (7)(12) \text{ in} \\ &= 84 \text{ in} \end{aligned}$$

More Examples (10)

Example 1: Convert 5 lb to ounces:

Solution: 16 oz = 1 lb

Step (1): 16 is the conversion factor
Step (2): Multiply by 16:

$$\begin{aligned} 5 \text{ lb} &= (5)(16) \text{ oz} \\ &= 80 \text{ oz} \end{aligned}$$

Example 2: Convert 4 hr to minutes:

Solution: 60 min = 1 hr

Step (1): 60 is the conversion factor
Step (2): Multiply by 60:

$$\begin{aligned} 4 \text{ hr} &= (4)(60) \text{ min} \\ &= 240 \text{ min} \end{aligned}$$

Example 3: Convert 3 gal to quarts:

Solution: 4 qt = 1 gal

Step (1): 4 is the conversion factor
Step (2): Multiply by 4

$$\begin{aligned} 3 \text{ gal} &= (3)(4) \text{ qt} \\ &= 12 \text{ qt} \end{aligned}$$

Study Exercise One (11)

Perform the following conversions. You may use the table of measures in frame 7.

1. Convert 5 gal to quarts.
2. Convert 3 short tons to pounds.
3. Convert 3 mi to feet.
4. Convert 4 da to hours.
5. Convert 5 qt to pints.
6. Convert 3 long tons to pounds.
7. Convert 30 yd to feet.
8. Convert 2 yr to days.
9. Convert 2 yr to weeks.

Converting to Larger Units of Measure (12)

When converting to a larger unit of measure, follow these two steps:

Step (1): Determine the conversion factor by finding the number of units of the smaller measurement which is equivalent to one unit of the larger measurement.
Step (2): Divide by this conversion factor.

Example: Convert 120 min to hours:

Solution: 60 min = 1 hr

Step (1): The conversion factor is 60
Step (2): Divide by 60

$$120 \text{ min} = \frac{120}{60} \text{ hr}$$
$$= 2 \text{ hr}$$

More Examples (13)

Example 1: Convert 43 oz to the nearest tenth of a pound:

Solution: 16 oz = 1 lb

Step (1): The conversion factor is 16
Step (2): Divide by 16:

```
     2.68
16)43.00
   32
   11 0
    9 6
    1 40
    1 28
      12
```

Therefore, 43 oz = 2.7 lb (to nearest tenth of a pound).

Example 2: Convert 40 mo to years and months:

Solution: 12 mo = 1 yr

Step (1): The conversion factor is 12
Step (2): Divide by 12:

```
    3 ←— years
12)40
   36
    4 ←— months
```

Therefore, 40 mo = 3 yr 4 mo.

More Examples (Continued)

Example 3: Convert 18 qt to the nearest tenth of a gallon:

Solution: 4 qt = 1 gal

Step (1): The conversion factor is 4
Step (2): Divide by 4:

$$\begin{array}{r} 4.5 \\ 4\overline{)18.0} \\ \underline{16} \\ 2\,0 \\ \underline{2\,0} \end{array}$$

Therefore, 18 qt = 4.5 gal

Study Exercise Two (14)

Perform the following conversions. If answers do not come out even, then round to the nearest tenth. You may use the table of measures in frame 7.

1. Convert 24 in to feet.
2. Convert 16 qt to gallons.
3. Convert 51 oz to pounds.
4. Convert 200 sec to minutes.
5. Convert 21 pt to quarts and pints.
6. Convert 29 ft to yards and feet.
7. Convert 250 min to hours.
8. Convert 42 hr to days and hours.

Simplified Form of Measurements (15)

A measurement is in simplified form when it has been converted to the largest possible unit of measure involving a whole number.

Example: Simplify 30 in.

Solution: Convert to feet and inches:

Step (1): The conversion factor is 12 since 12 in = 1 ft
Step (2): Divide by 12:

$$\begin{array}{r} 2 \longleftarrow \text{feet} \\ 12\overline{)30} \\ \underline{24} \\ 6 \longleftarrow \text{inches} \end{array}$$

Therefore, 30 in simplifies to 2 ft 6 in.

Another Example (16)

Sometimes we must take more than one step to simplify a measurement.

Example: Simplify 15 pt:

Solution: Convert to gallons, quarts, and pints:

Part I: Convert 15 pt to quarts and pints.

Line (a): 2 pt = 1 qt
Line (b): The conversion factor is 2

Another Example (Continued)

Line (c): Divide by 2:

$$\begin{array}{r} 7 \leftarrow \text{quarts} \\ 2\overline{)15} \\ \underline{14} \\ 1 \leftarrow \text{pints} \end{array}$$

Line (d): 15 pt = 7 qt (1 pt)

Part II: Convert 7 qt to gallons and quarts:

Line (e): 4 qt = 1 gal
Line (f): The conversion factor is 4
Line (g): Divide by 4:

$$\begin{array}{r} 1 \leftarrow \text{gallons} \\ 4\overline{)7} \\ \underline{4} \\ 3 \leftarrow \text{quarts} \end{array}$$

Line (h): 7 qt = 1 gal 3 qt

Therefore, 15 pt = 1 gal 3 qt 1 pt ←

Study Exercise Three (17)

Simplify the following measurements:

1. 174 in **2.** 320 sec **3.** 31 pt
4. 93 oz **5.** 33 qt

Addition of Denominate Numerals (18)

Denominate numerals may be added by summing together the like measurements which occur in each column.

Example: Add:

3 lb 7 oz
4 lb 6 oz

Solution: Add the like measurement in each column:

3 lb 7 oz
4 lb 6 oz
7 lb 13 oz

An Example Requiring A Simplification Step (19)

Example: Add:

5 lb 12 oz
7 lb 8 oz

Solution: Add the like measurements and simplify the result:

5 lb 12 oz
7 lb 8 oz

Line (a): 12 lb 20 oz (20 oz = 1 lb 4 oz) ⟵ simplification step
1 lb

Line (b): 13 lb 4 oz

Another Example Using Simplification (20)

Example: Add:

4 yd	2 ft	3 in
6 yd	2 ft	10 in
7 yd	1 ft	9 in

Solution: Add the like measurements and simplify the result:

	4 yd	2 ft	3 in	
	6 yd	2 ft	10 in	
	7 yd	1 ft	9 in	
Line (a):	17 yd	5 ft	22 in	
		1 ft ↙	↓	(22 in = 1 ft 10 in) ⟵ simplification step
Line (b):	17 yd	6 ft	10 in	
	2 yd ↙	↓		(6 ft = 2 yd 0 ft) ⟵ simplification step
Line (c):	19 yd	0 ft	10 in	
Line (d):	19 yd	10 in		

Study Exercise Four (21)

Add the following and simplify:

1.
2 ft 5 in
4 ft 10 in

2.
3 yd 2 ft 8 in
4 yd 2 ft 10 in

3.
1 gal 2 qt 1 pt 11 fl oz
2 gal 1 qt 1 pt 8 fl oz
1 gal 1 qt 1 pt 2 fl oz

4.
5 hr 20 min 51 sec
3 hr 22 min 46 sec
8 hr 17 min 14 sec

5.
3 lb 14 oz
2 lb 12 oz
8 lb 5 oz

6.
3 wk 5 da 7 hr
2 wk 2 da 13 hr
1 wk 3 da 14 hr

Subtraction of Denominate Numerals (22)

Example: Subtract:

5 yd	2 ft	10 in
2 yd	1 ft	8 in

Solution: Subtract the measures in each column:

5 yd	2 ft	10 in
2 yd	1 ft	8 in
3 yd	1 ft	2 in

Note: The answer may be checked by addition.

An Example Using Borrowing (23)

Example: Subtract:

9 hr	10 min
4 hr	50 min

An Example Using Borrowing (Continued)

Solution: Borrow 1 hr and convert it to minutes:

$$\begin{array}{l} 9\text{ hr } 10\text{ min} \\ \underline{4\text{ hr } 50\text{ min}} \end{array} \longrightarrow \begin{array}{l} 8\text{ hr } 60\text{ min } 10\text{ min} \\ \underline{4\text{ hr } \qquad\quad 50\text{ min}} \end{array} \longrightarrow \begin{array}{l} 8\text{ hr } 70\text{ min} \\ \underline{4\text{ hr } 50\text{ min}} \\ 4\text{ hr } 20\text{ min} \end{array}$$

Note: Often the middle conversion step can be done mentally.

Another Example Using Borrowing (24)

Example: Subtract:

$$\begin{array}{l} 4\text{ da } 7\text{ hr } 20\text{ min} \\ \underline{2\text{ da } 9\text{ hr } 50\text{ min}} \end{array}$$

Solution: Borrow 1 hr and convert it to minutes. Next, borrow 1 da and convert it to hours:

$$\begin{array}{l} 4\text{ da } 7\text{ hr } 20\text{ min} \\ \underline{2\text{ da } 9\text{ hr } 50\text{ min}} \end{array} \longrightarrow \begin{array}{l} 4\text{ da } 6\text{ hr } 80\text{ min} \\ \underline{2\text{ da } 9\text{ hr } 50\text{ min}} \end{array} \longrightarrow \begin{array}{l} 3\text{ da } 30\text{ hr } 80\text{ min} \\ \underline{2\text{ da } \ \ 9\text{ hr } 50\text{ min}} \\ 1\text{ da } 21\text{ hr } 30\text{ min} \end{array}$$

Study Exercise Five (25)

Subtract the following:

1. $\begin{array}{l} 5\text{ gal } 3\text{ qt} \\ \underline{1\text{ gal } 2\text{ qt}} \end{array}$

2. $\begin{array}{l} 3\text{ gal } 1\text{ qt} \\ \underline{1\text{ gal } 3\text{ qt}} \end{array}$

3. $\begin{array}{l} 3\text{ qt } 1\text{ pt } 4\text{ fl oz} \\ \underline{1\text{ qt } 1\text{ pt } 6\text{ fl oz}} \end{array}$

4. $\begin{array}{l} 4\text{ lb } 2\text{ oz} \\ \underline{1\text{ lb } 7\text{ oz}} \end{array}$

5. $\begin{array}{l} 8\text{ yd } 2\text{ ft } \ \ 9\text{ in} \\ \underline{4\text{ yd } 2\text{ ft } 11\text{ in}} \end{array}$

6. $\begin{array}{l} 6\text{ da } 12\text{ hr } 40\text{ min} \\ \underline{2\text{ da } 15\text{ hr } 55\text{ min}} \end{array}$

Multiplication Involving a Denominate Numeral (26)

Example: Multiply 5 lb 4 oz by 3:

Solution: Multiply 3 times each term of the denominate numeral:

$$\begin{array}{rr} 5\text{ lb} & 4\text{ oz} \\ & 3 \\ \hline 15\text{ lb} & 12\text{ oz} \end{array}$$

An Example Requiring Simplification Steps (27)

Example: Multiply 4 gal 2 qt 1 pt by 7:

Solution: Multiply 7 times each term of the denominate numeral; then simplify the result:

$$\begin{array}{lrrrl} & 4\text{ gal} & 2\text{ qt} & 1\text{ pt} & \\ & & & 7 & \\ \hline \textit{Line (a):} & 28\text{ gal} & 14\text{ qt} & 7\text{ pt} & \\ & & \underline{3\text{ qt}} \swarrow & \downarrow & (7\text{ pt} = 3\text{ qt } 1\text{ pt}) \longleftarrow \text{simplification step} \\ \textit{Line (b):} & 28\text{ gal} & 17\text{ qt} & 1\text{ pt} & \\ & \underline{4\text{ gal}} \swarrow & \downarrow & & (17\text{ qt} = 4\text{ gal } 1\text{ qt}) \longleftarrow \text{simplification step} \\ \textit{Line (c):} & 32\text{ gal} & 1\text{ qt} & 1\text{ pt} & \end{array}$$

Study Exercise Six (28)

Multiply the following and simplify:

1. 7 lb 2 oz × 6
2. 7 lb 2 oz × 9
3. 3 yd 2 ft 7 in × 5
4. 7 gal 3 qt × 4
5. 8 yr 3 mo × 6
6. 2 hr 40 min 35 sec × 2

Division Involving Denominate Numerals (29)

Example: Divide 12 lb 8 oz by 4:

Solution: Use the long division process to divide 4 into each term of the denominate numeral:

Line (a): 3 lb 2 oz
Line (b): 4)12 lb 8 oz
Line (c): 12 lb
Line (d): 0 lb 8 oz
Line (e): 8 oz
Line (f): 0 oz

An Example Requiring a Simplification Step (30)

Example: Divide 28 lb 4 oz by 5:

Solution: Use the long division process:

Line (a): 5 lb 10 oz
Line (b): 5)28 lb 4 oz
Line (c): 25 lb
Line (d): 3 lb 4 oz = 52 oz (3 lb = 48 oz) ⟵ simplification step
Line (e): 50 oz
Line (f): 2 oz remainder
Line (g): The answer is formed by putting the remainder over the divisor:

5 lb $10\frac{2}{5}$ oz

Another Example Using Simplification (31)

Example: Divide a length of 9 yd 2 ft 5 in into 4 equal parts:

Solution: Use the long division process to divide by 4:

Line (a): 2 yd 1 ft 4 in
Line (b): 4)9 yd 2 ft 5 in
Line (c): 8 yd ↓
Line (d): 1 yd 2 ft = 5 ft (1 yd = 3 ft) ⟵ simplification step
Line (e): 4 ft
Line (f): 1 ft 5 in = 17 in (1 ft = 12 in) ⟵ simplification step
Line (g): 16 in
Line (h): 1 in remainder
Line (i): The answer is 2 yd 1 ft $4\frac{1}{4}$ in

Study Exercise Seven (32)

Divide each of the following:

1. $2\overline{)26\text{ yd }2\text{ ft}}$

2. $4\overline{)26\text{ yd }2\text{ ft}}$

3. $10\overline{)31\text{ gal }2\text{ qt }1\text{ pt}}$

4. $8\overline{)35\text{ lb }15\text{ oz}}$

5. $3\overline{)5\text{ hr }17\text{ min }42\text{ sec}}$

REVIEW EXERCISES (33)

A. Perform the following conversions. You may use the table of measures in frame 7.

1. Convert 10 gal to quarts.
2. Convert 4 mi to feet.
3. Convert 8 lb to ounces.
4. Convert 2 pt to fluid ounces.
5. Convert 48 in to feet.
6. Convert 240 sec to minutes.
7. Convert 650 hr to weeks, days, and hours.
8. Convert 185 fl oz to quarts, pints, and fluid ounces.

B. Simplify the following measurements:

9. 30 in **10.** 100 oz (weight) **11.** 140 sec

C. Add the following and simplify:

12.
$$\begin{array}{r} 6\text{ yd }2\text{ ft }7\text{ in} \\ 4\text{ yd }1\text{ ft }9\text{ in} \\ \underline{2\text{ yd }2\text{ ft }10\text{ in}} \end{array}$$

13.
$$\begin{array}{r} 4\text{ hr }20\text{ min }30\text{ sec} \\ 3\text{ hr }10\text{ min }50\text{ sec} \\ \underline{8\text{ hr }30\text{ min }40\text{ sec}} \end{array}$$

14.
$$\begin{array}{r} 2\text{ gal }3\text{ qt }1\text{ pt} \\ 6\text{ gal }2\text{ qt }1\text{ pt} \\ \underline{4\text{ gal }1\text{ qt }0\text{ pt}} \end{array}$$

D. Subtract the following:

15.
$$\begin{array}{r} 10\text{ lb }4\text{ oz} \\ \underline{4\text{ lb }10\text{ oz}} \end{array}$$

16.
$$\begin{array}{r} 3\text{ qt }1\text{ pt }11\text{ fl oz} \\ \underline{1\text{ qt }1\text{ pt }12\text{ fl oz}} \end{array}$$

17.
$$\begin{array}{r} 6\text{ yd }1\text{ ft }7\text{ in} \\ \underline{4\text{ yd }2\text{ ft }10\text{ in}} \end{array}$$

E. Multiply the following and simplify:

18.
$$\begin{array}{r} 4\text{ gal }2\text{ qt }1\text{ pt} \\ \underline{5} \end{array}$$

19.
$$\begin{array}{r} 2\text{ hr }32\text{ min }3\text{ sec} \\ \underline{4} \end{array}$$

20.
$$\begin{array}{r} 2\text{ yd }1\text{ ft }5\text{ in} \\ \underline{3} \end{array}$$

F. Divide the following:

21. $3\overline{)27\text{ yr }9\text{ mo}}$ **22.** $6\overline{)22\text{ lb }14\text{ oz}}$ **23.** $4\overline{)22\text{ gal }3\text{ qt }1\text{ pt}}$

G. Solve the following problems:

24. Twenty years ago the height of a tree was 4 ft 8 in. Today its height is 63 ft 4 in. How much did it grow?

25. Find the total contents of a dozen jars if the capacity of each jar is 1 pt 10 fl oz.

26. An airplane carries enough fuel for a 5 hr 30 min flight. How much flight time remains after the plane flies for 3 hr 48 min?

27. A piece of lumber measuring 9 ft 5 in is to be cut into 4 equal parts. How long is each piece?

28. The first floor of a ten-story building is 21 ft 4 in high. The other 8 floors are each 12 ft 4 in high and the last floor is 14 ft 7 in high. How high is the building? (Give the answer in terms of feet and inches.)

29. Each can of fruit juice holds 1 qt 14 fl oz. What is the total contents of 5 cans?

REVIEW EXERCISES (Continued)

30. An airplane flew for 3 hr 40 min between cities A and B. It stayed at city B for 1 hr 22 min, and then flew on to city C in 4 hr 13 min. What was the total time from city A to city C?

31. An airplane is flying at an altitude of 31,680 ft. How high is this in miles?

32. How much ribbon is left on a 4 yd spool, if 62 in was used to trim a tablecloth? (Give the answer in terms of yards, feet and inches.)

33. At birth a baby weighed 7 lb 5 oz. Today she weighs 19 lb 3 oz. How much weight did she gain?

Solutions to Review Exercises (34)

A. **1.** 10 gal = (10)(4) qt = 40 qt **2.** 4 mi = (4)(5,280) ft = 21,120 ft **3.** 8 lb = (8)(16) oz = 128 oz **4.** 2 pt = (2)(16) fl oz = 32 fl oz

5. 48 in $=\frac{48}{12}$ ft $=4$ ft **6.** 240 sec $=\frac{240}{60}$ min $=4$ min **7.** 650 hr=3 wk 6 da 2 hr **8.** 185 fl oz=5 qt 1 pt 9 fl oz

B. **9.** 30 in = 2 ft 6 in **10.** 100 oz = 6 lb 4 oz **11.** 140 sec = 2 min 20 sec

C. **12.**

6 yd 2 ft 7 in
4 yd 1 ft 9 in
2 yd 2 ft 10 in
12 yd 5 ft 26 in
2 ft (26 in = 2 ft 2 in)
12 yd 7 ft 2 in
2 yd (7 ft = 2 yd 1 ft)
14 yd 1 ft 2 in

13.

4 hr 20 min 30 sec
3 hr 10 min 50 sec
8 hr 30 min 40 sec
15 hr 60 min 120 sec
2 min (120 sec = 2 min 0 sec)
15 hr 62 min 0 sec
1 hr (62 min = 1 hr 2 min)
16 hr 2 min

14.

2 gal 3 qt 1 pt
6 gal 2 qt 1 pt
4 gal 1 qt 0 pt
12 gal 6 qt 2 pt
1 qt (2 pt = 1 qt)
12 gal 7 qt 0 pt
1 gal (7 qt = 1 gal 3 qt)
13 gal 3 qt 0 pt

D. **15.**

10 lb 4 oz ⟶ 9 lb 16 oz 4 oz ⟶ 9 lb 20 oz
4 lb 10 oz   4 lb 10 oz   4 lb 10 oz
5 lb 10 oz

16.

3 qt 1 pt 11 fl oz ⟶ 2 qt 3 pt 11 fl oz ⟶ 2 qt 2 pt 27 fl oz
1 qt 1 pt 12 fl oz   1 qt 1 pt 12 fl oz   1 qt 1 pt 12 fl oz
1 qt 1 pt 15 fl oz

17.

6 yd 1 ft 7 in ⟶ 5 yd 4 ft 7 in ⟶ 5 yd 3 ft 19 in
4 yd 2 ft 10 in   4 yd 2 ft 10 in   4 yd 2 ft 10 in
1 yd 1 ft 9 in

E. **18.**

4 gal 2 qt 1 pt
5
20 gal 10 qt 5 qt
2 qt (5 pt = 2 qt 1 pt)
20 gal 12 qt 1 pt
3 gal (12 qt = 3 gal)
23 gal 0 qt 1 pt

19.

2 hr 32 min 3 sec
4
8 hr 128 min 12 sec
2 hr (128 min = 2 hr 8 min)
10 hr 8 min 12 sec

Solutions to Review Exercises (Continued)

20.

```
 2 yd  1 ft   5 in
                3
 ----------------
 6 yd  3 ft  15 in
       1 ft ↙ ↓          (15 in = 1 ft 3 in)
 6 yd  4 ft   3 in
 1 yd ↙ ↓                (4 ft = 1 yd 1 ft)
 ----------------
 7 yd  1 ft   3 in
```

F. 21.

```
     9 yr  3 mo
 3)27 yr  9 mo
   27 yr
   -----
          9 mo
          9 mo
          ----
          0 mo
```

22.

```
     3 lb  13 oz
 6)22 lb  14 oz
   18 lb
   -----
    4 lb  14 oz = 78 oz
                  78 oz
                  -----
                   0 oz
```

23.

```
     5 gal  2 qt            1 pt
 4)22 gal  3 qt            1 pt
   20 gal
   ------
    2 gal  3 qt = 11 qt
                   8 qt
                   ----
                   3 qt  1 pt = 7 pt
                                4 pt
                                ----
                                3 pt (remainder)
```

Answer: 5 gal 2 qt $1\frac{3}{4}$ pt

G. 24. Use subtraction:

```
 63 ft  4 in  ⟶  62 ft  16 in
  4 ft  8 in       4 ft   8 in
                  -----------
                  58 ft   8 in
```

The tree grew 58 ft 8 in over the 20-yr period.

25. Use multiplication:

```
  1 pt   10 fl oz
             12
 ----------------
 12 pt  120 fl oz
  7 pt ↙ ↓            (120 fl oz = 7 pt 8 fl oz)
 ----------------
 19 pt    8 fl oz
```

The total contents are 19 pt 8 fl oz

26. Use subtraction:

```
 5 hr 30 min  ⟶  4 hr 90 min
 3 hr 48 min       3 hr 48 min
                  -----------
                   1 hr 42 min
```

There is enough fuel for 1 hr 42 min of flight time.

27. Use division:

```
    2 ft  4 in
 4)9 ft  5 in
   8 ft
   ----
   1 ft  5 in = 17 in
                16 in
                -----
                 1 in (remainder)
```

Each piece would measure 2 ft $4\frac{1}{4}$ in

28. Use multiplication and addition:

```
 12 ft   4 in
          8
 -----------
 96 ft  32 in      (32 in = 2 ft 8 in)
  2 ft ↙ ↓
 -----------
 98 ft   8 in
```

The eight floors account for 98 ft 8 in. Therefore, add this to the height of the first floor and the last floor.

```
  98 ft   8 in
  21 ft   4 in     (first floor)
  14 ft   7 in     (last floor)
 ------------
 133 ft  19 in
   1 ft ↙ ↓
 ------------
 134 ft   7 in
```

29. Use multiplication:

```
 1 qt  14 fl oz
         5
 -------------
 5 qt  70 fl oz
 2 qt ↙ ↓          (70 fl oz = 4 pt 6 fl oz = 2 qt 6 fl oz)
 -------------
 7 qt   6 fl oz
```

The total contents are 7 qt 6 fl oz.

Solutions to Review Exercises (Continued)

30. Use addition:

```
3 hr 40 min
1 hr 22 min
4 hr 13 min
-----------
8 hr 75 min
1 hr ↙ ↓          (75 min = 1 hr 15 min)
-----------
9 hr 15 min
```

The total time was 9 hr 15 min.

31. Use division:
(1 mi = 5,280 ft)

```
          6 mi
     ______
5280)31680
     31680
```

The airplane's altitude is 6 mi.

32. Convert and use subtraction:
(62 in = 5 ft 2 in = 1 yd 2 ft 2 in)

```
4 yd           ⟶ 3 yd 3 ft      ⟶ 3 yd 2 ft 12 in
1 yd 2 ft 2 in   1 yd 2 ft 2 in   1 yd 2 ft  2 in
                                  ---------------
                                  2 yd 0 ft 10 in
```

There is 2 yd 10 in of ribbon left.

33. Use subtraction:

```
19 lb 3 oz ⟶ 18 lb 19 oz
 7 lb 5 oz     7 lb  5 oz
              -----------
              11 lb 14 oz
```

The baby gained 11 lb 14 oz.

SUPPLEMENTARY PROBLEMS

A. Perform the following conversions. You may use the table of measures in frame 7.

1. Convert 5 lb to ounces.
2. Convert 10 mi to feet.
3. Convert 15 gal to pints.
4. Convert 12 yd to feet.
5. Convert 60 in to feet.
6. Convert 48 oz to pounds.
7. Convert 100 min to the nearest tenth of an hour.
8. Convert 1,000 hr to weeks, days, and hours.

B. Simplify the following measurements:

9. 117 oz (weight) **10.** 385 sec **11.** 28 in

C. Add the following and simplify:

12.
```
5 lb 12 oz
7 lb  3 oz
2 lb  8 oz
----------
```

13.
```
7 hr 15 min 10 sec
2 hr 28 min  5 sec
4 hr 39 min 48 sec
------------------
```

14.
```
5 yd 2 ft 11 in
3 yd 1 ft  5 in
2 yd 2 ft
6 yd       8 in
---------------
```

15.
```
3 gal 3 qt 1 pt 12 fl oz
4 gal 1 qt       8 fl oz
6 gal      1 pt 10 fl oz
------------------------
```

D. Subtract the following:

16.
```
9 yd 1 ft 2 in
2 yd 2 ft 8 in
--------------
```

17.
```
4 hr 20 min 16 sec
2 hr 30 min 17 sec
------------------
```

SUPPLEMENTARY PROBLEMS (Continued)

18. 8 lb
 4 lb 10 oz
 ―――――

19. 10 gal 1 qt 1 pt
 2 gal 3 qt 1 pt
 ――――――――

E. Multiply the following and simplify:

20. 7 yd 2 ft
 × 4
 ―――――

21. 5 gal 2 qt
 × 6
 ―――――

22. 3 hr 15 min 14 sec
 × 4
 ――――――――

23. 10 lb 7 oz
 × 11
 ―――――

F. Divide the following:

24. $2\overline{)8 \text{ ft } 6 \text{ in}}$

25. $8\overline{)23 \text{ lb } 15 \text{ oz}}$

26. $3\overline{)7 \text{ hr } 12 \text{ min}}$

27. $3\overline{)7 \text{ yd } 2 \text{ ft } 9 \text{ in}}$

G. Solve the following problems:

28. An airplane had 6 hr and 30 min of fuel at take-off. How many hours of fuel remain after the plane flies for 2 hr and 45 min?

29. Find the total weight of 2 dozen cans of soup if each can weighs 1 lb 2 oz. Give the answer in terms of pounds and ounces.

30. A board 10 ft $9\frac{1}{8}$ in long must be cut into 4 equal parts. Allowing $\frac{1}{8}$ in for waste, find the length of each piece.

31. A woman works 8 hr on Monday, 8 hr 15 min on Tuesday, 9 hr 15 min on Wednesday, 8 hr 45 min on Thursday, and 7 hr 15 min on Friday. What was her total work time for the week?

32. At birth a baby weighed 8 lb 12 oz. Today he weighs 17 lb 8 oz. How much weight did he gain?

33. What is the difference in weight of two boxes of cereal if one weighs 2 lb 1 oz and the other weighs 1 lb 8 oz?

34. An airplane is flying at an altitude of 34,780 ft. To the nearest tenth, how high is this in miles?

35. How much ribbon is left on a 5 yd spool, if 54 in was used to trim a dress? (Give the answer in terms of yards, feet and inches.)

36. There are 24 cans of orange juice in a box. If each can holds 1 qt 14 fl oz, what is the total contents of the box?

37. An 8 lb 13 oz meat roast is to be divided into 3 equal smaller roasts. How much will each weigh?

38. In baseball the distance between each base is 90 ft. A player who hits a home run contacts 4 bases. What is this distance in feet? in yards?

39. A window is 7 ft long. A store has a sale on 72 in draperies. Will this size fit the window?

40. A sink faucet required 1 hr 20 min to install. A new design on the market claims it takes one fourth less time to install. How much time can be saved by using the new product? How long will the job take?

Solutions to Study Exercises (11A)

Study Exercise One (Frame 11)

1. 5 gal = (5)(4) qt
 = 20 qt

2. 3 short tons = (3)(2,000) lb
 = 6,000 lb

3. 3 mi = (3)(5,280) ft
 = 15,840 ft

4. 4 da = (4)(24) hr
 = 96 hr

5. 5 qt = (5)(2) pt
 = 10 pt

6. 3 long tons = (3)(2,240) lb
 = 6,720 lb

7. 30 yd = (30)(3) ft
 = 90 ft

8. 2 yr = (2)(365) da
 = 730 da

9. 2 yr = (2)(52) wk
 = 104 wk

Study Exercise Two (Frame 14) (14A)

1. 24 in = $\frac{24}{12}$ ft
   = 2 ft

2. 16 qt = $\frac{16}{4}$ gal
   = 4 gal

3. $16\overline{)51.00}$ = 3.18
   48
   3 0
   1 6
   1 40
   1 28
   12
   51 oz = 3.2 lb
   (to nearest tenth)

4. $60\overline{)200.00}$ = 3.33
   180
   20 0
   18 0
   2 00
   1 80
   20
   200 sec = 3.3 min
   (to nearest tenth)

5. $2\overline{)21}$ = 10 ← quarts
   20
   1 ← pint
   21 pt = 10 qt 1 pt

6. $3\overline{)29}$ = 9 ← yards
   27
   2 ← feet
   29 ft = 9 yd 2 ft

7. $60\overline{)250.00}$ = 4.16
   240
   10 0
   6 0
   4 00
   3 60
   40
   250 min = 4.2 hr
   (to nearest tenth)

8. $24\overline{)42}$ = 1 ← day
   24
   18 ← hours
   42 hr = 1 da 18 hr

Study Exercise Three (Frame 17) (17A)

1. *Part I:* Convert 174 in to feet and inches.
   $12\overline{)174}$ = 14 ← feet
   12
   54
   48
   6 ← inches
   174 in = 14 ft (6 in)

   Part II: Convert 14 ft to yards and feet.
   $3\overline{)14}$ = 4 ← yards
   12
   2 ← feet
   14 ft = 4 yd 2 ft
   Therefore, 174 in = 4 yd 2 ft 6 in

2. Convert 320 sec to minutes and seconds.
   $60\overline{)320}$ = 5 ← minutes
   300
   20 ← seconds
   Therefore, 320 sec = 5 min 20 sec

Study Exercise Three (Frame 17) (Continued)

3. *Part I:* Convert 31 pt to quarts and pints.

```
   15 ←— quarts
2)31
  2
  11
  10
   1 ←— pint
```

31 pt = 15 qt (1 pt)

Part II: Convert 15 qt to gallons and quarts

```
   3 ←— gallons
4)15
  12
   3 ←— quarts
```

15 qt = 3 gal 3 qt

Therefore, 31 pt = 3 gal 3 qt 1 pt

4. Convert 93 oz to pounds and ounces.

```
    5 ←— pounds
16)93
   80
   13 ←— ounces
```

Therefore, 93 oz = 5 lb 13 oz

5. Convert 33 quarts to gallons and quarts.

```
   8 ←— gallons
4)33
  32
   1 ←— quart
```

Therefore, 33 qt = 8 gal 1 qt

Study Exercise Four (Frame 21)

(21A)

1.

```
 2 ft  5 in
 4 ft 10 in
 ----------
 6 ft 15 in
 1 ft
 ----------
 7 ft  3 in
```

2.

```
 3 yd 2 ft  8 in
 4 yd 2 ft 10 in
 ---------------
 7 yd 4 ft 18 in
      1 ft
 ---------------
 7 yd 5 ft  6 in
 1 yd
 ---------------
 8 yd 2 ft  6 in
```

3.

```
 1 gal 2 qt 1 pt 11 fl oz
 2 gal 1 qt 1 pt  8 fl oz
 1 gal 1 qt 1 pt  2 fl oz
 ------------------------
 4 gal 4 qt 3 pt 21 fl oz
            1 pt
 ------------------------
 4 gal 4 qt 4 pt  5 fl oz
       2 qt
 ------------------------
 4 gal 6 qt 0 pt  5 fl oz
 1 gal
 ------------------------
 5 gal 2 qt 0 pt  5 fl oz
           or
 5 gal 2 qt 5 fl oz
```

4.

```
  5 hr 20 min  51 sec
  3 hr 22 min  46 sec
  8 hr 17 min  14 sec
 ---------------------
 16 hr 59 min 111 sec
        1 min
 ---------------------
 16 hr 60 min  51 sec
  1 hr
 ---------------------
 17 hr  0 min  51 sec
        or
 17 hr 51 sec
```

5.

```
  3 lb 14 oz
  2 lb 12 oz
  8 lb  5 oz
 -----------
 13 lb 31 oz
  1 lb
 -----------
 14 lb 15 oz
```

6.

```
 3 wk  5 da  7 hr
 2 wk  2 da 13 hr
 1 wk  3 da 14 hr
 ----------------
 6 wk 10 da 34 hr
       1 da
 ----------------
 6 wk 11 da 10 hr
 1 wk
 ----------------
 7 wk  4 da 10 hr
```

Study Exercise Five (Frame 25) (25A)

1. 5 gal 3 qt
   1 gal 2 qt
   ---
   4 gal 1 qt

2. 3 gal 1 qt ⟶ 2 gal 5 qt
   1 gal 3 qt   1 gal 3 qt
   ---
   1 gal 2 qt

3. 3 qt 1 pt 4 fl oz ⟶ 2 qt 2 pt 20 fl oz
   1 qt 1 pt 6 fl oz   1 qt 1 pt 6 fl oz
   ---
   1 qt 1 pt 14 fl oz

4. 4 lb 2 oz ⟶ 3 lb 18 oz
   1 lb 7 oz   1 lb 7 oz
   ---
   2 lb 11 oz

5. 8 yd 2 ft 9 in ⟶ 7 yd 4 ft 21 in
   4 yd 2 ft 11 in   4 yd 2 ft 11 in
   ---
   3 yd 2 ft 10 in

6. 6 da 12 hr 40 min ⟶ 5 da 35 hr 100 min
   2 da 15 hr 55 min   2 da 15 hr 55 min
   ---
   3 da 20 hr 45 min

Study Exercise Six (Frame 28) (28A)

1. 7 lb 2 oz
   6
   ---
   42 lb 12 oz

2. 7 lb 2 oz
   9
   ---
   63 lb 18 oz
   1 lb ↙ ↓
   ---
   64 lb 2 oz

3. 3 yd 2 ft 7 in
   5
   ---
   15 yd 10 ft 35 in
   2 ft ↙ ↓
   ---
   15 yd 12 ft 11 in
   4 yd ↙ ↓
   ---
   19 yd 0 ft 11 in
   or
   19 yd 11 in

4. 7 gal 3 qt
   4
   ---
   28 gal 12 qt
   3 gal ↙ ↓
   ---
   31 gal 0 qt
   or
   31 gal

5. 8 yr 3 mo
   6
   ---
   48 yr 18 mo
   1 yr ↙ ↓
   ---
   49 yr 6 mo

6. 2 hr 40 min 35 sec
   2
   ---
   4 hr 80 min 70 sec
   1 min ↙ ↓
   ---
   4 hr 81 min 10 sec
   1 hr ↙ ↓
   ---
   5 hr 21 min 10 sec

Study Exercise Seven (Frame 32) (32A)

1. 2)26 yd 2 ft
   Quotient: 13 yd 1 ft

2. 4)26 yd 2 ft
   Quotient: 6 yd 2 ft
   24 yd
   ---
   2 yd 2 ft = 8 ft
   8 ft
   ---
   0 ft

3. 10)31 gal 2 qt 1 pt
   Quotient: 3 gal 0 qt 1 pt
   30 gal
   ---
   1 gal 2 qt = 6 qt
   0 qt
   ---
   6 qt 1 pt = 13 pt
   10 pt
   ---
   3 pt

   Answer is: 3 gal $1\frac{3}{10}$ pt

4. 8)35 lb 15 oz
   Quotient: 4 lb 7 oz
   32 lb
   ---
   3 lb 15 oz = 63 oz
   56 oz
   ---
   7 oz

   Answer is: 4 lb $7\frac{7}{8}$ oz

5. 3)5 hr 17 min 42 sec
   Quotient: 1 hr 45 min 54 sec
   3 hr
   ---
   2 hr 17 min = 137 min
   135 min
   ---
   2 min 42 sec = 162 sec
   162 sec
   ---
   0 sec

UNIT

33 The Metric System of Measurement

OBJECTIVES (1)

By the end of this unit you should be able to:

1. WRITE THE MEANING OF EACH PREFIX USED IN THE METRIC SYSTEM.
2. THINK IN THE METRIC SYSTEM BY RELATING VARIOUS OBJECTS TO METRIC MEASUREMENTS.

A Unit in the Metric System for Length (2)

Length: Meter (m)

The meter is defined as 1,650,763.73 wavelengths in vacuum of the orange-red line of the spectrum of krypton 86.

1 m ≈ 1.09 yd

Meter

1 yd ≈ .91 m

Yard

A Unit in the Metric System for Volume (3)

Volume: Liter (L)*

The liter is defined as the space contained in a cube measuring 0.1 m on each side.

*Use a capital "L" to avoid confusing with the number "1".

A Unit in the Metric System for Volume (Continued)

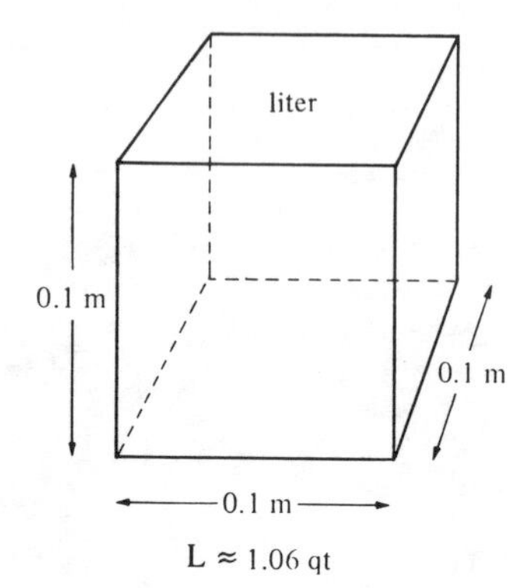

L ≈ 1.06 qt

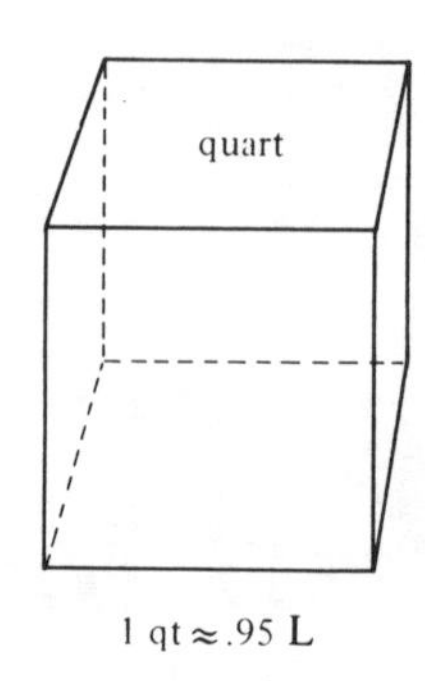

1 qt ≈ .95 L

A Unit in the Metric System for Weight (Mass) (4)

Mass or Weight: Kilogram (kg)

The standard for the kilogram is a cylinder of platinum-iridium alloy kept by the International Bureau of Weights and Measures in Paris. A duplicate is also kept by the National Bureau of Standards in the United States. This is the only base unit of the metric system which is still defined by an object made by humans.

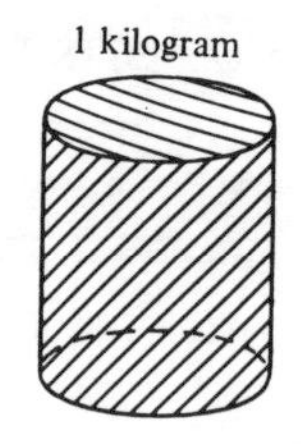

(relative sizes using same material)
1 kg ≈ 2.2 lb 1 lb ≈ .45 kg

Some Prefixes Used in the Metric System (5)

Prefix	Symbol	Meaning	
kilo	k	thousand	1,000
hecto	h	hundred	100
deka	da	ten	10
deci	d	tenth	.1 or $\frac{1}{10}$
centi	c	hundredth	.01 or $\frac{1}{100}$
milli	m	thousandth	.001 or $\frac{1}{1,000}$

m represents meter
L represents liter
g represents gram

Examples Using the Prefixes (6)

1. Kilometer (km) means 1,000 meters.
2. Hectometer (hm) means 100 meters.
3. Centimeter (cm) means .01 meter.
4. Kilogram (kg) means 1,000 grams.
5. Milligram (mg) means .001 gram.
6. Milliliter (mL) means .001 liter.

Remember (7)

Unit	Symbol	Measures	Replaces
meter	m	length, distance, thickness	inch, foot, yard, mile
gram	g	weight	ounce, pound
liter	L	volume	cup, pint, quart, gallon, fluid ounce

Study Exercise One (8)

Write the meaning for each of the following:

1. Kilometer **2.** Dekameter **3.** Millimeter **4.** Decimeter
5. Kilogram **6.** Centigram **7.** Milliliter **8.** Hectoliter

If "m" represents the meter family, "L" represents the liter family, and "g" the gram family, indicate by m, L, or g which family you would use for the following:

9. To put gasoline in your car.
10. To measure milk for a cake.
11. To measure the weight of a roast turkey.
12. To find the distance between two cities.
13. To find your weight.
14. To find your waistline.
15. To find your height.
16. To estimate the amount of soup in a pot.

Length in the Metric System (9)

meter

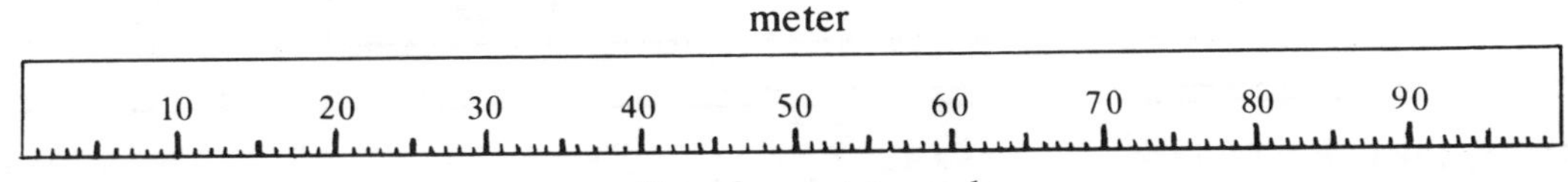

Drawing not to scale

A meter is subdivided into 100 centimeters.

1 km (kilometer)	=	1,000 m
1 hm (hectometer)	=	100 m
1 dam (dekameter)	=	10 m
1 dm (decimeter)	=	.1 m
1 cm (centimeter)	=	.01 m
1 mm (millimeter)	=	.001 m

Most Commonly Used Units of Length (10)

1. Kilometer: 1 km=1,000 m. One kilometer is approximately .62 or $\frac{5}{8}$ of a mile.
2. Meter: 1 m=100 cm. One meter is slightly longer than one yard.
3. Centimeter: 1 cm = .01 m. One centimeter is approximately .4 or $\frac{2}{5}$ of an inch.

Examples Involving Kilometers (11)

Kilometers are used for measuring long distances. Since 1 km is approximately .62 of a mile, 100 km would be about 62 mi.

The basic unit of speed in the metric system is kilometers per hour (km/hr) rather than miles per hour (mph). For comparison purposes, it is good to remember that 100 km/hr is about 62 mph. The following table shows equivalent speeds.

Comparison of Metric and English Speeds
(Approximate)

km/hr	mph
120	75
110	68
100	62
90	56
80	50
70	43
60	38
50	31
40	25
30	19
20	12
10	6

Examples Involving Meters (12)

A meter is slightly longer than a yard (1 m $\approx$ 1.09 yd).

Example 1: A football field would measure slightly less than 100 m.

Example 2: A basketball player measuring 2 m would be about 6 ft 6 in.

Example 3: A medium-size room might measure about 4 m wide by 5 m long.

Examples Involving Centimeters (13)

Centimeters are used for measuring small distances. A centimeter is about .4 or $\frac{2}{5}$ of an inch. Or, stated another way, 1 inch = 2.54 centimeters.

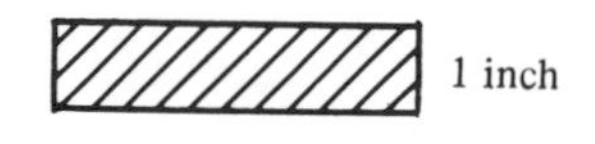

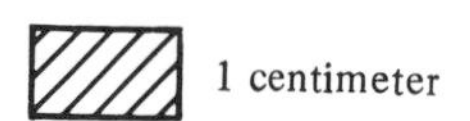

Examples Involving Centimeters (Continued)

Example 1: A nickel (5-cent piece) is about 2 cm in diameter.

Example 2: A quarter (25-cent piece) is about 2.5 cm in diameter.

Example 3: A new lead pencil is about 20 cm long.

Example 4: An individual 6 ft tall would measure 183 cm.

Study Exercise Two (14)

A. Multiple choice—select the one correct answer:

1. The width of a standard door would be about _________.
(a) 1 cm **(b)** 1 m **(c)** 1 km

2. A car traveling 100 km/hr would be traveling about _________.
(a) 100 mph **(b)** 62 mph **(c)** 162 mph

3. The width of your little finger would be about _________.
(a) 1 cm **(b)** 10 cm **(c)** 1 m

B. True or false:

4. A car traveling 60 km/hr would be breaking the 55 mph speed limit.

5. An individual who is 6 ft tall would measure less than 2 m.

6. A length of 2 in would measure about 5 cm.

Volume in the Metric System (15)

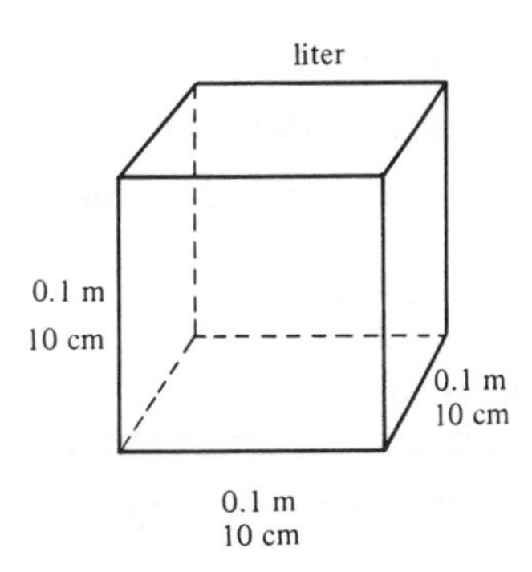

Drawing not to scale

The liter is defined as the space contained in a cube measuring 0.1 m or 10 cm on each side.

1 liter = (10 cm)(10 cm)(10 cm)
1 liter = 1,000 cubic centimeters (cm^3 or cc)
1 liter = 1,000 cm^3

Note: Cubic centimeter is abbreviated as cm^3 or cc.

Volume Relationships (16)

1 kL (kiloliter) = 1,000 liters
1 hL (hectoliter) = 100 liters
1 daL (dekaliter) = 10 liters
1 dL (deciliter) = .1 liter
1 cL (centiliter) = .01 liter
1 mL (milliliter) = .001 liter

Relating Cubic Centimeters to Liters (17)

Remember, a liter is defined as the space contained in a cube measuring 0.1 m or 10 cm on each side. This means that

$$1 \text{ liter} = (10 \text{ cm})(10 \text{ cm})(10 \text{ cm}) \text{ or } 1{,}000 \text{ cm}^3$$

$$1 \text{ L} = 1{,}000 \text{ cm}^3 \text{ (cc)}$$
$$1 \text{ mL} = 1 \text{ cm}^3 \text{ (cc)}$$

Most Commonly Used Units of Volume (18)

1. Liter: 1 L = 1,000 mL. One liter is slightly larger than 1 quart (1 liter ≈ 1.06 quart).
2. Milliliter: 1 mL = .001 L. One milliliter is approximately equal to $\frac{1}{5}$ of a teaspoon. Or, 1 milliliter is approximately $\frac{1}{30}$ of a fluid ounce.
3. Cubic centimeter: 1 cc = 1 mL. One cubic centimeter is exactly equal to one milliliter and the two units are used interchangeably.

Examples Involving Liters (19)

The liter is the basic unit used for measuring such things as gasoline, milk, paint, etc.

Example 1: 1 L of milk would be slightly more than a quart.

Example 2: 4 L of gasoline would be slightly more than a gallon.

Example 3: An 8 oz glass of water is about $\frac{1}{4}$ L.

Example 4: A 20 gal gasoline tank holds about 75 L.

Examples Involving Milliliters and Cubic Centimeters (20)

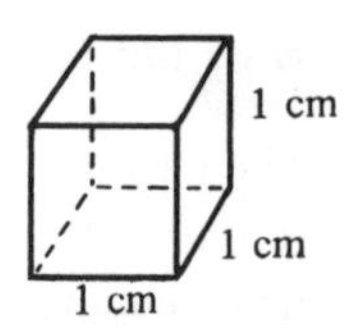

One Cubic Centimeter (1 cm^3 or 1 cc)

$1 \text{ cm}^3 = 1 \text{ mL}$

Milliliters and cubic centimeters are used to measure small volumes. The medical profession especially uses milliliters or cubic centimeters.

Example 1: One fluid ounce is about 30 milliliters or 30 cubic centimeters.

Example 2: An 8-ounce glass of water is about 240 milliliters or 240 cubic centimeters.

Example 3: A tablespoon holds about 15 milliliters or 15 cubic centimeters.

Study Exercise Three (21)

1. Which is a better buy, 1 gal of gasoline for $1.50 or 4 L of gasoline for $1.50?
2. Would a person become intoxicated upon drinking a milliliter of beer?
3. Would two normal glasses of water be about $\frac{1}{2}$L?
4. Which is larger, 1 fl oz or 1 cm^3?

Mass and Weight in the Metric System (22)

1 kg (kilogram) = 1,000 g
1 hg (hectogram) = 100 g
1 dag (dekagram) = 10 g
1 dg (decigram) = .1 g
1 cg (centigram) = .01 g
1 mg (milligram) = .001 g

The metric system also uses ton. One metric ton is defined as 1,000 kg.

Most Commonly Used Units of Mass and Weight (23)

1. Kilogram: 1 kg = 1,000 g. One kilogram weighs approximately 2.2 lb.
2. Gram: $1 \text{ g} = \frac{1}{1{,}000}$ kg. One gram weighs about $\frac{1}{500}$ lb. Or, 28 g weigh approximately 1 oz.

Examples Involving Kilograms (24)

The kilogram is used in place of pounds.

Example 1: A 220 lb football player weighs about 100 kg.

Example 2: A 5 lb sack of flour weighs about $2\frac{1}{4}$ kg.

Example 3: 1 qt of water weighs a little less than 1 kg.

Example 4: Here is a conversion table of pounds to kilograms. Find your weight in pounds and convert it to kilograms.

Pounds	Kilograms	Pounds	Kilograms
100	45	180	82
105	48	185	84
110	50	190	86
115	52	195	88
120	54	200	91
125	57	205	93
130	59	210	95
135	61	215	98
140	64	220	100
145	66	225	102
150	68	230	104
155	70	235	107
160	73	240	109
165	75	245	111
170	77	250	113
175	79	255	116

Examples Involving Grams (25)

The gram is a small unit of weight since 28 g weigh approximately one ounce.

Example 1: An ordinary paper clip weighs about 1 g.

Example 2: A nickel (5-cent piece) weighs about 5 g.

Example 3: 1 lb weighs about 455 g.

Study Exercise Four (26)

Multiple choice–Select the one correct answer:

1. A kilogram weighs approximately _________ .
(a) $\frac{1}{2}$ lb **(b)** 2.2 lb **(c)** 1,000 lb

2. A normal size pineapple weighs approximately _________ .
(a) 1 g **(b)** 1 kg **(c)** 10 kg

3. An average size candy bar weighs about _________ .
(a) 3 g **(b)** 30 g **(c)** 300 g

4. An ordinary size D flashlight battery weighs about _________ .
(a) 10 g **(b)** 100 g **(c)** 1 kg

REVIEW EXERCISES (27)

A. Complete the following table:

	Prefix	Symbol	Meaning
1.	kilo		
2.			hundred 100
3.	deka		
4.		d	
5.		c	
6.			thousandth .001

B. Write the meaning and the abbreviation for each of the following:

7. Kilogram. **8.** Milligram. **9.** Centimeter **10.** Decimeter.
11. Milliliter. **12.** Hectoliter. **13.** Dekameter.

C. Fill in the blanks:

14. 1 cubic centimeter = _________ milliliter.

15. 100 milliliters = _________ cubic centimeters.

REVIEW EXERCISES (Continued)

16. 1 kilometer = _________ meters.
17. 1 meter = _________ centimeters.
18. 1 liter = _________ milliliters.
19. 1 kilogram = _________ grams.

D. Multiple choice—Select the one correct answer:

20. One kilometer is about _________ miles.
(a) 62 **(b)** 6.2 **(c)** .62

21. The length of a football field is about _________.
(a) 10 m **(b)** 100 m **(c)** 1 km

22. A quarter is about _________ in diameter.
(a) 2.5 cm **(b)** 25 cm **(c)** 2.5 m

23. One liter is slightly more than _________.
(a) 1 gal **(b)** 1 qt **(c)** 1 pt

24. One fluid ounce is about _________.
(a) 30 L **(b)** 300 mL **(c)** 30 mL

25. A 220 lb football player weighs about _________.
(a) 100 g **(b)** 1 kg **(c)** 100 kg

26. A nickel (5-cent piece) weighs about _________.
(a) 5 kg **(b)** 50 g **(c)** 5 g

27. A softball bat is about _________ long.
(a) 1 L **(b)** 1 dam **(c)** 1 m

28. A 3 foot child is about _________ tall.
(a) 400 cm **(b)** 1 m **(c)** 200 cm

29. A small car weighs about _________ metric tons.
(a) 10 **(b)** 1 **(c)** .1

30. Ten cubic centimeters of medicine is the same amount as _________ milliliters of that medicine.
(a) 1 **(b)** 100 **(c)** 10

31. A $4\frac{1}{2}$ lb fish weighs about _________.
(a) 100 g **(b)** 2 kg **(c)** 1 m

32. A woman who weighs 110 lb weighs about _________ kilograms.
(a) 110 **(b)** 50 **(c)** 30

33. A 1.7 kg chicken weighs _________.
(a) More than 3 lb **(b)** less than 3 lb **(c)** exactly 3 lb

34. The ostrich, the largest living bird, reaches a height up to 2.5 m. This is _________.
(a) over 7 ft. **(b)** under 7 ft. **(c)** exactly 7 ft.

35. Ducks and geese can fly over 50 km/hr. This is about _________.
(a) 50 mph **(b)** 70 mph **(c)** 30 mph

36. In one year a typical family of four used 450 L of milk. This is about _________.
(a) 900 pt **(b)** 225 qt **(c)** 45 kL

37. Two size D flashlight batteries weigh about _________.
(a) 1 kg **(b)** 0.5 kg **(c)** 200 g

E. Assume that the items in column A and in column B are of the same quality and sell for the same price. Choose the item in each pair that gives you the most for your money.

	A	B
38.	4 L of gasoline	1 gal of gasoline
39.	1 square yd of cloth	1 square m of cloth

REVIEW EXERCISES (Continued)

40.	1 L of milk	1 qt of milk
41.	1 metric ton of grain	2000 lb of grain
42.	1 lb of meat	0.5 kg of meat
43.	3 L of apple juice	1 gal of apple juice
44.	750 mL of soda pop	1 L of soda pop
45.	50 tablets of vitamin C, each tablet containing 500 mg of vitamin C	30 tablets of vitamin C, each tablet containing 1 g of vitamin C

Solutions to Review Exercises (28)

A. **(1-6).** See Frame 5.

B. **7.** 1,000 grams; kg **8.** 0.001 gram; mg **9.** 0.01 meter; cm
10. 0.1 meter; dm **11.** 0.001 liter; mL **12.** 100 liters; hL
13. 10 meters; dam

C. **14.** 1 **15.** 100 **16.** 1,000 **17.** 100 **18.** 1,000 **19.** 1,000

D. **20.** **(c)** .62 mi **21.** **(b)** 100 m **22.** **(a)** 2.5 cm
23. **(b)** 1 qt **24.** **(c)** 30 mL **25.** **(c)** 100 kg
26. **(c)** 5g **27.** **(c)** 1 m **28.** **(b)** 1 m
29. **(b)** 1 metric ton **30.** **(c)** 10 mL **31.** **(b)** 2 kg
32. **(b)** 50 kg **33.** **(a)** more than 3 lb **34.** **(a)** over 7 ft
35. **(c)** 30 mph **36.** **(a)** 900 pt **37.** **(c)** 200 g

E. **38.** A **39.** B **40.** A **41.** A **42.** B
43. B **44.** B **45.** B

SUPPLEMENTARY PROBLEMS

A. Complete the following table:

	Prefix	Symbol	Meaning
1.		k	
2.	hecto		
3.			ten 10
4.	deci		
5.			hundredth .01
6.		m	

B. Write the meaning and the abbreviation for each of the following:

7. Decimeter. **8.** Centimeter. **9.** Millimeter. **10.** Kiloliter. **11.** Hectoliter.
12. Milliliter. **13.** Kilogram. **14.** Dekagram. **15.** Milligram.

SUPPLEMENTARY PROBLEMS (Continued)

C. Fill in the blanks:

16. 1 km = _________ m
17. 1 dam = _________ m
18. 1 cc = _________ mL
19. 1 kg = _________ g
20. 1 mg = _________ g
21. 1 L = _________ mL
22. 1 metric ton = _________ kg
23. 1 dL = _________ L
24. 1 daL = _________ L
25. 1 L = _________ cm^3 or cc
26. 1 hg = _________ g
27. 1 mg = _________ g
28. 100 cm = _________ m
29. 62 mph ≈ _________ km/hr
30. 1.09 yd ≈ _________ m
31. 1 kg ≈ _________ lb
32. .95L ≈ _________ qt.

D. Multiple choice—Select the one correct answer:

33. A car traveling 62 mph is traveling about _________.
(a) 62 km/hr **(b)** 100 km/hr **(c)** 1,000 km/hr

34. The height of an average ceiling is about _________.
(a) 2.5 m **(b)** 25 m **(c)** 25 km

35. A nickel (5-cent piece) is about _________ in diameter.
(a) 5 cm **(b)** 2 cm **(c)** 10 cm

36. A tablespoon will hold about _________.
(a) 15 liters **(b)** 150 cc **(c)** 15 cc

37. One m is slightly longer than _________.
(a) 1 ft **(b)** 1 rod **(c)** 1 yd

38. One cm is approximately _________ inch.
(a) .1 **(b)** .4 **(c)** .75

39. 31 mph would be about _________ km/hr.
(a) 100 **(b)** 70 **(c)** 50

40. One L is slightly larger than _________.
(a) 1 gal **(b)** 1 pt **(c)** 1 qt

41. One fl oz is about _________ mL or cc.
(a) 15 **(b)** 30 **(c)** 45

42. 240 mL or 240 cc of water would fill a glass that holds about _________ fl oz.
(a) 8 **(b)** 4 **(c)** 12

43. A tablespoon holds about _________ mL.
(a) 15 **(b)** 30 **(c)** 45

44. 0.5 kg would weigh about _________.
(a) 2.2 lb **(b)** 1.1 lb **(c)** 8 oz

45. One oz would weigh about _________ g.
(a) 28 **(b)** 100 **(c)** 56

46. In the Bible, Goliath is described as 6 cubits tall. This is about 9 ft, or slightly less than _________ m.
(a) 2 **(b)** 3 **(c)** 4

SUPPLEMENTARY PROBLEMS (Continued)

E. Assume that the items in column A and in column B are of the same quality and sell for the same price. Choose the item in each pair that gives you the most for your money.

	A	B
47.	2 lb of meat	1 kg of meat
48.	A 10 lb turkey	A 4 kg turkey
49.	A 4 lb fish	A 2 kg fish
50.	1 lb butter	0.5 kg butter
51.	A 3.75 lb chicken	A 1.5 kg chicken
52.	10 gal of gasoline	40 L of gasoline
53.	1000 lb of corn	0.5 metric ton of corn
54.	2 L of vinegar	3 bottles of vinegar, each containing 750 mL
55.	100 tablets of vitamin C, each tablet containing 200 mg of vitamin C	50 tablets of vitamin C, each tablet containing 500 mg of vitamin C

Solutions to Study Exercises

(8A)

Study Exercise One (Frame 8)

1. 1,000 meters
2. 10 meters
3. .001 meter
4. .1 meter
5. 1,000 grams
6. .01 gram
7. .001 liter
8. 100 liters
9. L
10. L
11. g
12. m
13. g
14. m
15. m
16. L

Study Exercise Two (Frame 14)

(14A)

A. **1.** **(b)** 1m **2.** **(b)** 62 mph **3.** **(a)** 1 cm

B. **4.** False; See frame 11, 60 km/hr ≈ 38 mph
5. True; 6 ft = 2 yd, and each yard is slightly less than a meter.
6. True; 1 in = 2.54 cm; therefore, 2 in = 5.08 cm

Study Exercise Three (Frame 21)

(21A)

1. The better buy is 4 L of gasoline for $1.50. Remember, a liter is slightly more than a quart, so 4 L would be more than a gallon.

2. No; a milliliter is a very minute quantity, about one-fifth of a teaspoonful. (See frame 18)

3. Yes; a normal glass of water is about 8 fl oz and would correspond to about $\frac{1}{4}$ L. Therefore, 2 glasses would be about $\frac{1}{2}$ L (see frame 19).

4. One fluid ounce is much larger than a cubic centimeter. One fluid ounce is about 30 cm^3 (see frame 20).

Study Exercise Four (Frame 26)

(26A)

1. **(b)** 2.2 lb **2.** **(b)** 1 kg **3.** **(b)** 30 g **4.** **(b)** 100 g

UNIT

34 Conversions Within the Metric System

OBJECTIVES (1)

By the end of this unit you should be able to:

1. CONVERT BETWEEN UNITS IN THE METRIC SYSTEM.
2. USE THE PREFIXES MEGA AND MICRO IN WRITING LARGE AND SMALL NUMBERS.

Summary of the Metric System (2)

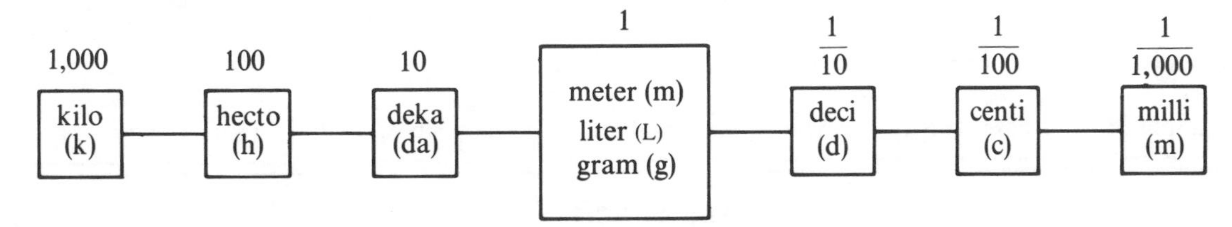

The metric system is based on powers of ten. In the chart, each prefix is ten times the one to its immediate right.

Example 1: "kilo" (1,000) corresponds to ten times a "hecto" (100).

Example 2: "centi" $\left(\frac{1}{100}\right)$ corresponds to ten times a "milli" $\left(\frac{1}{1,000}\right)$.

Example 3: our base units of meter, liter, and gram are each ten times a "deci" $\left(\frac{1}{10}\right)$.

Study Exercise One (3)

Each item in column A is ten times some item in column B. Make the correct match.

Example: one cg (1) is ten times one mg (f).

	A	"is ten times"	B
1.	cg	**(a)**	cm
2.	kL	**(b)**	dag
3.	dm	**(c)**	mL
4.	hm	**(d)**	mm
5.	m	**(e)**	dg
6.	cL	**(f)**	mg
7.	g	**(g)**	hm
8.	dg	**(h)**	dL
9.	L	**(i)**	hL
10.	cm	**(j)**	m
11.	dam	**(k)**	cg
12.	hg	**(l)**	dm
13.	dL	**(m)**	L
14.	km	**(n)**	dam
15.	daL	**(o)**	cL

The Chart—Left to Right **(4)**

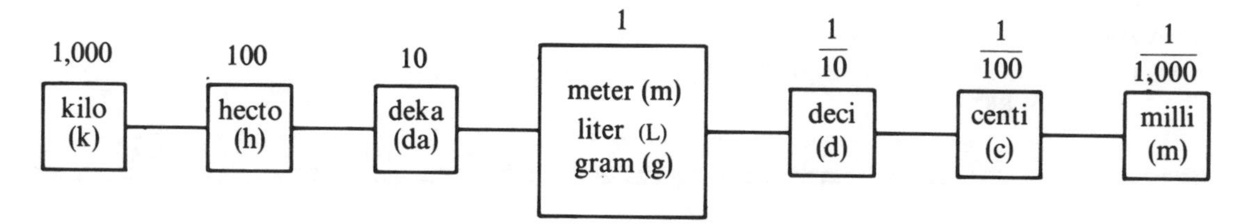

Notice as we move from left to right, each prefix is one tenth of the one to its immediate left.

Example 1: "hecto" (100) corresponds to one tenth of a "kilo" (1,000).

Example 2: "milli" $\left(\frac{1}{1{,}000}\right)$ corresponds to one tenth of a "centi" $\left(\frac{1}{100}\right)$.

Example 3: our base units of meter, liter, and gram are each one tenth of a "deka" (10).

Study Exercise Two **(5)**

Each item in column A is one tenth of some item in column B. Make the correct match.

Example: A mL (1) is one tenth of a cL (b).

	A	"is one tenth of"	B
1.	mL	**(a)**	m
2.	cg	**(b)**	cL
3.	dm	**(c)**	hm
4.	L	**(d)**	cg
5.	hm	**(e)**	dm
6.	dam	**(f)**	daL
7.	m	**(g)**	dag
8.	hg	**(h)**	dg
9.	mg	**(i)**	cm
10.	dL	**(j)**	kg
11.	cm	**(k)**	hg
12.	dag	**(l)**	dam
13.	g	**(m)**	dL
14.	cL	**(n)**	L
15.	mm	**(o)**	km

Using the Chart for Fast Conversions–Right to Left (6)

Example: Convert 65.5 centimeters to dekameters.

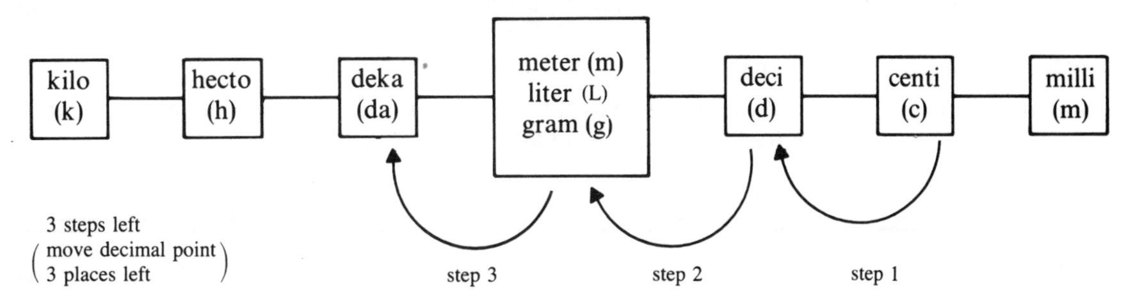

Solution: Each step left means we divide by a power of ten. Therefore, we divide by 10^3 or 1,000. This moves the decimal point 3 places to the left.

$$65.5 \text{ cm} = \frac{65.5}{1{,}000} \text{ or } .0655 \text{ dam}$$

Using the Chart for Fast Conversions–Left to Right (7)

Example: Convert .75 kilograms to decigrams.

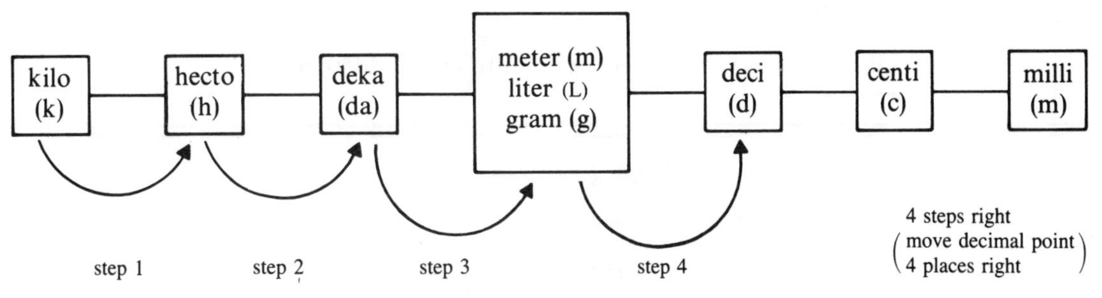

Solution: Each step right means we multiply by a power of ten. Therefore, we multiply by 10^4 or 10,000. This moves the decimal point 4 places to the right.

$$.75 \text{ kg} = (.75)(10{,}000) = 7{,}500 \text{ dg}$$

Basic Conversions–"Meter Family" (8)

Case (1): When converting to a larger unit of measure, use the operation of division. This means that we are going to move from right to left on our chart.

Example: Convert 1300 m to kilometers:

Solution: There are 3 steps to the left from meters to kilometers. Therefore, we divide by 10^3 or 1,000. This moves the decimal point 3 places to the left.

Divide by 1,000: thus, $1{,}300 \text{ m} = \frac{1{,}300}{1{,}000} \text{ km} = 1.3 \text{ km}$

Basic Conversions–"Meter Family" (Continued)

Case (2): When converting to a smaller unit of measure, use the operation of multiplication. This means that we are going to move from left to right on our chart.

Example: Convert 0.8 m to centimeters:

Solution: There are 2 steps to the right from meters to centimeters. Therefore, we multiply by 10^2 or 100. This moves the decimal point 2 places to the right.

Multiply by 100: thus, $0.8 \text{ m} = (100)(.8) \text{ cm} = 80 \text{ cm}$

More Examples (9)

Example 1: Convert 250 cm to meters:

Solution: There are 2 steps to the left from centimeters to meters.

Divide by 100:

(move decimal point 2 places left) $250 \text{ cm} = \frac{250}{100} \text{ m} = 2.5 \text{ m}$

Example 2: Convert 45 dam to meters:

Solution: There is one step to the right from dekameters to meters.

Multiply by 10:

(move decimal point 1 place right) $45 \text{ dam} = (10)(45) \text{ m} = 450 \text{ m}$

Study Exercise Three (10)

Use this chart to make the following conversions:

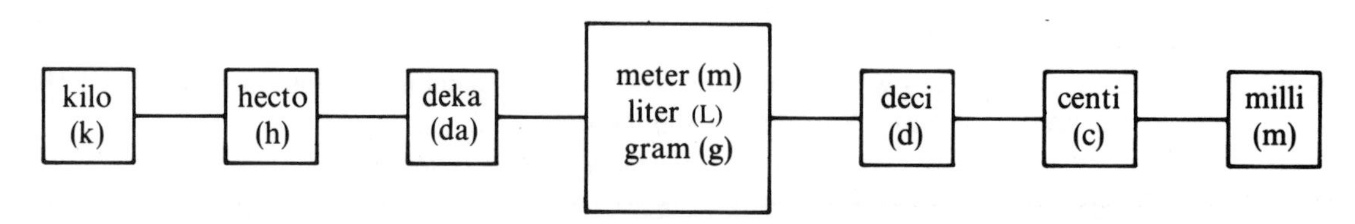

1. Convert 2,500 m to kilometers.
2. Convert 1.3 m to centimeters.
3. Convert 800 cm to meters.
4. Convert 76 cm to meters.
5. Convert 3.5 km to meters.

Basic Conversions–"Liter Family" (11)

Case (1): When converting to a larger unit of measure, use the operation of division. This means we are going to move from right to left on our chart.

Example: Convert 1,700 L to kiloliters:

Solution: There are 3 steps to the left from liters to kiloliters. Therefore, we divide by 10^3 or 1,000. This moves the decimal point 3 places to the left.

Divide by 1,000: $1{,}700 \text{ L} = \frac{1{,}700}{1{,}000} \text{ kL} = 1.7 \text{ kL}$

Basic Conversions–"Liter Family" (Continued)

Case (2): When converting to a smaller unit of measure, use the operation of multiplication. This means we are going to move from left to right on our chart.

Example: Convert .035 L to milliliters:

Solution: There are 3 steps to the right from liters to milliliters. Therefore, we multiply by 10^3 or 1,000. This moves the decimal point 3 places to the right.

Multiply by 1,000: thus, 0.035 L = (1,000)(.035) mL = 35 mL

More Examples (12)

Example 1: Convert 2,350 mL to liters.

Solution: There are 3 steps to the left from milliliters to liters.

Divide by 1,000:

$$\begin{pmatrix}\text{move decimal point} \\ \text{3 places left}\end{pmatrix} \qquad 2{,}350 \text{ mL} = \frac{2{,}350}{1{,}000} \text{ L} = 2.35 \text{ L}$$

Example 2: Convert 2.6 L to centiliters.

Solution: There are 2 steps to the right from liters to centiliters.

Multiply by 100:

$$\begin{pmatrix}\text{move decimal point} \\ \text{2 places right}\end{pmatrix} \qquad 2.6 \text{ L} = (100)(2.6) \text{ cL} = 260 \text{ cL}$$

Examples Using Cubic Centimeters (13)

Example 1: Convert 3,500 cc to liters.

Solution: Remember, 1 cc = 1 mL

There are 3 steps to the left from milliliters (cubic centimeters) to liters.

Divide by 1,000:

$$\begin{pmatrix}\text{move decimal point} \\ \text{3 places left}\end{pmatrix} \qquad 3{,}500 \text{ cc} = 3{,}500 \text{ mL} = \frac{3{,}500}{1{,}000} \text{ L} = 3.5 \text{ L}$$

Example 2: Convert 1.36 L to cubic centimeters.

Solution: Remember, 1 cc = 1 mL

There are 3 steps to the right from liters to milliliters (cubic centimeters).

Multiply by 1,000:

$$\begin{pmatrix}\text{move decimal point} \\ \text{3 places right}\end{pmatrix} \qquad 1.36 \text{ L} = (1{,}000)(1.36) \text{ mL} = 1{,}360 \text{ mL} = 1{,}360 \text{ cc}$$

Example 3: Convert 325 cc to milliliters.

Solution: Remember, 1 cubic centimeter (cc or cm^3) = 1 mL

Therefore, 325 cc = 325 mL

Study Exercise Four (14)

Use this chart to make the following conversions:

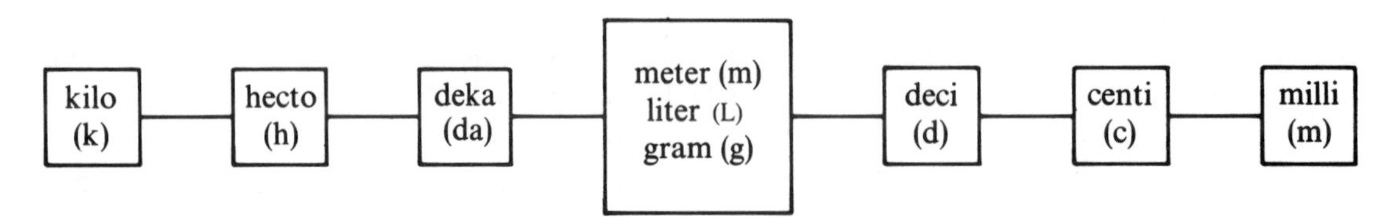

1. Convert 6,720 L to kiloliters.
2. Convert 0.025 L to milliliters.
3. Convert 1,265 mL to liters.
4. Convert 125 cL to liters.
5. Convert 1.5 L to cubic centimeters.
6. Convert 850 cm^3 to milliliters.
7. Convert 850 cc to liters.
8. Give the two abbreviations for cubic centimeter.

Basic Conversions—"Gram Family" (15)

Case (1): When converting to a larger unit of measure, use the operation of division. This means we are going to move from right to left on our chart.

Example: Convert 3,600 g to kilograms:

Solution: There are 3 steps to left from grams to kilograms. Therefore, we divide by 10^3 or 1,000. This moves the decimal point 3 places to the left.

Divide by 1,000: thus, $3{,}600\text{ g} = \frac{3{,}600}{1{,}000}\text{ kg} = 3.6\text{ kg}$

Case (2): When converting to a smaller unit of measure, use the operation of multiplication. This means we are going to move from left to right on our chart.

Example: Convert 2.3 g to centigrams:

Solution: There are 2 steps to the right from grams to centigrams. Therefore, we multiply by 10^2 or 100. This moves the decimal point 2 places to the right.

Multiply by 100: thus, $2.3\text{ g} = (100)(2.3)\text{ cg} = 230\text{ cg}$

More Examples (16)

Example 1: Convert .065 g to milligrams:

Solution: There are 3 steps to the right from grams to milligrams.

Multiply by 1,000:

(move decimal point 3 places right) $0.065\text{ g} = (1{,}000)(0.065)\text{ mg} = 65\text{ mg}$

Example 2: Convert 435 g to kilograms:

Solution: There are 3 steps to the left from grams to kilograms.

Divide by 1,000:

(move decimal point 3 places left) $435\text{ g} = \frac{435}{1{,}000}\text{ kg} = 0.435\text{ kg}$

Examples Using Metric Tons (17)

Example 1: Convert 5,650 kg to metric tons:

Solution: Remember, 1,000 kg = 1 metric ton

Divide by 1,000: $5{,}650 \text{ kg} = \frac{5{,}650}{1{,}000}$ metric tons = 5.65 metric tons

Example 2: Convert 2.5 metric tons to kilograms:

Solution: Remember, 1 metric ton = 1,000 kg

Multiply by 1,000: 2.5 metric tons = (1,000)(2.5) kg = 2,500 kg

Note: We did these problems by reduction and expansion, and did not use the chart.

Study Exercise Five (18)

Perform the following using the chart where possible:

1. Convert 2,750 g to kilograms.
2. Convert 0.57 g to milligrams.
3. Convert 85 cg to grams.
4. Convert 2.5 metric tons to kilograms.
5. Convert 3,800 kg to metric tons.
6. Convert 732 dag to kilograms.

The Prefix Mega (19)

In the metric system, *mega* represents one million (10^6 or 1,000,000). The symbol for mega is a capital M. Don't confuse this with the small m we use for milli.

Example 1: 1 megagram (1 Mg) = (1 g)(1,000,000) = 1,000,000 g (one million grams)

Example 2: 1 megaliter (1 ML) = (1L)(1,000,000) = 1,000,000 L (one million liters)

Example 3: 4,000,000 m (4 million meters) = (4 m)(1,000,000) = 4 Mm

The Prefix Micro (20)

In the metric system, *micro* represents one millionth $\left(\frac{1}{1{,}000{,}000} \text{ or } .000001\right)$. The symbol for micro is the Greek letter Mu, μ. You may also notice micro indicated by mc, particularly on vitamin labels.

Example 1: 1 microgram (1 μg or 1 mcg) = $(1 \text{ g})\left(\frac{1}{1{,}000{,}000}\right)$ = .000001 g (one millionth of a gram)

Example 2: 1 micrometer (1 μm) = (1 m)(.000001) = .000001 m (one millionth of a meter)

Example 3: .000003 m = $\left(\frac{1}{1{,}000{,}000}\right)(3 \text{ m})$ = 3 μm (3 micrometers)

(3 millionths of a meter)

More Examples (21)

Example 1: Convert 1 megagram to metric tons.

Solution: 1 Mg = (1 g)(1,000,000) = 1,000,000 g and 1,000,000 g = 1,000 kg (3 steps left on our chart) but 1,000 kg = 1 metric ton

Therefore, 1 megagram = 1 metric ton

More Examples (Continued)

Example 2: How many grams of vitamin B-12 are in a tablet containing 100 mcg (100 μg) of B-12?

Solution: 100 mcg = (100 g)(.000001) = .0001 g

Example 3: How many liters are there in 1,000 μL?

Solution: 1,000 μL = (1,000 L)(.000001) = .001 L

Study Exercise Six (22)

Fill in the blanks:

1. The prefix that means one million in the metric system is _________.
2. The prefix that means one millionth in the metric system is _________.
3. 1 Mg = _________ kg = _________ metric ton.
4. 2,000,000 L = _________ ML
5. 900 μg (mcg) = _________ g
6. .000002 m = _________ μm

Summary–Metric Prefixes We Have Studied (23)

Prefix	Symbol	Meaning
mega	M	1,000,000
kilo	k	1,000
hecto	h	100
deka	da	10
deci	d	.1 or $\frac{1}{10}$
centi	c	.01 or $\frac{1}{100}$
milli	m	.001 or $\frac{1}{1,000}$
micro	μ	.000001 or $\frac{1}{1,000,000}$

Summary–A Convenient Chart (24)

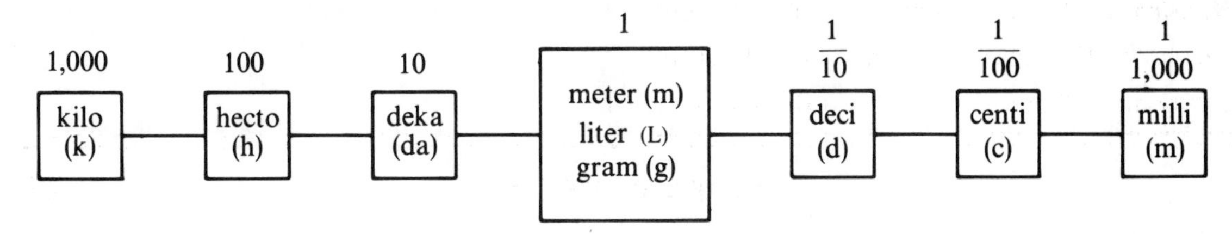

Note: We don't use the chart for mega and micro since the step from kilo to mega and micro to milli is more than just one multiple of 10.

REVIEW EXERCISES (25)

A. Complete the following table:

	Prefix	Symbol	Meaning
1.	mega		
2.		k	
3.			100
4.		da	
5.			.1 or $\frac{1}{10}$
6.	centi		
7.			.001 or $\frac{1}{1,000}$
8.		μ	

B. Write the meaning and the abbreviation for each of the following:

9. microgram
10. megaliter
11. megagram
12. micrometer
13. megameter
14. microliter

C. Perform the following conversions. You may use the chart in frame 24.

15. Convert 100 m to kilometers.
16. Convert 625 mm to centimeters.
17. Convert 0.36 L to milliliters.
18. Convert 62 mg to grams.
19. Convert 25 cm^3 to milliliters.
20. Convert 2.4 dm to millimeters.
21. Convert 1,800 cm to hectometers.
22. Convert 580 mL to liters.
23. Convert 1.3 L to cubic centimeters.
24. Convert 4 Mg to metric tons.
25. Convert 10,000μL to liters.
26. Convert 3 Mm to meters.

REVIEW EXERCISES (Continued)

27. Convert 3,000,000 meters to km.
28. Convert 2,100 m to kilometers.
29. Convert 3.1 cm to millimeters.
30. Convert 540 mL to liters.
31. Convert 54 hg to kilograms.
32. Convert 0.35 m to centimeters.
33. Convert 540 mL to cubic centimeters.
34. Convert 650 cg to dekagrams.

D. Assume that the items in column A and in column B are of the same quality and sell for the same price. Choose the item in each pair that gives you the most for your money.

	A	B
35.	12 L of gasoline	1 daL of gasoline
36.	0.4 kg of meat	450 g of meat
37.	1 dL of apple juice	12 cL of apple juice
38.	1.5 Mg of grain	1 metric ton of grain
39.	100 tablets of vitamin B-12, each tablet containing 100 mcg (100 μg) of vitamin B-12	80 tablets of vitamin B-12, each tablet containing 0.1 mg of vitamin B-12
40.	100 tablets of folic acid (a B vitamin), each tablet rated at 800 mcg (800 μg)	100 tablets of folic acid, each tablet rated at 0.6 mg

E. 41. Which automobile has the larger engine, an automobile with a 2500 cc engine or one with a 2.8 liter engine?

Solutions to Review Exercises (26)

A. **(1.-8.)** See Frame 23

B.
9. .000001 gram; μg (mcg)
10. 1,000,000 liters; ML
11. 1,000,000 grams; Mg
12. .000001 meter; μm (mcm)
13. 1,000,000 meters; Mm
14. .000001 liter; μL (mcL)

C.
15. $100 \text{ m} = \frac{100}{1{,}000} \text{ km} = 0.1 \text{ km}$
16. $625 \text{ mm} = \frac{625}{10} \text{ cm} = 62.5 \text{ cm}$
17. $0.36 \text{ L} = (1{,}000)(0.36) \text{ mL} = 360 \text{ mL}$
18. $62 \text{ mg} = \frac{62}{1{,}000} \text{ g} = 0.062 \text{ g}$
19. $25 \text{ cm}^3 = 25 \text{ mL}$
20. $2.4 \text{ dm} = (100)(2.4) \text{ mm} = 240 \text{ mm}$
21. $1{,}800 \text{ cm} = \frac{1{,}800}{10{,}000} \text{ hm} = 0.18 \text{ hm}$
22. $580 \text{ mL} = \frac{580}{1{,}000} \text{ L} = 0.58 \text{ L}$
23. $1.3 \text{ L} = (1{,}000)(1.3) \text{ mL} = 1{,}300 \text{ mL} = 1{,}300 \text{ cc}$
24. $4 \text{ Mg} = (1{,}000{,}000)(4\text{g}) = 4{,}000{,}000 \text{ g} = 4{,}000 \text{ kg} = 4$ metric tons
25. $10{,}000 \ \mu\text{L} = \left(\frac{1}{1{,}000{,}000}\right)(10{,}000 \text{ L}) = \frac{1}{100} \text{ L} = 0.01 \text{ L}$
26. $3 \text{ Mm} = (1{,}000{,}000)(3 \text{ m}) = 3{,}000{,}000 \text{ m}$
27. $3{,}000{,}000 \text{ m} = \frac{3{,}000{,}000}{1{,}000} \text{ km} = 3 \text{ km}$
28. $2{,}100 \text{ m} = \frac{2{,}100}{1{,}000} = 2.1 \text{ km}$
29. $3.1 \text{ cm} = (10)(3.1) = 31 \text{ mm}$
30. $540 \text{ mL} = \frac{540}{1{,}000} = .54$ liter
31. $54 \text{ hg} = \frac{54}{10} = 5.4 \text{ kg}$
32. $0.35 \text{ m} = (100)(0.35) = 35 \text{ cm}$
33. 540 mL = 540 cc
34. $650 \text{ mg} = \frac{650}{1{,}000} = .65 \text{ dag}$

D.
35. A since 1 daL = 10 L
36. B since 0.4 kg = 400 g
37. B since 12 cL = 1.2 dL

Solutions to Review Exercises (Continued)

38. A since 1 metric ton = 1 Mg
39. A since 100 mcg = 0.1 mg = .0001 g
40. A since 800 mcg = .0008 g and .6 mg = .0006 g

E. **41.** The 2.8 L engine is larger since 2.8 L = 2,800 mL = 2,800 cc or cm^3

SUPPLEMENTARY PROBLEMS

A. Complete the following table:

	Prefix	Symbol	Meaning
1.		M	
2.	kilo		
3.		h	
4.			10
5.	deci		
6.			.01 or $\frac{1}{100}$
7.		m	
8.			.000001 or $\frac{1}{1{,}000{,}000}$

B. Write the meaning and the abbreviation for each of the following:

9. megagram
10. microgram
11. micrometer
12. megaliter
13. megameter
14. microliter
15. centiliter

C. Fill in the blanks:

16. 1 km = ________ hm
17. 1 dam = ________ m
18. 1 cm = ________ mm
19. 1 kg = ________ g
20. 1 g = ________ mg
21. 1 liter = ________ mL
22. 1 mL = ________ cm^3 or cc
23. 1 liter = ________ cm^3 or cc
24. 1 Mg = ________ g
25. 1 μg = ________ g
26. 1 dL = ________ L
27. 1 hL = ________ kL
28. 1 metric ton = ________ kg
29. 1 metric ton = ________ Mg
30. 1 μm = ________ m

SUPPLEMENTARY PROBLEMS (Continued)

D. Perform the following conversions. You may use the chart in frame 24:

31. Convert 8 cm to millimeters.
32. Convert 7.25 dm to centimeters.
33. Convert 86 m to decimeters.
34. Convert 4.5 hm to dekameters.
35. Convert 4,875 m to kilometers.
36. Convert 32 cm^3 to milliliters.
37. Convert 22 mL to centiliters.
38. Convert 19.75 hL to liters.
39. Convert 29.8 daL to hectoliters.
40. Convert 5,870 L to kiloliters.
41. Convert 5.25 cg to milligrams.
42. Convert 43 dag to grams.
43. Convert 7 kg to grams.
44. Convert 4 g to decigrams.
45. Convert 452 mg to centigrams.
46. Convert 5 Mg to grams.
47. Convert 40,000 μg to grams.
48. Convert 5 Mm to meters.
49. Convert 5 Mg to metric tons.
50. Convert 6,000,000 m to megameters.

E. Assume that the items in column A and in column B are of the same quality and sell for the same price. Choose the item in each pair that gives you the most for your money.

	A	B
51.	1 daL of gasoline	9 L of gasoline
52.	375 g of meat	0.385 kg of meat
53.	0.9 Mg of grain	0.875 metric ton of grain
54.	100 tablets of vitamin B-12, each tablet containing 200 mcg (200 μg) of vitamin B-12	100 tablets of vitamin B-12, each tablet containing 0.175 mg of vitamin B-12

55. Which is the larger, a 3.5 L engine or a 3800 cc engine?

Solutions to Study Exercises

(3A)

Study Exercise One (Frame 3)

1. (f) **2.** (i) **3.** (a) **4.** (n) **5.** (l)
6. (c) **7.** (e) **8.** (k) **9.** (h) **10.** (d)
11. (j) **12.** (b) **13.** (o) **14.** (g) **15.** (m)

Study Exercise Two (Frame 5)

(5A)

1. (b) **2.** (h) **3.** (a) **4.** (f) **5.** (o)
6. (c) **7.** (l) **8.** (j) **9.** (d) **10.** (n)
11. (e) **12.** (k) **13.** (g) **14.** (m) **15.** (i)

Study Exercise Three (Frame 10)

(10A)

1. $2{,}500\text{ m}=\frac{2{,}500}{1{,}000}\text{ km}=2.5\text{ km}$

2. $1.3\text{ m}=(100)(1.3)\text{ cm}=130\text{ cm}$

3. $800\text{ cm}=\frac{800}{100}\text{ m}=8\text{ m}$

4. $76\text{ cm}=\frac{76}{100}\text{ m}=.76\text{ m}$

5. $3.5\text{ km}=(1{,}000)(3.5)\text{ m}=3{,}500\text{ m}$

Study Exercise Four (Frame 14)

(14A)

1. $6{,}720\text{ L} = \frac{6{,}720}{1{,}000}\text{ kL} = 6.72\text{ kL}$

2. $0.025\text{ L} = (1.000)(.025)\text{ mL} = 25\text{ mL}$

3. $1{,}265\text{ mL} = \frac{1{,}265}{1{,}000}\text{ L} = 1.265\text{ L}$

4. $125\text{ cL} = \frac{125}{100}\text{ L} = 1.25\text{ L}$

5. $1.5\text{ L} = (1.000)(1.5)\text{ mL} = 1{,}500\text{ mL} = 1{,}500\text{ cm}^3$

6. $850\text{ cm}^3 = 850\text{ mL}$

7. $850\text{ cc} = 850\text{ mL} = \frac{850}{1{,}000}\text{ L} = .85\text{ L}$

8. cm^3 or cc

Study Exercise Five (Frame 18) (18A)

1. $2{,}750 \text{ g} = \frac{2{,}750}{1{,}000} \text{ kg} = 2.75 \text{ kg}$
2. $.57 \text{ g} = (1{,}000)(0.57) \text{ mg} = 570 \text{ mg}$
3. $85 \text{ cg} = \frac{85}{100} \text{ g} = 0.85 \text{ g}$
4. 2.5 metric tons $= (1{,}000)(2.5) \text{ kg} = 2{,}500 \text{ kg}$
5. $3{,}800 \text{ kg} = \frac{3{,}800}{1{,}000}$ metric tons $= 3.8$ metric tons
6. $732 \text{ dag} = \frac{732}{100} \text{ kg} = 7.32 \text{ kg}$

Study Exercise Six (Frame 22) (22A)

1. mega
2. micro
3. 1 Mg=1,000,000 g=1,000 kg=1 metric ton
4. 2,000,000 L=(2 L)(1,000,000)=2 ML
5. 900 μg=(900 g)(.000001)=.0009 g
6. .000002 m=(2 m)(.000001)=2 μm

UNIT

35 Metric–English Conversions

OBJECTIVES (1)

By the end of this unit you should be able to:

1. USE CONVERSION TABLES TO CONVERT FROM THE METRIC SYSTEM TO THE ENGLISH SYSTEM.
2. USE CONVERSION TABLES TO CONVERT FROM THE ENGLISH SYSTEM TO THE METRIC SYSTEM.
3. CONVERT BETWEEN FAHRENHEIT AND CELSIUS TEMPERATURE SCALES.

Metric–English Conversion Tables (2)

These tables are *not* to be memorized.

Table I
Conversions of Length

Metric to English	English to Metric
1 km = .621 mi	1 mi = 1.61 km
1 m = 1.09 yd	1 yd = .914 m
= 3.28 ft	= 91.4 cm
= 39.37 in	1 ft = .305 m
1 cm = .394 in	= 30.5 cm
1 mm = .039 in	1 in = 2.54 cm

Metric–English Conversion Tables (Continued)

Table II Conversions of Volume	
Metric to English	English to Metric
1 liter = .264 gal = 1.06 qt = 2.11 pt 1 mL = .034 fl oz 1 cc = .034 fl oz	1 gal = 3.785 L 1 qt = .946 L 1 pt = .473 L 1 fl oz = 29.6 mL = 29.6 cc

Table III Conversions of Weight	
Metric to English	English to Metric
1 metric ton = 2,204.62 lb 1 kg = 2.20 lb = 35.3 oz 1 g = .035 oz	1 short ton = .907 met ton 1 lb = .454 kg 1 oz = 28.3 g

Converting from Metric to English (3)

To convert from the metric system to the English system, proceed to the left side of the appropriate table in frame 2 and multiply by the given conversion factor.

Example: Convert 5 km to miles:

Solution: Consult Table I and find that 1 km = .621 mi

Line (a): 5 km = (5)(.621 mi)

Line (b): = 3.105 mi

Computation

$$\begin{array}{r} .621 \\ \times 5 \\ \hline 3.105 \end{array}$$

More Examples (4)

Example 1: A bottle contains 737 g of catsup. How much is this in ounces? Round your answer to the nearest tenth of an ounce.

Solution: Consult Table III and find that 1 g = .035 oz

Line (a): 737 g = (737)(.035 oz)

Line (b): = 25.795 oz

Therefore, 737 g = 25.8 oz (to nearest tenth)

Computation

$$\begin{array}{r} 737 \\ \times .035 \\ \hline 3\ 685 \\ 22\ 11 \\ \hline 25.795 \end{array}$$

Example 2: A small bottle contains 55 mL of nasal spray. What is the contents in fluid ounces? **(5)**

Solution: Consult Table II and find that 1 mL = .034 fl oz

Line (a): 55 mL = (55)(.034 fl oz)

Line (b): = 1.87 fl oz

Computation

```
  .034
   ×55
   170
  1 70
 1.870
```

(6)

Example 3: A speed limit sign in Paris, France, says 30 km/hr. What is the speed limit to the nearest mile per hour?

Solution: Consult Table I and find that 1 km = .621 mi

Line (a): 30 km = (30)(.621 mi)

Line (b): = 18.63 mi

Therefore, 30 km/hr = 19 mph (to nearest mile per hour)

Computation

```
  .621
   ×30
18.630
```

Study Exercise One **(7)**

Perform the following conversions. Use the tables in frame 2.

1. A certain individual weighs 75 kg. What is his weight in pounds?
2. A girl is 155 cm tall. What is her height to the nearest inch?
3. You are driving a European sports car through a 25 mph speed zone. If your speedometer registers 35 km/hr, are you breaking the speed limit?
4. Which holds more fruit juice, a bottle containing 16 fl oz or a bottle containing 450 mL?
5. A certain motion picture camera uses 16 mm film. Give the film size in inches.
6. A can of corn is marked 200 grams. How much is this to the nearest whole ounce?

Converting from English to Metric **(8)**

To convert from the English system to the metric system, proceed to the right side of the appropriate table in frame 2 and multiply by the given conversion factor.

Example: Convert 2.5 fluid ounces to cubic centimeters.

Solution: Consult Table II and find that 1 fl oz = 29.6 cc

Line (a): 2.5 fl oz = (2.5)(29.6 cc)

Line (b): = 74 cc

Computation

```
  29.6
   2.5
 14 80
 59 2
 74.00
```

More Examples (9)

Example 1: A football player weighs 275 pounds. What is his weight in kilograms?

Solution: Consult Table III and observe that 1 lb = .454 kg

Line (a): 275 lb = (275)(.454 kg)

Line (b): = 124.85 kg

Computation

```
    .454
     275
  ------
   2 270
  31 78
  90 8
  ------
 124.850
```

Example 2: A girl is 5 ft 4 in tall. What is her height to the nearest centimeter? (10)

Solution: Convert 5 ft 4 in to inches; then consult Table I and find that 1 in = 2.54 cm.

Line (a): 5 ft 4 in = (5)(12 in) + 4 in

Line (b): = 60 in + 4 in

Line (c): = 64 in

Line (d): 64 in = (64)(2.54 cm)

Line (e): = 162.56 cm

Therefore, the girl is 163 cm tall.

Computation

```
   2.54
     64
 ------
  10 16
 152 4
 ------
 162.56
```

Example 3: A box contains 2 lb 4 oz of cereal. What is its weight to the nearest gram? (11)

Solution: Convert 2 lb 4 oz to ounces; then consult Table III and find that 1 oz = 28.3 g.

Line (a): 2 lb 4 oz = (2)(16 oz) + 4 oz

Line (b): = 32 oz + 4 oz

Line (c): = 36 oz

Line (d): 36 oz = (36)(28.3 g)

Line (e): = 1,018.8 g

Therefore, the box contains 1,019 grams of cereal.

Computation

```
    28.3
     3 6
 -------
   169 8
   849
 -------
 1,018.8
```

Study Exercise Two (12)

1. A drinking glass will hold 8 fl oz. What is its capacity in cubic centimeters?
2. A basketball player is 6 ft 11 in tall. What is his height to the nearest centimeter?
3. A candy bar weighs 3.5 oz. What does it weigh to the nearest gram?
4. An airplane is traveling 640 mph. What is its speed to the nearest kilometer per hour?
5. The capacity of a certain automobile's gas tank is 22 gal. What is the capacity to the nearest liter?
6. Which holds more orange juice, an 18 fl oz can or a 530 mL can?

Temperature (13)

The English system of measurement generally measures temperature by degrees Fahrenheit, °F. The metric system generally measures temperature by degrees Celsius (Centigrade), °C.

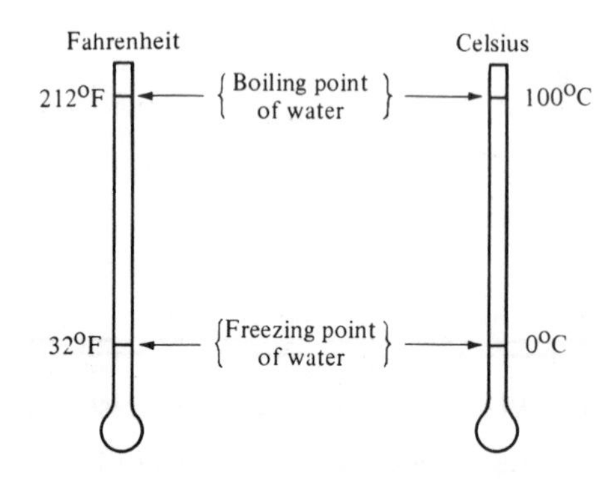

Comparison of Fahrenheit and Celsius Scales (14)

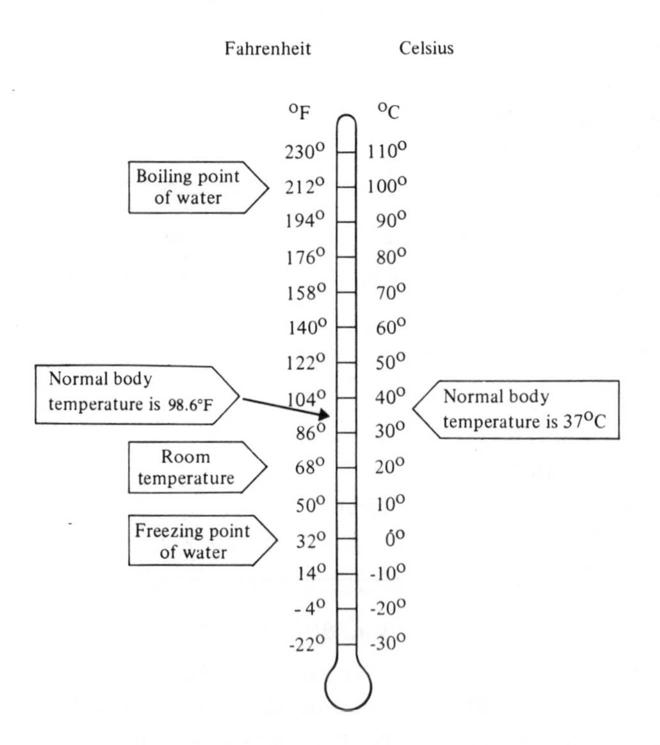

Conversion Between Fahrenheit and Celsius (15)

Case (1): To convert from Fahrenheit to Celsius:

$$C = \frac{5}{9}(F - 32)$$

Conversion Between Fahrenheit and Celsius (Continued)

Example: Convert 68°F to Celsius:

Solution: $C = \frac{5}{9}(68 - 32)$

$= \frac{5}{\cancel{9}_1} \cdot \cancel{36}^4$

$= 20°C$

Case (2): To convert from Celsius to Fahrenheit:

$$F = \frac{9}{5}C + 32$$

Example: Convert 30°C to Fahrenheit:

Solution: $F = \left(\frac{9}{5} \cdot 30\right) + 32$

$= 54 + 32$

$= 86°F$

Study Exercise Three (16)

A. Use the diagram in frame 14 to answer the following questions:

1. If the outdoor temperature were 40°C, would it be a chilly day or a hot day?
2. Will water freeze at a temperature of 32°C?
3. Would it snow at an outdoor temperature of 10°C?
4. A person's body temperature is 40°C. Is this person ill?

B. Use the conversion formulas in frame 15 to answer the following questions:

5. If an oven thermostat is set at 275°F, what is the setting in terms of degrees Celsius?
6. The melting point of iron is 1,535°C. What is the melting point in degrees Fahrenheit?
7. A metric cake recipe indicates that the cake is to be baked 30–35 min at 175°C. To the nearest 10 degrees, where should a Fahrenheit oven dial be set?

REVIEW EXERCISES (17)

A. Convert the following from metric to English. Use Tables I, II, and III in frame 2.

1. 12 km = _________ mi
2. 5 m = _________ ft
3. 20 cm = _________ in
4. 3 L = _________ gal
5. 100 mL = _________ fl oz
6. 30 cc = _________ fl oz
7. 5 metric ton = _________ lb
8. 60 kg = _________ lb
9. 2 kg = _________ oz
10. 200 g = _________ oz

B. Convert the following from English to metric. Use Tables I, II, and III in frame 2.

11. 5 mi = _________ km
12. 4 yd = _________ m
13. 2 ft = _________ cm
14. 6 in = _________ cm
15. 10 gal = _________ L
16. 3 qt = _________ L
17. 8 fl oz = _________ mL
18. 12 fl oz = _________ cc
19. 3 short ton = _________ metric ton
20. 5 oz = _________ g

REVIEW EXERCISES (Continued)

C. Convert the following temperatures using the formulas in frame 15:

21. Convert 113°F to Celsius. **22.** Convert 15°C to Fahrenheit.

D. Solve the following problems using Tables I, II, and III in frame 2:

23. The distance between two cities on a German road map is given as 60 km. How far is this in miles?

24. A speed limit sign in Mexico City says 40 km/hr. What is the speed limit to the nearest mile per hour?

25. A motion picture camera uses 8-mm film. What is the film size in inches?

26. A boy is 175 cm tall. What is his height to the nearest inch?

27. A bottle contains 250 mL of cough syrup. How much does it contain in fluid ounces?

28. Which has the greater capacity, a glass which holds 8 fl oz or a glass which holds 300 cm^3?

29. A girl weighs 100 lbs. What is her weight in kilograms?

30. A board is 2 ft 3 in long. What is its length in centimeters?

31. To the nearest whole liter, what is the capacity of a 12 gal gasoline tank?

32. Which is the better buy, gasoline at 40.9¢ per L or 154.9¢ per gal?

33. An automobile part is to be manufactured to a tolerance of .02 mm. Express this tolerance in inches.

34. It takes Jane the same amount of time to run a 1,000-meter race as it takes Mary to run a half-mile race. Who is the faster runner?

35. Make the following metric conversions for the game of baseball:

A. Distance between bases–90 ft = _________ m.

B. First, second, third bases–15 in by 15 in = _________ cm by _________cm.

C. Pitcher's plate to home base–60 ft 6 in = _________ m.

D. Batter's boxes–4 ft by 6 ft = _________ m by _________ m.

36. The following ingredients appear in an Austrian recipe for dumplings. Make the proper conversions.

A. 50 g butter = _________ oz.

B. 120 g sugar = _________ oz.

C. 150 g bread = _________ oz.

D. 20 g flour = _________ oz.

E. 300 g cottage cheese = _________ oz.

F. $\frac{1}{8}$L sour cream = _________ fl oz.

37. Bob found that it took 41.7 L of gasoline for his car to travel 231 miles. To the nearest whole number, how much gasoline did he use and how many miles did he get to the gallon?

Solutions to Review Exercises (18)

A. **1.** 12 km = (12)(.621 mi)
= 7.452 mi

2. 5 m = (5)(3.28 ft)
= 16.4 ft

3. 20 cm = (20)(.394 in)
= 7.88 in

4. 3 L = (3)(.264 gal)
= .792 gal

5. 100 mL = (100)(.034 fl oz)
= 3.4 fl oz

6. 30 cc = (30)(.034 fl oz)
= 1.02 fl oz

7. 5 metric ton = (5)(2204.62 lb)
= 11,023.1 lb

8. 60 kg = (60)(2.20 lb)
= 132 lb

9. 2 kg = (2)(35.3 oz)
= 70.6 oz

10. 200 g = (200)(.035 oz)
= 7 oz

Solutions to Review Exercises (Continued)

B. **11.** 5 mi = (5)(1.61 km)
= 8.05 km

12. 4 yd = (4)(.914 m)
= 3.656 m

13. 2 ft = (2)(30.5 cm)
= 61 cm

14. 6 in = (6)(2.54 cm)
= 15.24 cm

15. 10 gal = (10)(3.785 L)
= 37.85 L

16. 3 qt = (3)(.946 L)
= 2.838 L

17. 8 fl oz = (8)(29.6 mL)
= 236.8 mL

18. 12 fl oz = (12)(29.6 cc)
= 355.2 cc

19. 3 short ton = (3)(.907 metric ton)
= 2.721 metric ton

20. 5 oz = (5)(28.3 g)
= 141.5 g

C. **21.** $C = \frac{5}{9}(F - 32)$

$= \frac{5}{9}(113 - 32)$

$= \frac{5}{9} \cdot 81$

$= 45°C$

22. $F = \frac{9}{5}C + 32$

$= \left(\frac{9}{5} \cdot 15\right) + 32$

$= (9 \cdot 3) + 32$

$= 27 + 32$

$= 59°F$

D. **23.** 60 km = (60)(.621 mi)
= 37.26 mi

24. 40 km/m = (40)(.621 mph)
= 24.84 mph
= 25 mph (to nearest mph)

25. 8 mm = (8)(.039 in)
= .312 in

26. 175 cm = (175)(.394 in)
= 68.95 in
= 69 in (to nearest inch)

27. 250 mL = (250)(.034 fl oz)
= 8.5 fl oz

28. 8 fl oz = (8)(29.6 cc)
= 236.8 cc

A 300 cc glass has a greater capacity.

29. 100 lb = (100)(.454 kg)
= 45.4 kg

30. 2 ft 3 in = (2)(12 in) + 3 in
= 24 in + 3 in
= 27 in

27 in = (27)(2.54 cm)
= 68.58 cm

31. 12 gallons = (12)(3.785 L) = 45.42 L $\approx$ 45 L

32. 1 gal = 3.785 L; therefore, we must multiply 3.785 L of gasoline at 40.9¢ to find the price per gallon. (3.785)(40.9¢) = 154.8065¢. This is slightly better than 154.9¢ per gallon.

33. .02 mm = (.02)(.039 in) = 0.00078 in.

34. One half mile $= \frac{5{,}280}{2}$ ft = 2,640 ft

1,000 meters = (1,000)(3.28 ft) = 3,280 ft

Since Jane ran farther than Mary in the same amount of time, Jane is faster.

35. **A.** 90 ft = (90)(.305 m) = 27.45 m

B. 15 in × 15 in = (15)(2.54 cm) × (15)(2.54 cm) = 38.1 cm × 38.1 cm

C. 60 ft 6 in = 60.5 ft = (60.5)(.305 m) = 18.4525 m

D. 4 ft = (4)(.305 m) = 1.22 m
6 ft = (6)(.305 m) = 1.83 m

Solutions to Review Exercises (Continued)

36. **A.** butter; 50 g = (50)(.035 oz) = 1.75 oz
B. sugar; 120 g = (120)(.035 oz) = 4.2 oz
C. bread; 150 g = (150)(.035 oz) = 5.25 oz
D. flour; 20 g = (20)(.035 oz) = 0.7 oz
E. cottage cheese; 300 g = (300)(.035 oz) = 10.5 oz
F. sour cream; $\frac{1}{8}$L = $\left(\frac{1}{8}\right)$(1,000 mL) = 125 mL
125 mL = (125)(.034 fl oz) = 4.25 fl oz

37. 41.7 L = (41.7)(.264 gal) = 11.0088 gal ≈ 11 gal
231 miles ÷ 11 gal $\doteq$ 21 miles per gallon

SUPPLEMENTARY PROBLEMS

A. Convert the following from metric to English. Use Tables I, II, and III in frame 2.

1. 10 km = ________ mi **2.** 4 m = ________ yd **3.** 10 mm = ________ in
4. 5 L = ________ gal **5.** 10 mL = ________ fl oz **6.** 50 cc = ________ fl oz
7. 3 L = ________ pt **8.** 70 kg = ________ lb **9.** 3 kg = ________ oz
10. 6 g = ________ oz

B. Convert the following from English to metric. Use Tables I, II, and III in frame 2.

11. 2 mi = ________ km **12.** 5 yd = ________ cm
13. 4 ft = ________ cm **14.** 5 in = ________ cm
15. 2 gal = ________ L **16.** 2 qt = ________ L
17. 5 fl oz = ________ cm^3 **18.** 4 short ton = ________ metric ton
19. 5 lb = ________ kg **20.** 2 oz = ________ g

C. Solve the following problems using Tables I, II, and III in frame 2, or the temperature conversion formulas in frame 15:

21. A box of cereal weighs 325 g. What is its weight in ounces?
22. A room is 4.5 m in length. What is its length in feet?
23. A girl weighs 48 kg. What is her weight to the nearest pound?
24. Which is longer, the 1,000 meter or the half-mile run?
25. A candy bar weighs 2.5 oz. What is its weight to the nearest gram?
26. An airplane is traveling 500 mph. What is its speed in kilometers per hour?
27. A boy is 6 ft 2 in tall. How tall is he to the nearest centimeter?
28. To the nearest whole liter, what is the capacity of a 13 gal gasoline tank?
29. Which is the better buy, gasoline at 37.9¢ per L or 143.9¢ per gal?
30. An automobile part is to be manufactured to a tolerance of .03 mm. Express this tolerance in inches.
31. The distance between two cities on a French road map is given as 90 km. How far is this in miles (to the nearest whole mile)?
32. A speed limit sign in Mexico City says 50 km/hr. What is the speed limit to the nearest mile per hour?
33. A girl is 170 cm tall. What is her height to the nearest inch?
34. One automobile averages 17 mi per gallon, while another averages 8 km per liter. Which provides the better fuel economy?

SUPPLEMENTARY PROBLEMS (Continued)

35. The weight of a package is 1 lb 4 oz. What is its weight to the nearest gram?
36. The outside temperature is 15°C. What is the temperature in Fahrenheit?
37. A person has a fever of 104°F. What is the temperature in Celsuis?
38. A camera has a 35-mm lens. How many inches is this?
39. A golf drive was 165 yd. Change this to meters.
40. The thickness of some sheet metal is .0125 in. Change this to millimeters.
41. The radiator of an automobile holds 22 qt. What is its capacity in liters?
42. A man weighs 84 kilograms. What is his weight in pounds?
43. The Leaning Tower of Pisa is about 179 ft high. How high is it in meters?
44. The Eiffel Tower in Paris is about 300 m high. How high is this in feet?
45. Make the following metric conversion for the game of baseball:

A. Home base–17 in × 17 in = _________ cm by _________ cm.

B. Coaches' boxes–20 ft by 10 ft = _________ m by _________ m.

C. Distance from home base to batters boxes–6 in = _________ cm.

D. Pitcher's plate–24 in by 6 in = _________ cm by _________ cm.

E. Pitcher's plate to grass line–95 ft = _________ m.

46. The following ingredients appear in an Austrian recipe for apricot dumplings. Make the proper conversions.

A. 1 kg potatoes = _________ lb.

B. 350 g flour = _________ oz.

C. 10 g butter = _________ oz.

D. 500 g apricots = _________ oz.

E. 100 g bread crumbs = _________ oz.

F. 80 g sugar = _________ oz.

G. 15 mL rum = _________ fl oz.

Solutions to Study Exercises (7A)

Study Exercise One (Frame 7)

1. 75 kg = (75)(2.20 lb)
= 165 lb

2. 155 cm = (155)(.394 in)
= 61.07 in
= 61 in (to nearest inch)

3. 35 km/hr = (35)(.621 mph)
= 21.735 mph
No, you are not breaking the 25 mph speed limit.

4. 450 mL = (450)(.034 oz)
= 15.3 oz
The bottle containing 16 fl oz holds more fruit juice.

5. 16 mm = (16)(.039 in)
= .624 in

6. 200 g = (200)(.035 oz) = 7 oz.

Study Exercise Two (Frame 12) (12A)

1. 8 fl oz = (8)(29.6 cc)
= 236.8 cc

2. 6 ft 11 in = (6)(12 in) + 11 in
= 72 in + 11 in
= 83 in
83 in = (83)(2.54 cm)
= 210.82 cm
= 211 cm (to nearest centimeter)

Study Exercise Two (Frame 12) (Continued)

3. 3.5 oz = (3.5)(28.3 g)
= 99.05 g
= 99 g (to nearest gram)

4. 640 mph = (640)(1.61 km/hr)
= 1030.4 km/hr
= 1030 km/hr (to nearest km/hr)

5. 22 gal = (22)(3.785 L)
= 83.270 L
= 83 L (to the nearest liter)

6. 18 fl oz = (18)(29.6 mL) = 532.8 mL
The 18 fl oz can holds more orange juice.

Study Exercise Three (Frame 16) (16A)

A. 1. 40°C corresponds to 104°F. Thus, it would be a very hot day.
2. 32°C would be slightly more than 86°F. This is obviously far too warm for water to freeze.
3. 10°C corresponds to 50°F. This is far too warm for snow.
4. 40°C corresponds to 104°F. Since normal body temperature is 98.6°F, this person has a very high temperature and is quite ill.

B. 5. $C = \frac{5}{9}(F - 32)$

$= \frac{5}{9}(275 - 32)$

$= \frac{5}{9} \cdot 243$

$= 135°C$

6. $F = \frac{9}{5}C + 32$

$= \left(\frac{9}{5} \cdot 1{,}535\right) + 32$

$= 2{,}763 + 32$

$= 2{,}795°F$

7. $F = \frac{9}{5}C + 32$

$= \left(\frac{9}{5} \cdot 175\right) + 32$

$= 315 + 32$

$= 347°F$ To the nearest 10 degrees, the dial should be set at 350°F.

UNIT

36 Basic Geometry

OBJECTIVES (1)

By the end of this unit you should:

1. BE FAMILIAR WITH THE BASIC GEOMETRIC TERMS OF *POINT, LINE, LINE SEGMENT, RAY,* AND *ANGLE*.
2. BE ABLE TO CLASSIFY ANGLES AS ACUTE, RIGHT, OR OBTUSE.
3. BE ABLE TO RECOGNIZE POLYGONS AND CLASSIFY THEM ACCORDING TO THE NUMBER OF SIDES.
4. BE ABLE TO CLASSIFY TRIANGLES AS SCALENE, EQUILATERAL, ISOSCELES, OR RIGHT.
5. BE ABLE TO USE THE PYTHAGOREAN THEOREM TO FIND THE LENGTH OF THE HYPOTENUSE IN A RIGHT TRIANGLE.

Basic Geometric Terms (2)

1. Point

   Example: • A

2. Line

   Example: A B AB

3. Line segment

   Example:

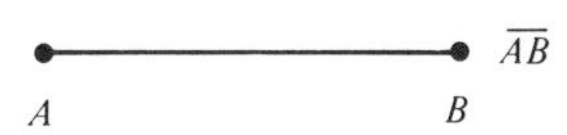

4. Ray

   Example:

Angles (3)

An angle is formed by two rays sharing the same endpoint, called the *vertex*.

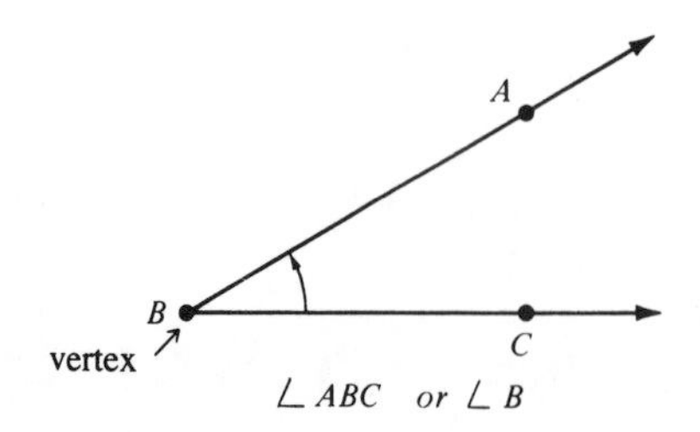

Measuring Angles (4)

Angles are measured in degrees, minutes, and seconds.

1 degree (1°) = 60 minutes $(60')$

1 minute $(1')$ = 60 seconds $(60'')$

One complete rotation of a ray about its endpoint is an angle of 360°.

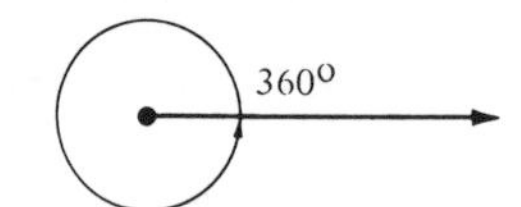

A straight angle contains 180°.

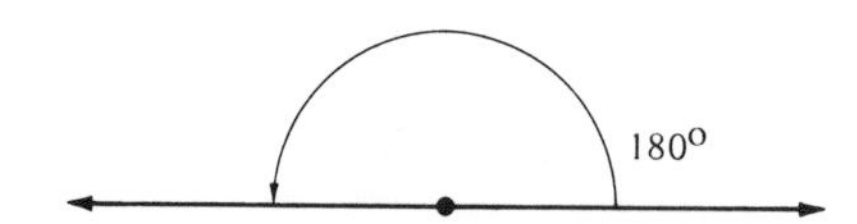

A right angle contains 90°. The sides are said to be *perpendicular.*

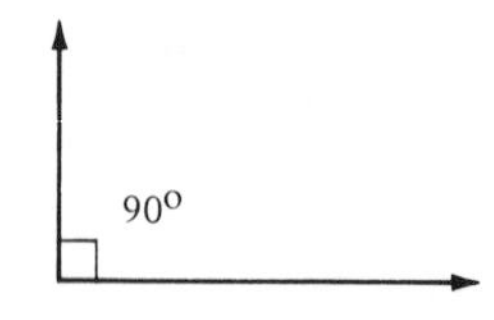

Types of Angles **(5)**

Angles may be classified as acute, right, or obtuse.

1. An acute angle measures less than 90°.

   Example:

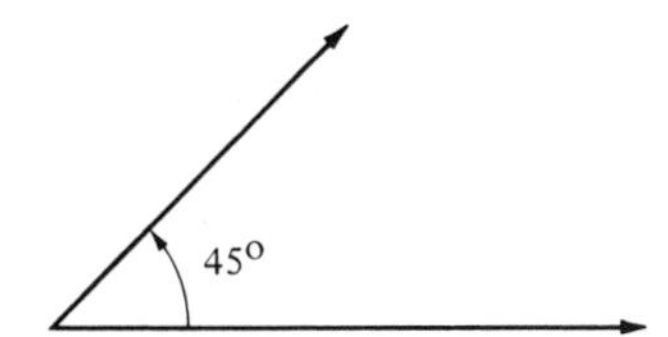

2. A right angle measures 90°.

   Example:

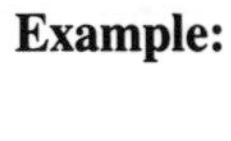

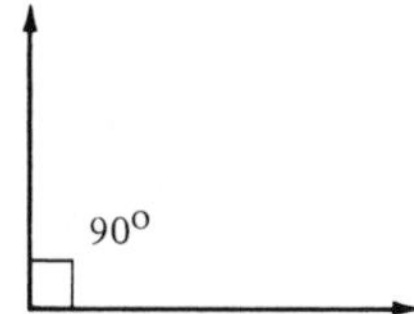

3. An obtuse angle measures more than 90° and less than 180°.

   Example:

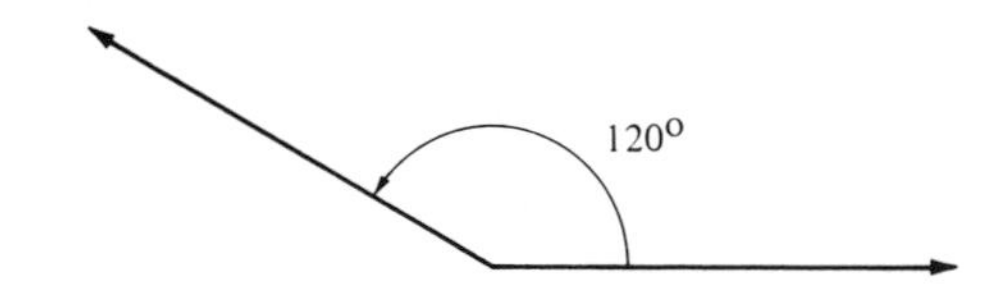

Study Exercise One **(6)**

A. Classify the following angles as acute, right, obtuse, or straight:

1.

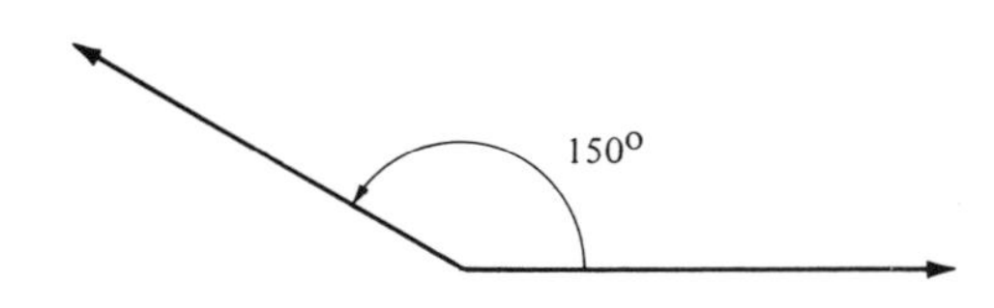

2.

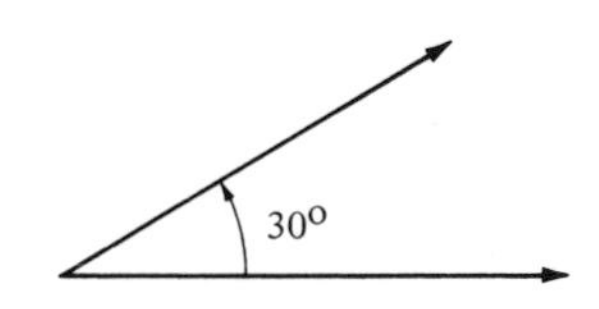

3.

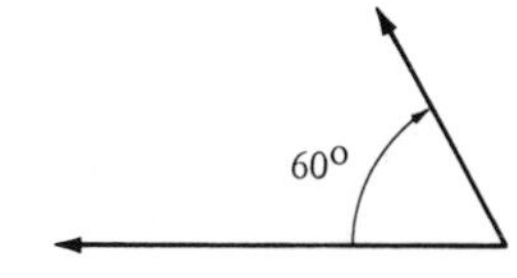

4.

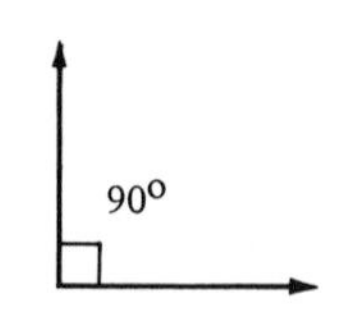

5.

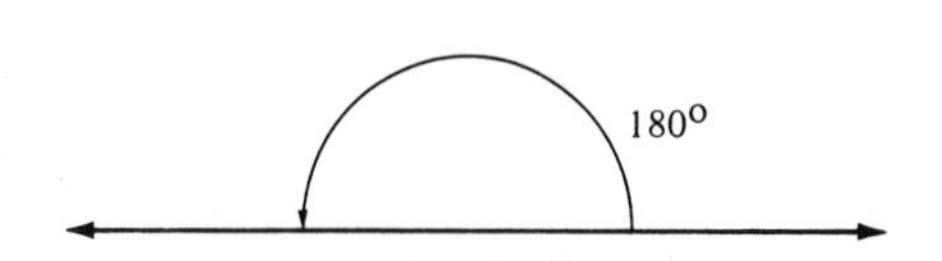

B. Complete the following statements:

6. In a right angle the sides are said to be _________ .

7. 1 degree = _________ minutes.

8. 5 degrees = _________ minutes.

***Study Exercise One* (Continued)**

9. An angle is formed by two ________ sharing the same endpoint, called the ________.
10. An acute angle measures less than ________ degrees.
11. An obtuse angle measures more than ________ degrees and less than ________ degrees.

Polygons (7)

A polygon is a plane figure composed of line segments which begin at a given point and end at the same point without crossing. Each of the following figures is a polygon:

Example 1:

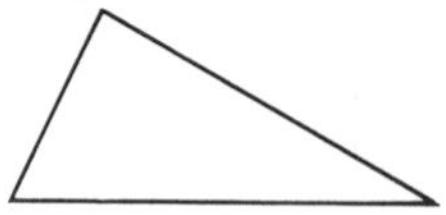

Example 2:

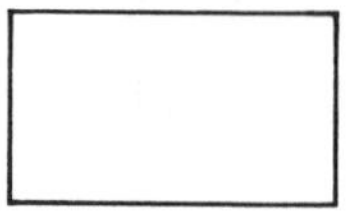

Example 3:

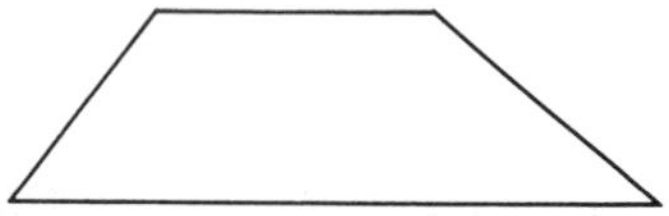

Example 4:

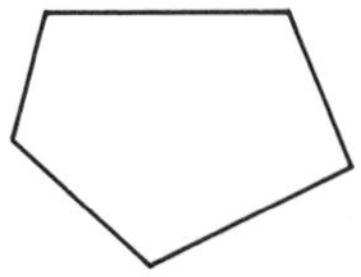

Examples of Figures that are *NOT* Polygons (8)

The following figures are not polygons:

Example 1: A circle is not composed of line segments.

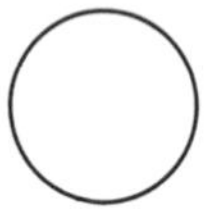

Example 2: A polygon cannot have crossing line segments.

Example 3: This figure is three dimensional and is therefore not a plane figure.

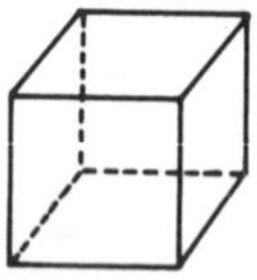

Example 4: This figure is not composed entirely of straight line segments.

Types of Polygons (9)

Polygons are classified according to the number of sides. The following are some of the more familiar polygons:

Number of Sides	Name
3	triangle
4	quadrilateral
5	pentagon
6	hexagon
8	octagon
10	decagon

Throughout the remainder of the unit, we will be especially interested in triangles and quadrilaterals.

Study Exercise Two (10)

A. Complete the following table referring to types of polygons:

	Number of Sides	Name
1.		quadrilateral
2.		hexagon
3.	10	
4.	3	
5.		octagon
6.	5	

B. Which of the following are polygons? (answer yes or no)

7.

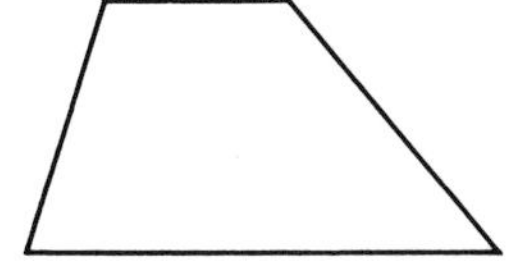

8.

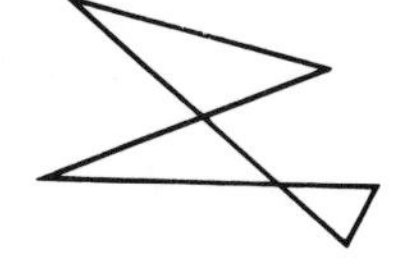

9.

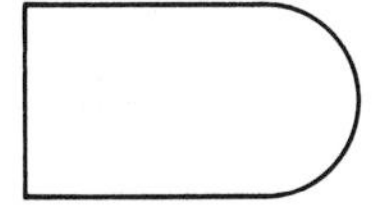

Types of Polygons (Continued)

10.

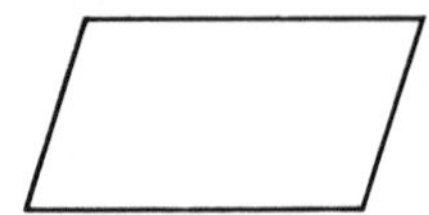

11.

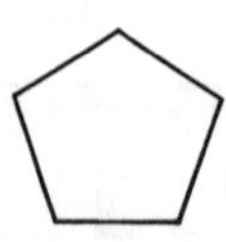

12.

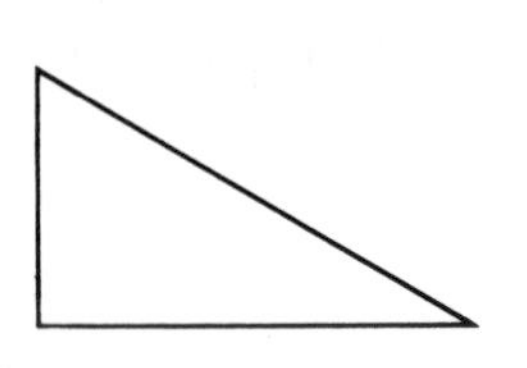

Triangles (11)

A triangle is a polygon having three sides and three interior angles.

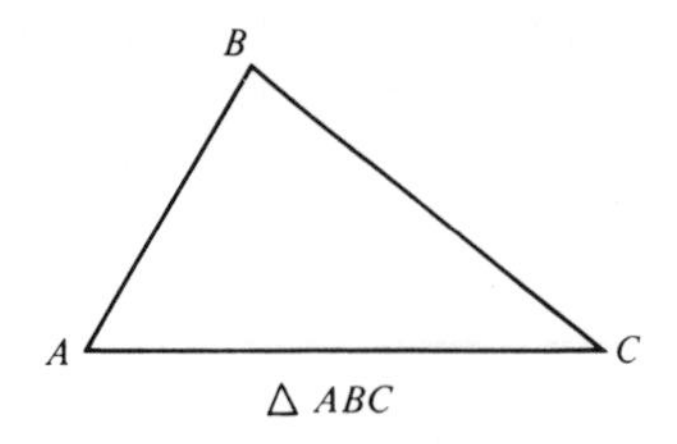

$\triangle ABC$

Points A, B, and C are said to be the *vertices* of the triangle.

Types of Triangles (12)

1. A scalene triangle has no equal sides and no equal angles.

   Example:

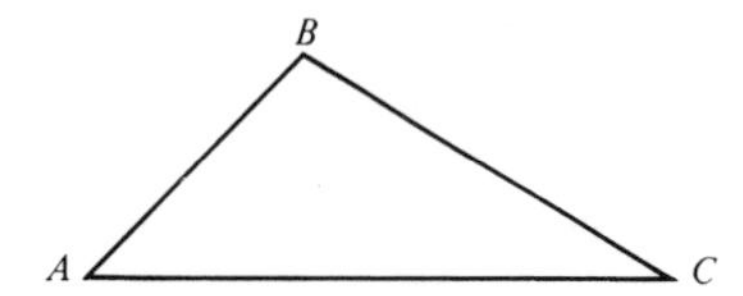

2. An equilateral triangle has three equal sides and three equal angles.

   Example:

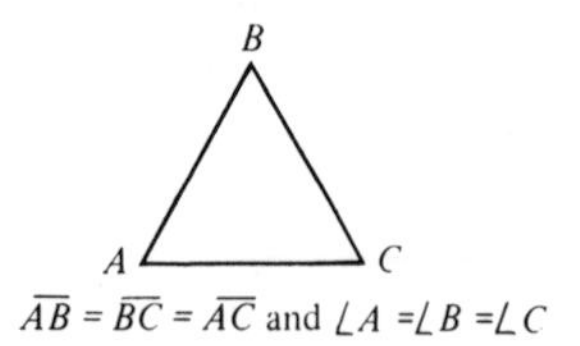

$\overline{AB} = \overline{BC} = \overline{AC}$ and $\angle A = \angle B = \angle C$

3. An isosceles triangle has two equal sides and, consequently, two equal angles.

   Example:

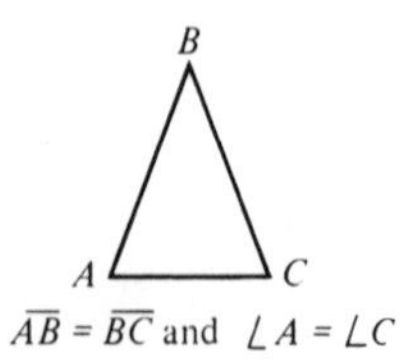

$\overline{AB} = \overline{BC}$ and $\angle A = \angle C$

Note: An equilateral triangle is also an isosceles triangle since it has at least two equal sides.

Triangles (Continued)

4. A right triangle has one right angle.

   Example:

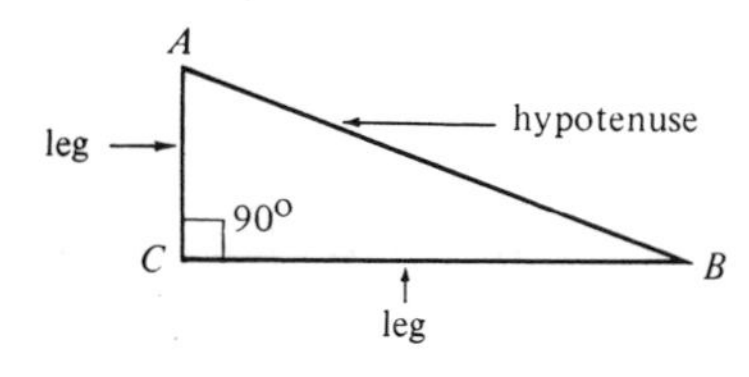

Sum of the Interior Angles (13)

The sum of the interior angles of a triangle is 180°.

Example 1: Find the measure of $\angle C$:

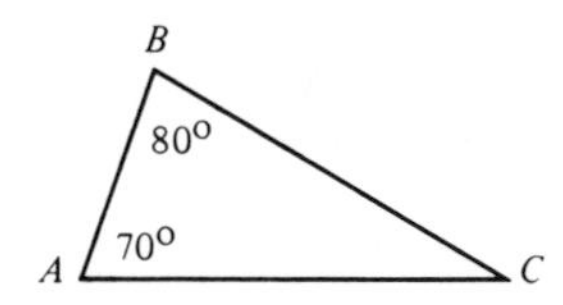

Solution: $\angle C = 180° - (80° + 70°) = 30°$

Example 2: Find the measure of $\angle A$ if $\triangle ABC$ is a right triangle:

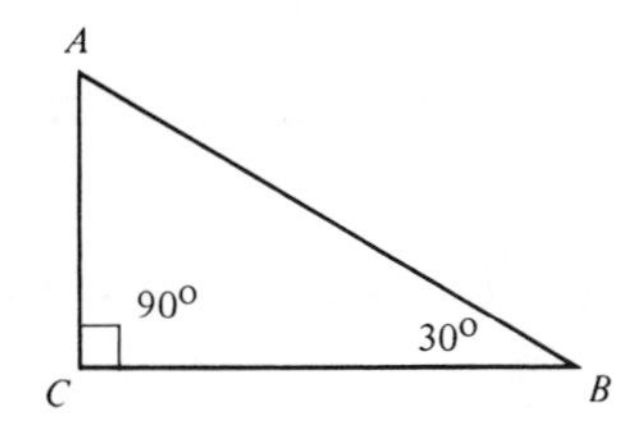

Solution: The sum of two acute angles in a right triangle is 90°. Therefore, $\angle A = 90° - 30° = 60°$.

Study Exercise Three (14)

A. Find the measure of the unknown angle:

1. Find the measure of $\angle A$:

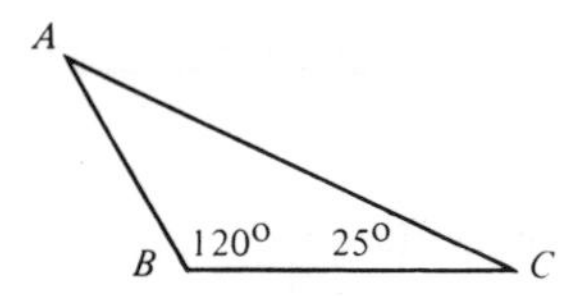

2. Find the measure of $\angle B$ if $\triangle ABC$ is a right triangle:

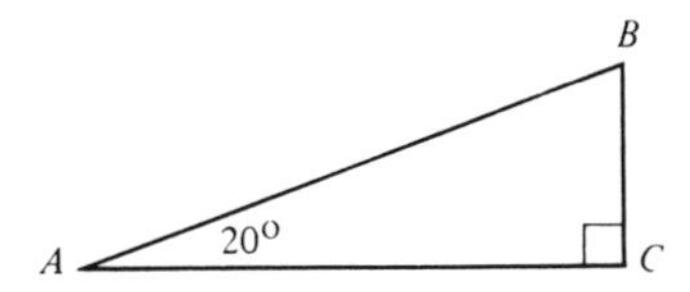

***Study Exercise Three* (Continued)**

B. Classify each of the following triangles as scalene, equilateral, isosceles, or right. (Some triangles may be classified as more than one type.)

3.

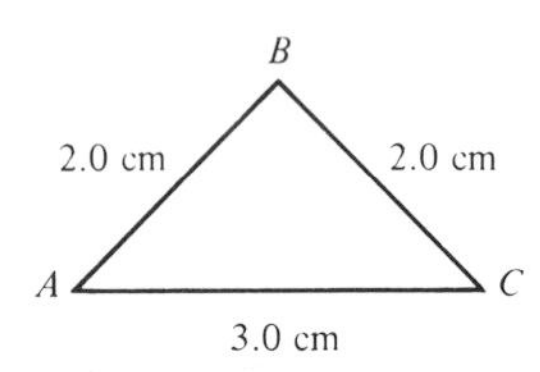

4.

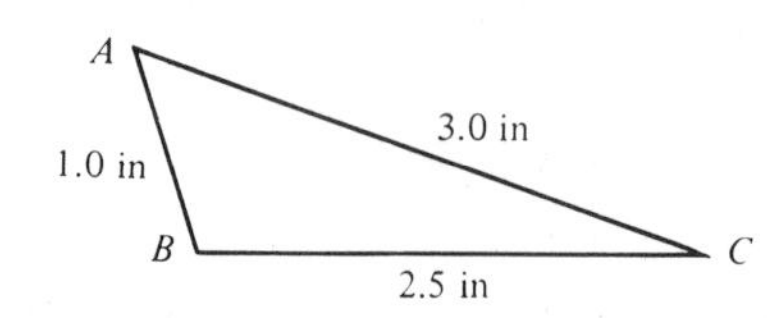

5.

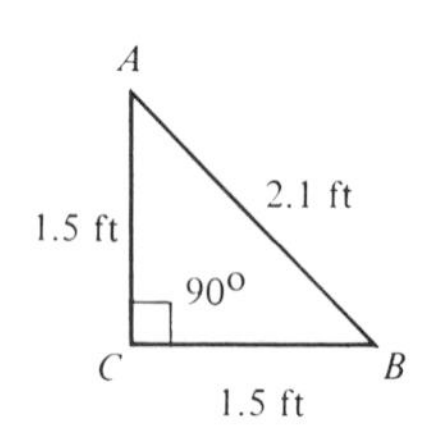

6.

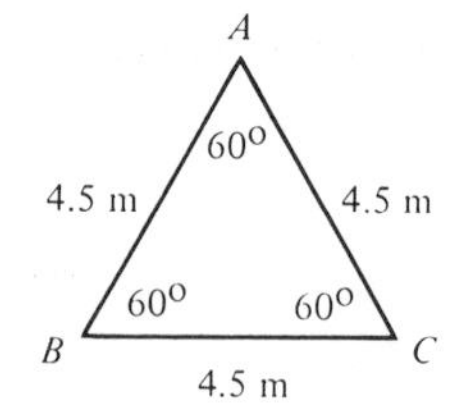

C. Complete the following statements:

7. A triangle with no equal sides and no equal angles is called a(n) __________ triangle.

8. Every equilateral triangle may also be considered a(n) __________ triangle.

9. The sum of the interior angles of a triangle is __________ degrees.

The Pythagorean Theorem (15)

In a right triangle, the square of the hypotenuse is equal to the sum of the squares of the two legs.

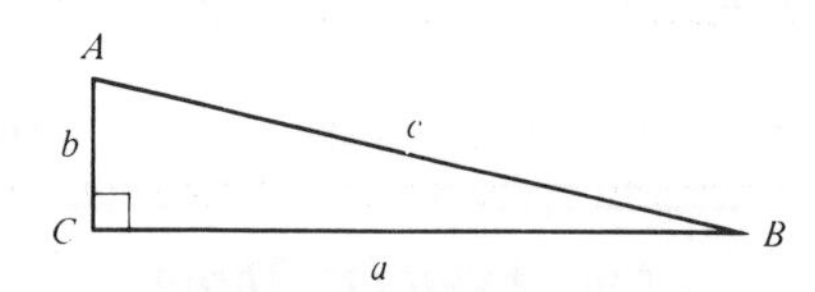

Pythagorean theorem: $c^2 = a^2 + b^2$

Remember, this theorem is valid only for right triangles.

An Example of the Pythagorean Theorem (16)

Example 1: Find the length of the hypotenuse for the following right triangle:

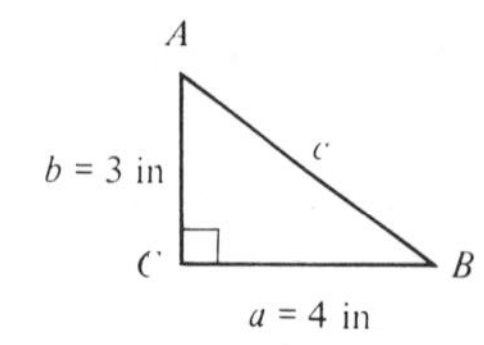

An Example of the Pythagorean Theorum (Continued)

Solution:

Line (a): $c^2 = a^2 + b^2$
Line (b): $= 4^2 + 3^2$
Line (c): $= 16 + 9$
Line (d): $= 25$
Line (e): $c = \sqrt{25}$
Line (f): $c = 5$ in

Therefore, the length of the hypotenuse is 5 inches.

An Example Using Metric Measurement (17)

Example 2: Find the length of the hypotenuse for the following right triangle:

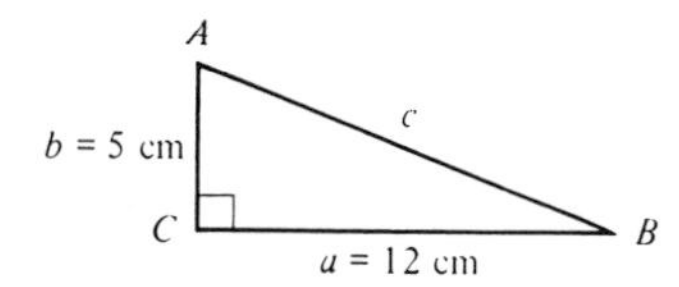

Solution:

Line (a): $c^2 = a^2 + b^2$
Line (b): $= 12^2 + 5^2$
Line (c): $= 144 + 25$
Line (d): $= 169$
Line (e): $c = \sqrt{169}$
Line (f): $c = 13$ cm

Therefore, the length of the hypotenuse is 13 centimeters.

An Example Requiring a Table (18)

Example 3: Find the length of the hypotenuse for the following right triangle:

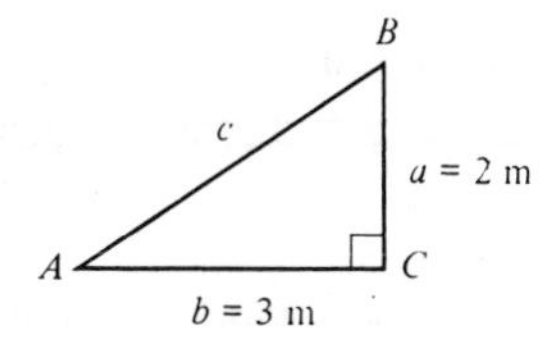

Solution:

Line (a): $c^2 = a^2 + b^2$
Line (b): $= 2^2 + 3^2$
Line (c): $= 4 + 9$
Line (d): $= 13$
Line (e): $c = \sqrt{13}$ (Consult Powers and Roots Table)*
Line (f): $c \approx 3.606$ m

Therefore, the length of the hypotenuse is 3.606 meters.

**Note:* Problems using the Pythagorean Theorem don't always result in perfect squares. Square roots of the whole numbers from 0 to 100 are found in the Powers and Roots Table in Appendix I.

Study Exercise Four (19)

Use the Pythagorean Theorem and the Powers and Roots Table to find the length of the hypotenuse for each of the following right triangles:

1.

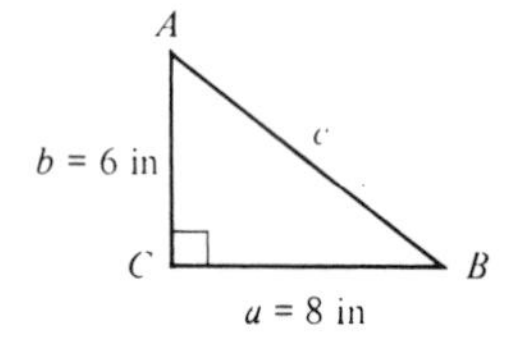

2.

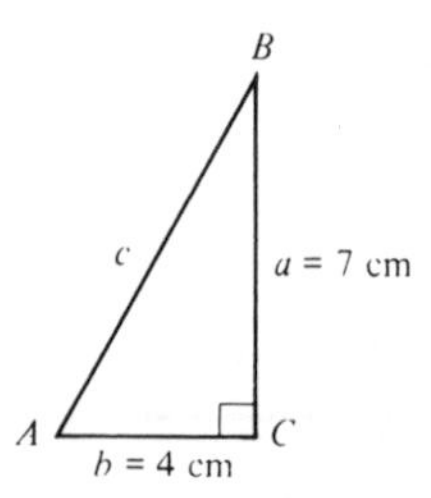

3.

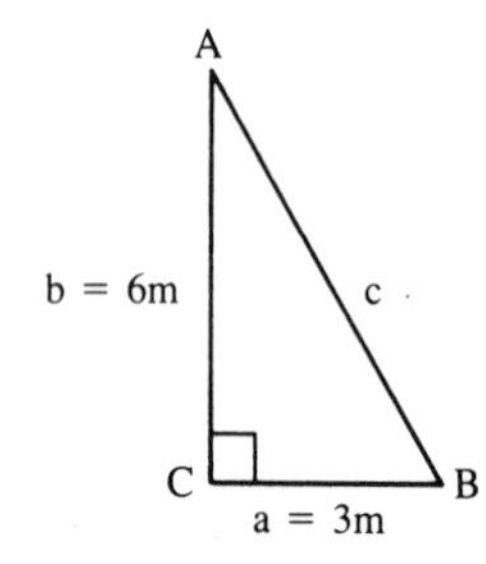

4.

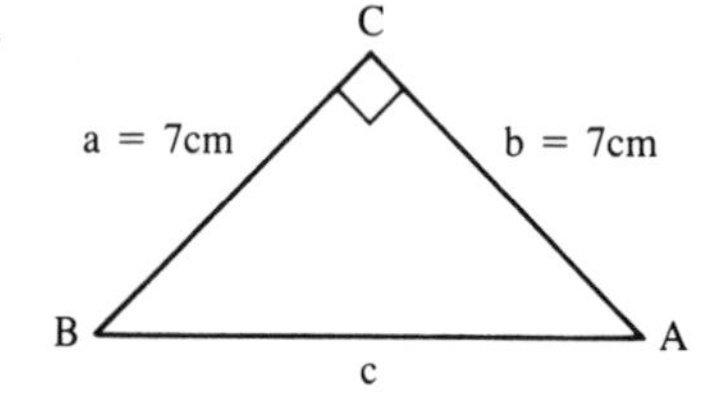

Quadrilaterals (20)

Recall that a quadrilateral is a polygon having four sides.

Example:

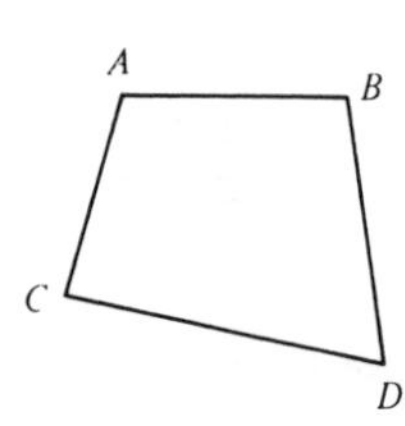

Points A, B, C, and D are said to be the *vertices* of the quadrilateral.

Special Quadrilaterals (21)

In the next unit we will be discussing areas and perimeters of triangles as well as the following special types of quadrilaterals.

1. Parallelogram:

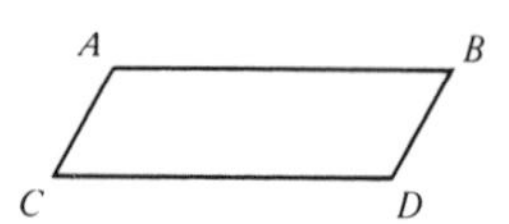

A parallelogram has two pairs of opposite sides which are parallel and equal.

2. Rectangle:

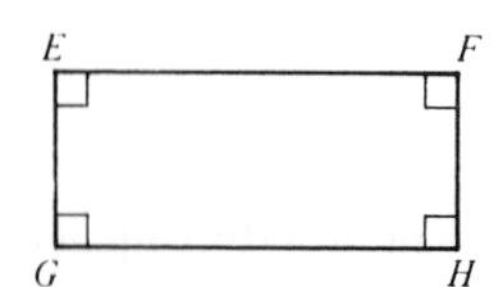

A rectangle has two pairs of opposite sides which are equal and parallel and four right angles.

Special Quadrilaterals (Continued)

3. Square:

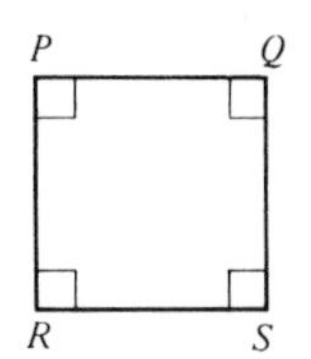

> A square has four equal sides with the opposite sides parallel and four right angles.

4. Trapezoid:

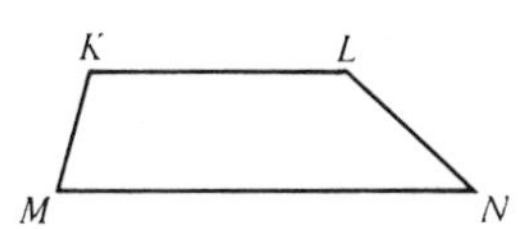

> A trapezoid has only one pair of opposite sides that are parallel.

Study Exercise Five (22)

True or false:

1. A trapezoid has four equal sides.
2. A rectangle has four equal angles.
3. A parallelogram has only one pair of parallel sides.
4. A parallelogram has four right angles.
5. A square has four equal sides.
6. Every polygon is a quadrilateral.
7. Every quadrilateral is a polygon.
8. A rectangle has four right angles.

REVIEW EXERCISES (23)

A. Classify the following angles as acute, right, obtuse, or straight:

1.

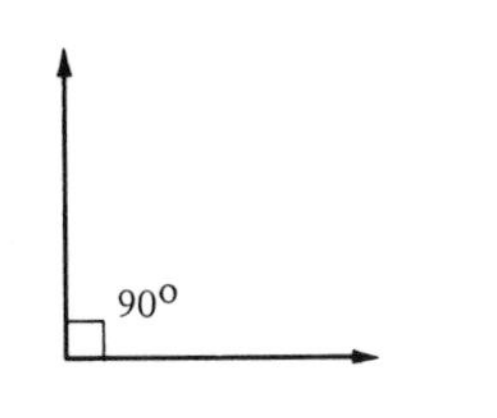

2.

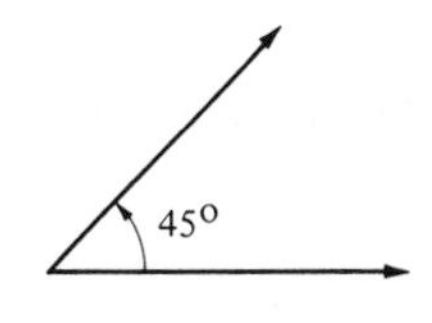

3.

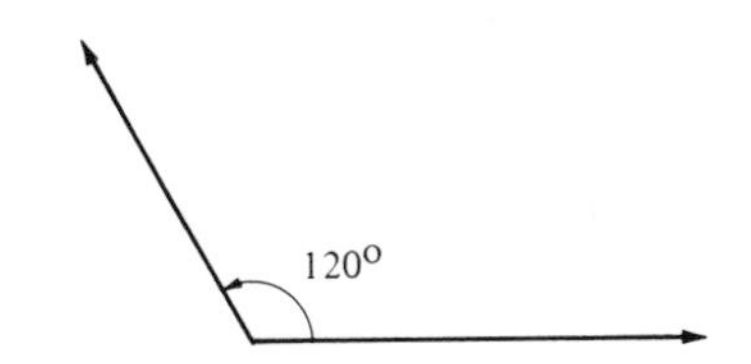

4.

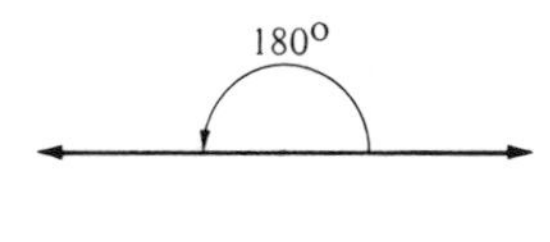

B. Match the number of sides in Column B with the figure in Column A:

Column A	Column B
5. Octagon	**(a)** 3
6. Pentagon	**(b)** 4
7. Hexagon	**(c)** 5
8. Triangle	**(d)** 6
9. Quadrilateral	**(e)** 8

REVIEW EXERCISES (Continued)

C. Which of the following figures are polygons? (answer yes or no)

10.

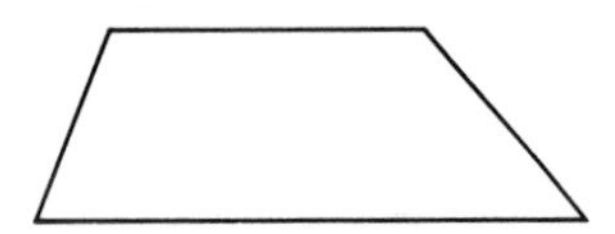

11.

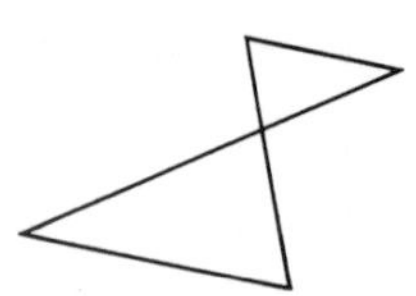

12.

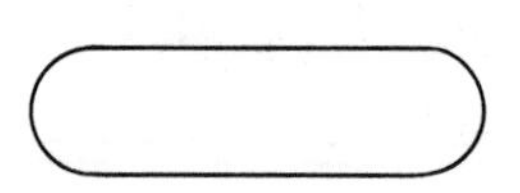

13.

D. Classify the following triangles as scalene, equilateral, isosceles, or right. (Some triangles may be classified as more than one type.)

14.

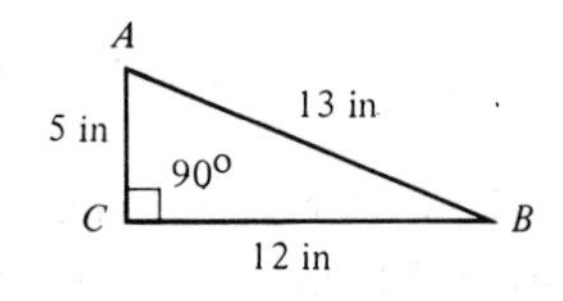

15.

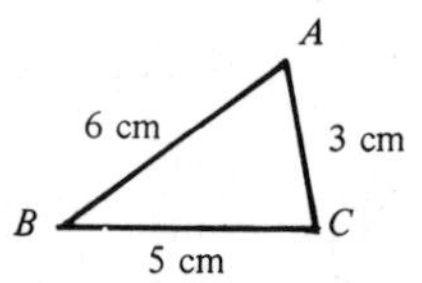

16.

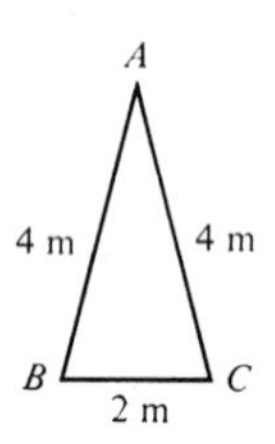

17.

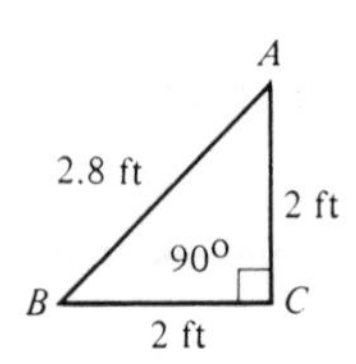

E. Find the measure of the unknown angle:

18.

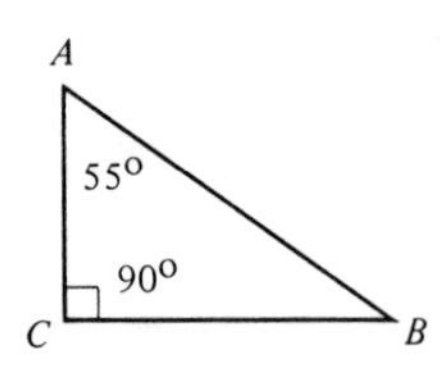

19.

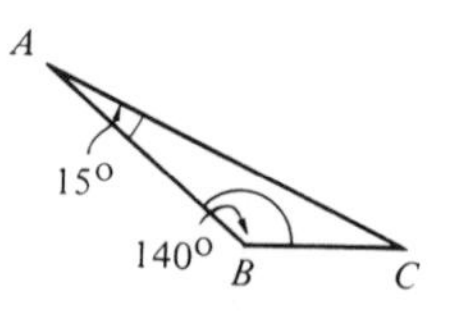

F. Use the Pythagorean theorem and the Powers and Roots Table to find the length of the hypotenuse for each of the following right triangles:

20.

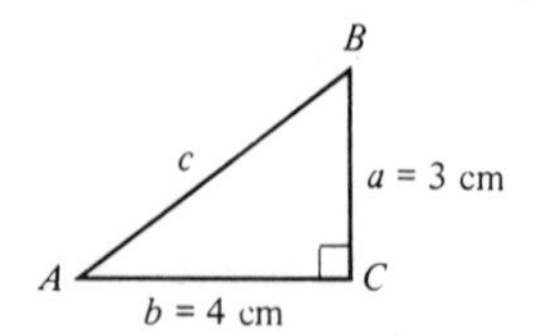

21.

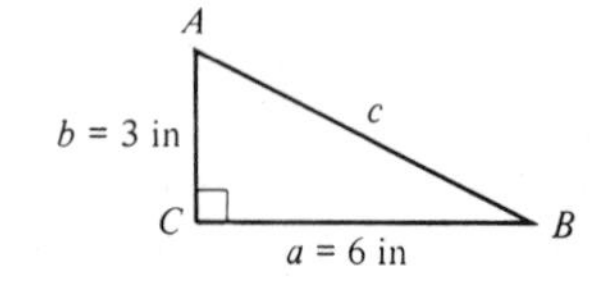

REVIEW EXERCISES (Continued)

G. Complete the following statements:

22. If the sides of an angle are perpendicular, then the angle is said to be a _________ angle.

23. 8 degrees = _________ minutes.

24. 1 minute = _________ seconds.

25. A quadrilateral which has only one pair of parallel sides is called a _________

26. A quadrilateral having its opposite sides equal and parallel is called a _________

H. Solve the following problems:

27. One of the equal angles of an isosceles triangle is 40°. Find the other two angles.

28. One acute angle of a right triangle is 37°. Find the other acute angle.

29. Two angles of a triangle are 17° and 43°. Find the third angle.

30. In a triangle the measures of angles A and B are 32°. Find angle C and classify this triangle.

31. The length of one side of a parallelogram is 12 m. What is the length of the opposite side?

32. One angle of a right triangle is 19°. Find the other two angles.

33. To the nearest tenth, what is the longest piece of pipe that can fit diagonally across a rectangular trailer bed that measures 4 ft by 8 ft?

34. To the nearest tenth, how long a board is needed to brace a square 3 ft gate, if the board is to be placed diagonally across the gate?

35. Will a 50 ft ladder reach the top of a 40 ft building if it is placed 30 ft from the building?

36. How much wire is needed to brace a tree, if the wire is to be tied 8 ft high on the tree and tied to the ground 2 ft away from the tree (to nearest hundredth)?

Solutions to Review Exercises (24)

A. **1.** Right. **2.** Acute. **3.** Obtuse **4.** Straight.

B. **5.** **(e)** 8 sides **6.** **(c)** 5 sides. **7.** **(d)** 6 sides. **8.** **(a)** 3 sides.
9. **(b)** 4 sides.

C. **10.** Yes. **11.** No. **12.** No. **13.** Yes.

D. **14.** Right and scalene. **15.** Scalene.
16. Isosceles. **17.** Right and isosceles.

E. **18.** $\angle B = 90° - 55°$
$= 35°$

19. $\angle C = 180° - (140° + 15°)$
$= 180° - 155°$
$= 25°$

F. **20.** $c^2 = a^2 + b^2$
$= 3^2 + 4^2$
$= 9 + 16$
$= 25$
$c = \sqrt{25}$
$c = 5$ cm

21. $c^2 = a^2 + b^2$
$= 6^2 + 3^2$
$= 36 + 9$
$= 45$
$c = \sqrt{45}$ (See Powers and Roots Table)
$c \approx 6.708$ in

G. **22.** Right. **23.** (8)(60) or 480 **24.** 60 **25.** Trapezoid. **26.** Parallelogram.

H. **27.** The other equal angle is 40°. 180° − (40° + 40°) = 180° − 80° = 100°

28. 90° − 37° = 53°

29. 180° − (17° + 43°) = 180° − 60° = 120°

30. 180° − (32° + 32°) = 180° − 64° = 116°. This is isosceles.

31. Opposite sides of parallelograms are equal. Therefore, the opposite side is also 12 m.

32. A right triangle must have exactly one 90° angle. 180° − (90° + 19°) = 180° − 109° = 71°. The angles are 90°, 71°, 19°.

Solutions to Review Exercises (Continued)

33.

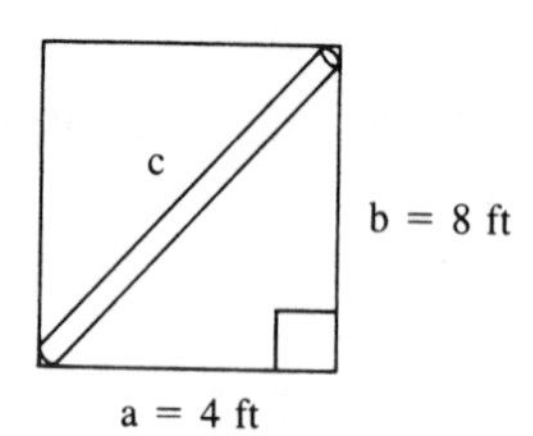

$c^2 = a^2 + b^2$
$= 4^2 + 8^2$
$= 16 + 64$
$= 80$
$c = \sqrt{80}$ (See Powers and Roots Table)
$c \approx 8.944$ or 8.9 ft

34.

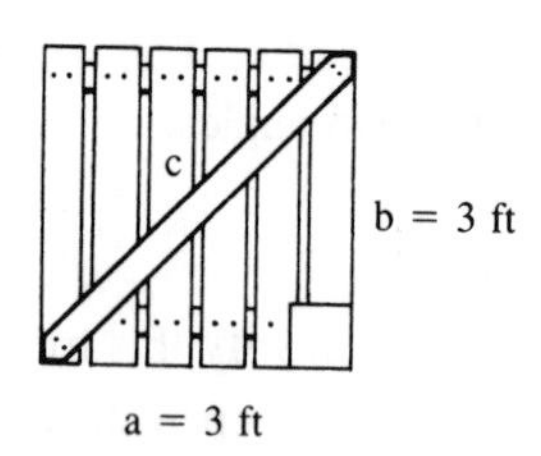

$c^2 = a^2 + b^2$
$= 3^2 + 3^2$
$= 9 + 9$
$= 18$
$c = \sqrt{18}$ (See Powers and Roots Table)
$c \approx 4.243$ or 4.2 ft

35.

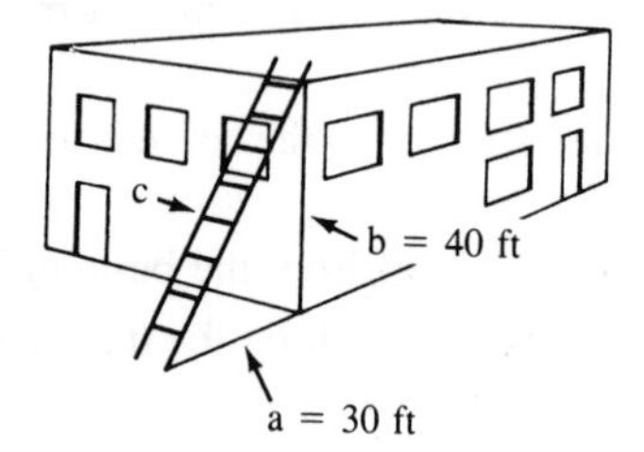

$c^2 = a^2 + b^2$
$= 30^2 + 40^2$
$= 900 + 1600$
$= 2500$
$c = \sqrt{2500}$
$c = 50$ ft (yes)

36.

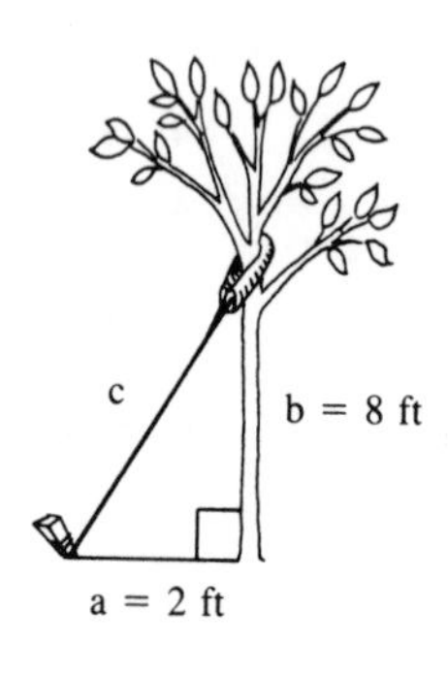

$c^2 = a^2 + b^2$
$= 2^2 + 8^2$
$= 4 + 64$
$= 68$
$c = \sqrt{68}$ (See Powers and Roots Table)
$c \approx 8.246$ or 8.25 ft

SUPPLEMENTARY PROBLEMS

A. Classify the following angles as acute, right, obtuse, or straight:

1.

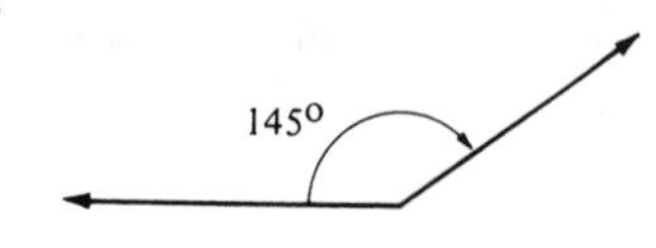

2.

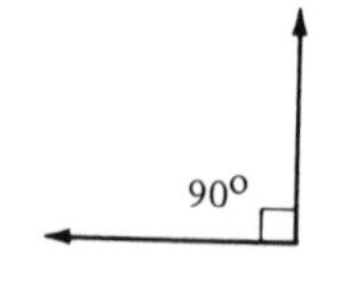

3.

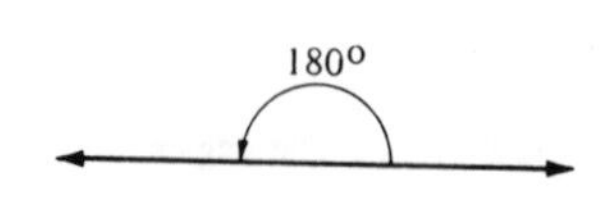

4.

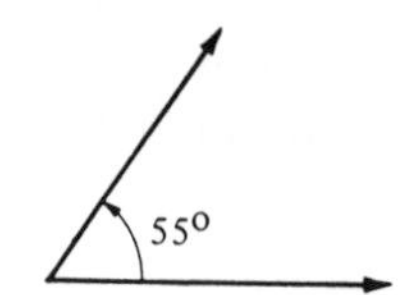

SUPPLEMENTARY PROBLEMS (Continued)

B. Complete the following table referring to types of polygons:

	Number of Sides	Name
5.	3	
6.	4	
7.		pentagon
8.		hexagon
9.	8	
10.		decagon

C. Which of the following figures are polygons? (answer yes or no)

11.

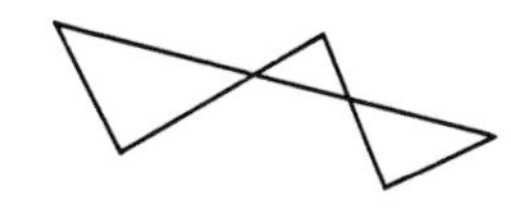

12.

13.

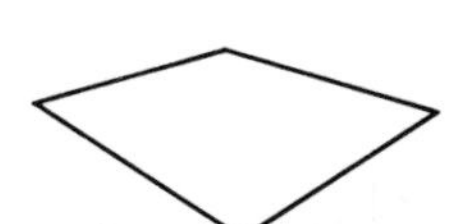

14.

D. Classify the following triangles as scalene, equilateral, isosceles, or right. (Some triangles may be classified as more than one type.)

15.

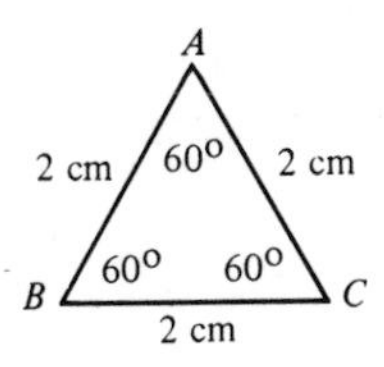

16.

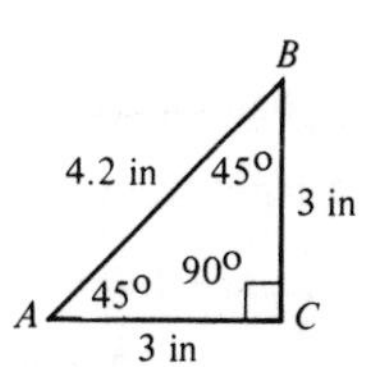

17.

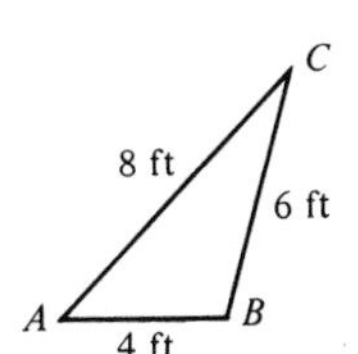

18.

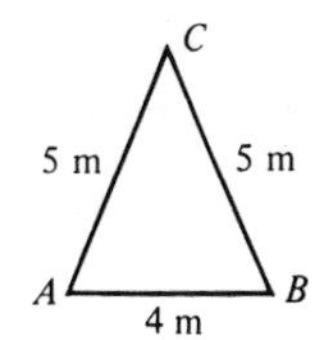

SUPPLEMENTARY PROBLEMS (Continued)

E. Find the measure of the unknown angle:

19.

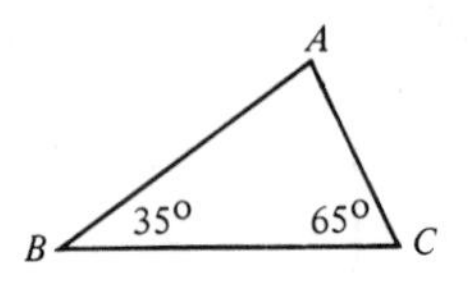

20.

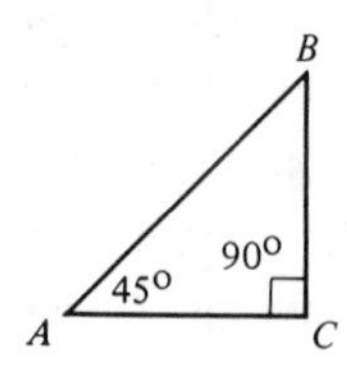

21.

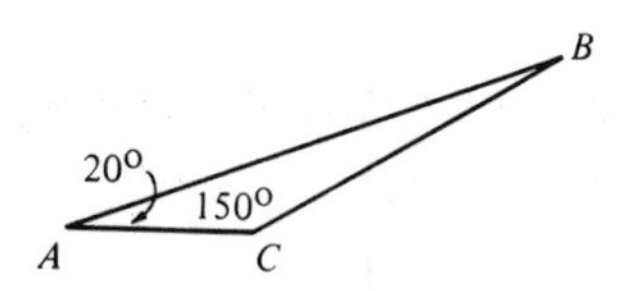

22.

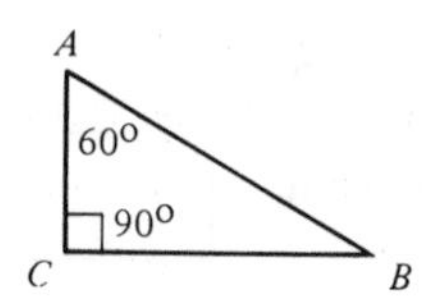

F. Use the Pythagorean theorem and the Powers and Roots Table to find the length of the hypotenuse for each of the following right angles:

23.

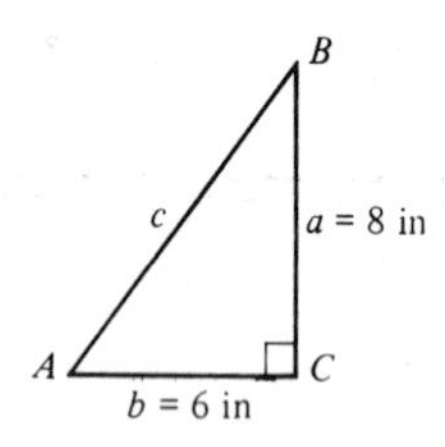

24.

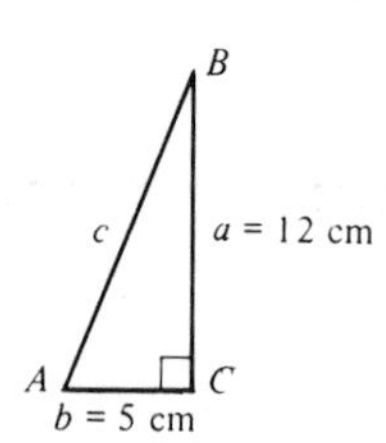

25.

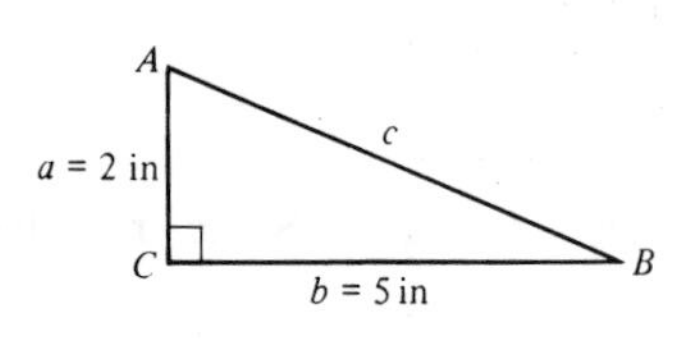

26.

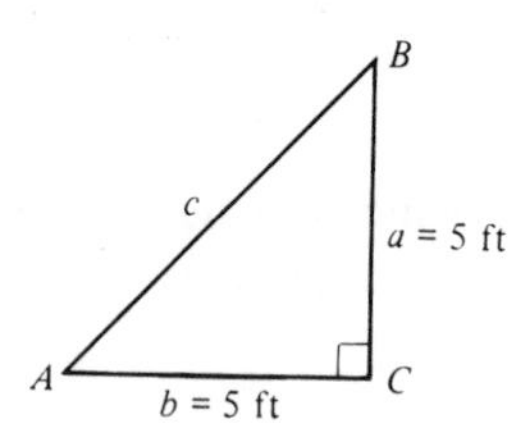

G. Complete the following statements:

27. The two legs of a right triangle are said to be _________

28. 1 degree = _________ minutes.

29. 10 degrees = _________ minutes.

30. 5 minutes = _________ seconds.

31. A square has four _________ sides.

32. Each angle of a rectangle measures _________ degrees.

33. In a parallelogram the opposite sides are _________ and _________

SUPPLEMENTARY PROBLEMS (Continued)

H. Solve the following problems:

34. One of the equal angles of an isosceles triangle is 55°. Find the other two angles.

35. One acute angle of a right triangle is 52°. Find the other acute angle.

36. Two angles of a triangle are 23° and 43°. Find the third angle.

37. In a triangle, the measures of A and B are 60°. Find angle C and classify this triangle.

38. One angle of a right triangle is 29°. Find the other two angles.

39. To the nearest whole number, what is the longest piece of pipe that can fit diagonally across a rectangular trailer bed that measures 4 ft by 7 ft?

40. To the nearest tenth, how long a board is needed to brace a square 4 ft gate, if the board is to be placed diagonally across the gate?

41. Will a 13 ft ladder reach the top of a 12 ft building if it is placed 5 ft from the building?

42. To the nearest whole foot, how much wire is needed to brace a tree, if the wire is to be tied 7 ft high on the tree and tied to the ground 3 ft away from the tree?

Solutions to Study Exercises

(6A)

Study Exercise One (Frame 6)

A. **1.** Obtuse. **2.** Acute. **3.** Acute. **4.** Right. **5.** Straight.

B. **6.** Perpendicular. **7.** 60 **8.** (5)(60) or 300 **9.** rays, vertex **10.** 90°

11. 90°, 180°

(10A)

Study Exercise Two (Frame 10)

A.

	Number of Sides	Name
1.	4	quadrilateral
2.	6	hexagon
3.	10	decagon
4.	3	triangle
5.	8	octagon
6.	5	pentagon

B. **7.** Yes. **8.** No; a polygon cannot have crossing line segments.

9. No; this figure is not composed entirely of straight line segments.

10. Yes. **11.** Yes. **12.** Yes.

(14A)

Study Exercise Three (Frame 14)

A. **1.**
$$\begin{aligned} \angle A &= 180° - (120° + 25°) \\ &= 180° - 145° \\ &= 35° \end{aligned}$$

2.
$$\begin{aligned} \angle B &= 90° - 20° \\ &= 70° \end{aligned}$$

B. **3.** Isosceles. **4.** Scalene.

5. Right and isosceles. **6.** Equilateral and isosceles.

C. **7.** Scalene **8.** Isosceles

9. 180°

Study Exercise Four (Frame 19) **(19A)**

1. $c^2 = a^2 + b^2$
$= 8^2 + 6^2$
$= 64 + 36$
$= 100$
$c = \sqrt{100}$
$c = 10$ in

2. $c^2 = a^2 + b^2$
$= 7^2 + 4^2$
$= 49 + 16$
$= 65$
$c = \sqrt{65}$ (Consult the Powers and Roots Table)
$c \approx 8.062$ cm

3. $c^2 = a^2 + b^2$
$= 3^2 + 6^2$
$= 9 + 36$
$= 45$
$c = \sqrt{45}$ (Consult the Powers and Roots Table)
$c = 6.708$ m

4. $c^2 = a^2 + b^2$
$= 7^2 + 7^2$
$= 49 + 49$
$= 98$
$c = \sqrt{98}$ (Consult the Powers and Roots Table)
$c = 9.899$ cm

Study Exercise Five (Frame 22) **(22A)**

1. False; a trapezoid has exactly one pair of parallel sides.
2. True; each angle of a rectangle measures 90°.
3. False; a parallelogram has two pair of parallel sides.
4. False; a rectangle has four right angles.
5. True.
6. False; see frame 9
7. True.
8. True.

UNIT

37 Area, Perimeter, and Circumference

OBJECTIVES (1)

By the end of this unit you should be able to:

1. FIND THE AREA AND PERIMETER OF A RECTANGLE, SQUARE, PARALLELOGRAM, TRIANGLE, AND A TRAPEZOID.
2. FIND THE AREA AND CIRCUMFERENCE OF A CIRCLE.

Area (2)

The area of the interior of any plane figure is given by the number of square units it contains.

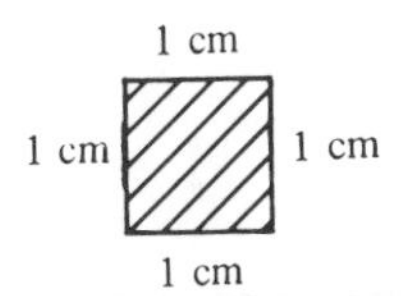

One square centimeter
($1\ cm^2$)

Example: This rectangle contains 10 of the square centimeters pictured above. Thus, its area equals $10\ cm^2$.

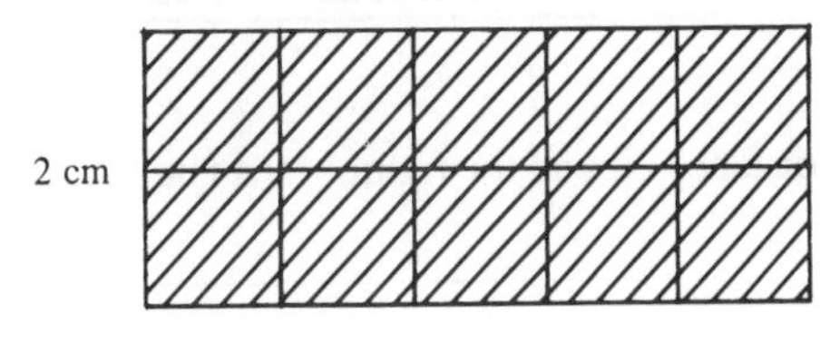

5 cm

Area may also be measured in terms of other square units, such as square meters (m^2), square inches (in^2), square feet (ft^2), or square miles (mi^2).

Area of a Rectangle (3)

The area of a rectangle is found by multiplying its length times its width. The formula for finding the area of a rectangle is $A = lw$.

Example 1: Find the area of the following rectangle whose length is 5 in and whose width is 4 in.

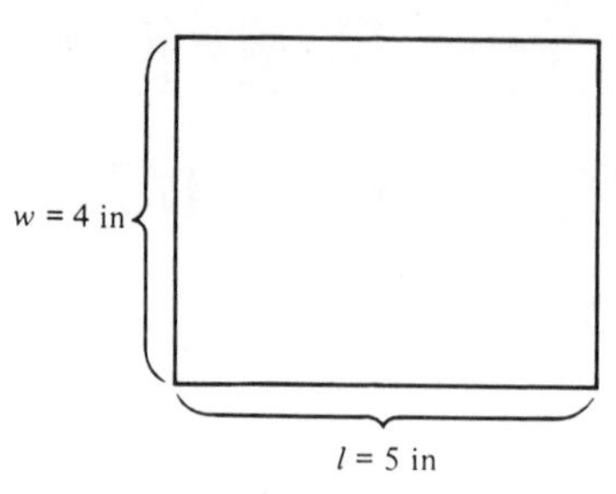

Solution: Use the formula $A = lw$.

Line (a): $A = lw$

Line (b): $A = (5 \text{ in})(4 \text{ in})$

Line (c): $= 20 \text{ in}^2$

Example 2: How many square yards of carpet is needed to carpet a rectangular room measuring 15 ft by 21 ft?

Solution: Convert to yards and use the formula $A = lw$:

Line (a): $15 \text{ ft} = \frac{15}{3}$ or 5 yd

Line (b): $21 \text{ ft} = \frac{21}{3}$ or 7 yd

Line (c): $A = lw$

Line (d): $= (5 \text{ yd})(7 \text{ yd})$

Line (e): $= 35 \text{ yd}^2$

Perimeter (4)

The perimeter of a polygon is the sum of the lengths of its sides. It is the distance around the polygon.

Perimeter of a Rectangle (5)

The perimeter of a rectangle is the distance around the figure. It can be found by doubling the sum of the lengths of two adjacent sides.

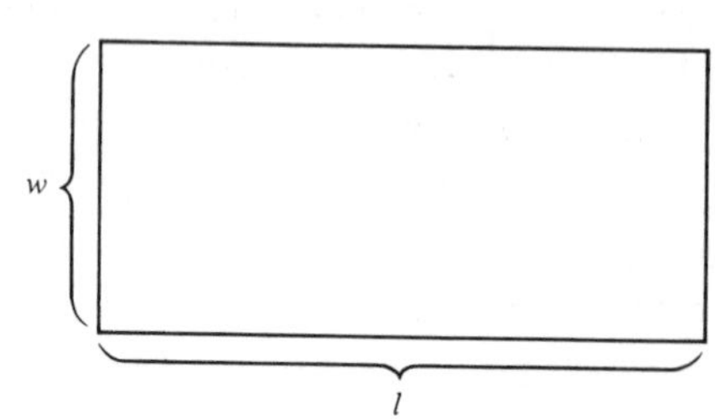

The formula is $P = 2(l + w)$:

UNIT 37

Perimeter of a Rectangle (Continued)

Example 1: Find the perimeter of a rectangle whose length is 20 cm and whose width is 15 cm.

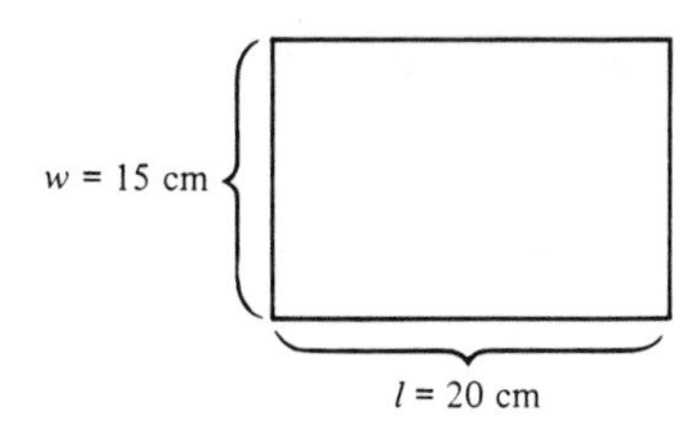

Solution: Use the formula $P = 2(l + w)$:

Line (a): $P = 2(l + w)$

Line (b): $P = 2(20 \text{ cm} + 15 \text{ cm})$

Line (c): $= 2(35 \text{ cm})$

Line (d): $= 70 \text{ cm}$

Example 2: A piece of rectangular carpet measures 9 ft by 12 ft. The edge of the carpet must be bound at a cost of 20¢ a foot. Find the total cost to bind the carpet.

Solution: Find the perimeter and multiply it by 20¢ or \$.20:

Line (a): $P = 2(l + w)$

Line (b): $= 2(12 \text{ ft} + 9 \text{ ft})$

Line (c): $= 2(21 \text{ ft})$

Line (d): $= 42 \text{ ft}$

Line (e): $\text{Cost} = (42)(.20)$

Line (f): $= \$8.40$

Study Exercise One (6)

A. Find the area and the perimeter for each of the following rectangles:

1.

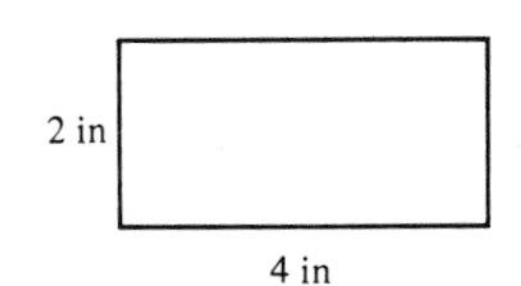

2.

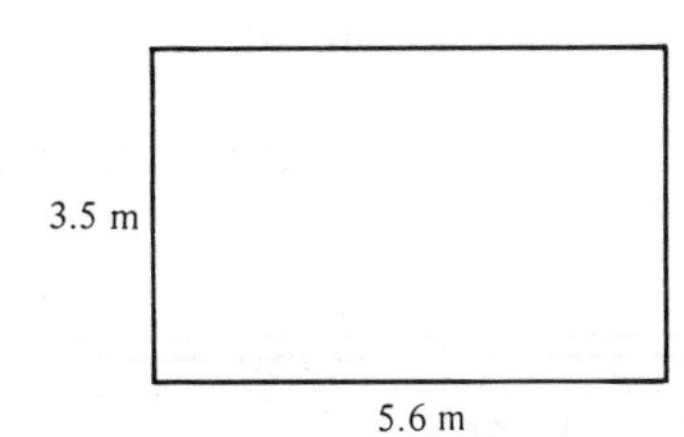

B. Solve the following problems:

3. Find the picture area of a rectangular television screen measuring 18.5 in by 14.5 in.

4. Find the cost of constructing 70 ft of sidewalk 36 in wide at \$3.20 per square foot.

5. A table top 62 in long and 40 in wide requires an aluminum molding around its edge. If the molding costs 5 cents per inch, what is the total cost of the molding?

6. How many feet of fencing is required to surround a garden 38 ft long and 23 ft wide?

7. How many yards of fringe is needed for a border on a 72 in by 76 in bedspread, if one of the 76 in sides is not to have a fringe?

Area of a Square (7)

A square is a rectangle where all four sides are of equal length.

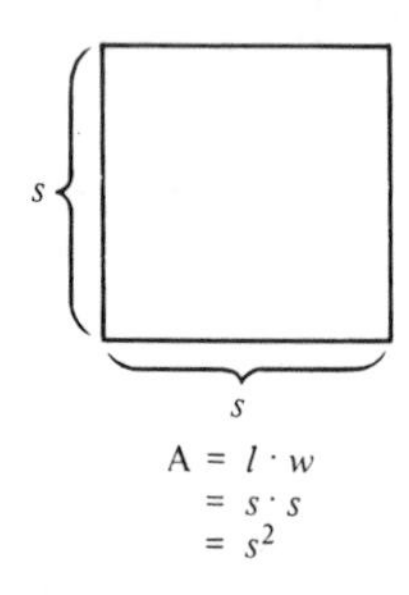

Thus, the area of a square can be found by the formula $A = s^2$.

Example 1: Find the area of a square where each side measures 9 cm:

Solution: Use the formula $A = s^2$:

Line (a): $A = s^2$

Line (b): $= (9 \text{ cm})^2$

Line (c): $= 81 \text{ cm}^2$

Example 2: Find the area of a square where each side measures 2 ft 3 in:

Solution: Convert to inches and use the formula $A = s^2$:

Line (a): 2 ft 3 in = (2)(12 in) + 3 in

Line (b): = 24 in + 3 in

Line (c): = 27 in

Line (d): $A = s^2$

Line (e): $= (27 \text{ in})^2$

Line (f): $= 729 \text{ in}^2$

Computation

```
  27
  27
 ---
 189
 54
 ---
 729
```

Perimeter of a Square (8)

The perimeter of a square is the sum of the lengths of its sides.

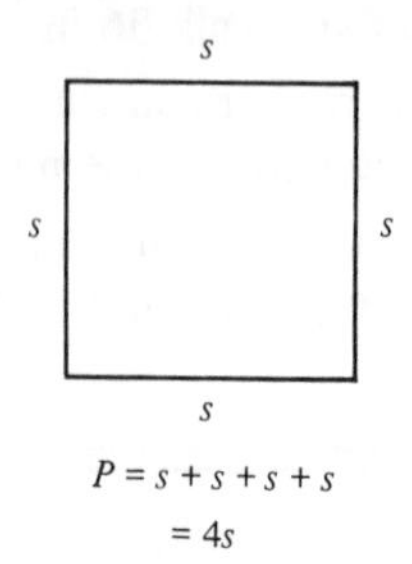

Thus, the perimeter of a square can be found by the formula $P = 4s$.

Perimeter of a Square (Continued)

Example 1: Find the perimeter of a square where each side measures 6 cm:

Solution: Use the formula $P = 4s$:

Line (a): $P = 4s$

Line (b): $= (4)(6 \text{ cm})$

Line (c): $= 24 \text{ cm}$

Example 2: Find the perimeter of a square where each side measures 6 yd 2 ft. (Leave answer in terms of yards and feet.)

Solution: Use the formula $P = 4s$:

Line (a): $P = 4s$

Line (b): $= (4)(6 \text{ yd } 2 \text{ ft})$

Line (c): $= 26 \text{ yd } 2 \text{ ft}$

Computation

6 yd 2 ft
×4
24 yd 8 ft
2 yd
26 yd 2 ft

Study Exercise Two (9)

A. Find the area and perimeter for each of the following squares:

1.

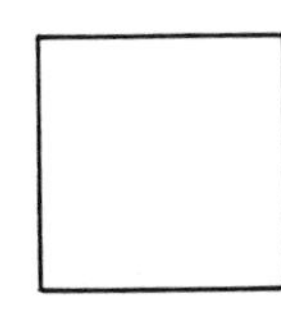

2.

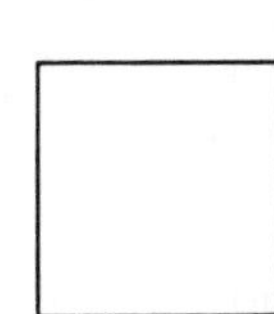

B. Solve the following problems:

3. Find the area in square inches of a square where each side measures 1 ft 7 in.

4. Find the perimeter of a square where each side measures 3 yd 1 ft. (Leave the answer in terms of yards and feet.)

5. If the distance between bases on a softball diamond is 60 ft, how many feet does a batter run on a home run?

6. What is the area in square feet of a lot that measures 141 ft by 141 ft?

7. How many feet of fence is needed to surround a square garden 33 ft by 33 ft?

Area of a Parallelogram (10)

We will now develop the formula for the area of a parallelogram. This will be done by changing a parallelogram into a corresponding rectangle.

1.

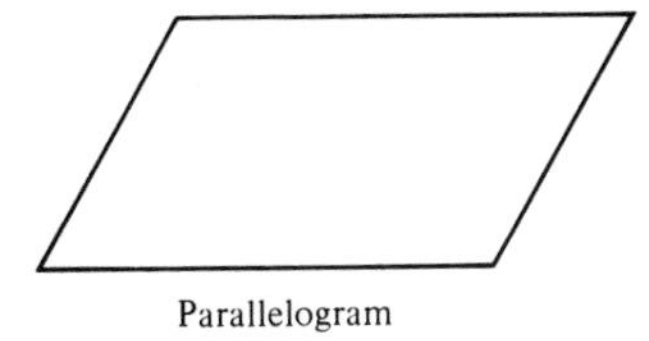
Parallelogram

2.

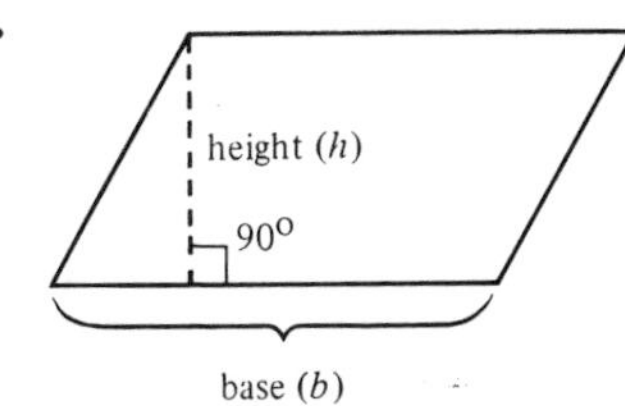

Area of a Parallelogram (Continued)

3.

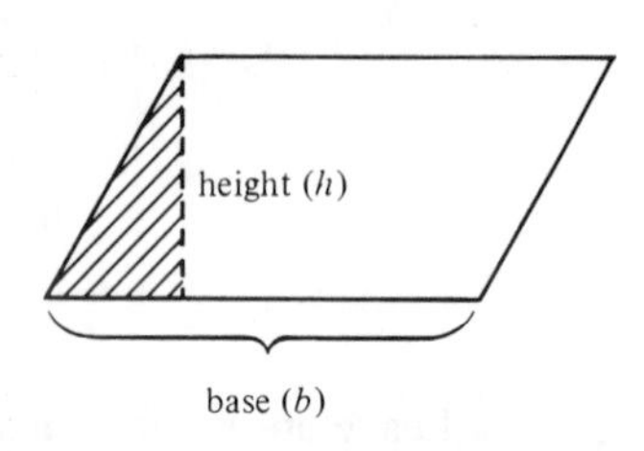

4.

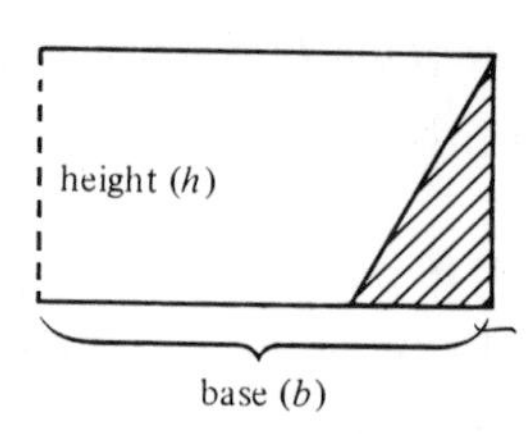

Line (a): Area of a rectangle: $A = \underbrace{(\text{length})} \cdot \underbrace{(\text{width})}$

Line (b): $A = (\text{base}) \cdot (\text{height})$

Line (c): Area of a parallelogram: $A = bh$

Using the Formula $A = bh$ (11)

Example 1: Find the area of the following parallelogram whose base measures 20 cm and whose height measures 7 cm:

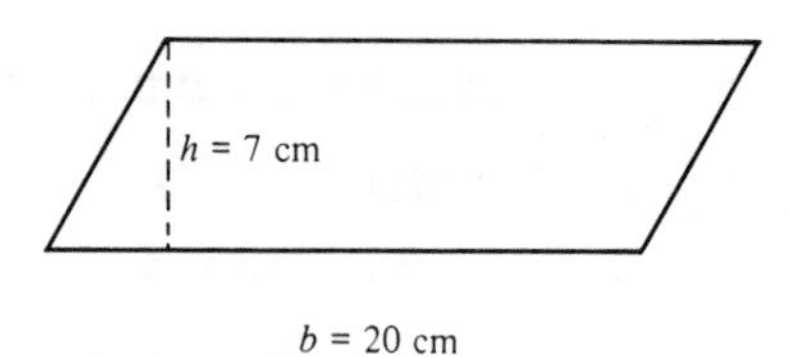

Solution: Use the formula $A = bh$:

Line (a): $A = bh$

Line (b): $= (20 \text{ cm})(7 \text{ cm})$

Line (c): $= 140 \text{ cm}^2$

Example 2: Find the area of the following parallelogram whose base measures 6 yd and whose height measures 2 ft:

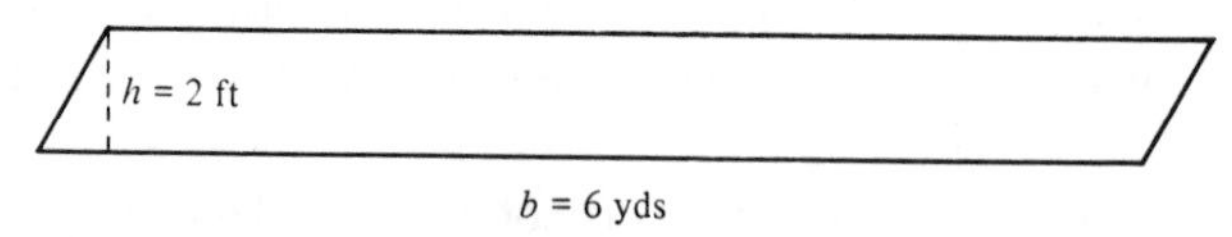

Solution: Convert 6 yd to feet and use the formula $A = bh$:

Line (a): $6 \text{ yd} = (6)(3 \text{ ft})$

Line (b): $= 18 \text{ ft}$

Line (c): $A = bh$

Line (d): $= (18 \text{ ft})(2 \text{ ft})$

Line (e): $= 36 \text{ ft}^2$

Perimeter of a Parallelogram (12)

The perimeter of a parallelogram is the distance around the figure. As with the rectangle, it can be found by doubling the sum of the lengths of two adjacent sides.

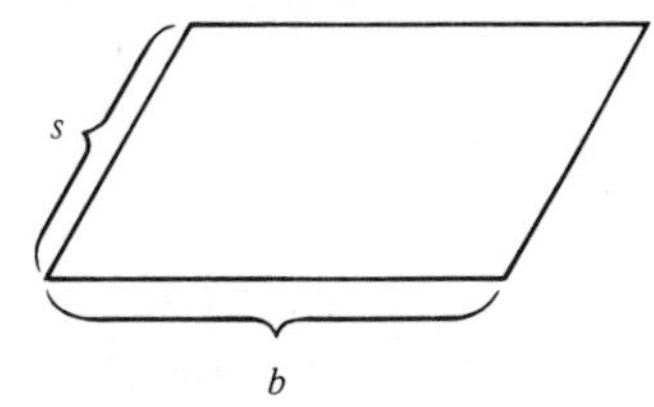

The formula is $P = 2(b + s)$.

Example: Find the perimeter of the following parallelogram whose base measures 15 cm and whose adjacent side measures 10 cm:

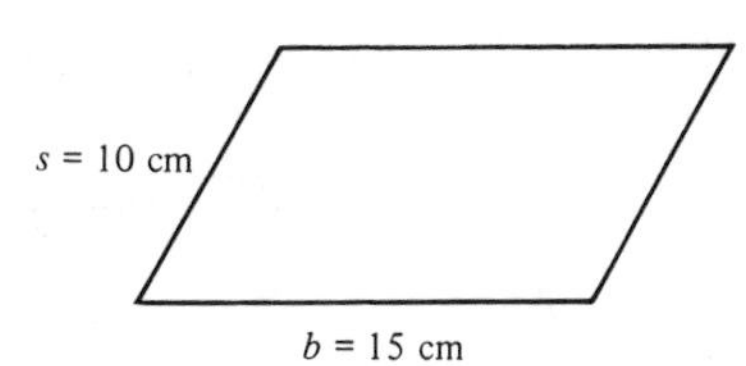

Solution: Use the formula $P = 2(b + s)$.

Line (a): $P = 2(b + s)$
Line (b): $= 2(15 \text{ cm} + 10 \text{ cm})$
Line (c): $= 2(25 \text{ cm})$
Line (d): $= 50 \text{ cm}$

Study Exercise Three (13)

A. Find the area of each of the following parallelograms:

1.

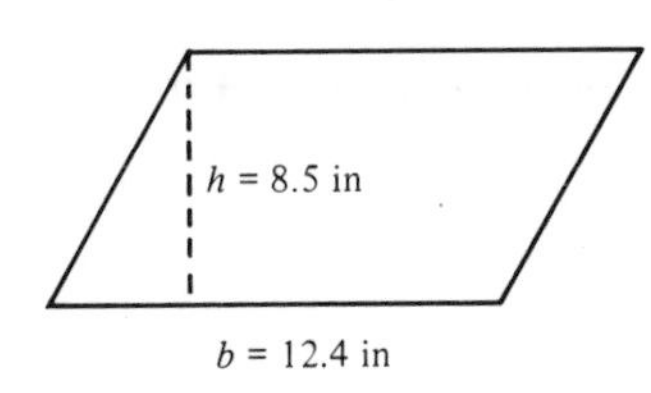

2.

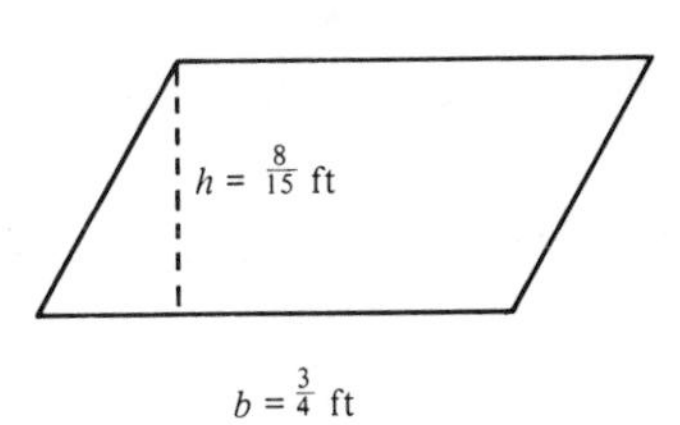

B. Find the perimeter of each of the following parallelograms:

3.

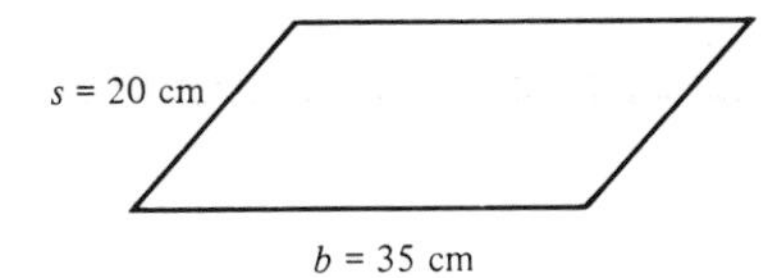

4.

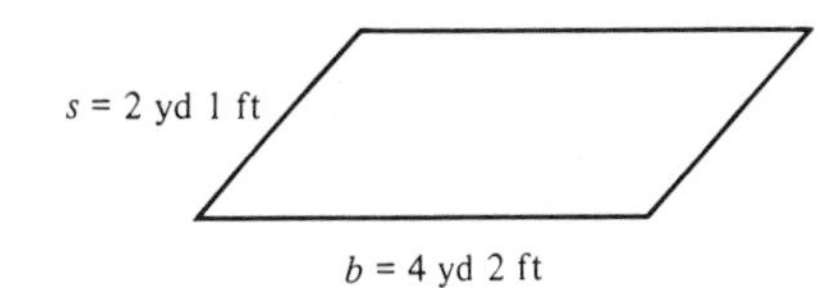

***Study Exercise Three* (Continued)**

C. Solve the following problem:

5. A parking lot contains 62 parking spaces. Each space is in the shape of a parallelogram with a base of 8 ft and a height of 16 ft. What is the total area covered by these spaces?

Area of a Triangle (14)

We will now develop the formula for the area of a triangle. This will be done by showing that the area within a triangle is one-half the area within a corresponding parallelogram.

1.

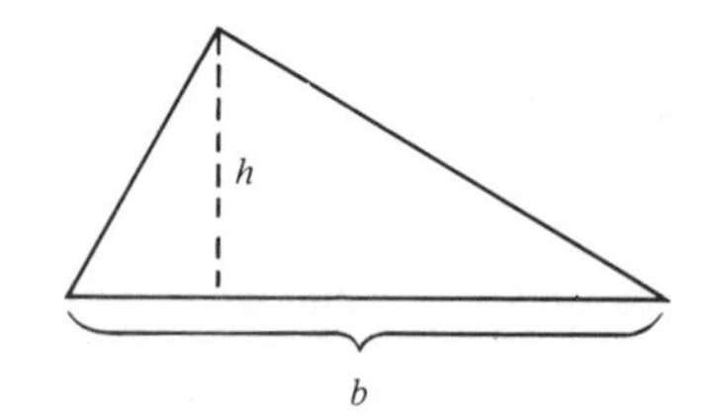

2.

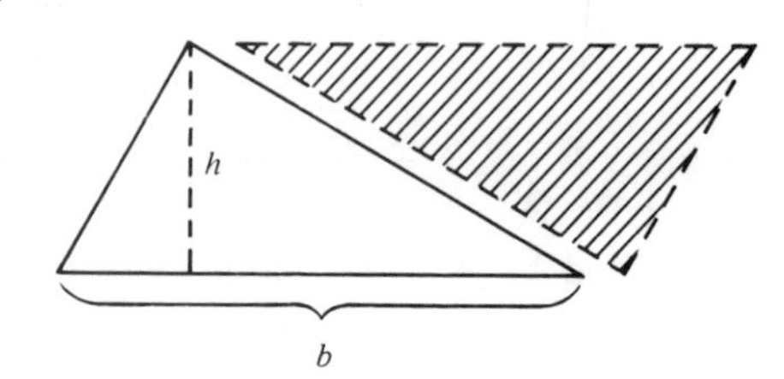

3.

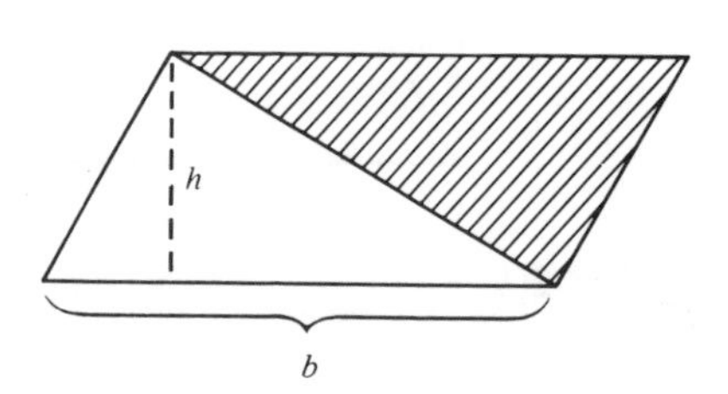

Area of the parallelogram is: $A = bh$

Area of the triangle is: $A = \frac{1}{2}bh$ or $\frac{bh}{2}$

Using the Formula $A = \frac{1}{2}bh$ or $A = \frac{bh}{2}$ (15)

Example 1: Find the area of the following triangle having a base of 24 cm and a height of 6 cm:

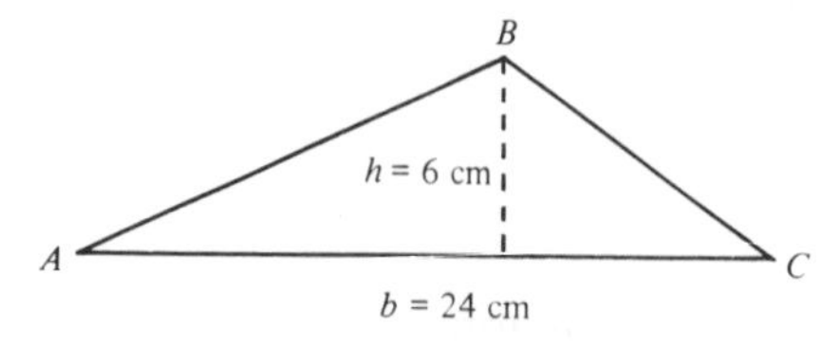

Solution: Use the formula $A = \frac{1}{2}bh$:

Line (a): $A = \frac{1}{2}bh$

Line (b): $= \frac{1}{2} \cdot (24 \text{ cm})(6 \text{ cm})$

Line (c): $= \frac{1}{\cancel{2}_1} \cdot (\cancel{24}^{12} \text{ cm})(6 \text{ cm})$

Line (d): $= 72 \text{ cm}^2$

Using the Formula $A = \frac{1}{2}bh$ or $A = \frac{bh}{2}$ (Continued)

Example 2: Find the area of the following right triangle having a base of 2 ft 3 in and a height of 9 in:

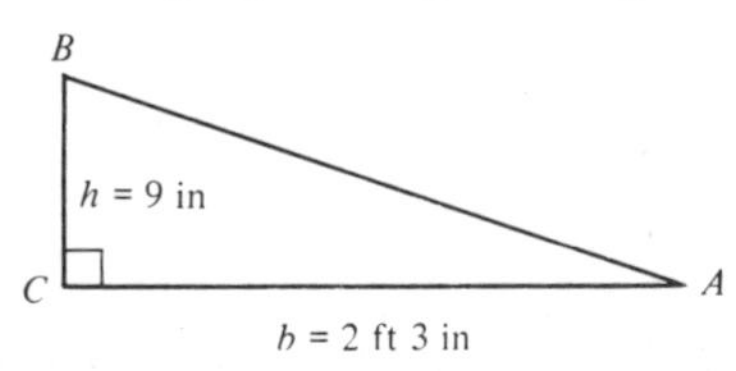

Solution: Convert 2 ft 3 in to inches and use the formula $A = \frac{bh}{2}$:

Line (a): $2 \text{ ft } 3 \text{ in} = 2(12 \text{ in}) + 3 \text{ in}$
Line (b): $= 24 \text{ in} + 3 \text{ in}$
Line (c): $= 27 \text{ in}$
Line (d): $A = \frac{bh}{2}$
Line (e): $= \frac{(27 \text{ in})(9 \text{ in})}{2}$
Line (f): $= \frac{243}{2} \text{ in}^2$
Line (g): $= 121\frac{1}{2}$ or 121.5 in^2

Perimeter of a Triangle (16)

The perimeter of a triangle is the sum of the lengths of the three sides.

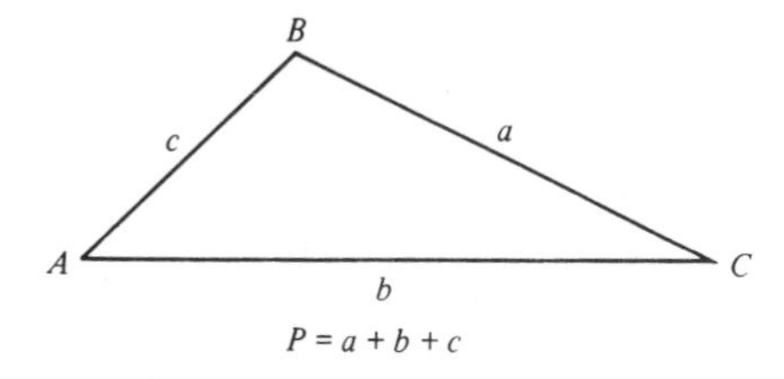

Examples (17)

Example 1: Find the perimeter of the following triangle:

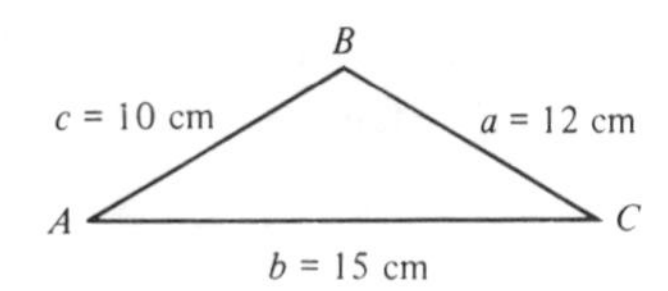

Solution: Use the formula $P = a + b + c$:

Line (a): $P = a + b + c$
Line (b): $= 12 \text{ cm} + 15 \text{ cm} + 10 \text{ cm}$
Line (c): $= 37 \text{ cm}$

Examples (Continued)

Example 2: Find the perimeter of the following right triangle:

Solution: Use the Pythagorean theorem to find the length of side c; then use the formula $P = a + b + c$:

Line (a): $c^2 = a^2 + b^2$

Line (b): $= 6^2 + 8^2$

Line (c): $= 36 + 64$

Line (d): $= 100$

Line (e): $c = \sqrt{100}$ or 10 ft

Line (f): $P = a + b + c$

Line (g): $= 6 \text{ ft} + 8 \text{ ft} + 10 \text{ ft}$

Line (h): $= 24 \text{ ft}$

Study Exercise Four (18)

A. Find the area for each of the following triangles:

1.

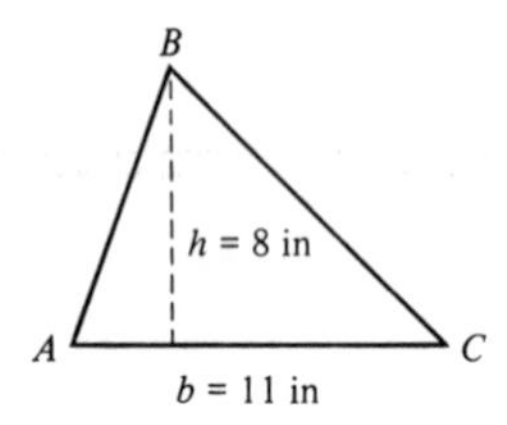

2.

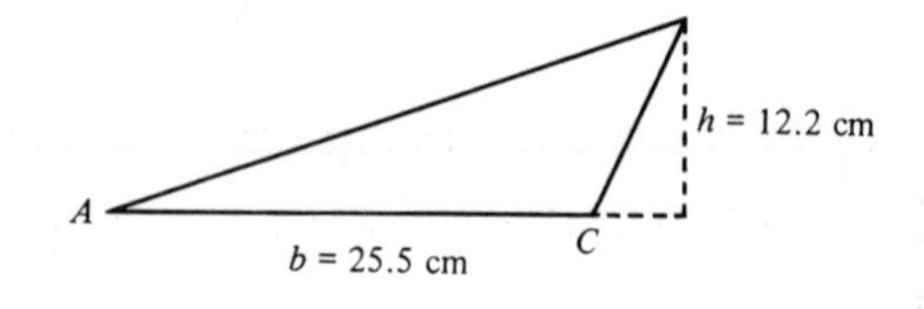

3.

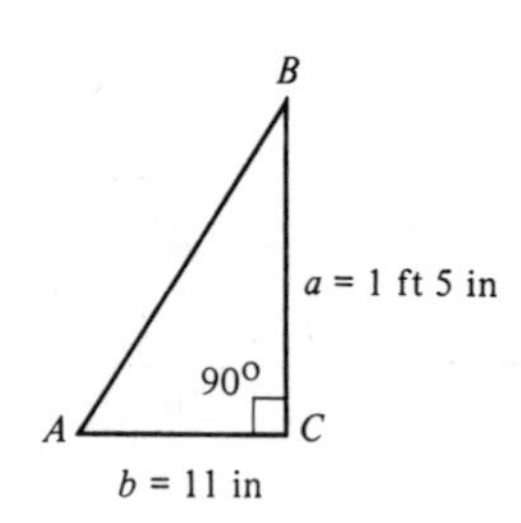

B. Find the perimeter for each of the following triangles:

4.

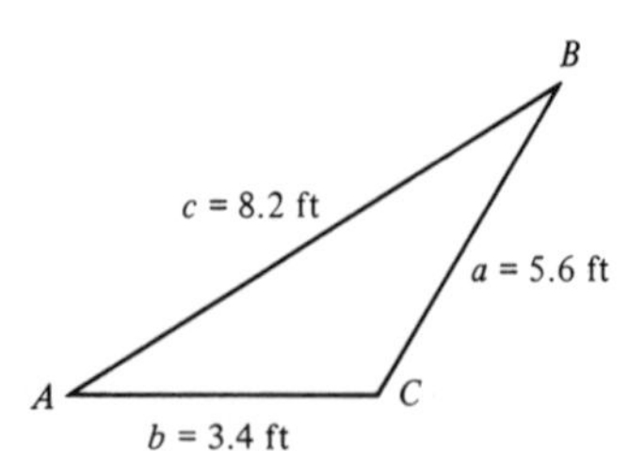

5.

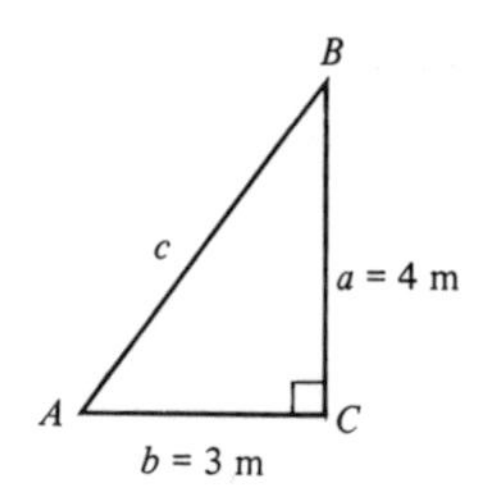

***Study Exercise Four* (Continued)**

C. Solve the following problems:

6. A triangular sail has a base of 12 ft and a height of 18 ft. How much area does one side of the sail expose to the wind?

7. Each side of an equilateral triangle measures 3 yds 2 ft. Find the perimeter in terms of yards and feet.

8. A roof of a house consists of four triangles, each with a base of 30 ft and a height of 20 ft. How many square feet of roofing material is needed to redo the roof on this house?

Area of a Trapezoid (19)

A trapezoid can be divided into two triangles as follows:

1.

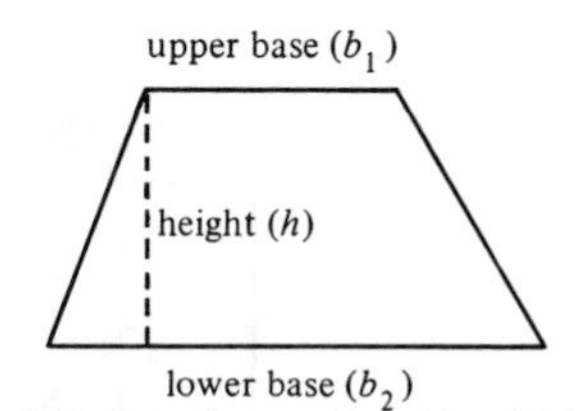

2.

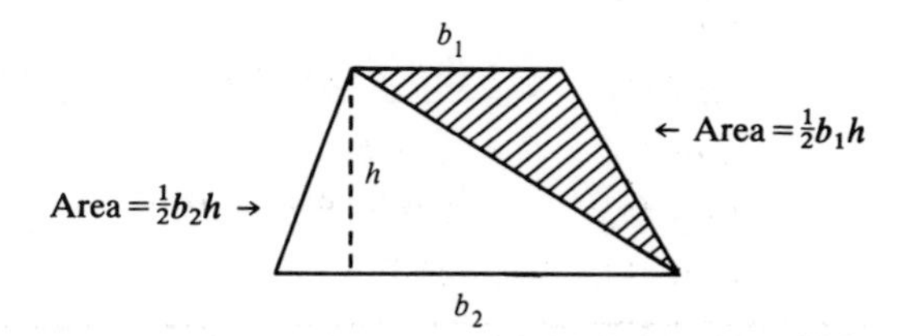

Using this fact, it can be shown using algebra that the area of a trapezoid is equal to one-half the height times the sum of the upper and lower bases.

$$A = \frac{h}{2}(b_1 + b_2) \quad \text{or} \quad \frac{h(b_1 + b_2)}{2}$$

Examples Involving the Area of a Trapezoid (20)

Example 1: Find the area of the following trapezoid:

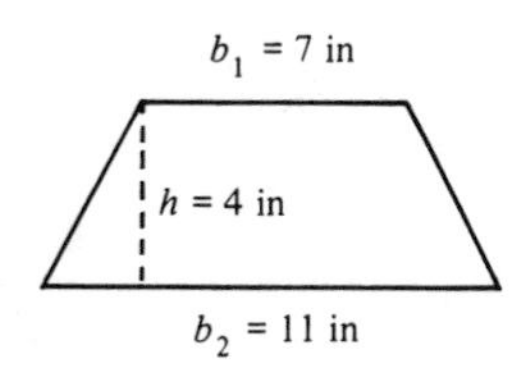

Solution: Use the formula $A = \frac{h(b_1 + b_2)}{2}$

Line (a): $A = \frac{h(b_1 + b_2)}{2}$

Line (b): $= \frac{(4 \text{ in})(7 \text{ in} + 11 \text{ in})}{2}$

Line (c): $= \frac{(4 \text{ in})(18 \text{ in})}{2}$

Line (d): $= \frac{(\not{4}^2 \text{ in})(18 \text{ in})}{\not{2}_1}$

Line (e): $= 36 \text{ in}^2$

Examples Involving the Area of a Trapezoid (Continued)

Example 2: Find the area of the following trapezoid:

$b_1 = 5.2$ cm

$h = 3.1$ cm

$b_2 = 8.3$ cm

Solution: Use the formula $A = \dfrac{h(b_1 + b_2)}{2}$

Line (a): $A = \dfrac{h(b_1 + b_2)}{2}$

Line (b): $= \dfrac{(3.1 \text{ cm})(5.2 \text{ cm} + 8.3 \text{ cm})}{2}$

Line (c): $= \dfrac{(3.1 \text{ cm})(13.5 \text{ cm})}{2}$

Line (d): $= \dfrac{41.85}{2} \text{ cm}^2$

Line (e): $= 20.925 \text{ cm}^2$

Computation

$$\begin{array}{r} 13.5 \\ \times 3.1 \\ \hline 1\ 35 \\ 40\ 5 \\ \hline 41.85 \end{array}$$

Perimeter of a Trapezoid (21)

The perimeter of a trapezoid is found by simply adding the lengths of the four sides.

Example: Find the perimeter of the following trapezoid:

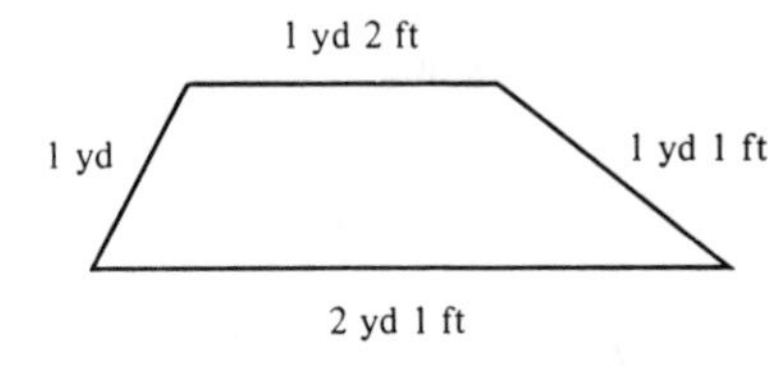

Solution: Add the lengths of the four sides:

$$\begin{array}{r} 1 \text{ yd } 2 \text{ ft} \\ 1 \text{ yd } 1 \text{ ft} \\ 2 \text{ yd } 1 \text{ ft} \\ 1 \text{ yd } 0 \text{ ft} \\ \hline 5 \text{ yd } 4 \text{ ft} \\ 1 \text{ yd} \phantom{\text{ 4 ft}} \\ \hline 6 \text{ yd } 1 \text{ ft} \end{array}$$

Therefore, the perimeter is 6 yd 1 ft.

Study Exercise Five (22)

A. Find the area of the following trapezoids:

1.

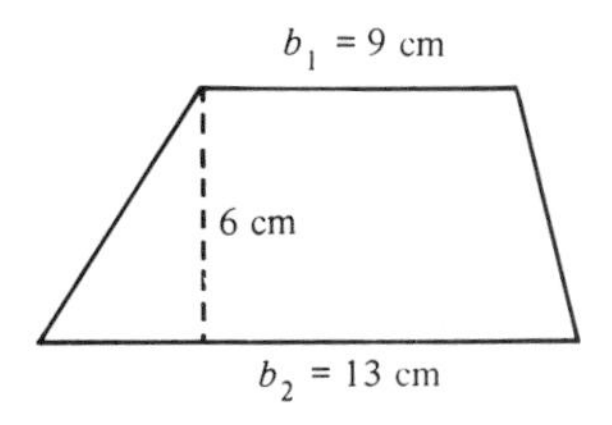

2.

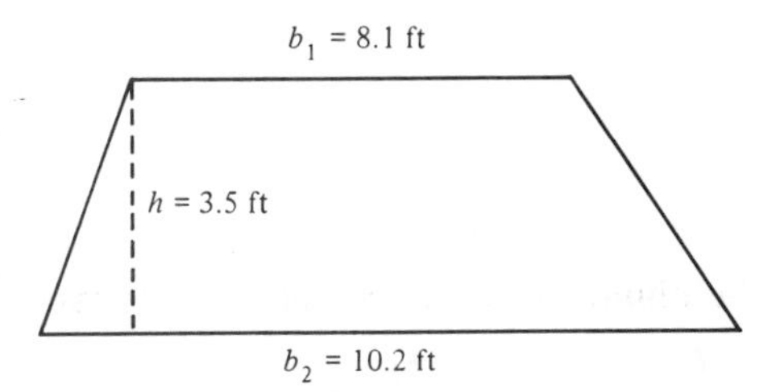

B. Solve the following problems:

3. Find the perimeter of this trapezoid:

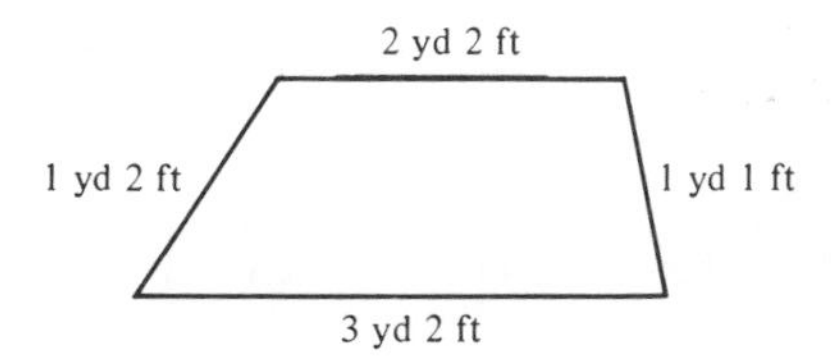

4. A roof is in the shape of a trapezoid. Its height is 9 ft. The bases are 14 ft and 24 ft. If a gallon of roofing tar covers 10 ft^2, how many gallons is needed to tar the entire roof?

Circles (23)

A circle is a set of points in a plane which are equidistant from a fixed point called the *center.*

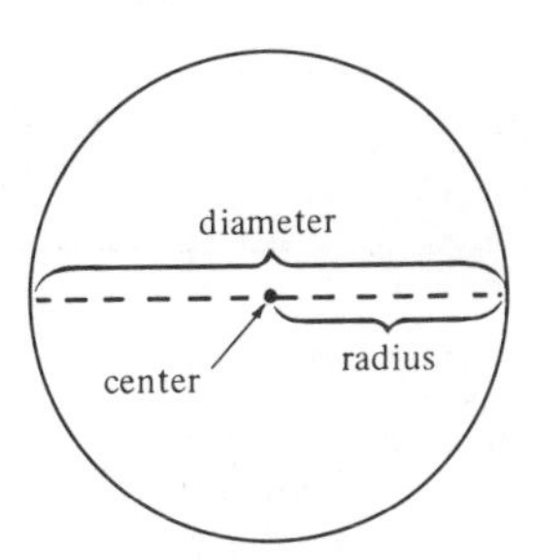

The radius of a circle is equal to one-half the diameter.

$$r = \frac{1}{2}d \quad \text{or} \quad \frac{d}{2}$$

Area of a Circle (24)

The area of a circle is the number of square units contained within the circle. It is given by this formula:

$$A = \pi r^2 \text{ where } \pi = 3.1415926\ldots$$

Unless instructed otherwise we will use 3.14 as our approximation for π.*

**Note:* The symbol "π" (read "pie") represents a number whose decimal expansion goes on at random and never terminates.

Area of a Circle (Continued)

Example 1: Find the area of a circle having a radius of 2 cm:

Solution: Use the formula $A = \pi r^2$ where $\pi = 3.14$:

Line (a): $A = \pi r^2$

Line (b): $= (3.14)(2\text{ cm})^2$

Line (c): $= (3.14)(4\text{ cm}^2)$

Line (d): $= 12.56\text{ cm}^2$

Example 2: Find the area of a circle having a diameter of 6 in:

Solution: Use the formula $A = \pi r^2$ where the radius is 3 in:

Line (a): $A = \pi r^2$

Line (b): $= (3.14)(3\text{ in})^2$

Line (c): $= (3.14)(9\text{ in}^2)$

Line (d): $= 28.26\text{ in}^2$

Circumference of a Circle (25)

The circumference of a circle is the distance around the circle.

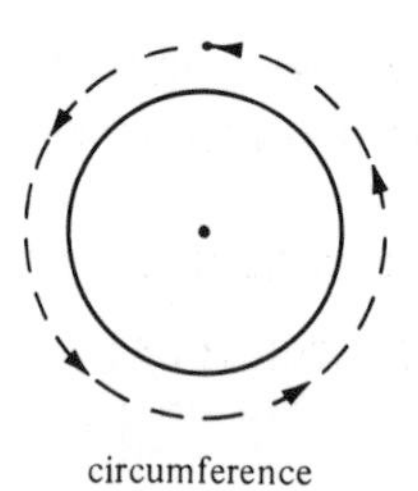

circumference

The circumference of a circle is found by using this formula:

$c = \pi d$ where π is approximated as 3.14*

Example 1: Find the circumference of a circle whose diameter is 5 in:

Solution: Use the formula $c = \pi d$ where $\pi = 3.14$:

Line (a): $c = \pi d$

Line (b): $= (3.14)(5\text{ in})$

Line (c): $= 15.7\text{ in}$

Example 2: Find the circumference of a circle whose radius is 7.5 cm:

Solution: Use the formula $c = \pi d$ where the diameter is 15 cm:

Line (a): $c = \pi d$

Line (b): $= (3.14)(15\text{ cm})$

Line (c): $= 47.1\text{ cm}$

Computation

$$\begin{array}{r} 3.14 \\ \underline{15} \\ 1570 \\ \underline{314} \\ 47.10 \end{array}$$

**Note:* Since the diameter of a circle is 2 times its radius you might also see this formula written as $c = 2\pi r$

Study Exercise Six (26)

A. Find the area of the following circles:

1.

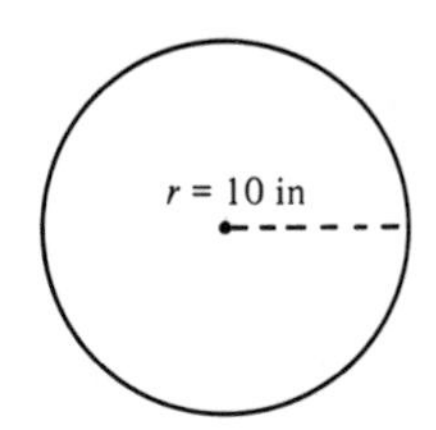

2.

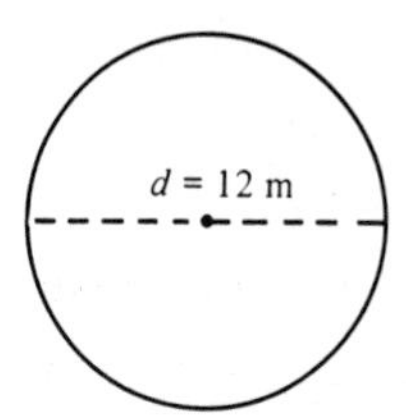

B. Find the circumference of the following circles:

3.

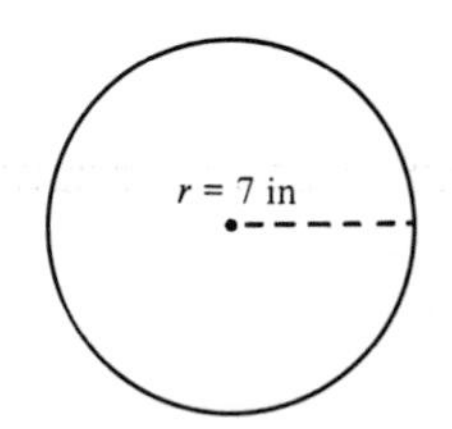

4.

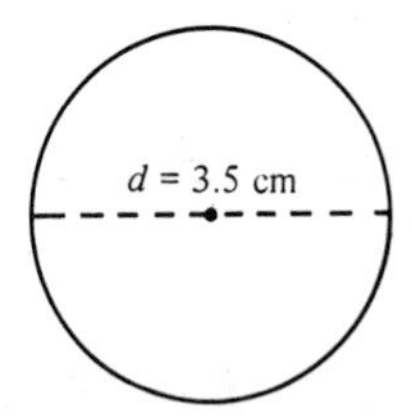

C. Solve the following problems:

5. What distance do you travel on one turn of a Ferris wheel if you sit 20 ft from the center?

6. A television station can be received for a distance of 50 mi in all directions. What area does the station cover?

Summary of Area, Perimeter, and Circumference (27)

1. Rectangle

$$A = lw$$
$$P = 2(l + w)$$

2. Square

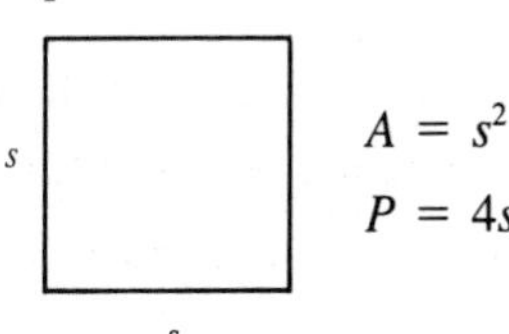

$$A = s^2$$
$$P = 4s$$

3. Parallelogram

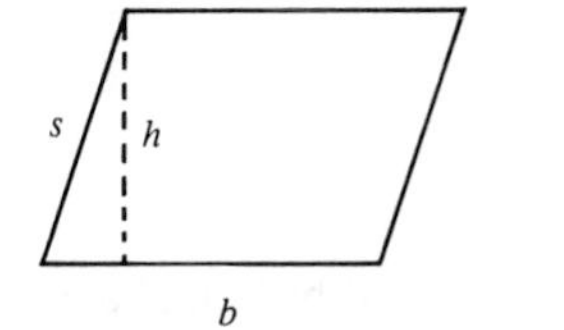

$$= bh$$
$$= 2(b + s)$$

4. Triangle

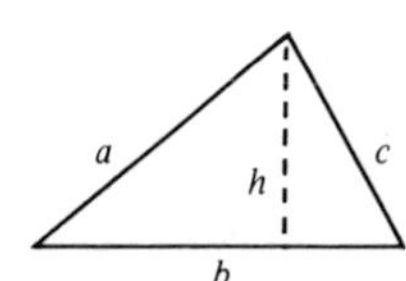

$$A = \frac{1}{2}bh \text{ or } \frac{bh}{2}$$
$$P = a + b + c$$

5. Trapezoid

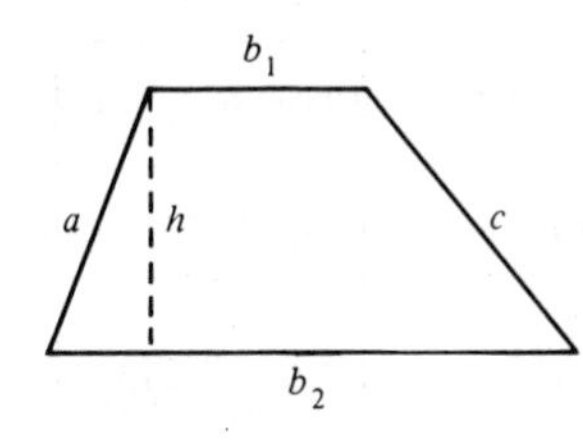

$$A = \frac{h}{2}(b_1 + b_2) \text{ or } \frac{h(b_1 + b_2)}{2}$$
$$P = a + b_1 + c + b_2$$

6. Circle

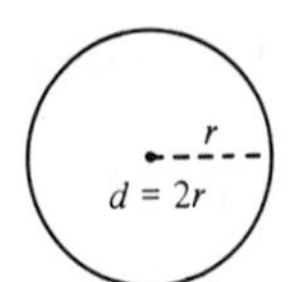

$$\left.\begin{aligned} A &= \pi r^2 \\ c &= \pi d \end{aligned}\right\} \pi \text{ is approximately } 3.14$$

REVIEW EXERCISES (28)

A. Find the area and perimeter of each of the following polygons:

1. Rectangle.

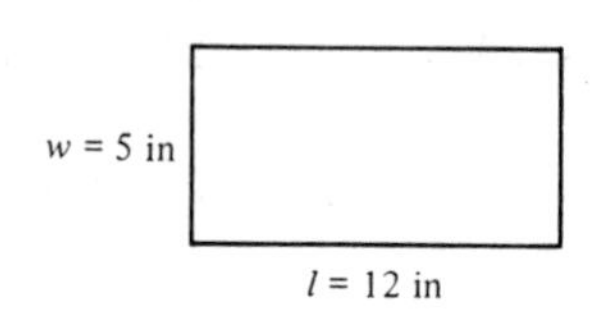

2. Square.

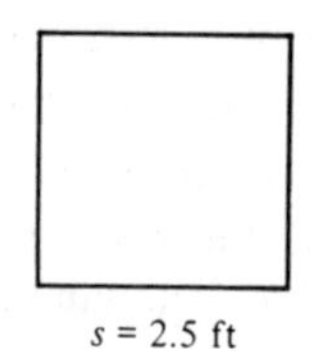

3. Parallelogram.

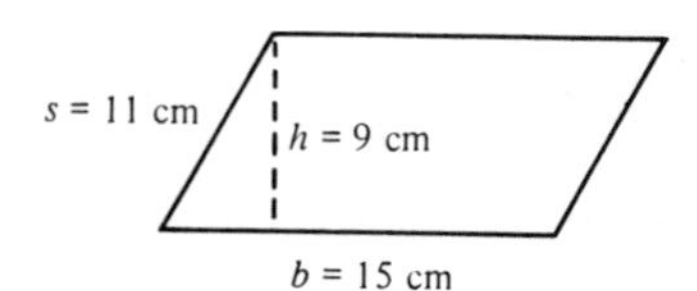

4. Triangle.

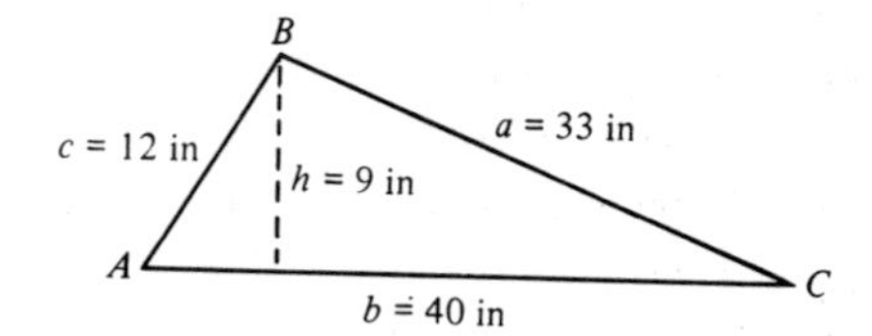

5. Right Triangle.

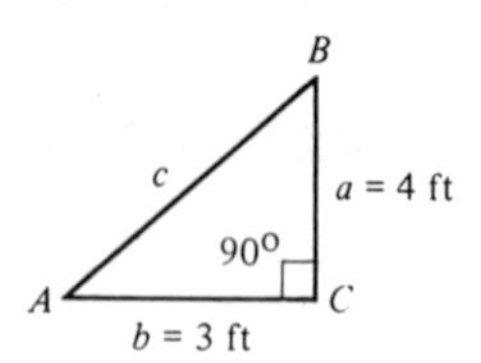

6. Trapezoid.

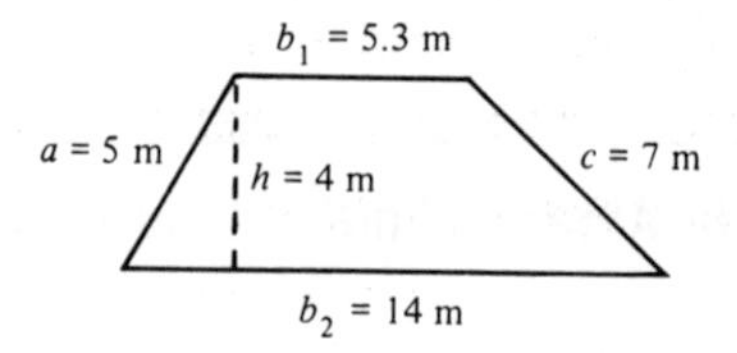

B. Find the area and circumference of the following circles:

7.

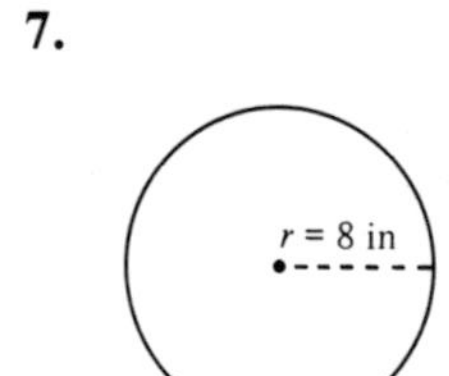

8.

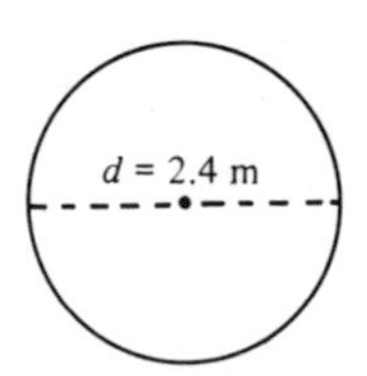

C. Solve the following problems:

9. Find the cost of constructing 50 ft of sidewalk 48 in wide at \$3.10 per ft^2.

10. A piece of rectangular carpet measures 8 ft by 10 ft. The edge of the carpet must be bound at a cost of 30¢ per foot. Find the total cost to bind the carpet.

11. A square room measures 15 ft on each side. If carpeting costs \$12 per square yard, what does it cost to carpet the room?

12. Find the perimeter of a square measuring 5 yds 2 ft on each side. Leave your answer in yards and feet.

13. Find the total area of a parking lot which contains 40 parking spaces in the form of parallelograms with bases of 16 ft and heights of 8 ft.

14. A triangular sail has a base of 10 ft and a height of 16 ft. How much area does one side of the sail expose to the wind?

REVIEW EXERCISES (Continued)

15. Each side of an equilateral triangle measures 2 ft 5 in. Find the perimeter in feet and inches.

16. A trapezoidal airplane wing has parallel sides measuring 25 ft and 30 ft respectively. The sides are 9 ft apart. Find the area of one side of the wing.

17. An automobile tire has a radius of 11 in. What is the distance around the outside edge?

18. A revolving sprinkler sprays a lawn for a distance of 14 ft. How many square feet is watered in one revolution if the spray pattern is a circle with a radius of 14 ft?

19. Two trapezoidal windows are to be replaced with stained glass. What is the total area involved, if the windows have bases of 4 ft and 6 ft and a height of 2 ft?

20. How much paint is needed to cover a rectangular 50 ft by 15 ft wall, if the paint covers at a rate of 400 square ft to 1 gal (nearest whole gallon)?

21. A lot 70 ft by 110 ft is to be fenced. How much will it cost to fence the lot if fencing costs $4.10 per foot?

22. How many yards of fringe is needed for a border on a bedspread 72 in by 102 in, if one of the 102 in sides is not to have a fringe (nearest whole yard)?

23. A rectangular lawn 42 ft by 52 ft is to be fertilized by a material that covers at the rate of 1 lb for every 200 square ft. To the nearest lb, how much fertilizer is needed?

24. To the nearest yd, how much fringe is needed for the border of a circular tablecloth, if the diameter is 44 in?

25. Which contains more pizza: a round pizza with a 12 inch diameter, or a square pizza 10 inches on each side?

26. A jewelry manufacturer makes circular medallions with diameters of 14 mm. What is the total area involved if both sides are to be gold plated?

Solutions to Review Exercises (29)

A.

1.
$$A = lw = (12\text{ in})(5\text{ in}) = 60\text{ in}^2$$
$$P = 2(l + w) = 2(12\text{ in} + 5\text{ in}) = 2(17\text{ in}) = 34\text{ in}$$

2.
$$A = s^2 = (2.5\text{ ft})^2 = 6.25\text{ ft}^2$$
$$P = 4s = 4(2.5\text{ ft}) = 10\text{ ft}$$

3.
$$A = bh = (15\text{ cm})(9\text{ cm}) = 135\text{ cm}^2$$
$$P = 2(b + s) = 2(15\text{ cm} + 11\text{ cm}) = 2(26\text{ cm}) = 52\text{ cm}$$

4.
$$A = \frac{bh}{2} = \frac{(40\text{ in})(9\text{ in})}{2} = \frac{360}{2}\text{ in}^2 = 180\text{ in}^2$$
$$P = a + b + c = 33\text{ in} + 40\text{ in} + 12\text{ in} = 85\text{ in}$$

5.
$$A = \frac{bh}{2} = \frac{(3\text{ ft})(4\text{ ft})}{2} = \frac{12}{2}\text{ ft}^2 = 6\text{ ft}^2$$
$$c^2 = a^2 + b^2 = 16 + 9 = 25$$
$$c = \sqrt{25}\text{ or }5\text{ ft}$$
$$P = a + b + c = 4\text{ ft} + 3\text{ ft} + 5\text{ ft} = 12\text{ ft}$$

6.
$$A = \frac{h(b_1 + b_2)}{2} = \frac{(4\text{ m})(5.3\text{ m} + 14\text{ m})}{2} = \frac{(\cancel{4}^{2}\text{ m})(19.3\text{ m})}{\cancel{2}_{1}} = 38.6\text{ m}^2$$
$$P = a + b_1 + c + b_2 = 5\text{ m} + 5.3\text{ m} + 7\text{ m} + 14\text{ m} = 31.3\text{ m}$$

B.

7.
$$A = \pi r^2 = (3.14)(8\text{ in})^2 = (3.14)(64\text{ in}^2) = 200.96\text{ in}^2$$
$$c = \pi d = (3.14)(16\text{ in}) = 50.24\text{ in}$$

8.
$$A = \pi r^2 = (3.14)(1.2\text{ m})^2 = (3.14)(1.44\text{ m}^2) = 4.5216\text{ m}^2$$
$$c = \pi d = (3.14)(2.4\text{ m}) = 7.536\text{ m}$$

Solutions to Review Exercises (Continued)

C. **9.** $A = lw$
$= (50 \text{ ft})(4 \text{ ft})$
$= 200 \text{ ft}^2$
$\text{Cost} = (200)(3.10)$
$= \$620$

10. $P = 2(l + w)$
$= 2(10 \text{ ft} + 8 \text{ ft})$
$= 2(18 \text{ ft})$
$= 36 \text{ ft}$
$\text{Cost} = (36)(.30)$
$= \$10.80$

11. Change feet to yards.

$15 \text{ ft} = \frac{15}{3} \text{ or } 5 \text{ yd}$

$A = s^2$
$= (5 \text{ yd})^2$
$= 25 \text{ yd}^2$
$\text{Cost} = (25)(12)$
$= \$300$

12. $P = 4s$
$= 4(5 \text{ yd } 2 \text{ ft})$

5 yd 2 ft
×4
20 yd 8 ft
2 yd
22 yd 2 ft

13. Find area of each parallelogram and then multiply by 40.

$A = bh$
$= (16 \text{ ft})(8 \text{ ft})$
$= 128 \text{ ft}^2$
$\text{Total} = (40)(128 \text{ ft}^2)$
$= 5{,}120 \text{ ft}^2$

14. $A = \frac{bh}{2}$
$= \frac{(\cancel{10}^5 \text{ ft})(16 \text{ ft})}{\cancel{2}_1}$
$= 80 \text{ ft}^2$

15. 2 ft 5 in
×3
6 ft 15 in
1 ft
7 ft 3 in

16. $A = \frac{h(b_1 + b_2)}{2}$
$= \frac{(9 \text{ ft})(25 \text{ ft} + 30 \text{ ft})}{2}$
$= \frac{(9 \text{ ft})(55 \text{ ft})}{2}$
$= \frac{495}{2} \text{ ft}^2$
$= 247.5 \text{ ft}^2$

17. $c = \pi d$
$= (3.14)(22 \text{ in})$
$= 69.08 \text{ in}$

18. $A = \pi r^2$
$= (3.14)(14 \text{ ft})^2$
$= (3.14)(196 \text{ ft}^2)$
$= 615.44 \text{ ft}^2$

19. $A = \frac{h(b_1 + b_2)}{2}$
$= \frac{(2 \text{ ft})(4 \text{ ft} + 6 \text{ ft})}{2}$
$= \frac{(2 \text{ ft})(10 \text{ ft})}{2}$
$= \frac{20}{2} \text{ ft}^2$
$= 10 \text{ ft}^2$ for one window
$(2)(10 \text{ ft}^2) = 20 \text{ ft}^2$ for two windows

20. $A = lw$
$= (15 \text{ ft})(50 \text{ ft})$
$= 750 \text{ ft}^2$

Divide to find the amount of paint needed.

$\frac{750}{400} = 1.875 \approx 2 \text{ gal}$

21. $P = 2(l + w)$
$= 2(110 \text{ ft} + 70 \text{ ft})$
$= 2(180 \text{ ft})$
$= 360 \text{ ft}$
$\text{Cost} = (360)(4.10)$
$= \$1{,}476$

22. $P = 72 \text{ in} + 72 \text{ in} + 102 \text{ in}$
$= 246 \text{ in}$

To change to yards divide by 36.

$\frac{246}{36} \approx 6.8 \text{ or } 7 \text{ yd}$

Solutions to Review Exercises (Continued)

23. $A = lw$
$= (52 \text{ ft})(42 \text{ ft})$
$= 2184 \text{ ft}^2$

Since each 200 ft² requires 1 lb, we divide.

$$\frac{2184}{200} = 10.92 \approx 11 \text{ lb}$$

24. $c = \pi d$
$= (3.14)(44 \text{ in})$
$= 138.16 \text{ in}$

Divide by 36 to find yards.

$$\frac{138.16}{36} \approx 3.8 \text{ or } 4 \text{ yd}$$

25. $A = \pi r^2$
$= (3.14)(6 \text{ in})^2$
$= (3.14)(36 \text{ in}^2)$
$= 113.04 \text{ in}^2$

$A = s^2$
$= (10 \text{ in})^2$
$= 100 \text{ in}^2$

The circular pizza is larger.

26. $A = \pi r^2$
$= (3.14)(7 \text{ mm})^2$
$= (3.14)(49 \text{ mm}^2)$
$= 153.86 \text{ mm}^2$ for one side

$(2)(153.86 \text{ mm}^2) = 307.72 \text{ mm}^2$

SUPPLEMENTARY PROBLEMS

A. Find the area and perimeter of each of the following polygons:

1. Rectangle.

$w = 8$ ft

$l = 12$ ft

2. Square.

$s = 6.2$ in

3. Parallelogram.

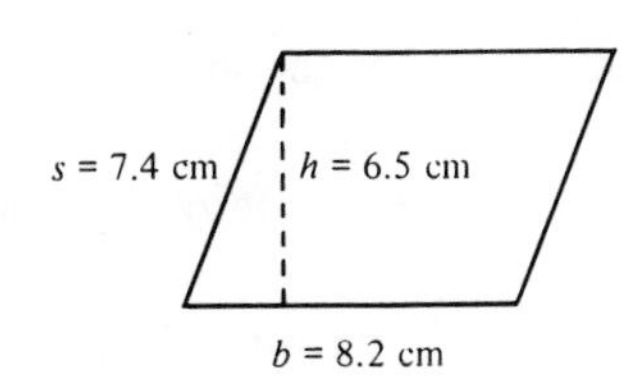

4. Triangle.

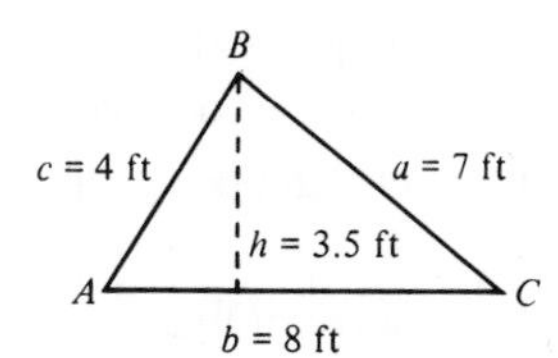

5. Right Triangle.

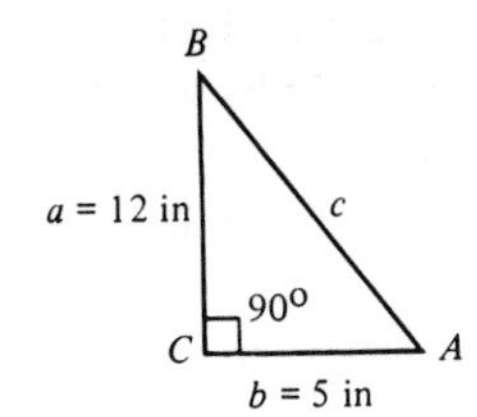

6. Trapezoid.

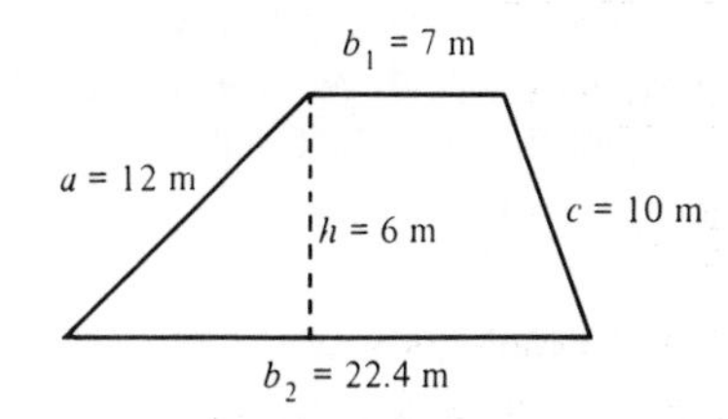

SUPPLEMENTARY PROBLEMS (Continued)

B. Find the area and circumference of the following circles:

7.

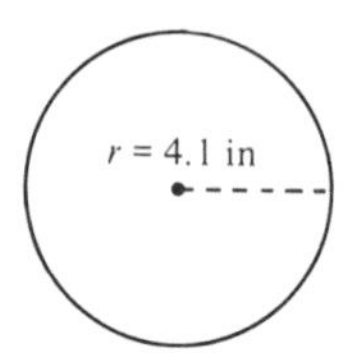

8.

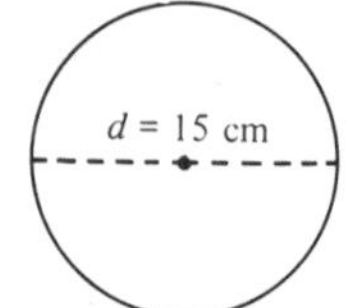

9. A room measures 15 ft by 21 ft. If carpeting costs $15 per square yard, what does it cost to carpet the room?

10. Find the cost of sodding a rectangular lawn measuring 35 ft by 18 ft at 30¢ a square foot.

11. How many feet of fencing is required to enclose a field measuring 120 ft by 70 ft?

12. A standard sheet of typing paper is 8.5 in by 11 in. What is its perimeter?

13. Find the area of a square table top measuring 32 in on each side.

14. The 55 parking spaces in a parking lot are in the form of parallelograms. Each has a base of 18 ft and a height of 9 ft. What is the total area covered by these spaces?

15. How many square feet of surface does one side of a triangular sail expose if it has a base of 8 ft and a height of 14 ft 6 in?

16. Each side of an equilateral triangle measures 1 ft 9 in. Find the perimeter in terms of feet and inches.

17. A roof is shaped in the form of a trapezoid. Its height is 9 ft and the bases are 14 ft and 24 ft. If a gallon of paint covers 300 ft^2, will there be enough to do two coats?

18. Over what distance will the tip of the minute hand of the clock "Big Ben" in London travel in one hour? The minute hand is 11 ft long.

19. A television station can televise programs for a distance of 60 mi in all directions. What area in square miles does it cover?

20. How many more square inches of pizza do you get in a 14-in diameter pan than a 12-in diameter pan?

21. Which has the greater area, a square measuring 10 in on each side or a circle having a diameter of 10 in?

22. Square 6 in by 6 in patio tiles sell for 27¢ each. How many are needed for one square foot, and how much would one square foot cost?

23. If 1 qt of liquid fertilizer covers 150 ft^2, how many gallons is needed for a 50 ft by 90 ft rectangular garden?

24. A room with a triangular fireplace in one corner is to be tiled. The room measures 15 ft by 14 ft and the fireplace triangle has a 6 ft base and 6 ft height. Compute the number of square feet to be tiled, if no tile is to be used for the fireplace area.

25. The walls of a room with a vaulted ceiling are to be covered with wallpaper. Two of the walls are identical trapezoids with a height of 30 ft and bases of 8 ft and 15 ft. The other two walls are rectangles 8 ft by 16 ft and 15 ft by 16 ft. Find the area to be wallpapered.

26. Two eaves of a garage are to be painted. Each eave is a rectangle 22 ft by 2 ft 6 in. If 1 qt of paint covers 100 ft^2, how much paint is needed?

27. From his lookout tower, a forest ranger can see a distance of 21 mi in all directions. How many square miles can he observe?

SUPPLEMENTARY PROBLEMS (Continued)

C. More difficult problems involving area. Find the area of each of the following composite figures:

28. Semicircular areas attached to a rectangle:

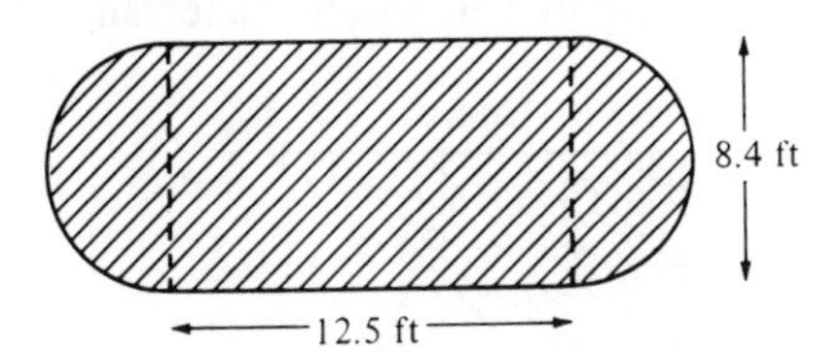

29. A machine part is in the shape of a trapezoid with two holes. Find the area of the shaded portion:

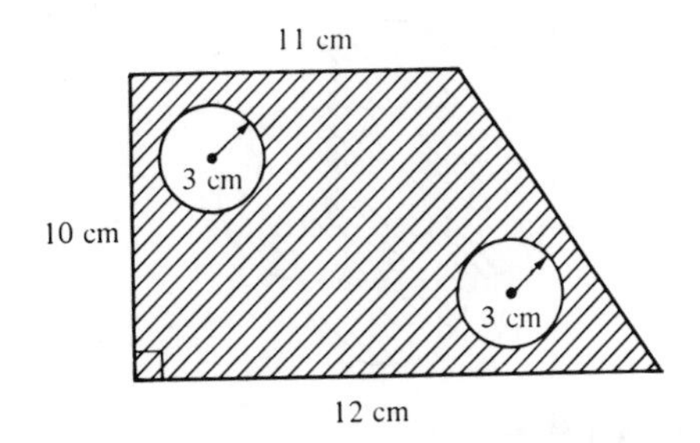

30. Find the area of the following ring:

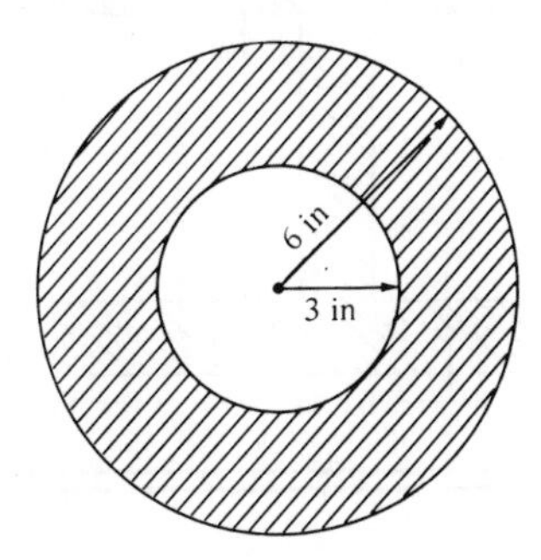

31. Find the area contained in the region outside the circle and inside the square:

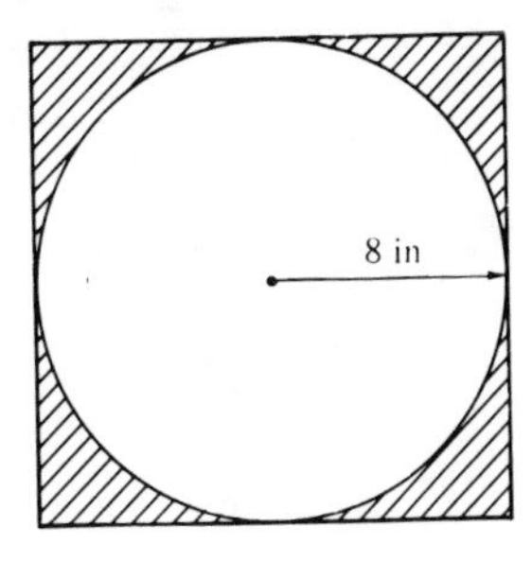

SUPPLEMENTARY PROBLEMS (Continued)

D. More difficult problems involving perimeter. Find the perimeter of each of the following composite figures:

32. Find the perimeter of the following figure composed of a rectangle with an inverted semicircular end:

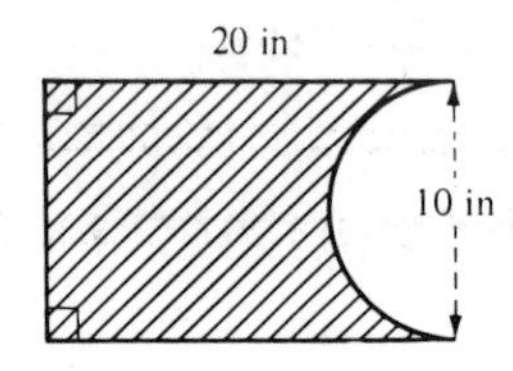

33. Find the perimeter of the following figure where semicircular regions have been attached to a square:

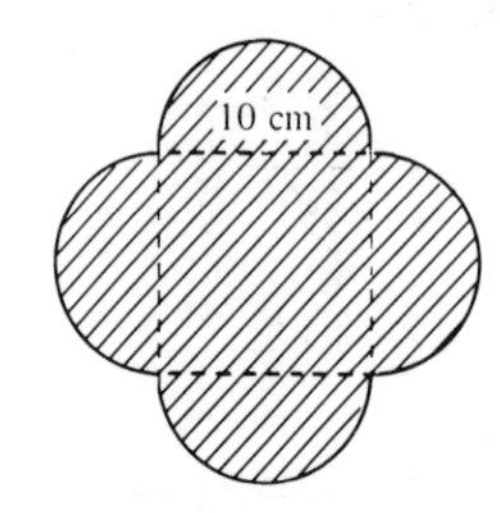

34. Find the perimeter of the given H-beam:

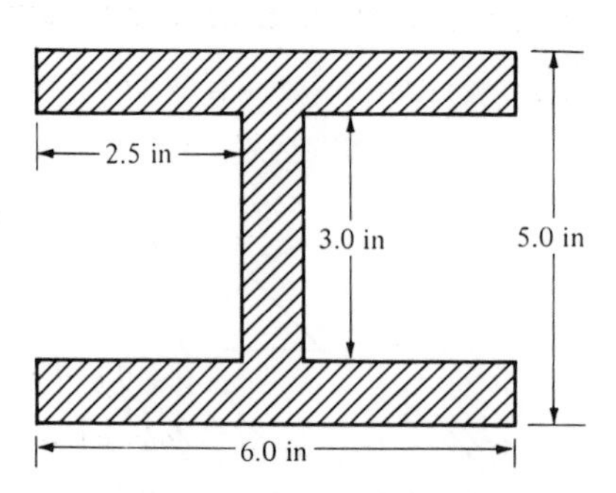

Solutions to Study Exercises (6A)

Study Exercise One (Frame 6)

A. 1. $A = lw$
$= (4 \text{ in})(2 \text{ in})$
$= 8 \text{ in}^2$

$P = 2(l + w)$
$= 2(4 \text{ in} + 2 \text{ in})$
$= 2(6 \text{ in})$
$= 12 \text{ in}$

2. $A = lw$
$= (5.6 \text{ m})(3.5 \text{ m})$
$= 19.6 \text{ m}^2$

$P = 2(l + w)$
$= 2(5.6 \text{ m} + 3.5 \text{ m})$
$= 2(9.1 \text{ m})$
$= 18.2 \text{ m}$

B. 3. $A = lw$
$= (18.5 \text{ in})(14.5 \text{ in})$
$= 268.25 \text{ in}^2$

4. Convert 36 in to feet.

$36 \text{ in} = \frac{36}{12}$ or 3 ft

$A = lw$
$= (70 \text{ ft})(3 \text{ ft})$
$= 210 \text{ ft}^2$

Cost $= (210)(3.20)$
$= \$672$

5. $P = 2(l + w)$
$= 2(62 \text{ in} + 40 \text{ in})$
$= 2(102 \text{ in})$
$= 204 \text{ in}$

Cost $= (204)(.05)$
$= \$10.20$

Solutions to Study Exercises (Continued)

6. $P = 2(l + w)$
$= 2(38 \text{ ft} + 23 \text{ ft})$
$= 2(61 \text{ ft})$
$= 122 \text{ ft}$

7. $P = 72 \text{ in} + 72 \text{ in} + 76 \text{ in}$
$= 220 \text{ in}$

To convert to yards divide by 36 in

$$\frac{220}{36} = 6\tfrac{1}{9} \text{ yd}$$

Study Exercise Two (Frame 9) **(9A)**

A. 1. $A = s^2$
$= (5 \text{ in})^2$
$= 25 \text{ in}^2$

$P = 4s$
$= (4)(5 \text{ in})$
$= 20 \text{ in}$

2. $A = s^2$
$= (3.2 \text{ m})^2$
$= 10.24 \text{ m}^2$

$P = 4s$
$= (4)(3.2 \text{ m})$
$= 12.8 \text{ m}$

B. 3. Convert to inches.

1 ft 7 in = 19 in

$A = s^2$
$= (19 \text{ in})^2$
$= 361 \text{ in}^2$

4. $P = 4s$

$$\begin{array}{r} 3 \text{ yd } 1 \text{ ft} \\ \times 4 \\ \hline 12 \text{ yd } 4 \text{ ft} \\ 1 \text{ yd} \phantom{ 1 \text{ ft}} \\ \hline 13 \text{ yd } 1 \text{ ft} \end{array}$$

5. $P = 4s$
$= (4)(60 \text{ ft})$
$= 240 \text{ ft}$

6. $A = s^2$
$= (141 \text{ ft})^2$
$= 19{,}881 \text{ ft}^2$

7. $P = 4s$
$= (4)(33 \text{ ft})$
$= 132 \text{ ft}$

Study Exercise Three (Frame 13) **(13A)**

A. 1. $A = bh$
$= (12.4 \text{ in})(8.5 \text{ in})$
$= 105.4 \text{ in}^2$

2. $A = bh$

$$= \left(\frac{\cancel{3}^{1}}{\cancel{4}_{1}} \text{ ft}\right)\left(\frac{\cancel{8}^{2}}{\cancel{15}_{5}} \text{ ft}\right)$$

$$= \frac{2}{5} \text{ ft}^2$$

B. 3. $P = 2(b + s)$
$= 2(35 \text{ cm} + 20 \text{ cm})$
$= 2(55 \text{ cm})$
$= 110 \text{ cm}$

4. $P = 2(b + s)$

Add the two sides, then multiply by 2

$$\begin{array}{l} 4 \text{ yd } 2 \text{ ft} \\ \underline{2 \text{ yd } 1 \text{ ft}} \\ 6 \text{ yd } 3 \text{ ft} = 7 \text{ yd} \end{array}$$

$P = 2(7 \text{ yd})$
$= 14 \text{ yd}$

C. 5. Find the area of each parallelogram, then multiply by 62

$A = bh$
$= (8 \text{ ft})(16 \text{ ft})$
$= 128 \text{ ft}^2$

Total $= (62)(128 \text{ ft}^2)$
$= 7{,}936 \text{ ft}^2$

Study Exercise Four (Frame 18) **(18A)**

A. 1. $A = \frac{1}{2}bh$

$= \frac{1}{2}(11 \text{ in})(8 \text{ in})$

$= \frac{1}{\not{2}_1}(11 \text{ in})(\not{8}^4 \text{ in})$

$= 44 \text{ in}^2$

2. $A = \frac{bh}{2}$

$= \frac{(25.5 \text{ cm})(12.2 \text{ cm})}{2}$

$= \frac{311.1}{2} \text{ cm}^2$

$= 155.55 \text{ cm}^2$

3. Convert 1 ft 5 in to inches.

1 ft 5 in = 17 in

$A = \frac{bh}{2}$

$= \frac{(11 \text{ in})(17 \text{ in})}{2}$

$= \frac{187}{2} \text{ in}^2$

$= 93.5 \text{ in}^2$

B. 4. $P = a + b + c$

$= 5.6 \text{ ft} + 3.4 \text{ ft} + 8.2 \text{ ft}$

$= 17.2 \text{ in}$

5. $c^2 = a^2 + b^2$

$= 16 + 9$

$= 25$

$c = \sqrt{25}$ or 5 m

$P = a + b + c$

$= 4 \text{ m} + 3 \text{ m} + 5 \text{ m}$

$= 12 \text{ m}$

6. $A = \frac{bh}{2}$

$= \frac{(12 \text{ ft})(\not{18}^9 \text{ ft})}{\not{2}_1}$

$= 108 \text{ ft}^2$

7. An equilateral triangle has three equal sides. Therefore, multiply 3 yd 2 ft by 3

```
 3 yd 2 ft
      ×3
 ---------
 9 yd 6 ft
 2 yd
 ---------
11 yd 0 ft or 11 yd
```

8. $A = \frac{1}{2}bh$

$= \frac{1}{2}(30 \text{ ft})(20 \text{ ft})$

$= \frac{1}{2}(600 \text{ ft}^2)$

$= 300 \text{ ft}^2$ area of one triangle

$4(300 \text{ ft}^2) = 1200 \text{ ft}^2$ area of entire roof

Study Exercise Five (Frame 22) **(22A)**

A. 1. $A = \frac{h(b_1 + b_2)}{2}$

$= \frac{(6 \text{ cm})(9 \text{ cm} + 13 \text{ cm})}{2}$

$= \frac{(6 \text{ cm})(\not{22}^{11} \text{ cm})}{\not{2}_1}$

$= 66 \text{ cm}^2$

2. $A = \frac{h(b_1 + b_2)}{2}$

$= \frac{(3.5 \text{ ft})(8.1 \text{ ft} + 10.2 \text{ ft})}{2}$

$= \frac{(3.5 \text{ ft})(18.3 \text{ ft})}{2}$

$= \frac{64.05}{2} \text{ ft}^2$

$= 32.025 \text{ ft}^2$

B. 3. Add the lengths of the four sides:

```
1 yd 2 ft
2 yd 2 ft
1 yd 1 ft
3 yd 2 ft
---------
7 yd 7 ft
2 yd
---------
9 yd 1 ft
```

4. Find the area of the roof; then since each gallon of tar covers 10 ft^2, divide by 10.

$A = \frac{h(b_1 + b_2)}{2}$

$= \frac{(9 \text{ ft})(14 \text{ ft} + 24 \text{ ft})}{2}$

$= \frac{(9 \text{ ft})(\not{38}^{19} \text{ ft})}{\not{2}_1}$

$= 171 \text{ ft}^2$

Therefore, the amount of tar is $\frac{171}{10}$ or 17.1 gal.

Study Exercise Six (Frame 26) (26A)

A. **1.** $A = \pi r^2$
$= (3.14)(10\text{ in})^2$
$= (3.14)(100\text{ in}^2)$
$= 314\text{ in}^2$

2. $A = \pi r^2$
$= (3.14)(6\text{ m})^2$
$= (3.14)(36\text{ m}^2)$
$= 113.04\text{ m}^2$

B. **3.** $c = \pi d$
$= (3.14)(14\text{ in})$
$= 43.96\text{ in}$

4. $c = \pi d$
$= (3.14)(3.5\text{ cm})$
$= 10.99\text{ cm}$

C. **5.** $c = \pi d$
$= (3.14)(40\text{ ft})$
$= 125.6\text{ ft}$

6. $A = \pi r^2$
$= (3.14)(50\text{ mi})^2$
$= 7{,}850\text{ mi}^2$

UNIT

38 Volume and Surface Area

OBJECTIVES (1)

By the end of this unit you should be able to:

1. FIND VOLUMES AND SURFACE AREAS FOR PRISMS (INCLUDING RECTANGULAR SOLIDS AND CUBES), AND SPHERES.
2. FIND THE VOLUME, LATERAL AREA, AND TOTAL AREA OF RIGHT CIRCULAR CYLINDERS.

Three-Dimensional Solids (2)

Three-dimensional solids are physical objects which take up space and have three measurements such as length, width, and height. In this unit we will learn to find the volume and surface area for these three-dimensional solids:

1. Prisms
2. Right circular cylinders
3. Spheres

Prisms (3)

A prism is any three-dimensional solid having two parallel bases that are identical polygons. The perpendicular distance between the bases is the *height*.

Example 1: A prism known as a *rectangular solid* (box).

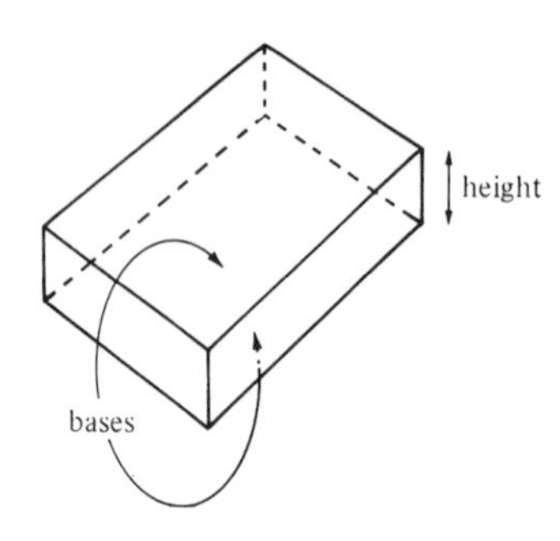

Prisms (Continued)

Example 2: A prism known as a *cube*. All sides are equal.

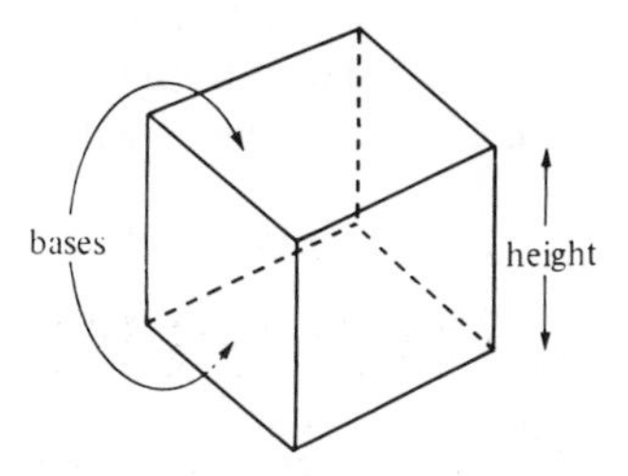

Example 3: A prism with triangular bases.

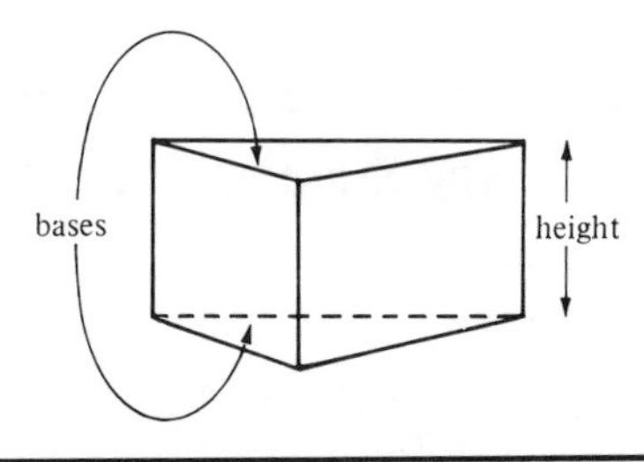

Right Circular Cylinders (4)

A right circular cylinder is any three-dimensional solid whose bases are parallel circles of the same size and whose sides are perpendicular to the bases.

Example 1: Right Circular Cylinder

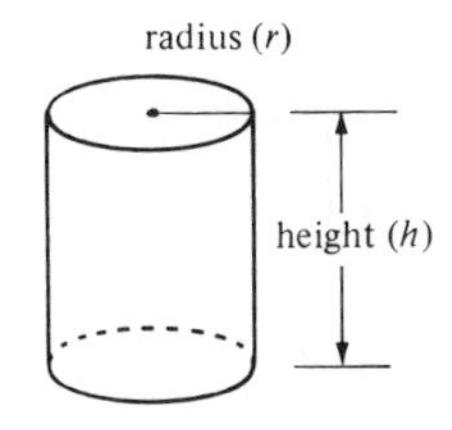

Example 2: Right Circular Cylinder

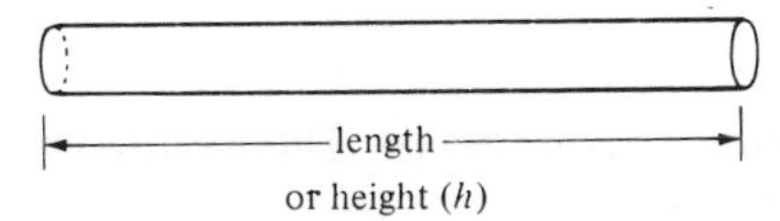

Example 3: Right Circular Cylinder

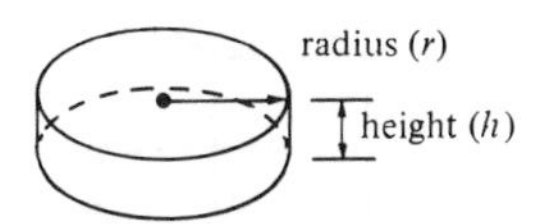

Spheres (5)

A sphere is a three-dimensional solid bounded by a closed surface such that every point in the surface is the same distance from a point called the *center.* This distance from the center to any point on the surface is called the *radius*. A circle, when rotated about one of its diameters, will generate a sphere.

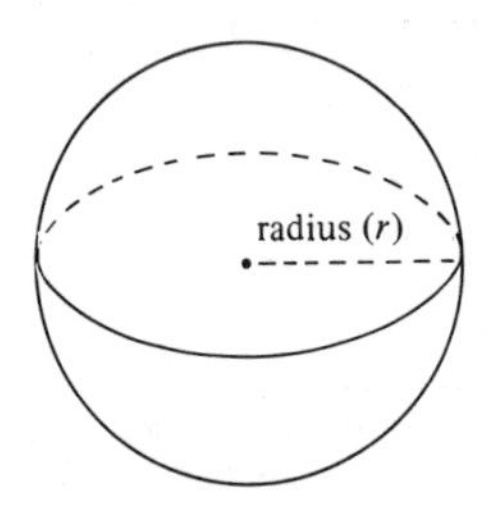

Study Exercise One (6)

Identify the following three-dimensional solids:

1.

2.

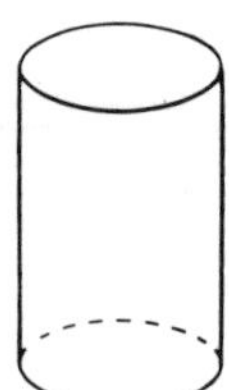

3.

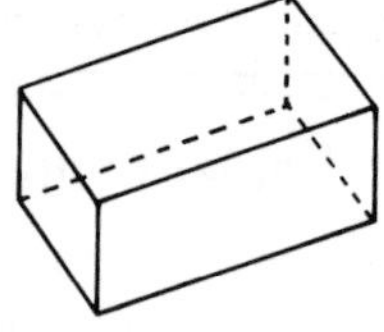

4.

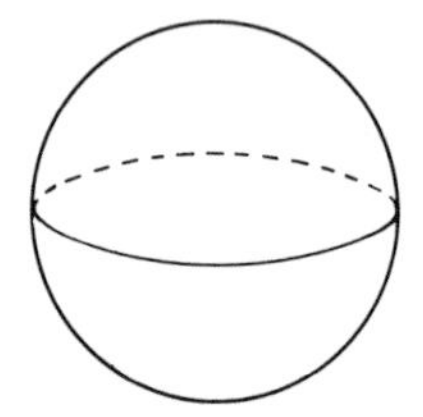

5.

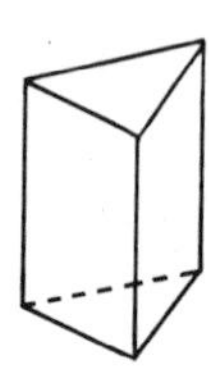

Volume (7)

Every three-dimensional object occupies a certain space. The measure of this space is called its *volume* and is given by the number of cubic units it contains.

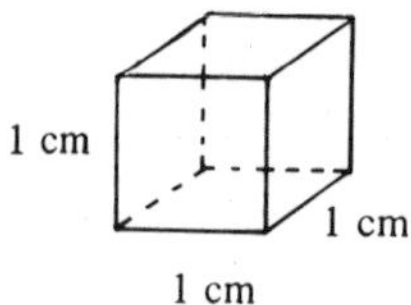

One cubic centimeter
(1 cm^3 or 1 cc)

Volume (Continued)

Example: This rectangular solid contains 18 of the cubic centimeters pictured above. Thus, its volume is 18 cm^3. This could also be written as 18 cc.

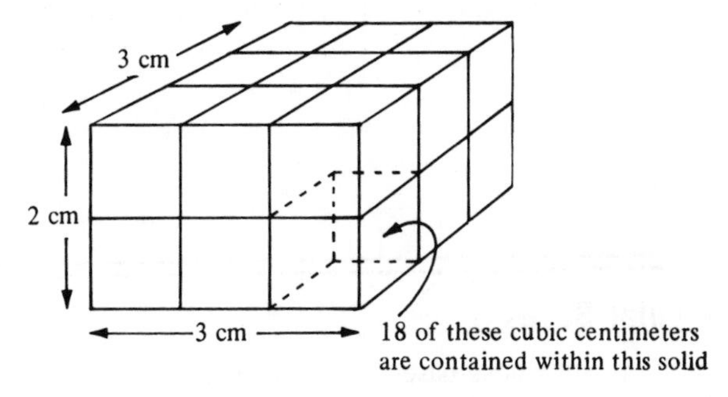

Volume may also be measured in other cubic units, such as cubic meters (m^3), cubic inches (in^3), cubic feet (ft^3), or cubic miles (mi^3).

Volume of a Prism (8)

The volume of a prism is found by multiplying the area of the base times the height.

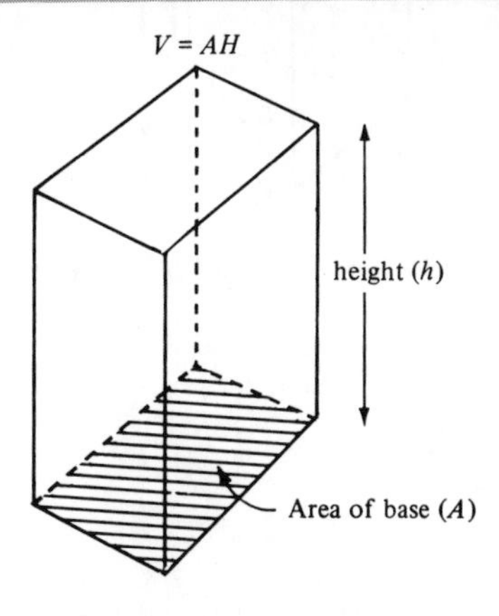

Examples Using the Formula $V = AH$ (9)

Example: Find the volume of a rectangular solid whose bases measure 2 in by 3 in and whose height measures 4 in:

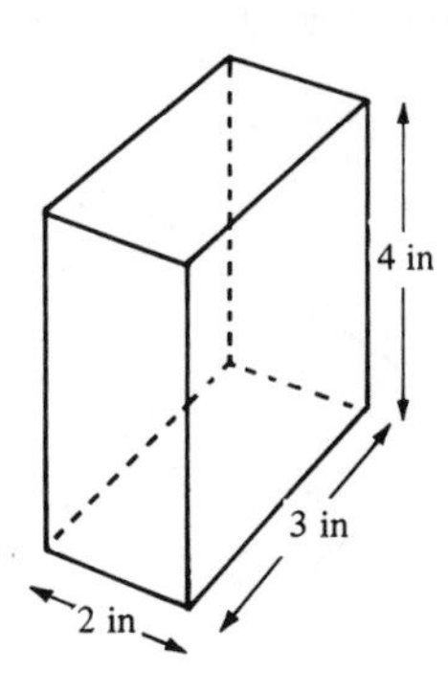

Examples Using the Formula $V = AH$ (Continued)

Solution: Find the area of the rectangular base and multiply it times the height.

Line (a): Area of base $= lw$

Line (b): $= (3\text{ in})(2\text{ in})$

Line (c): $= 6\text{ in}^2$

Line (d): $V = AH$

Line (e): $= (6\text{ in}^2)(4\text{ in})$

Line (f): $= 24\text{ in}^3$

A Prism with Right Triangular Bases (10)

Example: Find the volume of a prism whose bases are right triangles with legs measuring 3 cm and 4 cm. The height of the prism is 8 cm.

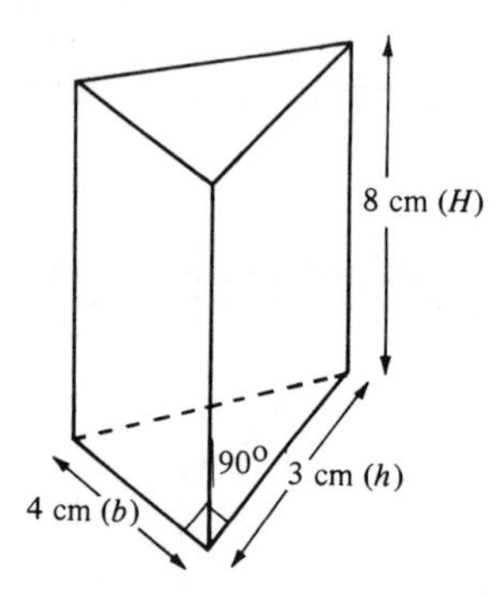

Solution: Find the area of the triangular base and multiply it times the height.

Line (a): Area of base $= \dfrac{bh}{2}$

Line (b): $= \dfrac{(\not{4}^{2}\text{ cm})(3\text{ cm})}{\not{2}_{1}}$

Line (c): $= 6\text{ cm}^2$

Line (d): $V = AH$

Line (e): $= (6\text{ cm}^2)(8\text{ cm})$

Line (f): $= 48\text{ cm}^3$

An Applied Problem (11)

Example: How many cubic yards of concrete is needed for a driveway measuring 21 ft long by 12 ft wide? The concrete is to be 4 in deep.

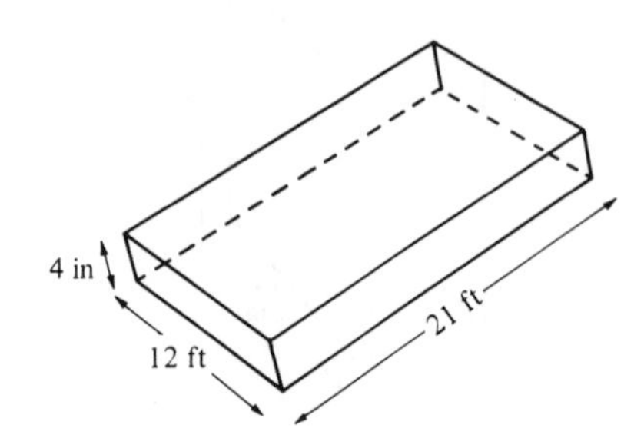

An Applied Problem (Continued)

Solution: Change all measurements to yards; then find the volume of the rectangular solid:

Line (a): 21 ft = 7 yd; 12 ft = 4 yd; and 4 in = $\frac{4}{36}$ yd or $\frac{1}{9}$ yd

Line (b): $V = AH$

Line (c): $= \overbrace{(7 \text{ yd})(4 \text{ yd})}\left(\frac{1}{9} \text{ yd}\right)$

Line (d): $= \frac{28}{9} \text{ yd}^3$

Line (e): $= 3\frac{1}{9} \text{ yd}^3$

Therefore, $3\frac{1}{9}$ yd^3 of concrete is needed to construct the driveway.

Special Volume Formulas (12)

1. The formula for the volume of a rectangular solid can also be written as $V = lwh$ where l represents the length, w represents the width, and h represents the height. This is true because lw is the area of the base.

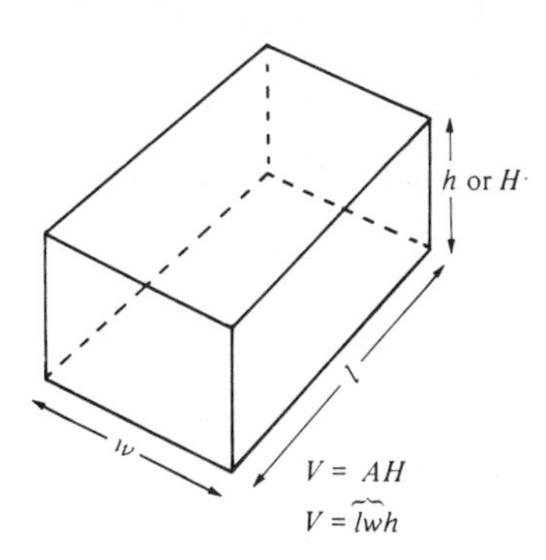

Example: Find the volume of a rectangular solid 3 in long, 2 in wide, and 1 in high:

Solution: $V = lwh$

$= (3 \text{ in})(2 \text{ in})(1 \text{ in})$

$= 6 \text{ in}^3$

2. The formula for the volume of a cube can also be written as $V = s^3$, where s represents the length of each side.

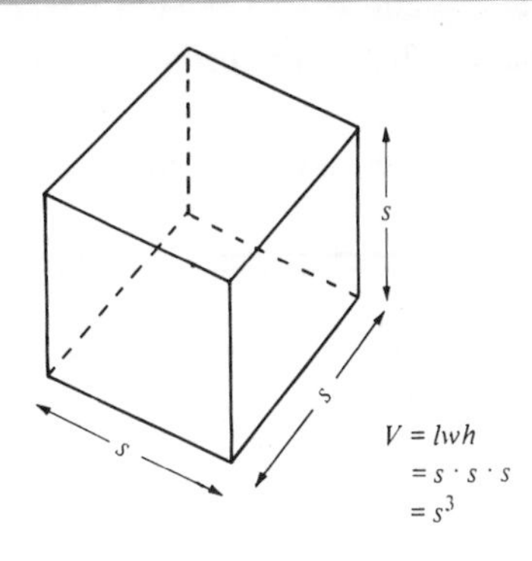

Special Volume Formulas (Continued)

Example: Find the volume of a cube whose sides each measure 2 cm:

Solution: $V = s^3$
$= (2 \text{ cm})^3$
$= 8 \text{ cm}^3$

Study Exercise Two (13)

1. Find the volume of the following prism whose bases are right triangles with legs measuring 5 in and 12 in. The height of the prism is 15 in.

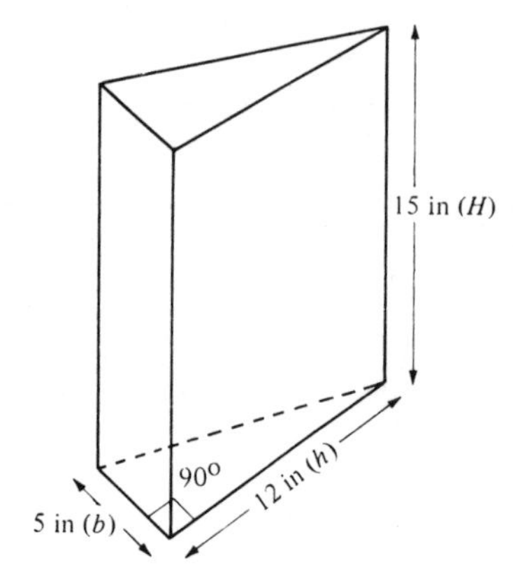

2. Find the volume of the following rectangular solid measuring 3.2 in long, 2.1 in wide, and 1.5 in high:

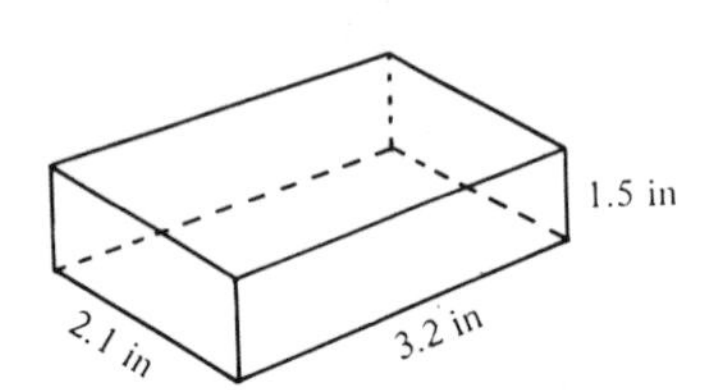

3. Find the volume of the following cube where each side measures 3 ft:

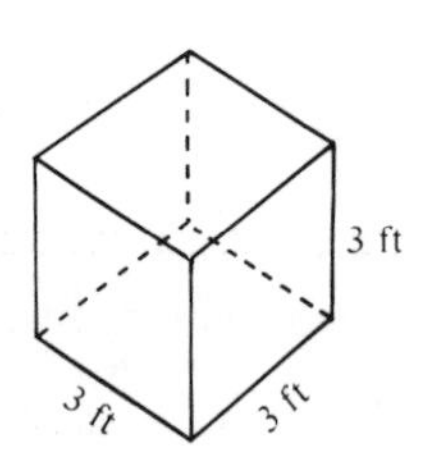

4. Dirt fill costs $4.50 per cubic yard. What is the cost to fill a rectangular hole measuring 15 ft long, 12 ft wide, and 9 ft deep?
5. How many cubic yards of concrete is needed for a patio measuring 15 ft long by 12 ft wide? The concrete is to be 3 in deep.

Volume of a Right Circular Cylinder (14)

As with a prism, the volume of a right circular cylinder is found by multiplying the area of the base times the height.

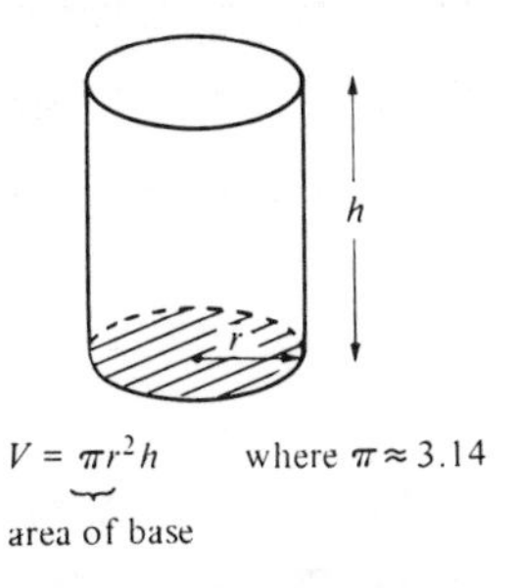

$V = \underbrace{\pi r^2}_{\text{area of base}} h$ where $\pi \approx 3.14$

Examples Using the Formula $V = \pi r^2 h$ (15)

Example: Find the volume of a right circular cylinder whose radius is 3 ft and whose height is 10 ft.

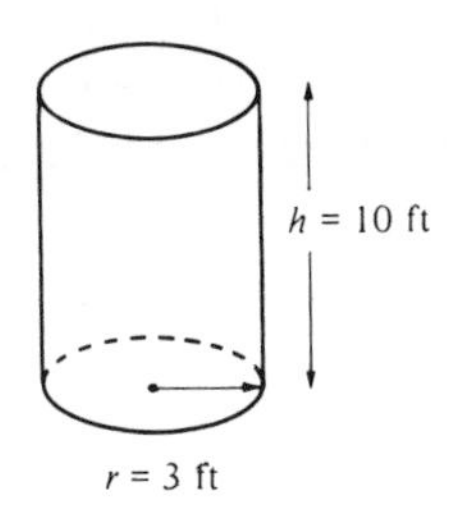

Solution:

Line (a): $V = \pi r^2 h$

Line (b): $= (3.14)(3 \text{ ft})^2(10 \text{ ft})$

Line (c): $= (3.14)\underbrace{(9 \text{ ft}^2)(10 \text{ ft})}$

Line (d): $= (3.14)(90 \text{ ft}^3)$

Line (e): $= 282.6 \text{ ft}^3$

Computation

3.14
90
282.60

An Applied Problem (16)

Example: A railroad tank car in the shape of a cylinder is used to transport LP gas. If the tank car is 2 yd wide and 50 ft long, what is its capacity in cubic feet?

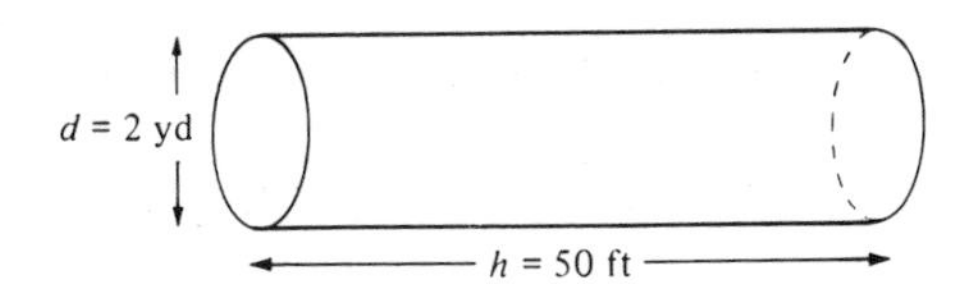

An Applied Problem (Continued)

Solution: Convert 2 yd to feet, find the radius, and use the formula $V = \pi r^2 h$:

Line (a): $2 \text{ yd} = (2)(3 \text{ ft}) \text{ or } 6 \text{ ft}$

Line (b): $r = \dfrac{6 \text{ ft}}{2} \text{ or } 3 \text{ ft}$

Line (c): $V = \pi r^2 h$

Line (d): $= (3.14)(3 \text{ ft})^2(50 \text{ ft})$

Line (e): $= (3.14)\underbrace{(9 \text{ ft}^2)(50 \text{ ft})}$

Line (f): $= (3.14)(450 \text{ ft}^3)$

Line (g): $= 1{,}413 \text{ ft}^3$

Computation

$$\begin{array}{r} 3.14 \\ \underline{4\ 50} \\ 157\ 00 \\ \underline{1256} \\ 1{,}413.00 \end{array}$$

Study Exercise Three (17)

1. Find the volume of a right circular cylinder whose radius is 2 m and whose height is 8.5 m.
2. A hole 4 yd in diameter has been dug 15 ft into the ground for a concrete footing. If concrete costs $5.00 a cubic yard, find the cost of the footing.
3. How many cubic feet of water does a tank hold if its diameter is 6 ft and its height is 12 ft?
4. To the nearest gallon, how many gallons does the tank in problem 3 hold if 7.5 gal of water occupies 1 ft³?

Volume of a Sphere (18)

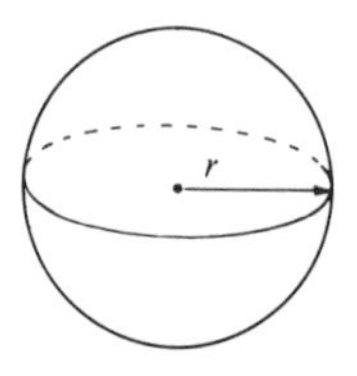

The volume of a sphere is given by this formula:

$$V = \frac{4}{3}\pi r^3 \text{ or } \frac{4\pi r^3}{3} \qquad \text{where } \pi \approx 3.14$$

Using the Formula $V = \dfrac{4\pi r^3}{3}$ (19)

Example: Find the volume of a sphere whose radius is 3 in:

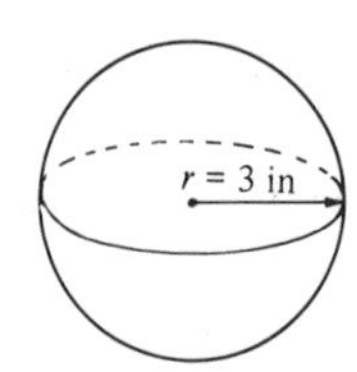

Using the Formula $V = \dfrac{4\pi r^3}{3}$ (Continued)

Solution:

Line (a): $V = \dfrac{4\pi r^3}{3}$

Line (b): $= \dfrac{(4)(3.14)(3\text{ in})^3}{3}$

Line (c): $= \dfrac{(4)(3.14)(\cancel{27}^{9}\text{ in}^3)}{\cancel{3}_1}$

Line (d): $= (3.14)(36\text{ in}^3)$

Line (e): $= 113.04\text{ in}^3$

Computation

```
  3.14
    36
 18 84
 94 2
113.04
```

An Applied Problem (20)

Example: How many cubic inches of rubber is needed to manufacture a solid toy rubber ball having a diameter of 2 in? Round the answer to the nearest hundredth.

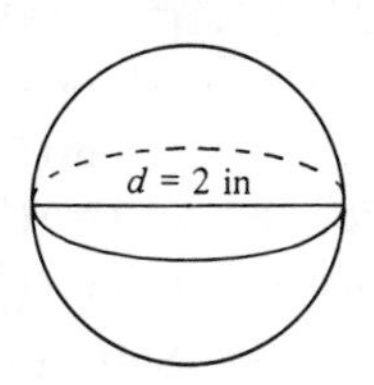

Solution: Find the radius and then use the formula $V = \dfrac{4\pi r^3}{3}$:

Line (a): $r = \dfrac{2\text{ in}}{2}$ or 1 in

Line (b): $V = \dfrac{4\pi r^3}{3}$

Line (c): $= \dfrac{(4)(3.14)(1\text{ in})^3}{3}$

Line (d): $= \dfrac{(4)(3.14)(1\text{ in}^3)}{3}$

Line (e): $= \dfrac{12.56\text{ in}^3}{3}$

Line (f): $= 4.19\text{ in}^3$ (to the nearest hundredth)

Computation

```
   4.186
3)12.56
  12
    5
    3
    26
    24
     20
     18
```

Study Exercise Four (21)

1. Find the volume of a sphere whose radius is 2 cm (round to nearest tenth).
2. A natural gas storage tank is in the shape of a sphere with a diameter of 60 ft. How many cubic yards of gas does it contain (round to nearest tenth)?
3. How many cubic centimeters of helium is needed to blow up a balloon having a radius of 10 cm? Round the answer to nearest whole cm^3.
4. Express the answer to problem 3 in milliliters and liters.

Surface Area (22)

The surface area of a three-dimensional object is the area of the entire surface.

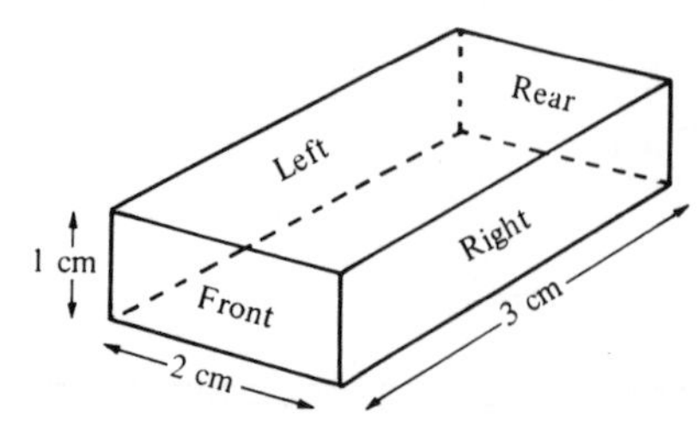

For example, this rectangular solid has six faces: top and bottom, front and rear, left and right. The surface area is the sum of these six areas.

Area of top face = (2 cm)(3 cm) or 6 cm^2
Area of bottom face = (2 cm)(3 cm) or 6 cm^2
Area of front face = (1 cm)(2 cm) or 2 cm^2
Area of rear face = (1 cm)(2 cm) or 2 cm^2
Area of left face = (1 cm)(3 cm) or 3 cm^2
Area of right face = (1 cm)(3 cm) or 3 cm^2

Surface area = 22 cm^2

Surface area may also be measured in terms of other square units, such as square meters (m^2), square inches (in^2), square feet (ft^2), or square miles (mi^2).

Remember, volume is measured in terms of cubic units, while surface area is measured in terms of square units.

Surface Area of a Prism (23)

The surface area of a prism is the sum of the areas of each face.

Rectangular Solids (24)

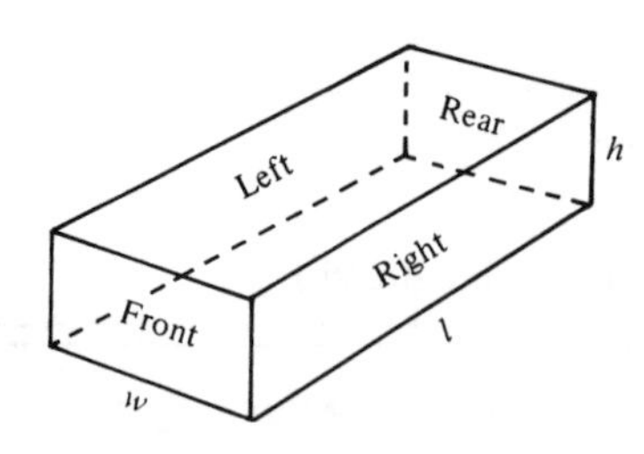

Area of top and bottom faces = $2lw$
Area of left and right faces = $2lh$
Area of front and rear faces = $2wh$

$$A = 2lw + 2lh + 2wh$$

Rectangular Solids (Continued)

Example: Find the surface area of a rectangular solid having a length of 4 in, a width of 3 in, and a height of 2 in:

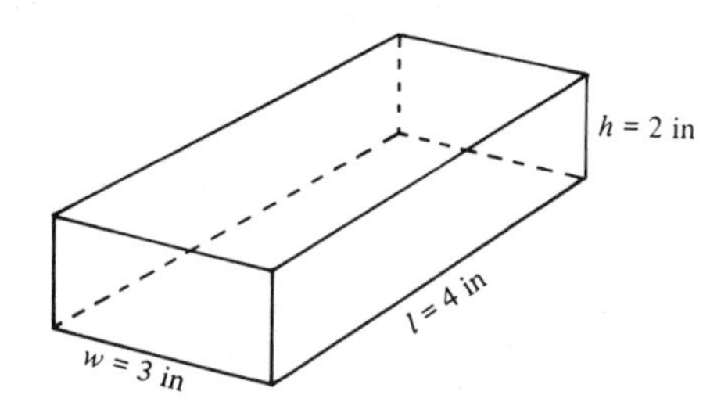

Solution: Use the formula $A = 2lw + 2lh + 2wh$:

Line (a): $A = 2lw + 2lh + 2wh$

Line (b): $= \underbrace{(2)(4 \text{ in})(3 \text{ in})} + \underbrace{(2)(4 \text{ in})(2 \text{ in})} + \underbrace{(2)(3 \text{ in})(2 \text{ in})}$

Line (c): $= 24 \text{ in}^2 + 16 \text{ in}^2 + 12 \text{ in}^2$

Line (d): $= 52 \text{ in}^2$

Cubes (25)

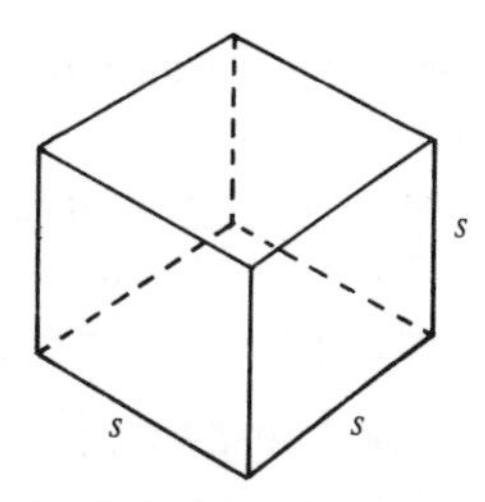

A cube has six faces. The area of each face is s^2. Therefore, the formula for the surface area of a cube is $A = 6s^2$.

Example: Find the surface area of a cube where each side measures 3 cm:

Solution: Use the formula $A = 6s^2$:

Line (a): $A = 6s^2$

Line (b): $= 6(3 \text{ cm})^2$

Line (c): $= 6(9 \text{ cm}^2)$

Line (d): $= 54 \text{ cm}^2$

Other Prisms (26)

The surface area of other types of prisms can be determined by adding together the area of each of the faces.

Example: Find the surface area of the following prism whose bases are right triangles with legs measuring 3 cm and 4 cm. The height is 10 cm.

Other Prisms (Continued)

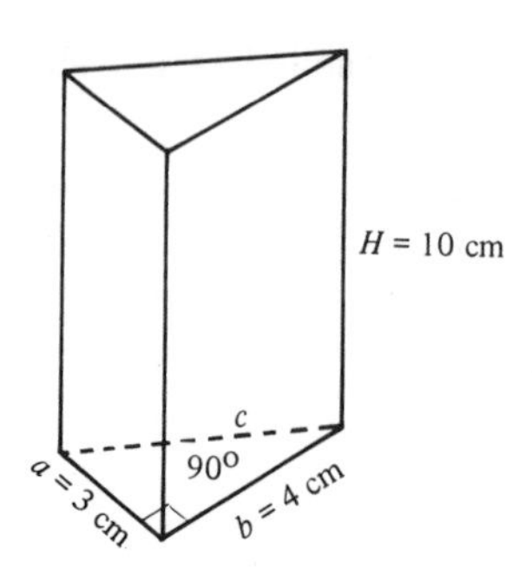

Solution: Use the Pythagorean theorem to find the length of the hypotenuse (c). Then find the area of each of the five faces and add them together.

Line (a): $c^2 = a^2 + b^2$

Line (b): $= 9 + 16$

Line (c): $= 25$

Line (d): $c = \sqrt{25}$ or 5 cm

Area of bottom face $= \dfrac{(3 \text{ cm})(4 \text{ cm})}{2} = 6 \text{ cm}^2$

Area of upper face $= \dfrac{(3 \text{ cm})(4 \text{ cm})}{2} = 6 \text{ cm}^2$

Area of rear face $= (5 \text{ cm})(10 \text{ cm}) = 50 \text{ cm}^2$

Area of left face $= (3 \text{ cm})(10 \text{ cm}) = 30 \text{ cm}^2$

Area of right face $= (4 \text{ cm})(10 \text{ cm}) = 40 \text{ cm}^2$

Surface area $= 132 \text{ cm}^2$

Study Exercise Five (27)

1. Find the surface area of a rectangular solid having a length of 7 in, a width of 5 in, and a height of 3 in:

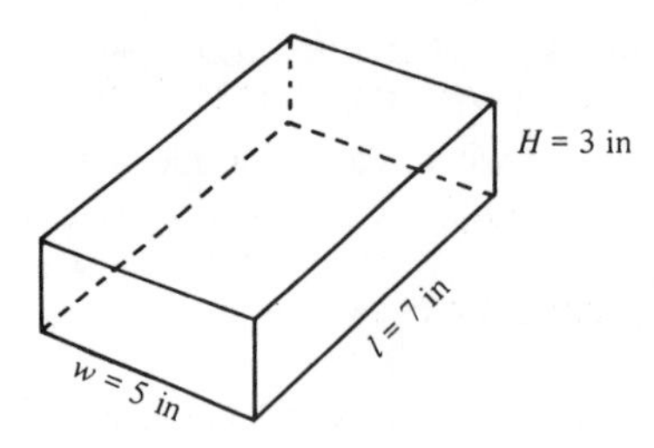

2. Find the surface area of a cube where each side measures 2 cm:

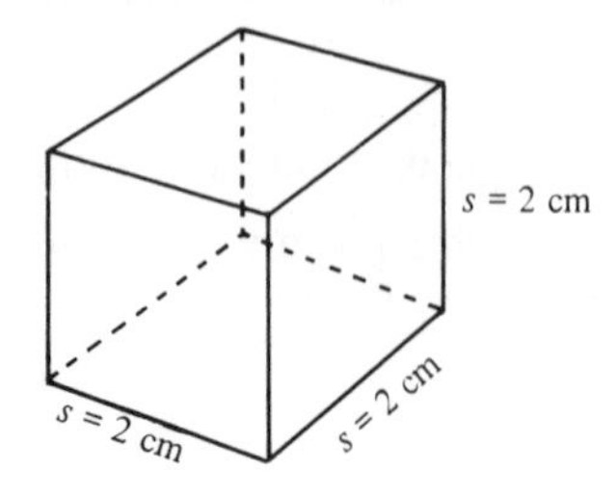

Study Exercise Five **(Continued)**

3. Find the surface area of a prism whose bases are right triangles with legs measuring 6 in and 8 in. The height is 15 in.

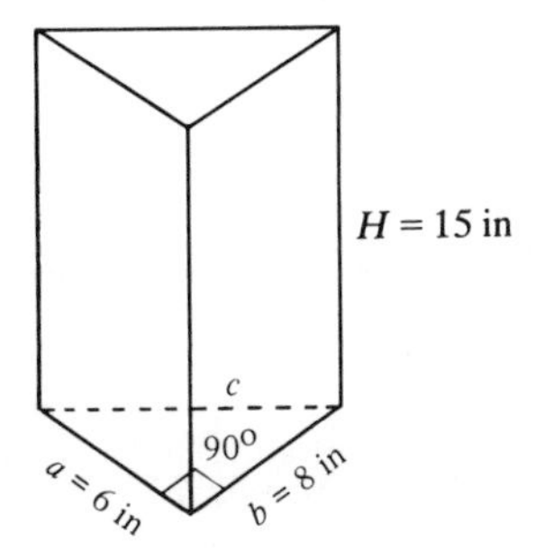

4. How many square feet of wood is needed to make a packing box having a length of 6 ft, a width of 5 ft, and a height of 3 ft?

5. If the wood used in problem 4 can be purchased at 40¢ per square foot, what will be the total cost of the wood?

Lateral Area of Right Circular Cylinders (28)

The lateral area of a right circular cylinder is the surface area, not considering the top and bottom bases.

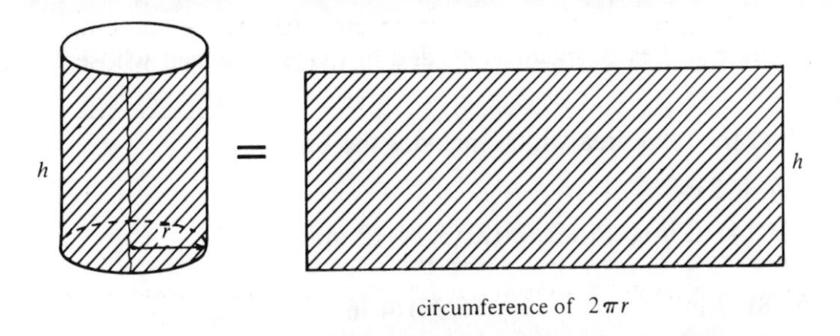

The lateral area of a right circular cylinder can be found by multiplying the circumference of the base ($2\pi r$) times the height (h).

$$A_L = \underbrace{2\pi r}_{\text{circumference}} h \qquad \text{where } \pi \approx 3.14$$

Example: Find the lateral area of a right circular cylinder whose radius is 3 in and whose height is 5 in:

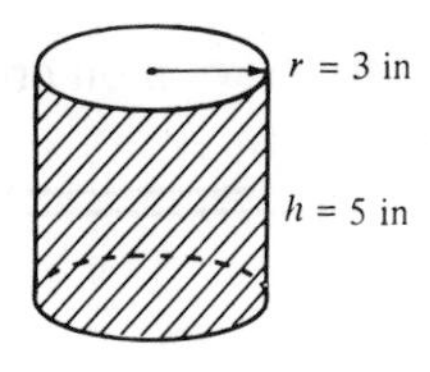

Lateral Area of Right Circular Cylinders (Continued)

Solution:

Line (a): $A_L = 2\pi rh$

Line (b): $= \underbrace{(2)(3.14)}\underbrace{(3\text{ in})(5\text{ in})}$

Line (c): $= (6.28)(15\text{ in}^2)$

Line (d): $= 94.2\text{ in}^2$

Total Area of a Right Circular Cylinder (29)

The total area of a right circular cylinder is the total surface area, including the two bases.

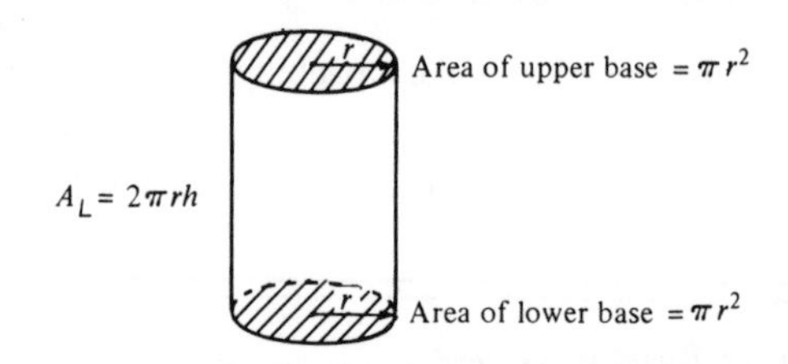

Therefore, the total area is the sum of the lateral area and the area of the two bases.

$$A = 2\pi rh + 2\pi r^2$$

Example: How many square inches of metal is needed to make a tin can whose radius is 2 in and whose height is 4.5 in?

Solution: We must find the total area which includes the bases.

Line (a): $A = 2\pi rh + 2\pi r^2$

Line (b): $= \underbrace{(2)(3.14)}\underbrace{(2\text{ in})(4.5\text{ in})} + \underbrace{(2)(3.14)}\underbrace{(2\text{ in})^2}$

Line (c): $= \underbrace{(6.28)(9\text{ in})} + \underbrace{(6.28)(4\text{ in}^2)}$

Line (d): $= 56.52\text{ in} + 25.12\text{ in}^2$

Line (e): $= 81.64\text{ in}^2$

Study Exercise Six (30)

1. Find the lateral area of a right circular cylinder whose radius is 4 in and whose height is 10 in.
2. Find the total area of a right circular cylinder whose radius is 3 cm and whose height is 8.5 cm.
3. How many square inches of paper is needed to make a label for a can of tomatoes if the can is 9 in high and has a diameter of 4 in?
4. How many square feet of metal is needed to make an oil drum whose height is 3 ft and whose radius is 12 in?

Surface Area of a Sphere (31)

The surface area of a sphere is the area of its entire outside surface.

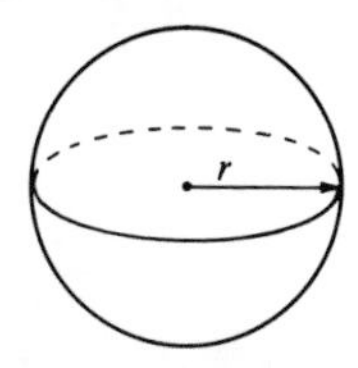

The surface area for a sphere is given by this formula:

$$A = 4\pi r^2 \qquad \text{where } \pi \approx 3.14$$

Example: What is the surface area of a globe measuring 14 inches in diameter?

Solution: Find the radius and use the formula $A = 4\pi r^2$.

Line (a): $r = \dfrac{14 \text{ in}}{2}$ or 7 in

Line (b): $A = 4\pi r^2$

Line (c): $= (4)(3.14)(7 \text{ in})^2$

Line (d): $= \underbrace{(4)(3.14)}(49 \text{ in}^2)$

Line (e): $= (12.56)(49 \text{ in}^2)$

Line (f): $= 615.44 \text{ in}^2$

Computation

```
  12.56
     49
 ------
 113 04
 502 4
 ------
 615.44
```

Study Exercise Seven (32)

1. Find the surface area of a sphere whose radius is 10 in.
2. Find the surface area of a globe whose diameter is 2 ft.
3. A natural gas storage tank is in the shape of a sphere with an outside diameter of 30 ft. If the outside surface is to be painted, how much paint will it take if each gallon of paint covers 200 ft^2?

Summary of Volume and Surface Area (33)

1. Prisms in General

   $V = AH$

   $A =$ Sum of the area of the faces

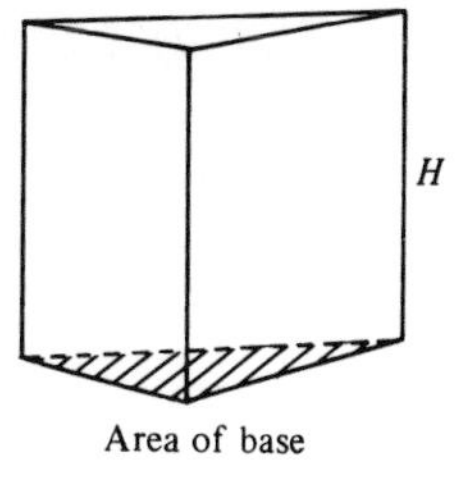

2. Rectangular Solid

   $V = lwh$

   $A = 2lw + 2lh + 2wh$

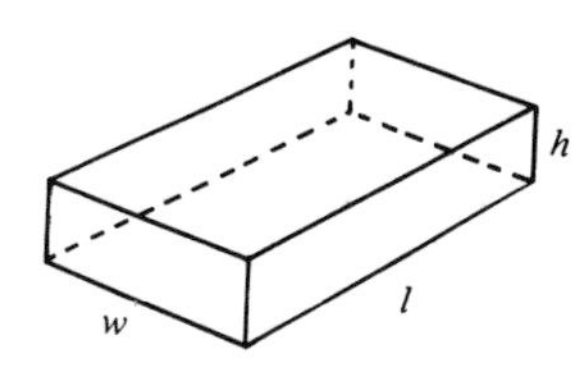

Summary of Volume and Surface Area (Continued)

3. Cube

$V = s^3$

$A = 6s^2$

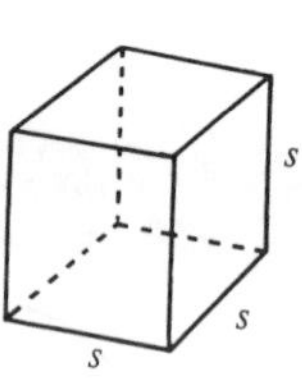

4. Right Circular Cylinder

$V = \pi r^2 h$

$A_L = 2\pi rh$

$A = 2\pi rh + 2\pi r^2$

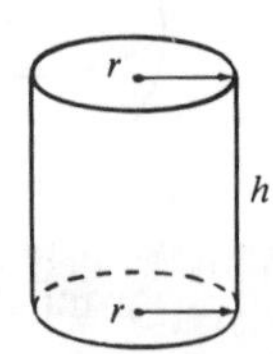

5. Sphere

$V = \frac{4}{3}\pi r^3$ or $\frac{4\pi r^3}{3}$

$A = 4\pi r^3$

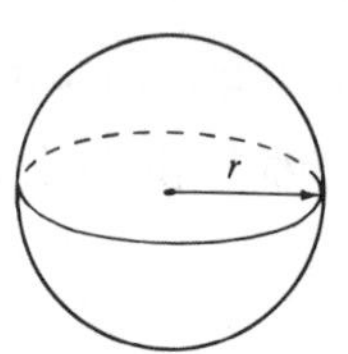

Remember, volume is measured in cubic units, while surface area is measured in square units.

REVIEW EXERCISES (34)

A. Find the volume and total surface area for each of the following solids:

1. Rectangular Solid

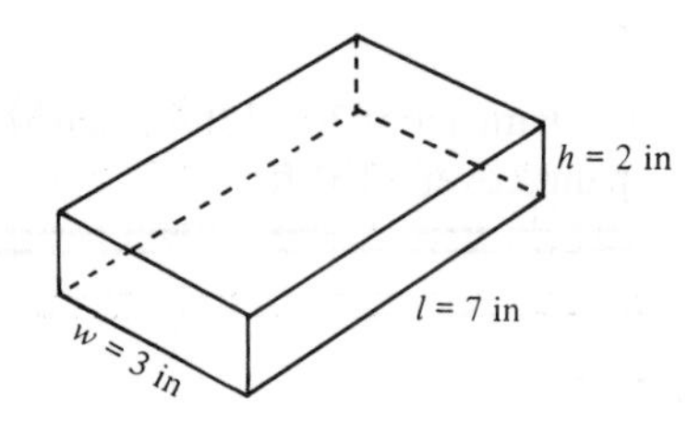

2. Cube

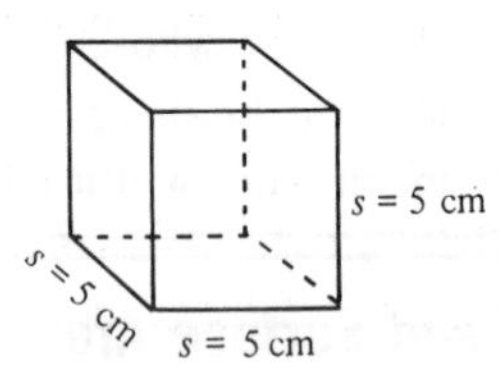

3. Prism

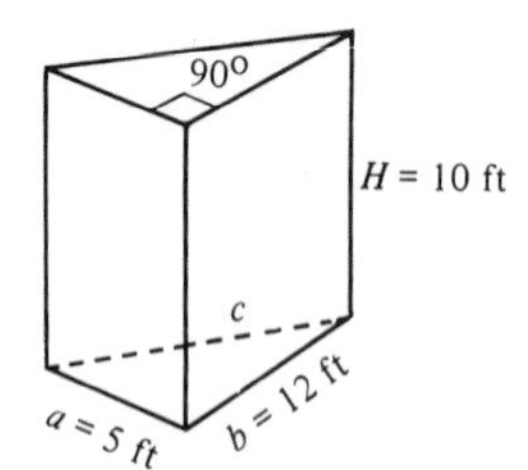

4. Right Circular Cylinder

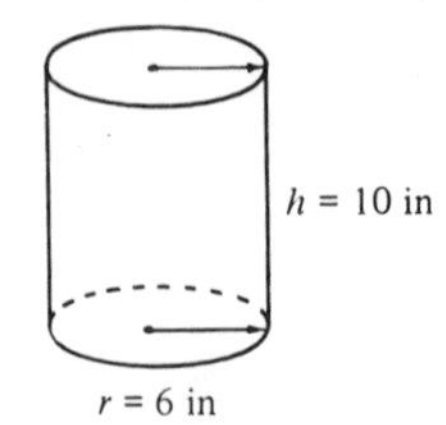

REVIEW EXERCISES (Continued)

5. Sphere

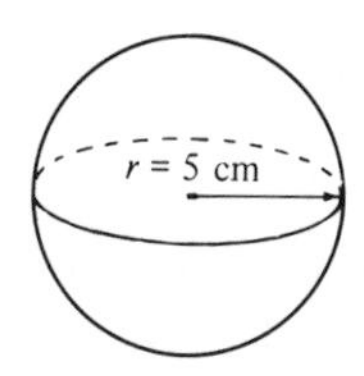

B. Solve the following problems:

6. How many cubic yards of dirt fill is needed to fill a rectangular hole measuring 78 ft long, 24 ft wide, and 12 ft deep?
7. How many cubic yards of dirt must be removed to dig a circular well 6 ft in radius and 75 ft deep?
8. How many square inches of aluminum is needed to make a furnace pipe 6 inches in diameter and 3 ft long? Neglect the size of the seam.
9. A water tank is in the shape of a right circular cylinder having a radius of 2 ft and a height of 5 ft. If 7.5 gal of water occupies one cubic foot, find the capacity of the tank.
10. Find the surface area of a globe whose diameter is 6 in.
11. How many square inches of metal is needed to manufacture a tin can having a radius of 2 in and a height of 4.5 in?
12. Find the surface area of a cubical box where each side measures 9 in.
13. How many cubic yards of concrete is needed to construct a circular patio 12 ft in diameter and 4 in thick (round your answer to the nearest tenth of a cubic yard)?
14. A storage tank for natural gas is in the shape of a sphere. Its inside diameter is 40 ft. How many cubic feet of gas is stored in the tank (round your answer to the nearest cubic foot)?
15. The outside of a water tank in the shape of a sphere is to be painted. Its diameter is 30 ft. How much paint is needed if each gallon covers 300 ft^2?
16. A schoolroom is 40 ft long by 20 ft wide by 15 ft high. Allowing 200 ft^3 per student, what is the maximum number of students that should be in this room?
17. If steel weighs 490 lb per ft^3, what is the weight of a steel plate that is 12 ft by 10 ft by 2 in?
18. To the nearest whole gallon, how many gallons of gasoline may be stored in a right circular cylinder tank with a diameter of 20 ft and a length of 40 ft, if 7.5 gal of gasoline occupies 1 ft^3?

Solutions to Review Exercises (35)

A. 1.
$$\begin{aligned} V &= lwh \\ &= (7 \text{ in})(3 \text{ in})(2 \text{ in}) \\ &= 42 \text{ in}^3 \end{aligned}$$

$$\begin{aligned} A &= 2lw + 2lh + 2wh \\ &= (2)(7 \text{ in})(3 \text{ in}) + (2)(7 \text{ in})(2 \text{ in}) + (2)(3 \text{ in})(2 \text{ in}) \\ &= 42 \text{ in}^2 + 28 \text{ in}^2 + 12 \text{ in}^2 \\ &= 82 \text{ in}^2 \end{aligned}$$

2.
$$\begin{aligned} V &= s^3 \\ &= (5 \text{ cm})^3 \\ &= 125 \text{ cm}^3 \end{aligned}$$

$$\begin{aligned} A &= 6s^2 \\ &= (6)(5 \text{ cm})^2 \\ &= (6)(25 \text{ cm}^2) \\ &= 150 \text{ cm}^2 \end{aligned}$$

Solutions to Review Exercises (Continued)

3. $A = \dfrac{(5\text{ ft})(12\text{ ft})}{2}$

$A = 30\text{ ft}^2$

$V = AH$

$= (30\text{ ft}^2)(10\text{ ft})$

$= 300\text{ ft}^3$

$c^2 = a^2 + b^2$

$= 25 + 144$

$= 169$

$c = \sqrt{169} = 13\text{ ft}$

Area of bottom face =	30 ft²
Area of upper face =	30 ft²
Area of rear face =	130 ft²
Area of left face =	50 ft²
Area of right face =	120 ft²
A =	360 ft²

4. $V = \pi r^2 h$

$= (3.14)(6\text{ in})^2(10\text{ in})$

$= (3.14)(36\text{ in}^2)(10\text{ in})$

$= 1130.4\text{ in}^3$

$A = 2\pi rh + 2\pi r^2$

$= (2)(3.14)(6\text{ in})(10\text{ in}) + (2)(3.14)(6\text{ in})^2$

$= 376.8\text{ in}^2 + (6.28)(36\text{ in}^2)$

$= 376.8\text{ in}^2 + 226.08\text{ in}^2$

$= 602.88\text{ in}^2$

5. $V = \dfrac{4\pi r^3}{3}$

$= \dfrac{(4)(3.14)(5\text{ cm})^3}{3}$

$= \dfrac{(4)(3.14)(125\text{ cm}^3)}{3}$

$= 523\tfrac{1}{3}\text{ cm}^3$

$A = 4\pi r^2$

$= (4)(3.14)(5\text{ cm})^2$

$= (4)(3.14)(25\text{ cm}^2)$

$= 314\text{ cm}^2$

B. **6.** 78 ft = 26 yd

24 ft = 8 yd

12 ft = 4 yd

$V = lwh$

$= (26\text{ yd})(8\text{ yd})(4\text{ yd})$

$= 832\text{ yd}^3$

7. 6 ft = 2 yd

75 ft = 25 yd

$V = \pi r^2 h$

$= (3.14)(2\text{ yd})^2(25\text{ yd})$

$= (3.14)(4\text{ yd}^2)(25\text{ yd})$

$= 314\text{ yd}^3$

8. 3 ft = 36 in

$r = \dfrac{6\text{ in}}{2}$ or 3 in

$A = 2\pi rh$

$= (2)(3.14)(3\text{ in})(36\text{ in})$

$= 678.24\text{ in}^2$

9. $V = \pi r^2 h$

$= (3.14)(2\text{ ft})^2(5\text{ ft})$

$= (3.14)(4\text{ ft}^2)(5\text{ ft})$

$= 62.8\text{ ft}^3$

Amount of water = (62.8)(7.5)

= 471 gal

10. $r = \dfrac{6\text{ in}}{2}$ or 3 in

$A = 4\pi r^2$

$= (4)(3.14)(3\text{ in})^2$

$= (4)(3.14)(9\text{ in}^2)$

$= 113.04\text{ in}^2$

11. $A = 2\pi rh + 2\pi r^2$

$= (2)(3.14)(2\text{ in})(4.5\text{ in}) + (2)(3.14)(2\text{ in})^2$

$= 56.52\text{ in}^2 + (2)(3.14)(4\text{ in}^2)$

$= 56.52\text{ in}^2 + 25.12\text{ in}^2$

$= 81.64\text{ in}^2$

Solutions to Review Exercises (Continued)

12. $A = 6s^2$
$= (6)(9 \text{ in})^2$
$= (6)(81 \text{ in}^2)$
$= 486 \text{ in}^2$

13. $12 \text{ ft} = 4 \text{ yd}$

$4 \text{ in} = \frac{4}{36} \text{ yd or } \frac{1}{9} \text{ yd}$

$r = \frac{4 \text{ yd}}{2} \text{ or } 2 \text{ yd}$

$V = \pi r^2 h$
$= (3.14)(2 \text{ yd})^2\left(\frac{1}{9} \text{ yd}\right)$
$= (3.14)(4 \text{ yd}^2)\left(\frac{1}{9} \text{ yd}\right)$
$= \frac{12.56}{9} \text{ yd}^3$
$= 1.4 \text{ yd}^3$ (to the nearest tenth of a cubic yard)

14. $r = \frac{40 \text{ ft}}{2} \text{ or } 20 \text{ ft}$

$V = \frac{4\pi r^3}{3}$
$= \frac{(4)(3.14)(20 \text{ ft})^3}{3}$
$= \frac{(4)(3.14)(8{,}000 \text{ ft}^3)}{3}$
$= \frac{100{,}480 \text{ ft}^3}{3}$
$= 33{,}493 \text{ ft}^3$ (to the nearest cubic foot)

15. $r = \frac{30 \text{ ft}}{2} \text{ or } 15 \text{ ft}$

$A = 4\pi r^2$
$= (4)(3.14)(15 \text{ ft})^2$
$= (4)(3.14)(225 \text{ ft}^2)$
$= 2826 \text{ ft}^2$

Divide by 300

$\frac{2826}{300} = 9.42$ gal of paint

16. $V = lwh$
$= (40 \text{ ft})(20 \text{ ft})(15 \text{ ft})$
$= 12{,}000 \text{ ft}^3$

Divide by 200.

$\frac{12{,}000}{200} = 60$ students

17. $2 \text{ in} = \frac{2}{12} \text{ ft} = \frac{1}{6} \text{ ft}$

$V = lwh$
$= (12 \text{ ft})(10 \text{ ft})\left(\frac{1}{6} \text{ ft}\right)$
$= 20 \text{ ft}^3$

Multiply by 490.

$(20)(490) = 9{,}800$ lb

18. $r = \frac{20 \text{ ft}}{2} = 10 \text{ ft}$

$V = \pi r^2 h$
$= (3.14)(10 \text{ ft})^2(40 \text{ ft})$
$= 12{,}560 \text{ ft}^3$

Multiply by 7.5.

$(12{,}560)(7.5) = 94{,}200$ gal of gasoline

SUPPLEMENTARY PROBLEMS

A. Find the volume and total surface area for each of the following solids:

1. Rectangular Solid

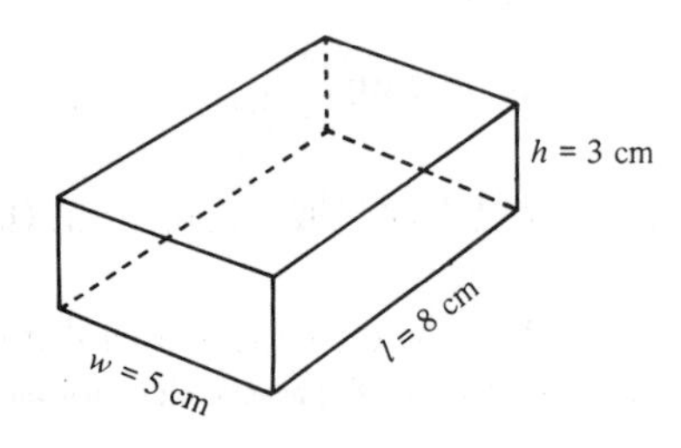

2. Prism

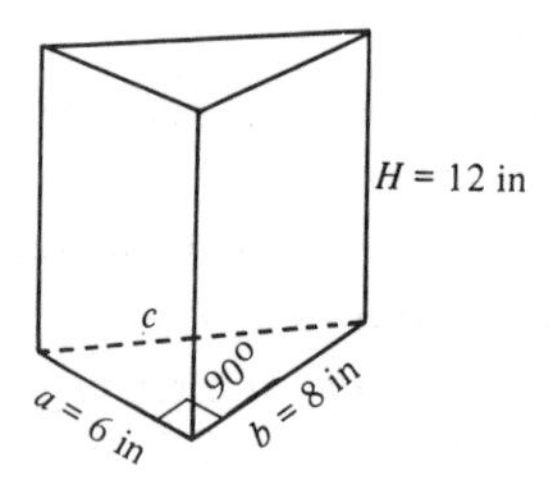

3. Cube

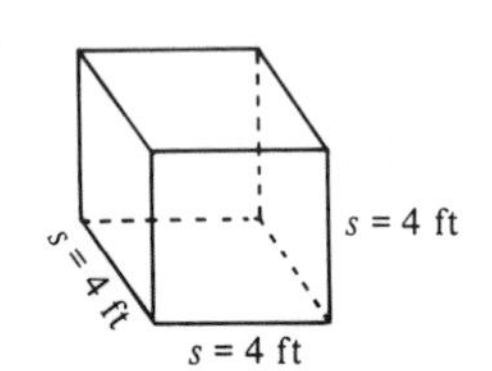

4. Sphere

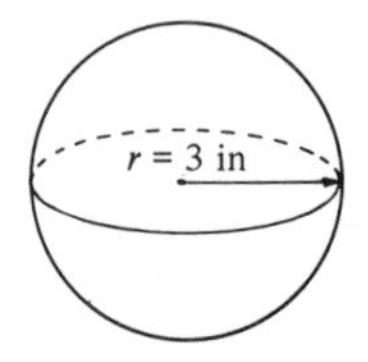

5. Right Circular Cylinder

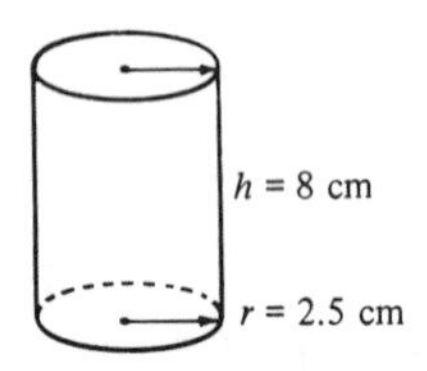

B. Solve the following problems:

6. How many cubic yards of concrete is needed to construct a sidewalk 72 ft long, 3 ft wide, and 4 in deep?

7. Dirt fill costs $5.00 per cubic yard. What is the cost to fill a rectangular hole 24 ft long, 15 ft wide, and 6 ft deep?

8. What is the capacity (volume) of a cubical safe where each interior side measures 1.5 ft?

9. A circular hole 2 yd in diameter has been dug 12 ft into the ground for a concrete footing. If concrete costs $45.00 per cubic yard, find the cost of the footing.

10. A railroad tank car in the shape of a cylinder is used to transport liquid ammonia. If the tank car is 2 yd wide and 42 ft long, find the capacity of the tank car to the nearest cubic foot.

11. A natural gas storage tank is in the shape of a sphere with a diameter of 60 ft. How many cubic feet of gas will it contain?

12. How many cubic inches of metal is needed to manufacture a ball bearing having a diameter of .6 of an inch?

13. Find the surface area of a rectangular box measuring 14 in long, 10 in wide, and 8 in high.

14. How many square inches of paper is needed to make a label for a can of peaches if the can is 5 in high and has a diameter of 3 in?

15. Find the surface area of a glass globe whose diameter measures 22 in.

SUPPLEMENTARY PROBLEMS (Continued)

16. What is the volume in cubic feet of a truck used to move furniture, if it is 20 ft long, 10 ft wide and 8 ft high?

17. What is the volume in liters of an aquarium that measures 70 cm by 30 cm by 40 cm?

18. How many cubic feet of water is needed to fill a circular wading pool that is 10 ft in diameter and 18 in deep? If 7.5 gal occupies 1 ft^3, how many gallons are needed to fill this pool (nearest whole gallon)?

19. How many cubic feet of food may be placed inside a freezer if its inside measurements are 48 in by 27 in by 20 in?

20. A silo in the shape of a right circular cylinder has a diameter of 12 ft and a height of 30 ft. How many cubic feet of silage will it hold?

21. A trailer in the shape of a rectangular solid is to be painted. What is the surface area if it measures 20 ft long, 8 ft wide and 7 ft high? What surface area will be painted if it is decided not to paint the roof and the bottom of the trailer?

C. More difficult problems involving volume and surface area of composite solids:

22. Find the volume and total surface area of this figure composed of a right circular cylinder surmounted by a hemisphere (one-half of a sphere):

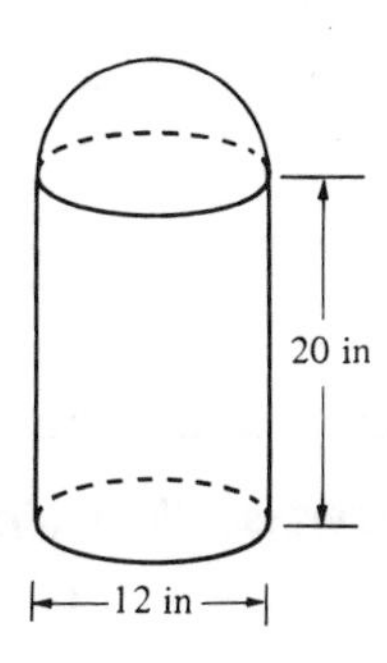

23. Find the volume and total surface area of this figure composed of a right circular cylinder and two hemispheres:

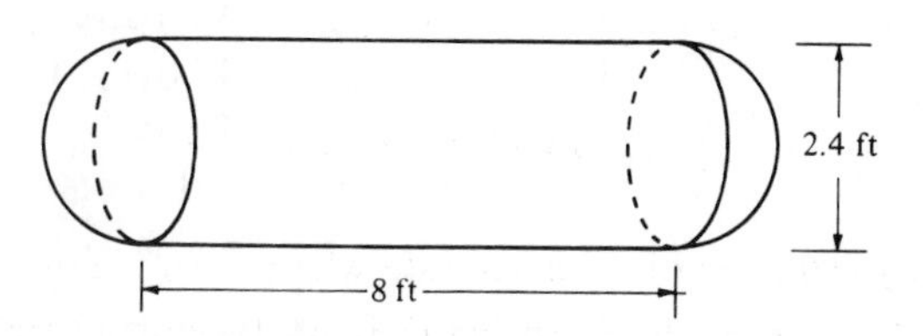

Solutions to Study Exercises (6A)

Study Exercise One (Frame 6)

1. Right circular cylinder. **2.** Right circular cylinder. **3.** Rectangular solid (prism).
4. Sphere. **5.** Prism with triangular bases.

Study Exercise Two (Frame 13) (13A)

1. Area of base $= \frac{(5 \text{ in})(\not{12}^{6} \text{ in})}{\not{2}_1} = 30 \text{ in}^2$

   $V = AH$
   $= (30 \text{ in}^2)(15 \text{ in})$
   $= 450 \text{ in}^3$

2. $V = lwh$
   $= (3.2 \text{ in})(2.1 \text{ in})(1.5 \text{ in})$
   $= 10.08 \text{ in}^3$

3. $V = s^3$
   $= (3 \text{ ft})^3$
   $= 27 \text{ ft}^3$

4. Convert to yards:

   15 ft = 5 yd; 12 ft = 4 yd; 9 ft = 3 yd

   $V = lwh$
   $= (5 \text{ yd})(4 \text{ yd})(3 \text{ yd})$
   $= 60 \text{ yd}^3$

   Cost = (\$4.50)(60)
   = \$270

5. Convert to yards:

   15 ft = 5 yd; 12 ft = 4 yd; 3 in $= \frac{1}{12}$ yd

   $V = lwh$

   $= (5 \text{ yd})(\not{4}^{1} \text{ yd})\left(\frac{1}{\not{12}_3} \text{ yd}\right)$

   $= \frac{5}{3} \text{ yd}^3$

   $= 1\frac{2}{3} \text{ yd}^3$

Study Exercise Three (Frame 17) (17A)

1. $V = \pi r^2 h$
   $= (3.14)(2 \text{ m})^2(8.5 \text{ m})$
   $= (3.14)(4 \text{ m}^2)(8.5 \text{ m})$
   $= 106.76 \text{ m}^3$

2. Convert to yards; 15 ft = 5 yd

   Find the radius; $r = \frac{4 \text{ yd}}{2}$ or 2 yd

   $V = \pi r^2 h$
   $= (3.14)(2 \text{ yd})^2(5 \text{ yd})$
   $= (3.14)(4 \text{ yd}^2)(5 \text{ yd})$
   $= 62.8 \text{ yd}^3$

   Cost = (\$5.00)(62.8)
   = \$314.00

3. Find the radius; $r = \frac{6 \text{ ft}}{2}$ or 3 ft

   $V = \pi r^2 h$
   $= (3.14)(3 \text{ ft})^2(12 \text{ ft})$
   $= 339.12 \text{ ft}^3$

4. Multiply by 7.5.

   (339.12)(7.5) = 2543.4 gal of water

Study Exercise Four (Frame 21) (21A)

1. $V = \dfrac{4\pi r^3}{3}$

$= \dfrac{(4)(3.14)(2\ \text{cm})^3}{3}$

$= \dfrac{(4)(3.14)(8\ \text{cm}^3)}{3}$

$= 33.5\ \text{cm}^3$ (to the nearest tenth)

2. Convert to yards; 60 ft = 20 yd

Find the radius; $r = \dfrac{20\ \text{yd}}{2}$ or 10 yd

$V = \dfrac{4\pi r^3}{3}$

$= \dfrac{(4)(3.14)(10\ \text{yd})^3}{3}$

$= \dfrac{(4)(3.14)(1{,}000\ \text{yd}^3)}{3}$

$= 4{,}186.7\ \text{yd}^3$ (to nearest tenth)

3. $V = \dfrac{4\pi r^3}{3}$

$= \dfrac{(4)(3.14)(10\ \text{cm})^3}{3}$

$= 4187\ \text{cm}^3$ (to nearest whole cm^3)

4. Remember, 1 cm^3 or cc = 1 mL;

therefore, 4,187 cm^3 = 4,187 mL

Remember, 1,000 mL = 1 L;

therefore, 4,187 mL = 4.187 L

Study Exercise Five (Frame 27) (27A)

1. $A = 2lw + 2lh + 2wh$

$= (2)(7\ \text{in})(5\ \text{in}) + (2)(7\ \text{in})(3\ \text{in}) + (2)(5\ \text{in})(3\ \text{in})$

$= 70\ \text{in}^2 + 42\ \text{in}^2 + 30\ \text{in}^2$

$= 142\ \text{in}^2$

2. $A = 6s^2$

$= (6)(2\ \text{cm})^2$

$= (6)(4\ \text{cm}^2)$

$= 24\ \text{cm}^2$

3. Use the Pythagorean theorem to find c; then find the area of each of the five faces and add them together.

$c^2 = a^2 + b^2$

$= 36 + 64$

$= 100$

$c = \sqrt{100}$ or 10 in

Area of bottom face $= \dfrac{(6\ \text{in})(8\ \text{in})}{2} = 24\ \text{in}^2$

Area of upper face $= \dfrac{(6\ \text{in})(8\ \text{in})}{2} = 24\ \text{in}^2$

Area of rear face $= (15\ \text{in})(10\ \text{in}) = 150\ \text{in}^2$

Area of left face $= (15\ \text{in})(6\ \text{in}) = 90\ \text{in}^2$

Area of right face $= (15\ \text{in})(8\ \text{in}) = 120\ \text{in}^2$

Surface Area $= 408\ \text{in}^2$

4. $A = 2lw + 2lh + 2wh$

$= 2(6\ \text{ft})(5\ \text{ft}) + 2(6\ \text{ft})(3\ \text{ft}) + 2(5\ \text{ft})(3\ \text{ft})$

$= 60\ \text{ft}^2 + 36\ \text{ft}^2 + 30\ \text{ft}^2$

$= 126\ \text{ft}^2$

5. Multiply by \$.40

(126)(\$.40) = \$50.40

Study Exercise Six (Frame 30) (30A)

1. $A_L = 2\pi rh$

$= (2)(3.14)(4\ \text{in})(10\ \text{in})$

$= 251.2\ \text{in}^2$

2. $A = 2\pi rh + 2\pi r^2$

$= (2)(3.14)(3\ \text{cm})(8.5\ \text{cm}) + (2)(3.14)(3\ \text{cm})^2$

$= 160.14\ \text{cm}^2 + (2)(3.14)(9\ \text{cm}^2)$

$= 160.14\ \text{cm}^2 + 56.52\ \text{cm}^2$

$= 216.66\ \text{cm}^2$

Study Exercise Six (Frame 30) (Continued)

3. Find the radius; $r = \frac{4 \text{ in}}{2}$ or 2 in

Find the lateral area.

$$\begin{aligned} A_L &= 2\pi rh \\ &= (2)(3.14)(2 \text{ in})(9 \text{ in}) \\ &= 113.04 \text{ in}^2 \end{aligned}$$

4. Convert to feet; 12 in = 1 ft

Find the total area.

$$\begin{aligned} A &= 2\pi rh + 2\pi r^2 \\ &= (2)(3.14)(1 \text{ ft})(3 \text{ ft}) + (2)(3.14)(1 \text{ ft})^2 \\ &= 18.84 \text{ ft}^2 + (2)(3.14)(1 \text{ ft}^2) \\ &= 18.84 \text{ ft}^2 + 6.28 \text{ ft}^2 \\ &= 25.12 \text{ ft}^2 \end{aligned}$$

Study Exercise Seven (Frame 32) **(32A)**

1.
$$\begin{aligned} A &= 4\pi r^2 \\ &= (4)(3.14)(10 \text{ in})^2 \\ &= (4)(3.14)(100 \text{ in}^2) \\ &= 1{,}256 \text{ in}^2 \end{aligned}$$

2. Find the radius; $r = \frac{2 \text{ ft}}{2}$ or 1 ft

$$\begin{aligned} A &= 4\pi r^2 \\ &= (4)(3.14)(1 \text{ ft})^2 \\ &= (4)(3.14)(1 \text{ ft}^2) \\ &= 12.56 \text{ ft}^2 \end{aligned}$$

4. Find the radius; $r = \frac{30 \text{ ft}}{2}$ or 15 ft

$$\begin{aligned} A &= 4\pi r^2 \\ &= (4)(3.14)(15 \text{ ft})^2 \\ &= (4)(3.14)(225 \text{ ft}^2) \\ &= 2{,}826 \text{ ft}^2 \end{aligned}$$

Amount of paint $= \frac{2{,}826}{200}$ or 14.13 gal

Module 6 Practice Test

Units 32–38

A. Do as directed:

1. Convert 6 da to hours.

2. Convert 31 ft to yards and feet.

3. Add and simplify:

5 hr 20 min 6 sec
3 hr 42 min 38 sec
2 hr 15 min 22 sec

4. Subtract:

5 lb 7 oz
2 lb 12 oz

5. Multiply and simplify:

3 yd 2 ft 2 in
×8

6. Divide:

6)39 lb 6 oz

B. You may use this chart to make the following conversions within the metric system.

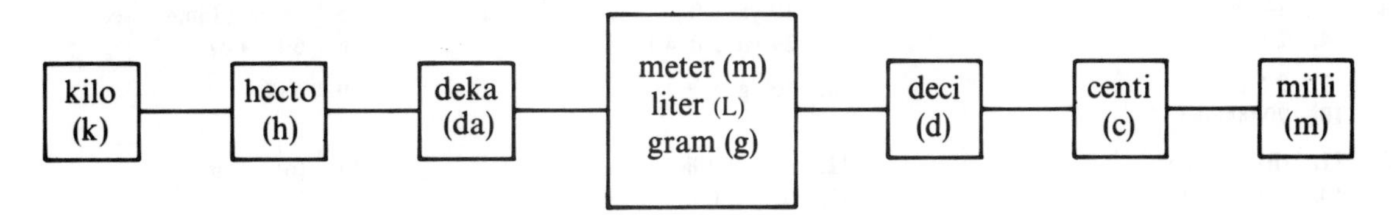

7. 250 cm = ________ m

8. 5 hg = ________ g

9. 600 mL = ________ dL

10. 2 km = ________ dm

C. Multiple choice.

11. A speed of 100 km/hr is approximately ________ mph.
(a) 100 **(b)** 62 **(c)** 35

12. A kilogram weighs approximately ________ pounds.
(a) 10 **(b)** 2.2 **(c)** $\frac{1}{2}$

13. A liter is approximately ________.
(a) 1 pt **(b)** 1 qt **(c)** 1 gal

14. Water boils at a temperature of ________.
(a) 100°F **(b)** 100°C **(c)** 212°C

15. Water freezes at a temperature of ________,
(a) 0°F **(b)** 0°C **(c)** 32°C

D. Do as directed:

16. Convert 50°F to Celsius (°C) by using the formula,

$$C = \frac{5}{9}(F - 32)$$

Module 6 Practice Test (Continued)

17. Convert 20°C to Fahrenheit (°F) by using the formula,

$$F = \frac{9}{5}C + 32$$

18. A boy weighs 80 kg. If 1 kg = 2.20 lb, find the boy's weight in pounds.

19. A girl is 5 ft 2 in tall. If 1 in = 2.54 cm, find her height to the nearest centimeter.

20. Find the perimeter of a square where each side measures 3 ft 4 in. Leave your answer in terms of feet and inches.

21. Find the cost of constructing a concrete sidewalk measuring 3 ft by 60 ft at $.90 per square foot.

22. What distance do you travel on one turn of a Ferris wheel if you sit 10 ft from the center?

23. A cereal box measures 8 in by 2 in by 11 in.
- **(a)** Find its surface area.
- **(b)** Find its volume.

24. A vegetable can is to be made with a height of 10 in and a radius of 3 in.
- **(a)** What area will the label of the can cover?
- **(b)** What total area of metal will be necessary?
- **(c)** What will the volume of the can be?

25. A solid rubber ball with a 3 in radius will contain what volume of rubber?

Answers to Module 6 Practice Test

A. **1.** 144 hr **2.** 10 yd 1 ft **3.** 11 hr 18 min 6 sec
4. 2 lb 11 oz **5.** 29 yd 2 ft 4 in **6.** 6 lb 9 oz

B. **7.** 2.5 m **8.** 500 g **9.** 6 dL
10. 20,000 dm

C. **11.** **(b)** 62 mph **12.** **(b)** 2.2 lb **13.** **(b)** 1 qt
14. **(b)** 100°C **15.** **(b)** 0°C

D. **16.** 10°C **17.** 68°F **18.** 176 lb
19. 157 cm **20.** 13 ft 4 in **21.** $162.00
22. 62.8 ft **23.** **(a)** 252 in^2 **(b)** 176 in^3 **24.** **(a)** 188.4 in^2 **(b)** 244.92 in^2 **(c)** 282.6 in^3
25. 113.04 in^3

APPENDIX I Tables

TABLE I POWERS AND ROOTS
TABLE II COMPOUND INTEREST
TABLE III TABLES OF MEASURE
TABLE IV METRIC PREFIXES

Table I—Powers and Roots

n	n^2	n^3	$\sqrt{n}$	$\sqrt[3]{n}$	n	n^2	n^3	$\sqrt{n}$	$\sqrt[3]{n}$
0	0	0	0.000	0.000	50	2 500	125 000	7.071	3.684
1	1	1	1.000	1.000	51	2 601	132 651	7.141	3.708
2	4	8	1.414	1.260	52	2 704	140 608	7.211	3.733
3	9	27	1.732	1.442	53	2 809	148 877	7.280	3.756
4	16	64	2.000	1.587	54	2 916	157 464	7.348	3.780
5	25	125	2.236	1.710	55	3 025	166 375	7.416	3.803
6	36	216	2.449	1.817	56	3 136	175 616	7.483	3.826
7	49	343	2.646	1.913	57	3 249	185 193	7.550	3.849
8	64	512	2.828	2.000	58	3 364	195 112	7.616	3.871
9	81	729	3.000	2.080	59	3 481	205 379	7.681	3.893
10	100	1 000	3.162	2.154	60	3 600	216 000	7.746	3.915
11	121	1 331	3.317	2.224	61	3 721	226 981	7.810	3.936
12	144	1 728	3.464	2.289	62	3 844	238 328	7.874	3.958
13	169	2 197	3.606	2.351	63	3 969	250 047	7.937	3.979
14	196	2 744	3.742	2.410	64	4 096	262 144	8.000	4.000
15	225	3 375	3.873	2.466	65	4 225	274 625	8.062	4.021
16	256	4 096	4.000	2.520	66	4 356	287 496	8.124	4.041
17	289	4 913	4.123	2.571	67	4 489	300 763	8.185	4.062
18	324	5 832	4.243	2.621	68	4 624	314 432	8.246	4.082
19	361	6 859	4.359	2.668	69	4 761	328 509	8.307	4.102
20	400	8 000	4.472	2.714	70	4 900	343 000	8.367	4.121
21	441	9 261	4.583	2.759	71	5 041	357 911	8.426	4.141
22	484	10 648	4.690	2.802	72	5 184	373 248	8.485	4.160
23	529	12 167	4.796	2.844	73	5 329	389 017	8.544	4.179
24	576	13 824	4.899	2.884	74	5 476	405 224	8.602	4.198
25	625	15 625	5.000	2.924	75	5 625	421 875	8.660	4.217
26	676	17 576	5.099	2.962	76	5 776	438 976	8.718	4.236
27	729	19 683	5.196	3.000	77	5 929	456 533	8.775	4.254
28	784	21 952	5.292	3.037	78	6 084	474 552	8.832	4.273
29	841	24 389	5.385	3.072	79	6 241	493 039	8.888	4.291
30	900	27 000	5.477	3.107	80	6 400	512 000	8.944	4.309
31	961	29 791	5.568	3.141	81	6 561	531 441	9.000	4.327
32	1 024	32 768	5.657	3.175	82	6 724	551 368	9.055	4.344
33	1 089	35 937	5.745	3.208	83	6 889	571 787	9.110	4.362
34	1 156	39 304	5.831	3.240	84	7 056	592 704	9.165	4.380
35	1 225	42 875	5.916	3.271	85	7 225	614 125	9.220	4.397
36	1 296	46 656	6.000	3.302	86	7 396	636 056	9.274	4.414
37	1 369	50 653	6.083	3.332	87	7 569	658 503	9.327	4.431
38	1 444	54 872	6.164	3.362	88	7 744	681 472	9.381	4.448
39	1 521	59 319	6.245	3.391	89	7 921	704 969	9.434	4.465
40	1 600	64 000	6.325	3.420	90	8 100	729 000	9.487	4.481
41	1 681	68 921	6.403	3.448	91	8 281	753 571	9.539	4.498
42	1 764	74 088	6.481	3.476	92	8 464	778 688	9.592	4.514
43	1 849	79 507	6.557	3.503	93	8 649	804 357	9.644	4.531
44	1 936	85 184	6.633	3.530	94	8 836	830 584	9.695	4.547
45	2 025	91 125	6.708	3.557	95	9 025	857 375	9.747	4.563
46	2 116	97 336	6.782	3.583	96	9 216	884 736	9.798	4.579
47	2 209	103 823	6.856	3.609	97	9 409	912 673	9.849	4.595
48	2 304	110 592	6.928	3.634	98	9 604	941 192	9.899	4.610
49	2 401	117 649	7.000	3.659	99	9 801	970 299	9.950	4.626
					100	10 000	1 000 000	10.000	4.642

Table II—Compound Interest Table Showing how much $1 Will Amount to at Various Rates

Number of periods	Interest rate per period ½%	1%	1½%	2%	2½%	3%	3½%	4%
1	1.0050	1.0100	1.0150	1.0200	1.0250	1.0300	1.0350	1.0400
2	1.0100	1.0201	1.0302	1.0404	1.0506	1.0609	1.0712	1.0816
3	1.0151	1.0303	1.0457	1.0612	1.0769	1.0927	1.1087	1.1249
4	1.0202	1.0406	1.0614	1.0824	1.1038	1.1255	1.1475	1.1699
5	1.0253	1.0510	1.0773	1.1041	1.1314	1.1593	1.1877	1.2167
6	1.0304	1.0615	1.0934	1.1262	1.1597	1.1941	1.2293	1.2653
7	1.0355	1.0721	1.1098	1.1487	1.1887	1.2299	1.2723	1.3159
8	1.0407	1.0829	1.1265	1.1717	1.2184	1.2668	1.3168	1.3686
9	1.0459	1.0937	1.1434	1.1951	1.2489	1.3048	1.3629	1.4233
10	1.0511	1.1046	1.1605	1.2190	1.2801	1.3439	1.4106	1.4802
11	1.0564	1.1157	1.1779	1.2434	1.3121	1.3842	1.4600	1.5395
12	1.0617	1.1268	1.1956	1.2682	1.3449	1.4258	1.5110	1.6010
13	1.0670	1.1381	1.2136	1.2936	1.3785	1.4685	1.5640	1.6651
14	1.0723	1.1495	1.2318	1.3195	1.4129	1.5126	1.6187	1.7317
15	1.0777	1.1610	1.2502	1.3459	1.4483	1.5580	1.6753	1.8009
16	1.0831	1.1726	1.2690	1.3728	1.4845	1.6047	1.7339	1.8730
17	1.0885	1.1843	1.2880	1.4002	1.5216	1.6528	1.7947	1.9479
18	1.0939	1.1961	1.3073	1.4282	1.5597	1.7024	1.8575	2.0258
19	1.0994	1.2081	1.3270	1.4568	1.5987	1.7535	1.9225	2.1068
20	1.1049	1.2202	1.3469	1.4859	1.6386	1.8061	1.9898	2.1911
25	1.1328	1.2824	1.4509	1.6406	1.8539	2.0938	2.3632	2.6658
50	1.2832	1.6446	2.1052	2.6916	3.4371	4.3839	5.5849	7.1067

Number of periods	Interest rate per period 5%	6%	7%	8%	9%	10%	11%	12%
1	1.0500	1.0600	1.0700	1.0800	1.0900	1.1000	1.1100	1.1200
2	1.1025	1.1236	1.1449	1.1664	1.1881	1.2100	1.2321	1.2544
3	1.1576	1.1910	1.2250	1.2597	1.2950	1.3310	1.3676	1.4049
4	1.2155	1.2625	1.3108	1.3605	1.4116	1.4641	1.5181	1.5735
5	1.2763	1.3382	1.4026	1.4693	1.5386	1.6105	1.6851	1.7623
6	1.3401	1.4185	1.5007	1.5869	1.6771	1.7716	1.8704	1.9738
7	1.4071	1.5036	1.6058	1.7138	1.8280	1.9787	2.0762	2.2107
8	1.4775	1.5938	1.7182	1.8509	1.9926	2.1436	2.3045	2.4760
9	1.5513	1.6895	1.8385	1.9990	2.1719	2.3579	2.5580	2.7731
10	1.6289	1.7908	1.9672	2.1589	2.3674	2.5937	2.8394	3.1058
11	1.7103	1.8983	2.1049	2.3316	2.5804	2.8531	3.1518	3.4785
12	1.7959	2.0122	2.2522	2.5182	2.8127	2.1384	3.4985	3.8960
13	1.8856	2.1329	2.4098	2.7196	3.0658	3.4523	3.8833	4.3635
14	1.9799	2.2609	2.5785	2.9372	3.3417	3.7975	4.3104	4.8871
15	2.0789	2.3966	2.7590	3.1722	3.6425	4.1772	4.7846	5.4736
16	2.1829	2.5404	2.9522	3.4259	3.9703	4.5950	5.3109	6.1304
17	2.2920	2.6928	3.1588	3.7000	4.3276	5.0545	5.8951	6.8660
18	2.4066	2.8543	3.3799	3.9960	4.7171	5.5599	6.5436	7.6900
19	2.5270	3.0256	3.6165	4.3157	5.1417	6.1159	7.2633	8.6128
20	2.6533	3.2071	3.8697	4.6610	5.6044	6.7275	8.0623	9.6463
25	3.3864	4.2919	5.4274	6.8485	8.6231	10.8347	13.5855	17.0001
50	11.4674	18.4202	29.4570	46.9016	74.3575	117.3909	184.5648	289.0022

Table III–Tables of Measure

Measure of Length
1 foot (ft) = 12 inches (in)
1 yard (yd) = 3 feet (ft)
= 36 inches (in)
1 rod (rd) = $16\frac{1}{2}$ feet (ft)
= $5\frac{1}{2}$ yards (yd)
1 mile (mi) = 5,280 feet (ft)
= 1,760 yards (yd)
= 320 rods (rd)

Measure of Area
1 square foot (ft^2) = 144 square inches (in^2)
1 square yard (yd^2) = 9 square feet (ft^2)
1 square rod (rd^2) = 30.25 square yards (yd^2)
1 acre (A) = 160 square rods (rd^2)
= 4,840 square yards (yd^2)
= 43,560 square feet (ft^2)
1 square mile (mi^2) = 640 acres (A)

Measure of Volume
1 cubic foot (ft^3) = 1,728 cubic inches (in^3)
1 cubic yard (yd^3) = 27 cubic feet (ft^3)

Dry Measure
1 quart (qt) = 2 pints (pt)
1 peck (pk) = 8 quarts (qt)
1 bushel (bu) = 4 pecks (pk)

Liquid Measure
1 tablespoon (tbsp) = 3 teaspoons (tsp)
= $\frac{1}{2}$ fluid ounce (fl oz)
1 cup (c) = 16 tablespoons (tbsp)
= 8 fluid ounces (fl oz)
1 pint (pt) = 2 cups (c)
= 16 fluid ounces (fl oz)
1 quart (qt) = 2 pints (pt)
1 gallon (gal) = 4 quarts (qt)

Measure of Time
1 minute (min) = 60 seconds (sec)
1 hour (hr) = 60 minutes (min)
1 day (da) = 24 hours (h)
1 week (wk) = 7 days (da)
1 year (yr) = 12 months (mo)
= 52 weeks (wk)
= 365 days (da)

Measure of Weight–Avoirdupois
1 pound (lb) = 16 ounces (oz)
1 short ton = 2,000 pounds (lb)
1 long ton = 2,240 pounds (lb)

Volume, Capacity, and Weight Equivalents
1 gallon (gal) = 231 cubic inches (in^3)
1 cubic foot (ft^3) = $7\frac{1}{2}$ gallons (gal)
1 ft^3 of fresh water weighs $62\frac{1}{2}$ pounds (lb)
1 ft^3 of sea water weighs 64 pounds (lb)

Angles and Arcs
1 circle = 360 degrees (°)
1 degree (°) = 60 minutes (′)
1 minute (′) = 60 seconds (″)

Summary of the Metric System

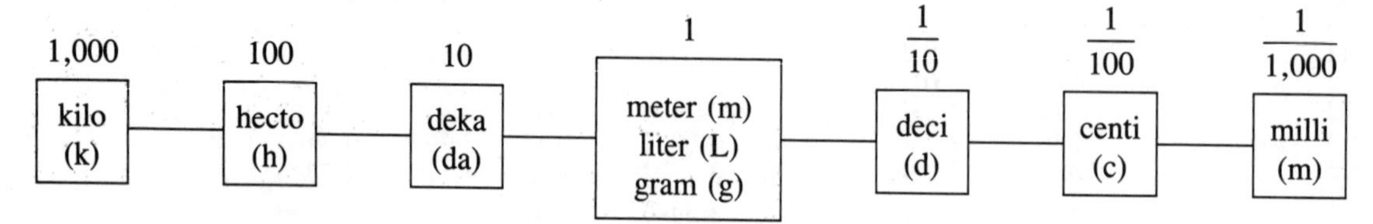

The metric system is based on powers of ten. In the chart, each prefix is ten times the one to its immediate right.

Length	Weight	Volume
1 km (kilometer) = 1000 m	1 kg (kilogram) = 1000 g	1 kL (kiloliter) = 1000 L
1 hm (hectometer) = 100 m	1 hg (hectogram) = 100 g	1 hL (hectoliter) = 100 L
1 dam (dekameter) = 10 m	1 dag (dekagram) = 10 g	1 daL (dekaliter) = 10 L

Table III (Continued)

1 dm (decimeter) =	.1 m	1 dg (decigram) =	.1 g	1 dL (deciliter) =	.1 L
1 cm (centimeter) =	.01 m	1 cg (centigram) =	.01 g	1 cL (centiliter) =	.01 L
1 mm (millimeter) =	.001 m	1 mg (milligram) =	.001 g	1 mL (milliliter) =	.001 L

Metric–English Equivalents

Conversions of Length

Metric to English		English to Metric	
1 km =	.621 mi	1 mi =	1.61 km
1 m =	1.09 yd	1 yd =	.914 m
=	3.28 ft	=	91.4 cm
=	39.37 in	1 ft =	.305 m
1 cm =	.394 in	=	30.5 cm
1 mm =	.039 in	1 in =	2.54 cm

Conversions of Volume

Metric to English		English to Metric	
1 L =	.264 gal	1 gal =	3.785 L
=	1.06 qt	1 qt =	.946 L
=	2.11 pt	1 pt =	.473 L
1 mL =	.034 fl oz	1 fl oz =	29.6 mL
1 cc =	.034 fl oz	=	29.6 cc

Conversions of Weight

Metric to English		English to Metric	
1 metric ton =	2204.62 lb	1 short ton =	.907 metric ton
1 kg =	2.20 lb	1 lb =	.454 kg
=	35.3 oz	1 oz =	28.3 g
1 g =	.035 oz		

Table IV–Metric Prefixes

Prefix	Symbol	Meaning
tera	T	1,000,000,000,000
giga	G	1,000,000,000
mega	M	1,000,000
kilo	k	1,000
hecto	h	100
deka	da	10
deci	d	0.1
centi	c	0.01
milli	m	0.001
micro	μ	0.000001
nano	n	0.000000001
pico	p	0.000000000001

APPENDIX

Answers to Odd-Numbered Supplementary Problems

Unit 1

1. 3, ten thousands. 0, tens.
3. 5, millions. 6, ten thousands. 1, hundreds. 9, units.
5. 9 tens + 3 units
7. 1 ten thousand + 2 thousands + 8 hundreds + 3 tens + 2 units
9. Fifty-six.
11. Five thousand, six.
13. Two hundred thousand, three hundred sixteen.
15. Ninety-five million.
17. Two billion, six hundred seventy-three million, four hundred ninety-eight thousand, two hundred twelve.
19. Six hundred thirty-five million, nine hundred two thousand, forty.
21. 3,429
23. 856,215,141
25. 13,000,000
27. 500,000
29. 100,602
31. 7,900,000
33. 7,936,000
35. 7,936,430
37. $68,000,000; $35,000,000; $32,000,000
39. Six thousand, seventy two.
41. Eight hundred, ninety-six.

Unit 2

1. Addends; sum.
3. Commutative.
5. 3 hundred thousands + 0 ten thousands + 2 thousands + 4 hundreds + 6 tens + 5 units.
7. 11
9. 273
11. 344,115
13. 52,809
15. 15,195
17. (a) 892 (b) 834; 58 (c) 476 (d) 86 (e) 562
19. 2,726
21. 5,887
23. 4,487
25. 1,253,666
27. 1,376 miles
29. $72 left
31. $9,460 profit
33. 32,776 seats
35. 10,799 books and magazines
37. $14,745
39. $8,257
41. Subtotal = $1,819; Total = $1,744
43. $6,376
45. $6,473
47. $7,200
49. $9,906
51. $19,188

Unit 3

1. $6 + 6 + 6 + 6 + 6 = 30$; $5 + 5 + 5 + 5 + 5 + 5 = 30$
3. $4 + 4 + 4 = 12$; $3 + 3 + 3 + 3 = 12$
5. 35, 8; 280
7. Associative.
9. 24
11. 32
13. 0
15. 0
17. 2,695
19. 269,500
21. 1,000
23. 100,000,000
25. 257,786
27. 116,888
29. 122,320,250
31. 207,757,160
33. $17,920
35. $1,100
37. $119,250
39. $4,096
41. 5,700 words
43. $260
45. 275 miles
47. 6,000 sheets
49. 414 seats

Unit 4

1. Dividend, 1,617; divisor, 3; quotient, 539; no remainder.
3. 124; 13; 7,205
5. Zero.
7. 5
9. 0
11. Impossible.
13. 3
15. 141
17. 90
19. 200 R8
21. 154 R 3
23. 12 R 29
25. 200 R 51
27. 30
29. 170 R 104
31. 23,949 R 94
33. 1,089 R 73
35. 1,130 R 592
37. $53 per sq ft
39. $225 a week
41. 5 days
43. 50,000 units per dose
45. 12 cents per ounce
47. $112 per month

Unit 5

1. 96¢
3. 65¢
5. 30 seconds
7. 135 miles
9. $960
11. 22 articles
13. 60 pounds
15. 56,310 feet
17. 1,358 articles
19. 6 inches
21. $2,970
23. 10,560 items
25. $702
27. 540 miles
29. $4,000

Unit 6

	Base	Exponent
1.	2	8
3.	1	5
5.	10	2
7.	3	4

9. 4^2
11. 8^1
13. $4^2 \times 5^3$
15. 16
17. 5
19. 36
21. 2,646
23. 100,000
25. 1,575
27. 12
29. 2
31. 0
33. 1
35. 13
37. 7

Unit 7

1. 3
3. 2
5. 5
7. Even.
9. Even.
11. Even.
13. Prime.
15. Composite.
17. Prime.
19. False.
21. True.
23. True.
25. False.
27. False.
29. $2^1 \times 7^1$
31. $3^1 \times 7^1$
33. $2^2 \times 3^2 \times 7^1$
35. $3^1 \times 5^1 \times 7^1 \times 11^1$
37. $5^2 \times 7^1 \times 13^1$
39. $2^3 \times 5^3$

Unit 8

1. Least common multiple.
3. 12
5. 7
7. 8; 24; 32; 48
9. 20; 40; 60
11. $2^2 \times 3^2 \times 5^1 = 180$
13. $2^3 \times 3^1 = 24$
15. $2^2 \times 3^1 \times 5^1 = 60$
17. $2^2 \times 5^2 = 100$
19. $2^2 \times 3^3 \times 7^2 = 5,292$
21. 77
23. 12
25. 24
27. 45
29. 30
31. 693

Unit 9

1. $\frac{4}{5}$
3. $\frac{4}{4}$
5. A fraction compares a portion to the whole. A fraction is an indicated division.
7. $\frac{4}{5}$
9. $\frac{1}{5}$
11. $\frac{35}{40}$
13. 35
15. 72
17. 9
19. 28
21. 0
23. True.
25. False.
27. True.
29. $\frac{16}{17}$

Unit 10

1. $\frac{12}{16}$
3. $\frac{7}{42}$
5. $\frac{60}{64}$
7. $\frac{10}{12}$
9. $\frac{2}{3}$
11. $\frac{9}{16}$
13. $\frac{3}{4}$
15. $\frac{4}{9}$
17. $\frac{2}{5}$
19. $\frac{3}{4}$
21. $\frac{2}{3}$
23. $\frac{14}{15}$
25. $\frac{5}{7}$
27. $\frac{1}{4}$
29. $3\frac{1}{4}$, $2\frac{11}{13}$
31. $1\frac{1}{2}$
33. 4
35. $2\frac{7}{8}$
37. 1
39. $4\frac{1}{2}$

Unit 11

1. 28
3. 105
5. 182
7. 90
9. 180
11. $\frac{3}{12}$ $\frac{10}{12}$
13. $\frac{20}{30}$ $\frac{18}{30}$ $\frac{25}{30}$
15. $\frac{20}{60}$ $\frac{45}{60}$ $\frac{24}{60}$
17. $\frac{2}{20}$ $\frac{15}{20}$
19. $\frac{20}{24}$ $\frac{9}{24}$
21. $\frac{15}{40}$ $\frac{28}{40}$
23. $\frac{2}{3}$
25. $\frac{1}{8}$
27. $\frac{3}{5}$ $\frac{2}{3}$ $\frac{3}{4}$
29. $\frac{3}{8}$ $\frac{1}{2}$ $\frac{5}{6}$

Unit 12

1. $\frac{3}{4}$
3. 3
5. $2\frac{1}{2}$
7. $1\frac{1}{10}$
9. $1\frac{7}{8}$
11. $5\frac{1}{10}$
13. $6\frac{1}{5}$
15. $2\frac{19}{20}$
17. $9\frac{11}{24}$
19. $\frac{1}{4}$
21. $1\frac{3}{5}$
23. $\frac{1}{4}$
25. $6\frac{1}{5}$
27. $4\frac{1}{2}$
29. $2\frac{5}{12}$
31. $1\frac{13}{24}$
33. $\frac{9}{10}$
35. $6\frac{1}{2}$
37. $2\frac{55}{72}$
39. $1\frac{9}{10}$

Unit 13

1. $3\frac{1}{3}$
3. 15
5. 3
7. $\frac{1}{10}$
9. 32
11. 60
13. $12\frac{1}{2}$
15. 80
17. $2\frac{1}{12}$
19. $\frac{3}{4}$
21. $\frac{1}{6}$
23. $2\frac{2}{3}$
25. $\frac{9}{40}$
27. $4\frac{1}{2}$
29. $\frac{49}{64}$
31. $12\frac{1}{4}$
33. $27\frac{1}{25}$
35. 1
37. $\frac{1}{8}$
39. $1\frac{61}{64}$
41. $\frac{5}{6}$
43. $1\frac{1}{7}$
45. $1\frac{1}{5}$
47. 2
49. 76¢
51. $49\frac{1}{3}$ pounds
53. $360
55. $6000

Unit 14

1. $1\frac{5}{8}$
3. $\frac{5}{18}$
5. $\frac{1}{7}$
7. Reciprocal.
9. $\frac{2}{9}$
11. $1\frac{2}{3}$
13. $\frac{2}{3}$
15. $\frac{1}{27}$
17. 12
19. 2
21. $1\frac{1}{5}$
23. $4\frac{1}{5}$
25. $\frac{7}{16}$
27. 1
29. $5\frac{2}{5}$
31. $1\frac{3}{32}$
33. $3

Unit 15

1. Food \$504 Housing \$315
3. 45¢
5. $49\frac{1}{3}$ pounds
7. 12 feet
9. $2\frac{1}{2}$ pounds
11. $14\frac{5}{32}$ inches
13. $1\frac{5}{7}$ minutes
15. $\frac{5}{24}$
17. $13\frac{3}{5}$
19. $\frac{15}{56}$
21. $\frac{19}{36}$
23. 1
25. 0
27. $\frac{8}{9}$
29. $\frac{5}{24}$
31. $\frac{1}{32}$
33. 15
35. $7\frac{1}{2}$

Unit 16

1. $\frac{4}{1}$
3. $\frac{1}{13}$
5. $\frac{3}{4}$
7. $\frac{2}{3}$
9. $\frac{2}{1}$
11. Not true.
13. True.
15. $n = 7$
17. $n = 8$
19. $n = 25$
21. $n = 240$
23. $n = 4$
25. 12 gallons
27. 44 grams
29. $18\frac{3}{4}$ cups
31. 16 hours
33. 154 pounds
35. 16 sets

Unit 17

1. 3 tenths plus 7 hundredths.
3. 2 and 0 tenths plus 0 hundredths plus 0 thousandths plus 7 ten-thousandths.
5. Five-tenths.
7. Six hundred nine thousandths.
9. Thirteen ten-thousandths.
11. Seven hundred eight and eighty thousandths.
13. Seventy-two hundred-thousandths.
15. Ten-thousandths.
17. Ten-thousandths.
19. Tenths.
21. Millionths.
23. .48
25. 5.093
27. .0083
29. .00406
31. $\frac{1}{4}$
33. $\frac{1}{20}$
35. (a) 6.4; (b) 6.4
37. (a) .1; (b) .1
39. (a) 1.205; (b) 1.21
41. (a) 23.16 (b) 23.165

Unit 18

1. $\frac{1}{20}$
3. $\frac{1}{125}$
5. $\frac{3}{40}$
7. $\frac{9}{20}$
9. $\frac{3}{2500}$
11. $12\frac{1}{1000}$
13. $4\frac{1}{50}$
15. $1\frac{1}{1000}$
17. 1.0875
19. .010125
21. 1.005
23. $\frac{11}{250}$
25. $2\frac{1}{3}$
27. $3\frac{9}{400}$
29. .344
31. 2.1
33. .61

Unit 19

1. 1.75
3. 7.678
5. 11.619
7. 19.339
9. 5.9178
11. 58.2676
13. 2.626
15. 5.887
17. 3.383
19. .02
21. 4.095
23. 2.79
25. \$1.38
27. 2.209 inches
29. (a) 28.5 (b) 27.2; 1.3 (c) 1.93 (d) .11 (e) 2.04 (f) .11
31. 10.7 gallons
33. \$12.01
35. \$11.25
37. Total \$504.73
Total Deposit \$479.23

Unit 20

1. 1.8
3. .006
5. .0006
7. .00328
9. 10.99
11. 3,340
13. 8.16
15. 273
17. 51.76
19. 63.336
21. .0961
23. .000027
25. .027
27. $102
29. 105.6 miles
31. $692.50
33. $7.84

Unit 21

1. 89
3. 180
5. .004
7. 860
9. 9,000
11. .6
13. .5625
15. .625
17. .35
19. 140
21. 27.8
23. .385
25. 177.1
27. .682
29. .0143
31. .002824

Unit 22

1. .016 inches
3. $1.20625
5. $3.50
7. .01
9. 5
11. .0625
13. 42.75
15. 70.5
17. 20.1
19. 60
21. .29
23. $(3.01)n = 12.04; n = 4$
25. $(7.3)n = 18.4; n = 2.5$
27. 6.5 kilometers
29. 7.68 min.

Unit 23

1. $\frac{3}{50}$
3. $\frac{13}{100}$
5. $\frac{23}{400}$
7. $1\frac{1}{5}$
9. $\frac{1}{8}$
11. $\frac{1}{16}$
13. $\frac{1}{6}$
15. $\frac{7}{1}$ or 7
17. .08
19. 1.528
21. .002
23. .375
25. .036
27. 33
29. 1.26
31. 9
33. $75,680

Unit 24

1. 83.27%
3. $37\frac{1}{2}\%$
5. .8%
7. $4\frac{3}{4}\%$
9. $18\frac{3}{4}\%$
11. $4\frac{1}{4}\%$
13. 24%
15. 15%
17. $\frac{1}{4}$ .25
19. $\frac{1}{3}$ $.33\frac{1}{3}$
21. $\frac{3}{4}$ .75
23. $\frac{2}{3}$ $.66\frac{2}{3}$
25. $\frac{1}{2}$ .5
27. $\frac{9}{10}$ .9
29. 20%
31. 45.5%

Unit 25

1. 6
3. 192
5. 1.125
7. 56.7
9. .2944
11. 36.72
13. 32
15. $22.40
17. 1,111
19. 6,800
21. $66\frac{2}{3}\%$
23. 120%
25. 15%
27. 19%
29. $5.25
31. $65.34
33. 24%
35. $54,000

Unit 26

1. $130; $1,130
3. $16; $616
5. $21.88; $646.88
7. 7%
9. 4%
11. 9%
13. 8%
15. Annual percentage rate.
17. $.63
19. $8.58
21. $.32
23. Total amount accumulated at the end of a given interval of time when interest has been compounded.
25. $81.21 or $81.20 using Table II.
27. $2,396.60
29. $4,291.90
31. $1075.12; $275.12
33. $225.30; $25.30
35. $536.05; $36.05

Unit 27

1. $12; 14%
3. $3; 15%
5. Amount of decrease.
7. 1,886
9. 336
11. $57
13. $196.80
15. $365
17. 8.7%
19. 1.59%

Unit 28

1. Amount of commission; sales amount.
3. Net proceeds.
5. Net proceeds.
7. $130; 25%
9. $432; 17%
11. $600
13. $429
15. $187,500
17. $94, $30
19. $221; $179
21. $1,095,000
23. $117,180
25. $143,000

Unit 29

1. Amount of discount.
3. Net price.
5. 100%–rate of discount.
7. 19%
9. $750.96
11. $54.56
13. $99
15. $450; $27
17. 22%; $76
19. 4%; $16
21. $1,100; $110
23. 13.9%
25. 21.1%
27. 6.86%

Unit 30

1. Original amount, amount of increase.
3. Original amount, amount of increase.
5. 21%
7. 23%
9. $4.74
11. $.72
13. 703 articles
15. 1,028,430
17. $8,200
19. $22; 10%
21. $49; 12%
23. $1,400; $1,582
25. 164%
27. 84°

Unit 31

1. Cost price.
3. Rate of profit.
5. 28%
7. $22.40
9. $4
11. $216
13. $8,300
15. $74.48; $25.48
17. $36; 8%
19. $1,940; 24%
21. $15
23. 128%
25. $240

Unit 32

1. 80 oz
3. 120 pt
5. 5 ft
7. 1.7 hr
9. 7 lb 5 oz
11. 2 ft 4 in
13. 14 hr 23 min 3 sec
15. 14 gal 1 qt 1 pt 14 fl oz
17. 1 hr 49 min 59 sec
19. 7 gal 2 qt
21. 33 gal
23. 114 lb 13 oz
25. 2 lb $15\frac{7}{8}$ oz
27. 2 yd 1 ft 11 in
29. 27 lb
31. 41 hr 30 min
33. 9 oz
35. 3 yd 1 ft 6 in
37. 2 lb 15 oz
39. No; 7 ft = 84 in

Unit 33

1. Kilo, thousand 1,000
3. Deka, da
5. centi, c
7. .1 meter, dm
9. .001 meter, mm
11. 100 liters, hL
13. 1,000 grams, kg
15. .001 gram, mg
17. 10
19. 1,000
21. 1,000
23. .1
25. 1,000
27. .001
29. 100
31. 2.2
33. b
35. b
37. c
39. c
41. b
43. a
45. a
47. b
49. b
51. a
53. b
55. b

Unit 34

1. mega, 1,000,000
3. hecto, 100
5. d, .1 or $\frac{1}{10}$
7. milli, .001 or $\frac{1}{1,000}$
9. 1,000,000 grams, Mg
11. .000001 meter, μm (mcm)
13. 1,000,000 meters, Mm
15. .01 liter, cL
17. 10
19. 1,000
21. 1,000
23. 1,000
25. .000001
27. 0.1
29. 1
31. 80 mm
33. 860 dm
35. 4.875 km
37. 2.2 cL
39. 2.98 hL
41. 52.5 mg
43. 7,000 g
45. 45.2 cg
47. .04 g
49. 5 metric tons
51. a
53. a
55. b

Unit 35

1. 6.21 mi
3. .39 in
5. .34 fl oz
7. 6.33 pt
9. 105.9 oz
11. 3.22 km
13. 122 cm
15. 7.57 L
17. 148 cm^3
19. 2.27 kg
21. 11.375 oz
23. 106 lb
25. 71 g
27. 188 cm
29. 37.9¢/L
31. 56 mi
33. 67 in
35. 566 g
37. 40°C
39. 150.81 m
41. 20.812 L
43. 54.595 m
45. A. 43.18 cm by 43.18 cm
B. 6.1 m by 3.05 m
C. 15.24 cm
D. 60.96 cm by 15.24 cm
E. 28.975 m

Unit 36

2. Obtuse.
3. Straight.
5. Triangle.
7. 5
9. Octagon.
11. No.
13. Yes.
15. Equilateral and isosceles.
17. Scalene.
19. 80°
21. 10°
23. 10 in
25. 5.385 in
27. Perpendicular.
29. 600
31. Equal.
33. Equal; parallel.
35. 38°
37. 60°, equilateral.
39. 8 ft
41. Yes.

Unit 37

1. 96 ft^2; 40 ft
3. 53.3 cm^2; 31.2 cm
5. 30 in^2; 30 in
7. 52.7834 in^2; 25.748 in
9. $525
11. 380 ft
13. 1,024 in^2
15. 58 ft^2
17. No.
19. 11,304 mi^2
21. Square.
23. 7.5 gal
25. 1058 ft^2
27. 1,384.74 square miles
29. 58.48 cm^2
31. 55.04 in^2
33. 62.8 cm

Unit 38

1. 120 cm^3; 158 cm^2
3. 64 ft^3; 96 ft^2
5. 157 cm^3; 164.85 cm^2
7. $400
9. $565.20
11. 113,040 ft^3
13. 664 in^2
15. 1519.76 in^2
17. 84 L
19. 15 ft^3
21. 712 ft^2; 392 ft^2
23. 43.40736 ft^3; 78.3744 ft^2

Index

INDEX

INDEX